BASIN ANALYSIS

BASIN ANALYSIS
PRINCIPLES AND APPLICATIONS

Philip A. Allen
Department of Earth Sciences,
University of Oxford

John R. Allen
British Petroleum Company plc,
London

b
Blackwell
Science

© 1990 by
Blackwell Scientific Publications
Editorial offices:
Osney Mead, Oxford OX2 0EL
25 John Street, London WC1N 2BL
23 Ainslie Place, Edinburgh EH3 6AJ
238 Main Street, Cambridge,
  Massachusetts 02142, USA
54 University Street, Carlton
  Victoria 3053, Australia

Other editorial offices:
Arnette Blackwell SA
1, rue de Lille, 75007 Paris
France

Blackwell Wissenschafts-Verlag GmbH
Kurfürstendamm 57
10707 Berlin, Germany

Blackwell MZV
Feldgasse 13, A-1238 Wien
Austria

First published 1990
Reprinted with corrections 1992
Reprinted 1993, 1995

Set by Times Graphics, Singapore
Printed and bound in Great Britain
by The Bath Press, Avon

DISTRIBUTORS

Marston Book Services Ltd
PO Box 87
Oxford OX2 0DT
(*Orders*: Tel: 01865 791155
       Fax: 01865 791927
       Telex: 837515)
USA
  Blackwell Science, Inc
  238 Main Street
  Cambridge, MA 02142
  (*Orders*: Tel: 800 215-1000
           617 876-7000
       Fax: 617 492-5263)
Canada
  Oxford University Press
  70 Wynford Drive
  Don Mills
  Ontario M3C 1J9
  (*Orders*: Tel: 416 441-2941)
Australia
  Blackwell Science Pty Ltd
  54 University Street
  Carlton, Victoria 3053
  (*Orders*: Tel: 03 347-5552)

British Library
Cataloguing in Publication Data

Allen, Philip A.
  Basin analysis: principles and applications.
  1. Sedimentary basins
  I. Title II. Allen, John R.
  552′.5

  ISBN 0-632-02423-2
  ISBN 0-632-02422-4 pbk

Library of Congress
Cataloging in Publication Data

Allen, P.A.
   Basin analysis: principles and applications/
     Philip A. Allen, John R. Allen.
       p.   cm.
     ISBN 0-632-02423-2 — ISBN 0-632-02422-4 (pbk.)
     1. Sedimentary basins. 2. Petroleum — Geology.
   I. Allen, John R. II. Title.
   QE571.A45   1990
   552′.5—dc20

Contents

Section 4: Evolution of the Basin-fill

Section 5: Application to Petroleum Play Assessment

Preface

Basin analysis is the integrated study of sedimentary basins as geodynamical entities. Sedimentary basins are worthy of study because they contain the sedimentary record of processes on the Earth's surface that have operated for countless millennia. They also contain in their geometry, tectonic evolution and stratigraphic history, valuable clues to the way in which the lithosphere deforms. They are therefore primary repositories of geological information. Sedimentary basins past and present are also the sites of almost all of the world's commercial hydrocarbons.

The investigation of sedimentary basins has brought about an unprecedented collaboration of earth scientists from a variety of disciplines, notably geophysicists, structural geologists, stratigraphers, sedimentologists and geochemists. This is not to say that communication is always easy between the different specialisms, but a general awareness of the importance of the contributing disciplines to the study of sedimentary basins is almost universal. Surely this integrated problem-solving approach is a significant omen for the development of the earth sciences over the next decade or two.

We write this book in an attempt to give a flavour of the multidisciplinary nature of basin analysis. We introduce subject matter such as the mechanics of the lithosphere, the formative mechanisms of the various genetic classes of basin, the stratigraphic patterns of the basin-fill, the typical depositional systems and facies, and the burial history in sufficient detail for the many links to be illuminated but at a level that is hopefully understandable and stimulating to an undergraduate at university, college or polytechnic. The concluding two chapters deal with the application of basin analysis in the arena where most professional geologists work – the hydrocarbon industry. We stress that the inclusion of this material is by no means an apologetic. It summarizes in an applied sense many of the principles proposed in the preceding chapters.

Authors of textbooks in rapidly developing subject areas are confronted with a major problem in keeping the book as up-to-date as possible and yet getting it finally published. Two important compilations of papers have been published during the writing of this text to which we have not been able to do full justice in overviewing. These are *Sea Level Changes: an Integrated Approach* published by the Society of Economic Paleontologists and Mineralogists, and *Sedimentary Basins and Basin-Forming Mechanisms* published by the Canadian Society of Petroleum Geologists. We were able, through availability of preprints and through participation in an Exxon sequence stratigraphy school by PAA (Esher, London 1988) to incorporate the new concepts on facies tracts in Chapter 6, and wish to thank Exxon, especially Dr Bernie Vining, for this opportunity.

We have taken the approach of emphasizing wherever possible the explanatory nature of the principles of basin analysis. This has necessarily been at the expense of a more descriptive, one might say encyclopaedic, approach to the topic. It has required the inclusion of a certain amount of theory. This has the effect of producing a somewhat more 'abstract' text, but the rewards are an enhanced understanding of geological processes in general. This knowledge can then be used to investigate a range of geological problems that remain obscure when tackled by the case-study or analogue approach.

Many insights into geological problems come indeed from standing back, looking at first principles and deriving a result using what is known about continuum mechanics. A book that took this approach to basin analysis across the board would be vast. However, to ignore the quantitative work that underpins our understanding of a particular phenomenon is often to fatally remove the physical insight. We have therefore compromised. We have, in places, provided enough in the form of mathematical development to enable readers to prove for themselves a particular result (the greatest mathematical knowledge required is of elementary calculus). Thus, it is possible to arrive at a solution for heat flow and thermal subsidence in a basin formed by uniform extension of the lithosphere, or to decompact a borehole record to obtain the observed subsidence as a function of time. For the less mathematically adept or physically minded, or the plain unwilling, this material, which is separated off from the main text, can be skipped over without disastrous consequences. We have also, however, for the sake of space and simplicity, explained certain phenomena purely qualitatively or summarized results graphically without a rigorous proof. For those who feel frustrated by this approach, we can merely recommend that they inspect more specialized texts or the literature cited for the derivation. A limited number of worked examples are provided which give a 'hands-on' feel to the problem solving.

We believe this to be the first undergraduate textbook to examine sedimentary basins as geodynamic entities responding closely to lithospheric and surface controls. We also espouse a modern basin analysis approach to hydrocarbon play assessment. The book should serve as a course text for basin analysis at university, college or polytechnic, as a primer for research students becoming involved in particular aspects of basin research, and as a refresher for professional geologists now separated from higher education.

We wrote the book as identical twins, one (JRA) firmly rooted in the petroleum exploration business, the other (PAA) transplanted at a tender age into academia. We still talk to each other. Such a twin authorship is no doubt unusual and more than a little curious. You must make of it what you will. You may agree with Tennyson who in *Locksley Hall* speaks of

'Nourishing a youth sublime
with fairy-tales of science,
and the long result of Time'.

Many have helped us to achieve the goal of publishing this modest volume. Our families have suffered terribly from absent stares and lunatic ravings – God bless them. We gratefully acknowledge the help of colleagues who have read chapters of the book for us: Hugh Sinclair and Philip England at Oxford, Tony Watts and Bernie Coakley at Lamont, Peter Molnar at MIT and Marc Helman at Queen's, New York. Chapters 10 and 11 are published with the kind permission of the British Petroleum Company. We also thank reviewers Christopher T. Baldwin at Boston University and John Grotzinger at MIT. Mary Marsland cheerfully wrestled with a difficult typescript and Claire Carlton, with Julia York's help, draughted the illustrations. We thank you all.

P.A.A., J.R.A.
Oxford

SECTION 1
THE FOUNDATIONS OF
SEDIMENTARY BASINS

1 Basins in their plate tectonic environment

Assumptions, hasty, crude and vain,
Full oft to use will Science deign;
The corks the novice plies today
The swimmer soon shall cast away

(A.H. Clough, Poem (1940))

SUMMARY

Sedimentary basins are regions of prolonged subsidence of the Earth's surface. The driving mechanisms of subsidence are ultimately related to processes within the relatively rigid, cooled thermal boundary layer of the Earth known as the lithosphere. The lithosphere is composed of a number of plates which are in motion with respect to each other. Sedimentary basins therefore exist in a background environment of plate motion.

The Earth's interior is composed of a number of compositional and rheological zones. The main compositional zones are between crust, mantle and core, the crust containing relatively low density rocks overlain by a sedimentary cover. The mechanical and rheological divisions do not necessarily match the compositional zones. A fundamental rheological boundary is between the lithosphere and the underlying aesthenosphere. The lithosphere is sufficiently rigid to comprise a relatively coherent plate. Its base is marked by a characteristic isotherm (c. 1330 °C) and is commonly termed the *thermal lithosphere*. The upper portion (c. 50 km thick) of the thermal lithosphere is able to store elastic stresses over long time scales and is referred to as the *elastic lithosphere*. The continental lithosphere has a strength profile with depth that suggests that a weak, ductile zone exists in the lower crust, separating a brittle upper crust and upper mantle, giving a jam-sandwich type structure. The oceanic lithosphere, however, lacks this low-strength layer, its strength increasing with depth to the brittle-ductile transition in the upper mantle.

The relative motion produces deformation and seismicity concentrated along plate boundaries. These are of three types
• divergent boundaries such as the mid-ocean ridge spreading centres of the ocean basins
• convergent boundaries associated with large amounts of shortening, such as continental collision zones
• conservative boundaries characterized by strike-slip deformation.

Sedimentary basins have been classified principally in terms of the type of lithospheric substratum (i.e. continental, oceanic, transitional), their position with respect to the plate boundary (intracratonic, plate margin) and type of plate margin nearest to the basin (divergent, convergent, transform). The widely cited basin classification of Bally and Snelson (1980) involves three distinct families of basins
• those located on rigid, relatively undeformed lithosphere unassociated with the formation of megasutures

- those associated with, but outside of, megasutures on rigid, relatively undeformed lithosphere
- those located upon and mostly contained within megasutures.

The formative mechanisms of sedimentary basins fall into three classes, although all three mechanisms may operate during the evolution of a basin:

1 *Purely thermal mechanisms*, such as the cooling and subsidence of the oceanic lithosphere as it moves away from spreading centres.

2 *Changes in crustal/lithospheric thickness*; thinning of the crust by mechanical stretching is accompanied by extensional fault-controlled subsidence, whereas thinning of the lithosphere produces thermal uplift.

3 *Loading* of the lithosphere causes a deflection or flexural deformation and therefore subsidence. An example is the subsidence in foreland basins.

From a genetic point of view there are two main groups of sedimentary basins; (1) basins due to lithospheric stretching, and (2) basins formed by flexure of continental and oceanic lithosphere. To this may be added (3) strike-slip or megashear-related basins which are characterized by local stretching in complex fault zones.

Inspection of any map showing hydrocarbon occurrences (e.g. St. John, Bally and Klemme 1984) reveals their clustered pattern. In general, provinces of hydrocarbon occurrence correspond to the locations of sedimentary accumulation greater than, say 1 km thick. These accumulations include sedimentary basins in the strict sense, implying zones of pronounced subsidence (Bally and Snelson 1980) but also carbonate bank build-ups on elevated oceanic crust, cratonic arches and so on which become fossilized in the geological record. In terms of production and potential production of hydrocarbons, however, sedimentary basins located at sites of prolonged and substantial subsidence are of overriding importance.

Historically, basin studies in terms of their rationale and methodology have developed from a number of distinct viewpoints such as that of stratigraphic sequences and their relation to sea level fluctuations (Sloss 1950, 1963), the geosyncline (Kay 1947, 1951, Aubouin 1965) and, more recently, the concept of plate tectonics. The location of sedimentary basins and their driving mechanisms are intimately associated with the motion of discrete, relatively rigid slabs which together represent the cooled thermal boundary layer of the Earth and with the convection system of the underlying mantle. The outer shell of the Earth comprises a relatively small number of these thin, rigid plates, and they are in a state of motion with respect to each other. Such motions set up plate boundary forces which may be transferred considerable distances into the interior of the plates, so that sedimentary basins exist in a background environment of stress set up by plate motion. Plates are also superimposed on a mantle that is undergoing slow thermal convection, so that the lithosphere providing the foundation of basins is also subject to differential thermal stresses along its base. Some basic ideas on plate tectonics and Earth structure are introduced in this chapter in so far as they help to explain the location and evolution of sedimentary basins. More exhaustive summaries can be found in Wyllie (1971), Cox (1973), Le Pichon, Francheteau and Bonnin (1973), Windley (1977), Smith (1976) and Cox and Hart (1986). A treatment of the physical properties of the Earth can be found in geophysical texts such as Garland (1971) and, more recently, Bott (1982).

The Earth's interior is composed of a number of essentially concentric zones which are defined on the basis of either

- compositional changes, or
- mechanical or rheological changes.

1.1 COMPOSITIONAL ZONATION OF THE EARTH

There are three main compositional units; the crust, mantle and core (Fig. 1.1).

1.1.1 Oceanic crust

The *crust* is an outer shell of relatively low density rocks. The oceanic crust is thin, ranging from approximately 4–20 km in thickness, 10 km being 'normal', and with an average density of about 2900 kg m^{-3}. It comprises a number of layers that reflect its mode of creation, an upper veneer (layer 1) of unconsolidated or poorly consolidated sediments, generally up to 0.5 km thick, an intermediate layer 2 of basaltic composition, consisting of

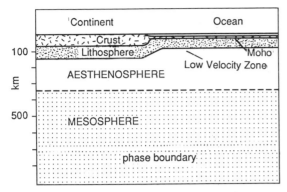

Fig. 1.1. The main compositional and rheological boundaries of the Earth. The most important compositional boundary is between the crust and the mantle, although there are certainly strong compositional variations within the continental crust. The base of the crust is marked by the Moho. The main rheological boundary is between the lithosphere and the aesthenosphere. The lithosphere is rigid enough to act as a coherent plate.

pillow lavas and associated products of submarine eruptions, and a layer 3 of gabbros and peridotites that may form the parent rocks which upon differentiation give rise to the basalts of layer 2. The oceanic crust has been thought to be distinctly layered in terms of velocity of seismic waves, but more recent views are that it possesses a more gradual and continuous increase in velocity with depth.

The lifetime of oceanic crust is short, despite the fact that it occupies about 60 per cent of the surface of the Earth (c. 3.2×10^9 km^2). This is because as the oceanic crust cools during ageing it becomes gravitationally unstable with respect to its substratum; as a result it is consumed. This explains why the oldest oceanic crust in today's oceans is Jurassic in age. Compared to the continents, the oceanic crust therefore has a very short lifespan.

1.1.2 Continental crust

The continental crust is thicker, ranging from 30 to 70 km, but with an 'average' thickness of perhaps 35 km. It was originally thought to be divided into two layers, each with a distinct composition and density: (1) an upper layer with physical properties similar to those of granites, granodiorites or di-

orites overlain by a thin veneer of sedimentary rocks. This so-called 'granitic layer' has a thickness of between 20–25 km and a density of 2500–2700 kg m^{-3}. The term 'granitic' is, however, misleading, since average densities are greater than that of granite. (2) A lower layer of primarily basaltic composition, but the pressure and temperature at depths in excess of 25 km imply that the rocks are granulites, or their high pressure, high temperature equivalents, eclogites or amphibolites. The density of this lower layer is 2800–3100 kg m^{-3}. These layers may not in reality be well defined, and instead a more continuous variation of composition with depth may exist.

In some regions, particularly the attenuated margins of continents, the crust is intermediate in character and thickness between typical oceanic and continental varieties. This may be due to the injection of dense intrusions, to metamorphism, or to other processes accompanying stretching.

Information on the density of crustal rocks has been obtained largely by observations on seismograms of the speed of seismic waves passing through the various layers, coupled with laboratory experiments on rock materials. The existence of a low velocity crust was discovered by the geophysicist

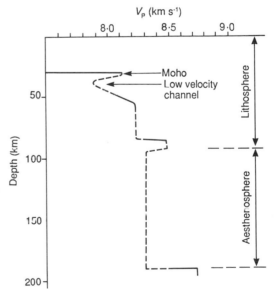

Fig. 1.2. Velocity–depth relationships for P waves beneath western Europe, showing a 'low velocity channel' underlying the Moho (after Hirn 1976).

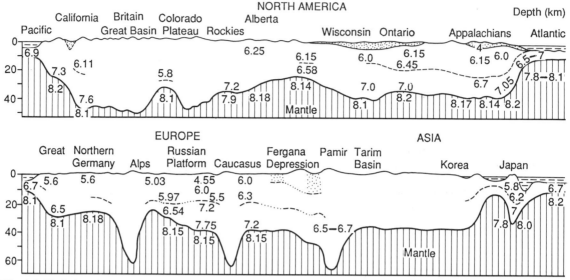

Fig. 1.3. Variations in crustal thickness across two continents (a) North America, (b) Eurasia, based on P wave velocities (after Holmes 1965). Mountains have well-developed crustal roots, as under the Alps, Caucasus and Pamir Range in the Eurasian section.

Mohorovičić shortly after the turn of the century. At the crust–mantle boundary, seismic P (longitudinal) wave velocities increase markedly; this abrupt increase in velocity may reflect a corresponding increase in rock density (Fig. 1.2). This horizon is known as the Mohorovičić discontinuity or Moho. The Moho varies in depth considerably, as can be seen from the traverses across Eurasia and North America (Fig. 1.3).

1.1.3 Mantle

The *mantle* is divided into two layers, the upper and lower mantle. The upper mantle extends to about 680 km ±20 km and is punctuated by phase transitions. The lower mantle extends to the outer limit of the core at 2900 km, with an increasing density with depth.

1.2 RHEOLOGICAL ZONATION OF THE EARTH

The mechanical or rheological divisions of the interior of the Earth do not necessarily match the compositional zones. One of the rheological zonations of interest to students of basin analysis is the

differentiation between the lithosphere and the aesthenosphere. This is because the vertical motions (subsidence, uplift) in sedimentary basins are ultimately a response to the deformation of this uppermost rheological zone of the Earth.

1.2.1 Lithosphere

The *lithosphere* is the rigid outer shell of the Earth, comprising the crust and the upper part of the mantle. It is of particular importance to note the difference between the *thermal* and *elastic* thicknesses of the lithosphere. It is generally believed (e.g. Parsons and Sclater 1977, Pollack and Chapman 1977) that the base of the lithosphere is represented by a characteristic isotherm (1100°C–1330°C) at which mantle rocks approach their solidus temperature. This defines the *thermal lithosphere*. Typical thicknesses of lithosphere under the oceans varies from *c.* 5 km at mid-ocean ridges to about 100 km in the coolest parts of the oceans. The lower boundary of the lithosphere is poorly defined under continents, depths of 100 km to 250 km being typical. The stepwise increases in velocities of S and P waves with depth through the lithosphere suggest that it contains compositional boundaries within it.

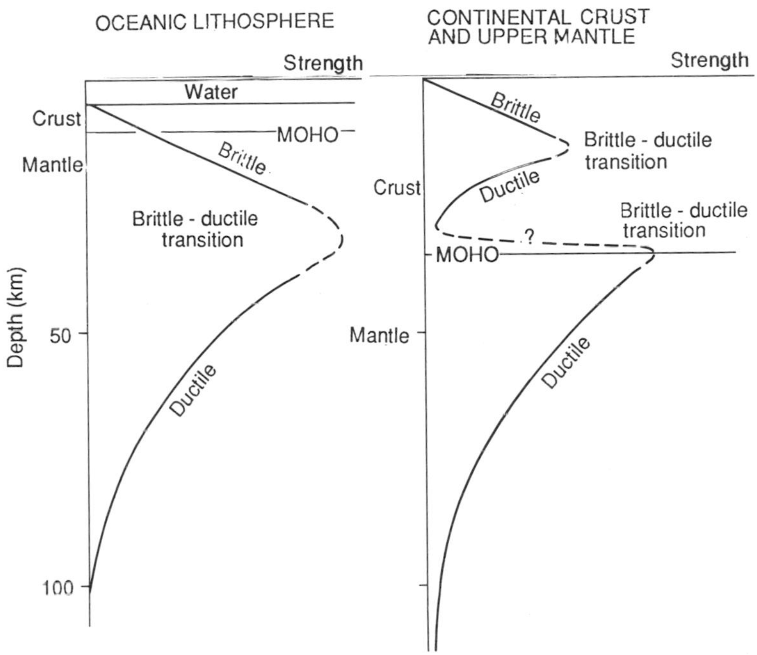

Fig. 1.4. The yield strength of the oceanic and continental lithosphere as a function of depth (after Molnar 1988). The oceanic lithosphere has a strong elastic core extending to depths of over 50 km, whereas the continental lithosphere appears to have a weak ductile layer in the lower continental crust. This gives a rheological layering like a jam sandwich. The elastic lithosphere is the upper portion that is able to store elastic stresses over geological periods of time. The base of the thermal lithosphere is a mechanical boundary separating the relatively strong outer shell of the lithosphere from the very weak aesthenosphere.

The rigidity of the lithosphere allows it to behave as a coherent plate but only the upper half of the lithosphere is sufficiently rigid to retain elastic stresses over geological time scales (say 10^9 years). Below this upper *elastic lithosphere* creep processes efficiently relax elastic stresses, so that there is a physical and conceptual difference between the elastic lithosphere and the thermal lithosphere. The lithosphere below the upper elastic portion must therefore be sufficiently soft to relax elastic stresses but sufficiently rigid to remain a coherent part of the surface plate.

The oceanic and continental lithosphere differ in their strength (Fig. 1.4). The strongest part of the oceanic lithosphere occurs in the mantle between 20 and 60 km depth, below which it becomes increasingly ductile. The continental lithosphere, however, appears to be markedly zoned rheologically.

In particular the upper seismically active brittle zone overlies a generally aseismic zone that may deform by ductile processes. This mid-lower crustal (?) ductile zone has been invoked as a level of detachment of major upper crustal faults (e.g. Kusznir and Park 1987) (see Sections 3.1.6.1 and 3.3.3). There is a second, deeper strong layer in the mantle part of the continental lithosphere where earthquakes occasionally occur (Chen and Molnar 1983).

There are also heterogeneities in the mantle part of the lithosphere, although they are small compared to the crust. Seismological studies of western Europe (Hirn 1976) suggest a highly stratified lithosphere beneath the Moho (Fig. 1.2). In particular, a 'channel' of reduced P wave velocities has been interpreted between 10 and 20 km below the Moho. This 10 km thick layer cannot be explained in terms of partial melting since the solidus temperature is far in excess of the actual temperature, and the hydration (serpentinization) of peridotites has been postulated as a possible mechanism. Whatever the cause, this upper low velocity channel may serve as a zone of decoupling of the upper lithosphere from the lower portion of the lithosphere when acted upon by tangential tectonic forces. There are few examples, however, where a process of decoupling can be unambiguously demonstrated at these levels.

The underlying region, the *aesthenosphere* is weaker than the lithosphere and is able to undergo deformation relatively easily by flow. The upper part of the aesthenosphere is known as the low

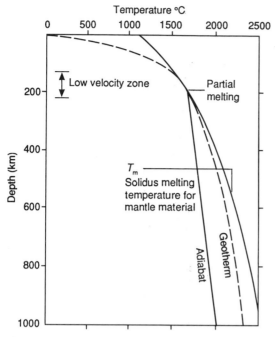

Fig. 1.5. Variation of temperature with depth, or geotherm, and the solidus melting temperature for mantle material. Where the solidus curve (T_m) and the geotherm become tangential, partial melting in the mantle is likely to take place, resulting in a zone of low seismic wave velocities (low velocity zone).

velocity zone where P and S wave transmission speeds drop markedly, presumably due to partial melting (Fig. 1.5).

The mechanical subdivisions of the mantle and core are of less direct concern in terms of sedimentary basin development.

1.3 PLATE MOTION

Plate tectonics can operate because the lithosphere is a coherent rheological 'plate'. The underlying concepts of relative plate motion come from studies of focal mechanism solutions of large earthquakes and from observations of the distribution of earthquake epicentres, and from studies of magnetic lineations in the ocean basins. The nature and rates of plate motion govern many aspects of the geodynamic environment of basins. We provide a brief background to plate motion here, but texts such as Le Pichon *et al.* (1973), Cox (1973)

and the review by Isacks, Oliver and Sykes (1968) give a great deal more information.

The global pattern of seismic activity is of continuous and narrow belts of high frequency of earthquakes, bounding extensive regions of stability (Barazangi and Dorman 1969). The narrow zones of earthquake activity define plate margins. Oceanic plate boundaries are very sharply defined whilst continental boundaries are rather more diffuse. The fact that earthquake epicentres occur at depths as great at 650 to 700 km along some plate boundaries suggests that a process exists that is capable of transferring brittle material to depths normally associated with deformation by flow. This process of plate subduction is responsible for both the relative youth of the oceanic crust and the distribution of earthquake epicentres. The fact that the interiors of plates experience only infrequent earthquake activity reflects the concentration of large relative motions of plates along their boundaries.

The lithospheric plates can be easily deformed by bending about a horizontal axis, but are highly resistant to torsion about steeply inclined axes. This latter property of strength allows the motion of plates over the Earth's surface to be modelled assuming no internal deformation, except at plate boundaries. But how do the oceanic and continental lithosphere compare in terms of strength? Different views exist on this problem. On the one hand, oceanic plates are stronger because they consist of more mafic mineral assemblages, whilst the continents contain quartz which shows ductile flow at lower temperatures than olivine (see Fig. 1.4). They also contain fewer intrinsic weaknesses such as old fundamental fault systems. However, the oceanic plates are thinner and hotter, and therefore bend more easily under an applied force system. Whether continental or oceanic lithosphere is stronger is therefore controversial and the strength must at least in part depend on parameters such as geothermal gradient and strain rate.

Three classes of plate boundary exist; divergent, convergent and conservative (Fig. 1.6).

Divergent boundaries are typified by the mid-ocean ridge spreading centres of the ocean basins. Here, the recognition of magnetic bands correlated with a magnetic reversal chronology (Vine and Matthews 1963) allows the rate of divergent plate motion to be estimated. Transform faults with strike-slip

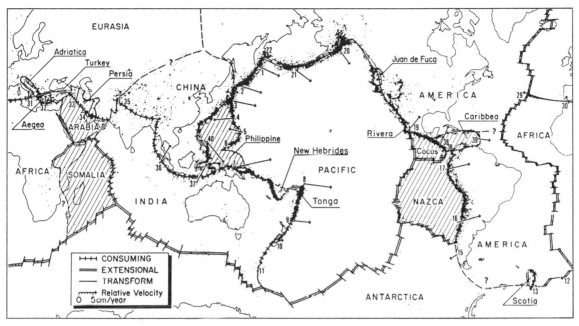

Fig. 1.6. Present distribution of lithospheric plates on the surface of the Earth showing the three main types of plate boundary (from Le Pichon *et al.* 1973).

displacement offset the divergent boundaries, producing a highly segmented pattern.

Convergent boundaries are of two classes.
1 Subduction boundaries where oceanic lithosphere constitutes the downgoing plate. Ocean–ocean boundaries, as for example, in the Mariana Islands, are characterized by a well-developed ocean trench and volcanic island arcs, whereas ocean–continent boundaries such as along the west of the Andes consist of an ocean trench with an associated continental magmatic arc with intense plutonic activity.
2 Collisional boundaries where continental lithosphere constitutes the downgoing plate. Where both plates are continental as in the Alps or Himalayan zones, the buoyancy of the downgoing plate resists subduction, leading to intense and widespread deformation. Less commonly, oceanic lithosphere may override continental lithosphere attached to subducting oceanic lithosphere, as in Taiwan.

Conservative boundaries occur where the adjoining plates are moving parallel to each other and are therefore dominated by transform faults.

The relative movement between plates causes earthquakes, a fact demonstrated by the concentration of seismic activity along plate boundaries. Earthquakes occur along trenches, ridges and transforms, but they are distinctly different along the three types of boundary;
• Ridges are characterized by small to moderate earthquakes generated at shallow depths of 10 km or less.
• Transforms experience larger earthquakes originating from depths of up to 20 km.
• Subduction zones are sites of very large and deep earthquakes, with foci occurring as deep as 700 km (Fig. 1.7).

The disappearance of earthquakes at relatively shallow levels along transforms and ridges is thought to be due to the change in rheology from brittle (capable of storing elastic stresses before rupture) to ductile (flowing by creep). This transition takes place in the range 600°C to 900°C, which corresponds to a depth of 20 to 30 km at transforms, but at shallower levels of about 10 km at ridges where temperatures are elevated. In contrast, at subduction zones, if the plate is descending quickly it remains cool relative to its surroundings and is capable of brittle deformation

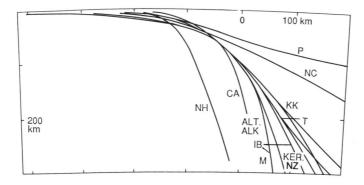

Fig. 1.7. Distribution of earthquake foci along Benioff zones (after Isacks and Barazangi 1977). [NH: New Hebrides; CA: Central America; ALT: Aleutians; ALK: Alaska; M: Marianas; IB: Izu-Bonin; KER: Kermadec; NZ: New Zealand; T: Tonga; KK: Kuril-Kamchatka; NC: Northern Chile; P: Peru]

to large depths. Hence, earthquake foci along subduction zones may be very deep. The rate of slab subduction may affect the depth of earthquakes, however. If the slab is only slowly subducted, it may heat up sufficiently to prevent the occurrence of earthquakes at great depths.

It is possible to study the type of motion that occurred during a particular earthquake and to work out the stresses released at the time of the earthquake, and thereby the direction of plate motion that gave rise to the stresses. These methods are called *first motion* or *focal mechanism* studies, since it is the first motion of the ground surface, whether it is up/away from the source of the earthquake or down/towards the focus. Distinct radiation patterns result from first motion on strike-slip, normal and reverse faults.

Sykes (1967) studied the first motion of earthquakes along mid-Atlantic ridge transforms (Fig. 1.8) and suggested that the motion was strike-slip, to the east on the northern blocks and to the west on the southern blocks. This is right lateral motion – opposite to that indicated by the offsets of the ridge, but the correct relative movement to support the plate tectonic interpretations of transforms as actively shearing only between the ridge segments and not beyond (Fig. 1.9). The earthquakes originating from spreading centres are quite different from those being produced at transforms. First motion studies suggest that faults in the mid-ocean ridge are dip-slip and extensional (Fig. 1.10). The situation at trenches is more complex. First motion studies of earthquakes along the *Benioff Zone* show that faulting takes place at roughly 45° to the inclined surface of the downgoing slab. At depths of greater than about 300 km, the focal mechanisms are compressional, but at shallower depths

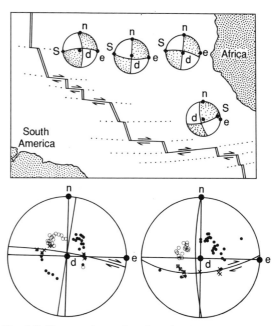

Fig. 1.8. First motions of earthquakes located on transforms that offset the mid-Atlantic ridge. Below are the first motion data used to determine the relative motion associated with the two most easterly earthquakes shown on the map. The solid circles are compressions, the open circles dilatations, and the crosses are stations located very close to the nodal plane representing the active fault plane (after Sykes 1967).

they are tensional. This pattern supports the view that in the lowermost part of the subducting slab, the plate experiences compression as it is forced into a zone of greater viscosity or strength at depth. The upper part, however, is in a state of tension

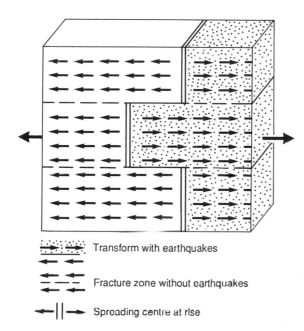

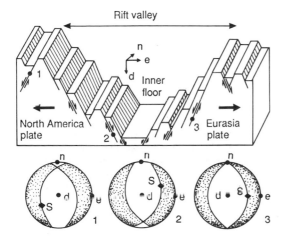

Fig. 1.9. Plate motion associated with a spreading ridge and transforms. Earthquakes only occur where the plates are actively shearing along transforms linking ridge segments (after Cox and Hart 1986).

Fig. 1.10. Focal mechanism solutions at spreading centres. The shapes of the beachballs, with nearly horizontal intersections of the nodal planes (termed null axes), suggest that dip-slip relative motion is taking place (after Cox and Hart 1986).

because of the gravitational body forces on the cool plate 'hanging' from its upper edge. This force constitutes slab-pull. Isacks, Oliver and Sykes (1968) provide a detailed analysis of the use of seismological studies at sites of subduction.

1.4 CLASSIFICATION SCHEMES OF SEDIMENTARY BASINS

Recent classification schemes of sedimentary basins based on plate tectonics have much in common. Dickinson (1974) emphasized the position of the basin in relation to the type of lithospheric substratum, the proximity of the basin to a plate margin and the type of plate boundary nearest to the basin (divergent, convergent, transform). The evolution of a basin could then be explained by changing plate settings and interactions. Dickinson (1974) recognized five major basin types on this basis:

1 Oceanic basins.
2 Rifted continental margins.
3 Arc-trench systems.
4 Suture belts.
5 Intracontinental basins.

Strike-slip or transform related basins were conspicuously missing as a distinct basin type in this classification, a deficiency corrected in Reading (1982).

Bally (1975) and Bally and Snelson (1980) differentiated three different families of sedimentary basins (Table 1.1):

1 Basins located on rigid, relatively undeformed lithosphere unassociated with the formation of megasutures.
2 Basins associated with but outside of megasutures on rigid, relatively undeformed lithosphere (perisutural).
3 Basins located upon and mostly contained within megasutures (episutural).

Megasutures in this context can be defined to include all the products of orogenic and igneous activity associated with predominantly compressional deformation. The boundaries of megasutures are often associated with subduction, whether it be of slabs of oceanic lithosphere (Benioff or B-type subduction) or of relatively buoyant continental lithosphere (Amferer or A-type subduction).

Table 1.1 Basin classification of Bally and Snelson (1980)

---

1 Basins located on rigid lithosphere, not associated with formation of megasutures
  1.1 Related to formation of oceanic crust
    1.1.1 *Rifts*
    1.1.2 *Oceanic transform fault associated basins*
    1.1.3 *Oceanic abyssal plains*
    1.1.4 *Atlantic-type passive margins (shelf, slope & rise) which straddle continental and oceanic crust*
      1.1.4.1 Overlying earlier rift systems
      1.1.4.2 Overlying earlier transform systems
      1.1.4.3 Overlying earlier backarc basins of (321) and (322) type
  1.2 Located on pre-Mesozoic continental lithosphere
    1.2.1 *Cratonic basins*
      1.2.1.1 Located on earlier rift grabens
      1.2.1.2 Located on former backarc basins of (321) type

2 Perisutural basins on rigid lithosphere associated with formation of compressional megasuture
  2.1 Deep sea trench or moat on oceanic crust adjacent to B-subduction margin
  2.2 *Foredeep and underlying platform sediments*, or moat on continental crust adjacent to A-subduction margin
    2.2.1 Ramp with buried grabens, but with little or no blockfaulting
    2.2.2 Dominated by block faulting
  2.3 *Chinese-type basins* associated with distal blockfaulting related to compressional megasuture and without associated A-subduction margin

3 Episutural basins located and mostly contained in compressional megasuture
  3.1 Associated with B-subduction zone
    3.1.1 *Forearc basins*
    3.1.2 *Circum Pacific backarc basins*
      3.1.2.1 Backarc basins floored by oceanic crust and associated with B-subduction (marginal sea *sensu stricto*).
      3.1.2.2 Backarc basins floored by continental or intermediate crust, associated with B-subduction
  3.2 Backarc basins, associated with continental collision and on concave side of A-subduction arc
    3.2.1 On continental crust or *Pannonian-type* basins
    3.2.2 On transitional and oceanic crust or *W. Mediterranean-type* basins
  3.3 Basins related to episutural megashear systems
    3.3.1 *Great basin-type* basin
    3.3.2 *California-type* basins

Boundaries of megasutures may also be the sites of important wrench tectonism along transform faults.

Industry-based classifications are typified by the scheme suggested by Halbouty *et al.* (1970) and later developed by Fischer (1975) and Klemme (1980). Klemme's scheme recognizes eight main types of basin based on their architectural characteristics such as linearity, asymmetry, cross-sectional geometry, which are themselves related to the tectonic setting and basin evolution.

Since Klemme's classification is based essentially on hydrocarbon characteristics rather than on fundamental geological properties, many sedimentary basins appear to change from one type to another type and yet another in their evolution. Whilst the idea of a polyhistory or multicyclic basin is a useful one, a classification which does not take into account *mechanisms* of formation and development has little to offer in the way of *explanation*.

The goal of categorizing a sedimentary basin and thereby gaining some predictive insights into frontier basins is common to industry classifications such as those of Huff (1978) and Klemme (1980). It is pursued by an Exxon group (Kingston, Dishroon and Phillips 1983a, b) to the extent of devising a formula for each basin, thereby facili

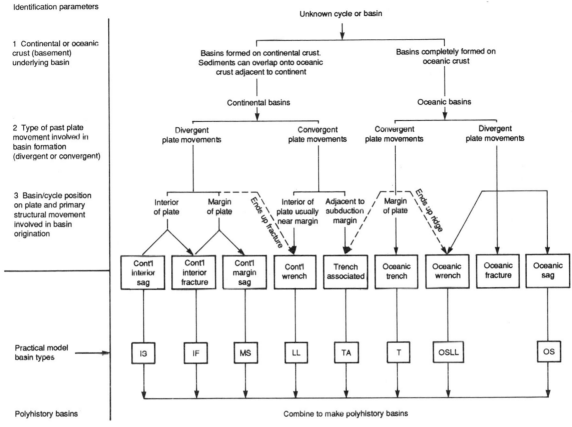

Fig. 1.11. Basin classification system of Kingston *et al.* (1983a) based on the type of lithospheric substrate, the relative plate motion and location on the plate.

tating easy comparisons between basins and providing an 'instant' idea of hydrocarbon potential. This classification system (Fig. 1.11) once again places basins primarily in their place tectonic setting (lithospheric substrate, type of plate motion and location on plate), reminiscent of Dickinson's analysis over a decade earlier, and categorizes a basin according to three critical factors:

1 The basin-forming tectonics.
2 The depositional sequences filling the basin.
3 The basin-modifying tectonics.

Once again, many basins are seen to be polyhistory and there is a distinct hydrocarbon potential for each basin type.

Whilst these classifications undoubtedly have their uses, particularly in predicting source presence, reservoir quality, availability of traps, etc., they have the effect of scrambling some of the essential differences and similarities between basins from the point of view of lithospheric mechanisms. There are three such mechanisms (Fig. 1.12):

1 Purely thermal mechanisms.
2 Changes in crustal/lithospheric thickness.
3 Loading and unloading.

Section 2.2.5 explains the importance of *cooling* of the oceanic lithosphere as it moves away from spreading centres in explaining the oceanic bathymetry. *Thinning* of the crust by subcrustal erosion, by thermal doming and subaerial erosion and by mechanical stretching have all been postulated and will be discussed in later sections. *Loading* of the lithosphere may take place on a small scale in the form of volcanoes or seamount chains, and on a large scale in the form of mountain belts, causing flexure and therefore subsidence. On top of all of these mechanisms of

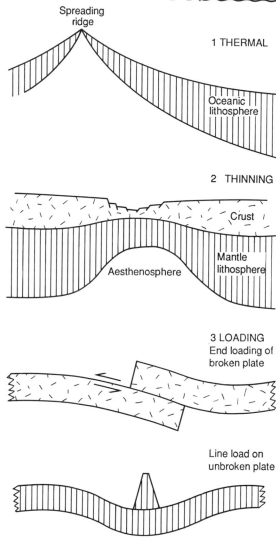

Fig 1.12 The three basic mechanisms for basin subsidence.

sequence (Veevers 1981). The early stages of the sequence correspond to the development of intra-cratonic rifts often associated with crustal doming. Such rifts may evolve into oceanic spreading centres or may be aborted to form failed rifts or aulacogens. With sea floor creation and drifting of the continental edge away from the spreading centre, passive margin basins develop. The sequence has been termed the rift-drift suite of sedimentary basins. The mechanisms of interest within this evolutionary sequence are therefore primarily the thermal and mechanical behaviour of the lithosphere under tension, and the thermal contraction of the lithosphere following stretching.

Basins formed by flexure fall into two groups. Flexure of oceanic lithosphere as it approaches subduction zones is responsible for the formation of deep oceanic trenches. It was the investigation of the deflection of the oceanic lithosphere at arc-trench boundaries which provided much of the framework for the general theory of lithospheric flexure. Flexure of the continental lithosphere in continental collision zones gives rise to foreland basins. The force system which causes flexure can be varied. In the case of ocean trenches, it is probably a combination of gravitational body forces on the downgoing oceanic slab and the excess in mass of the magmatic arc. In the case of foreland basins it is a combination of topographic loads represented by the mountain belt, lateral density variations of lithospheric material (caused, for example, by prior stretching of the continental lithosphere in the overridden plate or by obduction of dense mantle flakes) and horizontal forces set up by the shallow 'end-on' collision of buoyant continental lithosphere.

Basins in zones of strike-slip tectonics present difficulties in rational explanation. Every transform boundary involves some extension (transtension) and some compression (transpression). Segments of terrain caught up in the strike-slip zone may undergo considerable horizontal rotations and individual fault movement may open up small basins, or *pull-aparts* at points of change in curvature of the fracture or at fault divergences. As a result, basin formation can rarely be determined to result from one distinct mechanism. The formation of strike-slip related basins may involve the rotation of crustal blocks decoupled at great depth, or the thin-skinned extension of the crust.

subsidence must be added the additional driving force of the sediment load, amplifying the thermal and tectonic primary mechanisms. With this in mind, we shall discuss the formation of sedimentary basins under the following headings:

1 Basins due to lithospheric stretching.
2 Basins generated by flexure on continental and oceanic lithosphere.
3 Strike-slip or megashear-related basins.

Basins formed by stretching or thinning of the continental lithosphere fall within an evolutionary

2 Lithospheric mechanics

The kindly Earth shall slumber,
lapt in universal law

(Tennyson, *In Memoriam*)

SUMMARY

A knowledge of the behaviour of the lithosphere is essential if we are to understand the initiation and development of sedimentary basins. Lithospheric processes are responsible for the highly dynamic nature of the tectonic processes near the surface of the Earth. The Earth's outer layer or lithosphere can be regarded as a thermal boundary layer between the cool atmosphere or oceans and the hot interior. The lithosphere is therefore a thermal entity but it also has a physical significance. The upper half of the lithosphere is sufficiently rigid that it is able to store and transmit stresses. Because the deformations caused by applied forces are generally recoverable, this outer zone is known as the elastic lithosphere.

Applied forces of whatever origin cause stresses which result in deformation or strain. The simplest view of applied forces is of those due simply to the weight of an overlying rock column, known as *lithostatic stress*. The difference between the actual stress and the lithostatic stress is a tectonic contribution known as *deviatoric stress*. Deviatoric stresses can be either tensile or compressive.

Analysis of stress requires that a suitable coordinate system is established. A coordinate system is chosen whereby surfaces exist upon which no shear forces act. The normal stresses acting on these surfaces can be viewed as acting along one axis which is known as the *principal axis*. In a three-dimensional situation there are three principal axes and three principal stresses acting along these axes.

Deformation or strain leads to a volume change (dilatation), a displacement or solid body rotation. As in the case of stresses, a coordinate system can be erected so that principal axes of strain exist and shear strain components are zero.

In an elastic solid there is a clear relationship between the stresses and resultant strains. The exact relationship depends on material properties known as Young's modulus and Poisson's ratio. Where only one of the principal axes is non-zero, a *uniaxial* state of stress is said to occur and the relation between stress and strain is called Hooke's law. If there are two non-zero components of principal stress, we have the condition termed *plane stress*. In an analogous fashion, *uniaxial strain* and *plane strain* refer to the coordinate system of principal strains. In a state of stress where all the principal stresses are equal, an *isotropic* state of stress, the fractional volume change caused by isotropic compression is given by the bulk modulus or its reciprocal, the compressibility.

We know that the lithosphere is able to bend. The flexure of the lithosphere depends on its rigidity and the nature of the applied force or load causing the bending. Bending is accompanied by longitudinal stresses and bending moments in the plate which are related to the curvature of the plate and depth within the plate. The bending moment is related to the local radius of curvature by a coefficient called *flexural rigidity*.

Whereas the upper 25 to 50 km of the Earth is thought to behave more or less elastically over geological time scales, rocks below this level deform with a creep-like behaviour and thus relax stresses.

In order to understand the mechanical behaviour of the Earth it is necessary to understand its thermal structure, since this determines its rheology. In the lithosphere, heat transfer is predominantly by conduction, whereas in the mantle convection is extremely important.

For conductive heat transport, the heat flux is related to the temperature gradient by a coefficient, the *thermal conductivity*. Heat fluxes in the continents are determined primarily by conduction from radioisotopic heat sources whereas in the oceans they reflect cooling of newly created oceanic lithosphere. The variation of temperature with depth is known as the *geotherm*. Measurements of heat fluxes suggest that there are additional sources of heat apart from those causing conductive heat transport. Such sources are thought to be due primarily to mantle convection.

Temperature changes cause thermal stresses and strains which are additional to those of linear elasticity, but they may act in opposite senses. The relative magnitude of elastic and thermal strains is very much a function of geothermal gradient.

The temperature gradient in a convecting fluid should be *adiabatic*, that is, the temperature increase with depth is caused purely by compression due to the overlying rock column. The mantle follows an adiabatic gradient fairly closely but there are marked departures, one at 400 km thought to be due to an exothermic phase change from olivine to spinel and a deeper one at 650 km.

Convection is the natural result of differential heating above a critical Rayleigh Number and estimates for Earth's mantle strongly suggest that convection must be taking place. A thermal boundary layer develops along the upper surface of the convecting fluid because heat is lost by conduction to the surface. This cool boundary layer, the lithosphere, detaches and sinks due to gravitational instability along subduction zones.

Mantle convection is thought to take place by means of a thermally activated creep. *Diffusion creep* and *dislocation creep* are both strongly temperature dependent processes, but the latter is more rapid since it is driven by shear stresses acting on crystal imperfections. Crustal rocks may also behave in a ductile manner by *pressure-solution creep*, or they may deform in a brittle manner by fracturing. The rheology of the crust is complex because of its compositional heterogeneity. A low strength region in the middle crust may serve mechanically to decouple the upper crust from the lower crust and mantle lithosphere. The lower lithosphere may deform as an elastic solid on short time scales but viscously on longer time scales. The time scale of the viscous relaxation of stresses is not fully understood. In the case of a flexed plate, plastic deformation may take place if a critical elastic curvature is exceeded.

The formation and evolution of sedimentary basins takes place under a gross tectonic regime determined by lithospheric processes. This regime is manifested in a state of stress of the lithosphere which may result in a deformation or strain. The mechanical behaviour of the Earth is strongly dependent on its thermal properties, and a change in thermal structure will invariably result in a deformation. In the following sections therefore, we briefly examine the forces acting in the lithosphere, the heat flow through the lithosphere, and the resulting rheologies of lithospheric materials. These concepts provide a useful framework for the study of sedimentary basins.

This chapter is a brief overview of some of the continuum mechanics of importance to basin analysis. Detailed information can be found in a number of sources such as Turcotte and Schubert (1982), Hobbs, Means and Williams (1976), Ramsay and Huber (1983).

Stresses are forces per unit area that are transmitted by interatomic force fields. Those thatact perpendicular to a surface are *normal stresses* and those acting parallel to a surfacc are *shear stresses*. The pressure is the mean value of the normal stresses. Deformation of a solid results in *strain*. Normal strain is simply the ratio of the change in length of a solid to its original length. Shear strain is defined as one half of the decrease in a right angle in a solid when it is deformed.

2.1 STRESS AND STRAIN

2.1.1 Stresses in the lithosphere

Body forces on an element of a solid act throughout the volume of a solid and are directly proportional to its volume or mass. For example, the force of gravity per unit volume is the product of ρ, the density and g, the acceleration of gravity. The body forces on rocks within the Earth's interior depend on their densities, but density is itself a function of pressure. If we normalize rock densities to a zero-pressure value, typical mantle rocks would have densities of about 3250 kg m^{-3}, ocean crust (basalt and gabbro) would have densities of about 2950 kg m^{-3}, whereas continental granites and diorites would be in the range 2650 to 2800 kg m^{-3}. Sedimentary rocks are highly variable, ranging from 2100 kg m^{-3} for some shales to 2800 kg m^{-3} for vcry compact marbles (Table 2.1).

Surface forces act only on the surface area bounding a volume and arise from the interatomic stresses exerted from one side of the surface to the other. The magnitude of the force depends on the surface area over which the force acts and the orientation of the surface. The normal force per unit area on horizontal planes increases linearly with depth. That due to the weight of the rock overburden is known as *lithostatic stress* or *lithostatic pressure*. This concept of surface forces in the Earth's interior is made use of in considering the way in which hydrostatic equilibrium (Archimedes Principle) influences the support of the oceanic and continental plates by the mantle, an application known as *isostasy*.

Table 2.1 Common properties of rock

	Density kg m^{-3}	Young's modulus E 10^{11} Pa	Shear modulus G 10^{11} Pa	Poisson's ratio ν	Thermal conductivity K Wm^{-1}°K^{-1}	Coefficient of thermal expansion α 10^{-5}°K^{-1}
Sedimentary						
Shale	2100–2700	0.1–0.3	0.14		1.2–3	
Sandstone	2200–2700	0.1–0.6	0.04–0.3	0.2–0.3	1.5–4.2	3
Limestone	2200–2800	0.6–0.8	0.2–0.3	0.25–0.3	2–3.4	2.4
Dolomite	2200–2800	0.5–0.9	0.3–0.5		3.2–5	
Marble	2200–2800	0.3–0.9	0.2–0.35	0.1–0.4	2.5–3	
Metamorphic						
Gneiss	2700	0.04–0.7	0.1–0.35	0.04–0.15	2.1–4.2	
Amphibolite	3000		0.5–1.0	0.4	2.5–3.8	
Igneous						
Basalt	2950	0.6–0.8	0.3	0.25	1.3–2.9	
Granite	2650	0.4–0.7	0.2–0.3	0.1–0.25	2.4–3.8	2.4
Diabase	2900	0.8–1.1	0.3–0.45	0.25	1.7–2.5	
Gabbro	2950	0.6–1.0	0.2–0.35	0.15–0.2	1.9–2.3	1.6
Diorite	2800	0.6–0.8	0.3–0.35		2.8–3.6	
Pyroxenite	3250				4.1–5	
Anorthosite	2750	0.83	0.35	0.25	1.7–2.1	
Granodiorite	2700				2.6–3.5	
Mantle						
Peridotite	3250				2.3–3	2.4
Dunite	3250	1.4–1.6	0.6–0.7		3.7–4.6	
Miscellaneous						
Halite			0.3	0.15	5.4–7.2	13
Ice			0.092	0.033	2.2	5

(a)

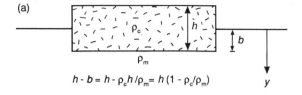

$$h - b = h - \rho_c h / \rho_m = h(1 - \rho_c/\rho_m)$$

(b)

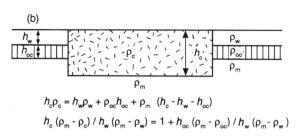

$$h_c \rho_c = h_w \rho_w + \rho_{oc} h_{oc} + \rho_m (h_c - h_w - h_{oc})$$

$$h_c (\rho_m - \rho_c) / h_w (\rho_m - \rho_w) = 1 + h_{oc} (\rho_m - \rho_{oc}) / h_w (\rho_m - \rho_w)$$

Fig. 2.1. Schematic diagrams to introduce concepts of isostasy, (a) continental block 'floating' in a fluid mantle, (b) continental block flanked by oceanic crust and overlying water column.

The surface force acting on a unit area at the base of a vertical column of rock is given by

$$\sigma_{yy} = \rho g y \qquad (2.1)$$

where ρ is the density of the rock column, g is the acceleration of gravity and y is the height of the column. For equilibrium to be maintained, we can equate the surface forces due to the differing rock columns under continental and oceanic lithosphere. The equilibrium for a continental block simply floating on a fluid mantle and for a similar case but with an ocean filled with water, is given in Fig. 2.1. A more familiar parameterization for isostasy is of a mountain belt with a 'root', the excess mass in the elevated continental crust being compensated by the mass deficit at depth of the continental root replacing mantle (Fig. 2.2). We shall examine a simple application of isostasy in studying the bending of lithosphere in Section 2.1.5.

Normal surface forces can also be exerted on vertical planes (Fig. 2.3). If the normal surface forces, σ_{xx}, σ_{yy} and σ_{zz} are all equal and they are also equal to the weight of overburden, the rock is said to be in a lithostatic state of stress. The normal surface forces σ_{xx}, σ_{yy} and σ_{zz} are rarely equal when a rock mass is being subjected to tectonic forces. In such a case the total horizontal surface force (normal stress) acting on a continent, for example, would be made up of two components, a lithostatic term and a tectonic contribution known as a *deviatoric stress* ($\Delta\sigma_{xx}$),

$$\sigma_{xx} = \rho_c g y + \Delta\sigma_{xx} \qquad (2.2)$$

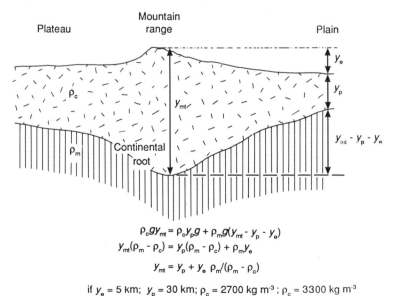

$$\rho_c g y_{mt} = \rho_c y_p g + \rho_m g (y_{mt} - y_p - y_e)$$

$$y_{mt}(\rho_m - \rho_c) = y_p(\rho_m - \rho_c) + \rho_m y_e$$

$$y_{mt} = y_p + y_e \, \rho_m/(\rho_m - \rho_c)$$

if $y_e = 5$ km; $y_p = 30$ km; $\rho_c = 2700$ kg m^{-3}; $\rho_c = 3300$ kg m^{-3}
then $y_{mt} = 57.5$ km

Fig. 2.2. Schematic diagram of isostasy for a continental mountain belt with a root.

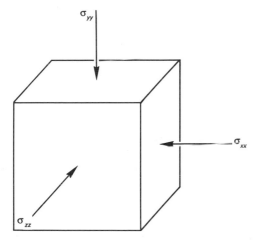

Fig. 2.3. Normal surface forces acting on vertical and horizontal planes.

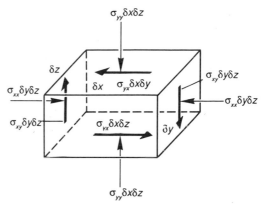

Fig. 2.4. Surface forces acting on a small rectangular element in a two-dimensional state of stress. The rectangular element has dimensions δx, δy and δz defined in accordance with the x, y, z Cartesian coordinate system. The state of stress is two-dimensional in the sense that there are no surface forces in the z direction, and none of the surface forces shown varies in the z direction. The normal forces are σ_{xx}, σ_{yy} and σ_{zz}, and the tangential or shear forces are σ_{xy} and σ_{yx}. The notation therefore immediately shows the orientation of the force – the second subscript on σ gives the direction of the force and the first subscript gives the direction of the normal to the surface on which the force acts. Each individual tangential force exerts a moment on the rectangular element, tending to rotate it about the z axis. However, for equilibrium, the sum of these moments is zero.

where ρ_c is the density of the continent. Normal stresses can be either *tensile* when they tend to pull on planes or *compressive* when they push on planes. Horizontal deviatoric stresses may result from uplift producing excess potential energy or may be transmitted from plate boundaries (see Chapter 3 and Section 6.2).

Surface forces acting parallel to a surface are known as *shear stresses*. Examples are provided by a thrust sheet with a lower fault plane that experiences a frictional resistance, or the gravitational sliding of a rock mass down an inclined plane.

Stress components can be generalized to any point in a material by using the x, y, z Cartesian coordinate system. In a two-dimensional system (Fig. 2.4), three independent components of stress σ_{xx}, σ_{yy} and σ_{xy} are required in order to determine the state of stress. However, the state of stress is dependent on the orientation of the coordinate system. In a coordinate system x', y' inclined at an angle Θ to the x, y system, we can determine normal (pressure) forces $\sigma_{x'x'}$ and tangential (shear) forces $\sigma_{x'y'}$, and there are surfaces which are orientated such that no shear forces act upon them. These surfaces are determined by the magnitudes of the independent components of stress, so that

$$\tan 2\Theta = \frac{2\sigma_{xy}}{(\sigma_{xx} - \sigma_{yy})} \tag{2.3}$$

and Θ defines the *principal axis of stress*; two other principal axes exist as orthogonals to that defined by Θ. No shear stresses exist on surfaces orientated

perpendicular to the principal axes, and normal stresses in the principal axis coordinate system are known as *principal stresses*. It is now possible to use the principal coordinate system as a reference with axes

$$\sigma_1 = \sigma_{xx}, \ \sigma_2 = \sigma_{yy} \text{ and } \sigma_{xy} = 0.$$

The angle at which the shear stress $\sigma_{x'y'}$ is a maximum is given by

$$\tan 2\Theta = (\sigma_{yy} - \sigma_{xx})/2\sigma_{xy} \tag{2.4}$$

A comparison of (2.3) with (2.4) shows that $\tan 2\Theta$ for the principal axis orientation and for the maximum shear stress orientation are negative reciprocals, so 2Θ differs by 90° between the two cases. Thus the axes which maximize the shear stress lie at 45° to those which minimize it. By letting $2\Theta = \pi/2$ (equivalent in radians to 90°)

$$(\sigma_{xy})_{max} = \frac{1}{2}(\sigma_1 - \sigma_2) \tag{2.5}$$

In other words, the maximum shear stress is half the difference of the principal stresses.

The same approach can be used to describe stress in three dimensions, only this time there are three principal stresses σ_1, σ_2 and σ_3 acting along three principal axes, and the sum of the shear stress components is always zero using this coordinate system. If the principal stresses σ_1, σ_2 and σ_3 are identical, the state of stress is isotropic and any principal stress is equal to the pressure. In such a case, any set of orthogonal axes qualifies as a principal axis coordinate system. This is known as a *hydrostatic state of stress*. Where the state of stress is not isotropic, the pressure is equal to the mean of the normal stresses

$$p = \frac{1}{3}(\sigma_1 + \sigma_2 + \sigma_3) \tag{2.6}$$

or, for any coordinate system

$$p = \frac{1}{3}(\sigma_{xx} + \sigma_{yy} + \sigma_{zz}) \tag{2.7}$$

Subtraction of the mean stress (that is, pressure) from the normal stress component reveals the *deviatoric normal stresses*.

As in the two-dimensional case, the orientation of the plane on which shear stresses are a maximum can be defined. The normal to this plane bisects the angle between the maximum (σ_1) and minimum (σ_3) principal stresses and the maximum value of shear stress is half the difference of the maximum and minimum principal stresses, that is $(\sigma_1 - \sigma_3)/2$.

2.1.2 Strain in the lithosphere

Strain is the deformation of a solid caused by the application of stress. We can define the components of strain by considering a rock volume with sides δ_x, δ_y and δ_z which changes in dimensions but not in shape, so that the new lengths of the sides after deformation are $\delta x - \varepsilon_{xx}\delta x$, $\delta y - \varepsilon_{yy}\delta y$ and $\delta z - \varepsilon_{zz}\delta z$ where ε_{xx}, ε_{yy} and ε_{zz} are the strains in the x, y and z directions (Fig. 2.5). As long as the deformation on the volume element is relatively small, the volume change, or *dilatation*, is simply the sum of the strain components ($\varepsilon_{xx} + \varepsilon_{yy} + \varepsilon_{zz}$). Volume elements may also change their position without changing their shape, in which case the strain components are due to *displacement*.

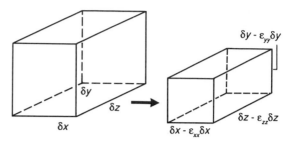

Fig. 2.5. A rectangular block which changes its dimensions but not its shape; this is a deformation involving no shear.

Shear strains, however, may distort the shape of an element of a solid. In the case of a two-dimensional rectangular element which is distorted to a parallelogram (Fig. 2.6), the shear strain is dependent on the amount of rotation of the sides of the rectangular element. Thus, shear strain in the two dimensions x and y is determined by the angles through which the sides of the rectangle are rotated (Fig. 2.6). The shear strain ε_{xy} is negative if the original right angle becomes acute when it is strained.

If the angles through which the sides of the rectangle are rotated (ϕ_1 and ϕ_2 in Fig. 2.6) are not equal, *solid body rotation* is said to have occurred. The magnitude of the solid body rotation is also dependent on the angles ϕ_1 and ϕ_2. Solid body rotations do not involve changes in the distances

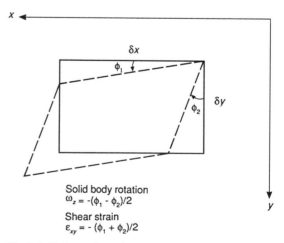

Solid body rotation
$\omega_z = -(\phi_1 - \phi_2)/2$
Shear strain
$\varepsilon_{xy} = -(\phi_1 + \phi_2)/2$

Fig. 2.6. Deformation of a rectangle into a parallelogram by a strain field involving shear.

between neighbouring elements of a solid and therefore do not reflect strain.

The deformation of any element can now be described according to the shear strain and the solid body rotation. If no solid body rotation occurs, $\phi_1 = \phi_2$ and the deformation is a result only of shear strains, it is known as *pure shear*. If there is solid body rotation but $\phi_1 = 0$, the element has undergone *simple shear*. Figure 2.7 illustrates the two circumstances. We shall see in Chapter 4 how models of pure shear (e.g. uniform extension with depth) and simple shear (e.g. asymmetrical extension associated with translithospheric shear zones) have been applied to the formation of rift basins.

Once again the two-dimensional case can be generalized to three dimensions by the addition of extra strain components. A pure shear strain in the xz plane has an associated shear strain component ε_{xz} and that in the yz plane has an associated shear strain ε_{yz}.

As in the case of shear stresses, shear strains can be described with reference to a coordinate system which is orientated such that shear strain components are zero. Such a system contains the *principal axes of strain*. In a directly analogous fashion to equation (2.3), the orientation of the principal axes can be found from the magnitudes of the independent components of strain:

$$\tan 2\Theta = 2\varepsilon_{xy}/(\varepsilon_{xx} - \varepsilon_{yy}) \qquad (2.8)$$

Here Θ is the angle between the Cartesian coordinate system x, y and the new rotated system x', y', and defines a principal axis of strain. Axes at right angles to this principal axis of strain are also principal axes of strain. The fractional changes in length along the directions of the principal strain axes are the principal strains. In the three-dimensional case the condition of isotropic strain is satisfied by

$$e = (\varepsilon_{xx} + \varepsilon_{yy} + \varepsilon_{zz})/3 = \Delta/3 \qquad (2.9)$$

where Δ is the dilatation and e is the mean normal strain.

Deviatoric strain components are strains that are the difference between the actual strain and the mean normal strain. Deviatoric strains invariably result from the operation of tectonic processes. Their analysis therefore greatly aids the interpretation of lithospheric deformation.

2.1.3 Introductory remark on elastic and plastic deformation

Elastic materials deform when they are subjected to a force and regain their original shape and volume when the force is removed. For relatively low temperatures, pressures and applied forces, almost all solid materials behave elastically. The relation between stress and elastic strain is linear. However, at high temperatures and pressures or high levels of stress rocks do not behave elastically (see Section 2.3). In near-surface regions where temperatures and pressures are low, rocks deform by brittle fracture at high levels of stress. Deeper in the Earth, high temperatures and pressures cause the rock to deform plastically under an applied force, with no fracturing (Sibson 1983). Brittle materials which have exceeded their yield strength and plastic materials do not regain their original shape when the force is removed.

Because much of the lithosphere behaves as a strong material over geological (i.e. long) periods of time, it is able to bend under surface loads, to store

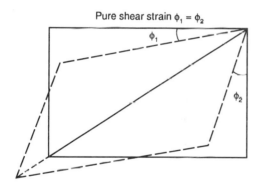

Pure shear strain $\phi_1 = \phi_2$

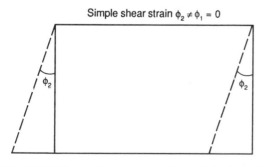

Simple shear strain $\phi_2 \neq \phi_1 = 0$

Fig. 2.7. Difference between pure shear strain (no solid body rotation) and simple shear strain (solid body rotation is $\phi_2/2$).

the elastic stresses responsible for earthquakes, and to transmit stresses over large horizontal distances. This fundamental property of the lithosphere is crucial to an understanding of the formation of sedimentary basins.

2.1.4 Linear elasticity

The theory of linear elasticity underpins a very great deal of thought on lithospheric mechanics and often constitutes the basic assumption in models of lithospheric behaviour.

In a linear, isotropic, elastic solid the stresses are linearly proportional to strains, and mechanical properties have no preferred orientation. The principal axes of stress and the principal axes of strain coincide. The relation between the principal strain and the components of principal stress can be stated as follows.

$$\varepsilon_1 = \frac{\sigma_1}{E} - \frac{\nu\sigma_2}{E} - \frac{\nu\sigma_3}{E} \qquad (2.10)$$

$$\varepsilon_2 = -\frac{\nu\sigma_1}{E} + \frac{\sigma_2}{E} - \frac{\nu\sigma_3}{E} \qquad (2.11)$$

$$\varepsilon_3 = -\frac{\nu\sigma_1}{E} - \frac{\nu\sigma_2}{E} + \frac{\sigma_3}{E} \qquad (2.12)$$

The exact partioning of stresses to give a resultant strain is clearly strongly influenced by E and ν, which are material properties known as *Young's modulus* and *Poisson's ratio* respectively. In general terms, a principal stress produces a strain component σ/E along the same axis and strain components $-\nu\sigma/E$ along the two other orthogonal axes.

Where only one of the principal stresses is non-zero (*uniaxial stress*), a shortening in the direction of the applied compressive stress will be accompanied by an extension in the two orthogo-nal directions (Fig. 2.8), and vice versa. Under these conditions where, let us say $\sigma_2 = \sigma_3 = 0$ and $\sigma_1 \neq 0$, there is a simple relation along the axis of uniaxial stress

$$\sigma_1 = E\varepsilon_1 \qquad (2.13)$$

This well-known relationship is called *Hooke's Law*.

There is a fractional volume change or dilatation due to uniaxial stress, but the contraction in the direction of uniaxial stress is compensated by expansion by half as much in the two other orthogonal directions in incompressible materials.

Uniaxial strain is where there is only one non-zero component of principal strain, that is, assuming the non-zero axis to be ε_1, $\varepsilon_2 = \varepsilon_3 = 0$.

If forces are applied not just along one axis but instead result in two non-zero components of principal stress, the condition is termed *plane stress* (Fig. 2.9). Such a system, where one of the principal stresses is zero, is suggestive of the horizontal stresses in the lithosphere caused by tectonic processes.

Plane strain is where only one of the principal strain components is zero, for example $\varepsilon_3 = 0$ and ε_1 and ε_2 are non-zero. It, too, is a common starting point for studying lithospheric deformation, since it can be assumed that the strain in the direction of an infinite plate will be zero (see as an example Section 2.1.4).

In an isotropic state of stress, all the principal stresses are equal ($\sigma_1 = \sigma_2 = \sigma_3$) and any principal stress is equal to the pressure. The principal strains developed in an isotropic state of stress are also equal. Each principal strain is equal to one third of the dilatation ($\varepsilon_1 = \varepsilon_2 = \varepsilon_3 = 1/3 \Delta$). The pressure under these conditions is related to the dilatation by K, the *bulk modulus* or its reciprocal

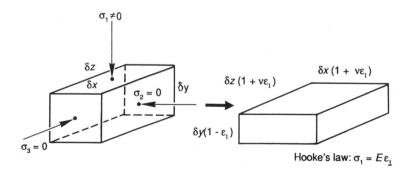

Fig. 2.8. Deformation under a uniaxial stress. Contraction in the direction of the compressive stress σ_1 is compensated by extension in the two orthogonal directions.

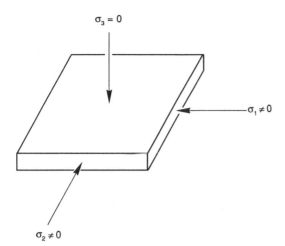

Fig. 2.9. Two non-zero principal axes of stress give the condition of plane stress.

β, the *compressibility*. These parameters therefore give the fractional volume change during isotropic compression under a given pressure. If there is a volume change, in order that matter is conserved, there must be an increase in density. Such a density increase δρ is given simply by

$$\delta\rho = \rho\beta p \qquad (2.14)$$

where p is the pressure, ρ is the density of the solid element and β the compressibility. In terms of the previously defined Young's modulus E and Poisson ratio v,

$$K = \frac{1}{\beta} = \frac{E}{3(1 - 2v)} \qquad (2.15)$$

showing that as v approaches 1/2 the bulk modulus tends to infinity, that is, the material becomes essentially incompressible.

2.1.5 Flexure in two dimensions

Because the lithosphere behaves elastically, it is able to bend when force systems or loads are applied to it. We shall return to this topic in considerable detail when considering the initiation and maintenance of foreland basins (Chapter 4). The aim here is to provide a brief theoretical background of the way the lithosphere responds by flexure to these applied force systems. A full analysis is provided by Turcotte and Schubert (1982). The concepts involved in the flexure of an

elastic solid may be briefly summarized as follows:
1 Flexure results from vertical forces, horizontal forces and torques (bending moments) in any combination. Horizontal loads are commonly neglected in geodynamical problems.
2 The bending moment is the integration of the fibre (normal) stresses on cross-sections of the plate acting over the distance to the midline of the plate. The bending moment is related to the local radius of curvature of the plate by a coefficient called the *flexural rigidity*. Flexural rigidity is proportional to the cube of the *elastic thickness*.
3 A general flexural equation can be derived which expresses the deflection of the plate in terms of the vertical and horizontal loads, bending moment and flexural rigidity. This equation can readily be adapted for use in the study of geological problems (Section 2.1.6 and Chapter 4).

In the simplest terms the flexure of a plate depends on its thickness, elastic properties and the nature of the applied load. Imagine a plate of thickness h and width L which is fixed at both ends and which is subjected to a line load at its midpoint (Fig. 2.10). In order to attain a force balance, the vertical line load V_a must be counteracted by vertical forces $V_a/2$ at both ends. Assuming that both plate thickness h and deflection w are small compared to the width of the plate L (as required by linear elastic theory), we can study the forces and torques on a small element of the plate. A downward force per unit area $q(x)$ is exerted on the plate by the applied load and on the end sections there is a net shear force per unit length V and horizontal force P per unit length, the latter being independent of x

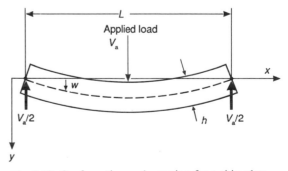

Fig. 2.10. Configuration and notation for a thin plate, pinned at its ends, bending under an applied load.

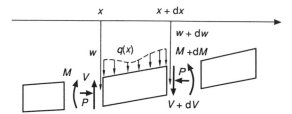

Fig. 2.11. Forces on a small element of a flexed plate. The forces can be balanced vertically and in terms of their tendency to rotate the element (moments or torques).

(Fig. 2.11). A bending moment M also acts on the end section related to the effects of normal stresses on the cross-section, σ_{xx}. These normal stresses are known as *fibre stresses* (Fig. 2.12).

The bending moment M is related to the curvature of the plate, since forces on the end section exert a torque about the midpoint of the plate. If the force on an element of thickness dy on the end of the plate is $\sigma_{xx}\, dy$, then this force will exert a torque about the midpoint ($y = 0$) of $\sigma_{xx}\, y\, dy$. Integrated over the entire end section we obtain the bending moment

$$M = \int_{-h/2}^{h/2} \sigma_{xx}\, y\, dy \qquad (2.16)$$

The fibre stresses σ_{xx} result in longitudinal strains in the plate, contractional in the upper half and extensional in the lower half (Fig. 2.12). In the two-dimensional case we assume that the plate is infinite in the plane normal to the figure, so there is no strain in the direction perpendicular to the xy

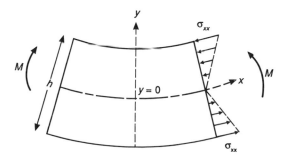

Fig. 2.12. Normal stresses on the end section of a flexed plate. These normal or fibre stresses exert torques about the midpoint of the plate which when integrated over the end section (from $-h/2$ to $+h/2$) give the bending moment.

plane, that is, $\varepsilon_{zz} = 0$. This is the situation of plane strain (Section 2.1.2). If the plate is thin, it is possible to make the further assumption that stresses normal to the plate's surface are zero ($\sigma_{yy} = 0$). The relations between stress and strain can then be restated as (cf. eq. 2.10)

$$\varepsilon_{xx} = \frac{1}{E}(\sigma_{xx} - \nu\sigma_{zz}), \qquad (2.17)$$

and from equation (2.11)

$$\varepsilon_{zz} = \frac{1}{E}(\sigma_{zz} - \nu\sigma_{xx}) \qquad (2.18)$$

If the bending is two-dimensional (there is no strain in the direction perpendicular to the page in Fig. 2.13), $\varepsilon_{zz} = 0$, then (2.17) and (2.18) give

$$\sigma_{xx} = \frac{E}{(1 - \nu^2)}\varepsilon_{xx} \qquad (2.19)$$

which is the relation between fibre stresses and longitudinal strains. The bending moment can now be related to longitudinal strains by substitution of (2.19) in (2.16)

$$M = \frac{E}{(1 - \nu^2)}\int_{-h/2}^{h/2} \varepsilon_{xx}\, y\, dy \qquad (2.20)$$

The bending moment M can also be related to the deflection w. The local radius of curvature R is inversely proportional to the rate of change in slope of the deflection, or $- d^2w/dx^2$ (if strains are small and $dw/dx \ll 1$, and where the negative sign simply states that w is positive downwards) (Fig. 2.13a). In addition, the longitudinal strain is also related to the radius of curvature. This relation can be derived by geometrical similarity (Fig. 2.13b), for the length of the plate is dependent on the local radius of curvature and ϕ ($l = R\phi$) and the change in length of the plate (Δl) is determined by the distance from the midline of the plate (y) and ϕ ($\Delta l = \phi y$). Using this result, the longitudinal strains ($\Delta l/l$) can be expressed in terms of the deflection,

$$\varepsilon_{xx} = \frac{\Delta l}{l} = \frac{y}{R} = \frac{-y\, d^2w}{dx^2} \qquad (2.21)$$

The bending moment can therefore be rewritten

$$M = \frac{-E}{(1 - \nu^2)}\frac{d^2w}{dx^2}\int_{-h/2}^{+h/2} y^2\, dy \qquad (2.22)$$

which when evaluated between the limits $-h/2$ and $h/2$ gives

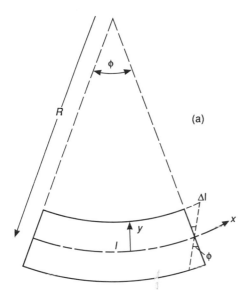

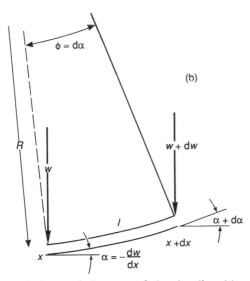

Fig. 2.13. Geometrical aspects of plate bending. (a) Longitudinal strain (extension above the midplane and contraction below the midplane of the plate) is a function of the distance from the midplane of the plate y, and the angle ϕ. (b) Notation to show that the second derivative of the deflection (d^2w/dx^2) gives the rate of change of slope of the plate; this is inversely related to the local radius of curvature R of the plate.

$$M = \frac{-Eh^3}{12(1-v^2)} \frac{d^2w}{dx^2} \qquad (2.23)$$

The coefficient of $-d^2w/dx^2$ is defined as the *flexural rigidity, D*. In other words,

$$M = -D\frac{d^2w}{dx^2} = \frac{D}{R} \qquad (2.24)$$

The force balance (Fig. 2.11) on a flexed plate with a downward force caused by loading $q(x)$ and horizontal forces (P) on the end sections is as follows.

Equating the forces in a vertical direction we have

$$q(x)dx = -dV$$

$$q(x) = \frac{-dV}{dx} \qquad (2.25)$$

Balancing torques we have in the counterclockwise direction a force P acting over a moment arm of $-dw$ and the torque dM. In the clockwise sense we have ($V + dV$) acting over a moment arm of dx. Since the term dV is small we can write

$$-P\,dw + dM = Vdx \qquad (2.26)$$

Differentiating (2.26) twice, so that (2.25) can be substituted into (2.26) gives

$$-P\frac{d^2w}{dx^2} + \frac{d^2M}{dx^2} = -q(x) \qquad (2.27)$$

Since we have already expressed the bending moment in terms of the flexural rigidity and the local curvature, (2.27) can be rewritten as

$$D\frac{d^4w}{dx^4} = q(x) - P\frac{d^2w}{dx^2} \qquad (2.28)$$

which is the general flexure equation for the deflection of a plate.

In the context of an elastic plate overlying a fluid-like mantle, this equation must be modified to account for a restoring force (buoyancy) acting upwards on the deflected plate (see below).

There will be different bending stresses under situations where the plate is pinned at one or both ends and according to whether it is point loaded (for example, at its free end) or uniformly loaded along its length, or loaded in some other fashion. However, the general flexural equation provides the basic starting point for more specific analyses.

When an applied load flexes a plate, the deflected region is filled either with water, as in the case of oceanic lithosphere or a starved continental basin, or with sediment, as in the case of most basins adjacent to hinterlands undergoing erosion. This infilling material has a smaller density than the mantle which is being replaced (Fig. 2.14, 2.15). The density difference can be denoted by $\Delta\rho$. The magnitude of the restoring force on the base of the deflected plate can be estimated by considering a balance of pressure ($\rho g h$) under the region of maximum deflection and under the unaffected region. This upward restoring force is $\Delta\rho g w$, and the net vertical force acting on the plate is the applied load less the restoring hydrostatic force. The general flexural equation (2.28) therefore becomes

$$D\frac{\mathrm{d}^4w}{\mathrm{d}x^4} + P\frac{\mathrm{d}^2w}{\mathrm{d}x^2} + \Delta\rho g w = q_a(x) \qquad (2.29)$$

where $\Delta\rho$ is ($\rho_m - \rho_w$) for a purely water-filled basin (Fig. 2.14) and $\Delta\rho$ is ($\rho_m - \rho_s$) for a fully sediment-filled basin (Fig. 2.15).

The lithosphere has a different flexural response according to the spatial distribution of the load. If the wavelength of a load of a certain mass, say excess topography, is sufficiently short, the vertical deflection of the lithosphere is small, and the lithosphere can be regarded as infinitely rigid for loads of this scale. However, if the wavelength of a load of the same mass is sufficiently long there is an effective isostatic response towards hydrostatic equilibrium and the lithosphere appears to have no rigidity. These two situations must be regarded as end members. The degree of compensation of the topographic load is the ratio of the deflection of the lithosphere to its maximum or hydrostatic deflection.

The way in which the lithosphere responds to sediment loads in evolving sedimentary basins is discussed in Section 8.3.4.

For reasonable values of plate thickness and flexural rigidity it is found that horizontal forces applied at the end of a plate are generally inadequate to cause buckling. Horizontal forces as buckling agents may, however, be much more important where the lithosphere has been strongly thinned in regions of high heat flow. The effects of horizontal stresses in amplifying existing lithospheric deformations are described in Section 6.2.

q_h = Upward hydrostatic force = $(\rho_m - \rho_w)gw$
q = Net force = $q_a - (\rho_m - \rho_c)gw$

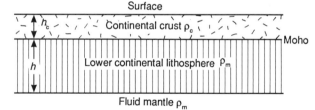

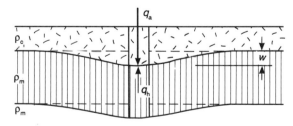

Fig. 2.15. Model for calculating the hydrostatic restoring force on the base of some continental crust where the deflection caused by the applied load q_a is assumed to be filled with material of the same density as the continental crust. This therefore approximates the case of a sediment-filled sedimentary basin on continental lithosphere.

q_h = Upward hydrostatic force = $(\rho_m - \rho_w)gw$
q = Net force = $q_a - (\rho_m - \rho_w)gw$

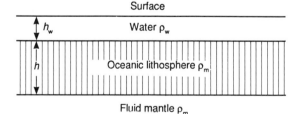

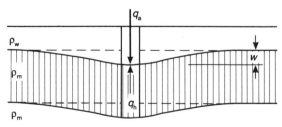

Fig. 2.14. Model for calculating the upward-acting hydrostatic restoring force on an oceanic plate, overlain by water, deflected by an applied force q_a.

The general features of basins such as oceanic trenches and foreland basins adjacent to mountain belts can be explained by flexural models. Applications of the foregoing discussion of elasticity and flexure to sedimentary basin analysis will be given in Chapter 4.

2.2 HEAT FLOW: CONDUCTION AND CONVECTION

2.2.1 Fundamentals

Empirical results from flexure observations in the oceans and on the continents (see Chapter 4) suggest that the upper 25 km to 50 km of the Earth behaves elastically on geological time scales. Rocks below this level deform with a fluid or creep-like behaviour and therefore relax stresses. To understand the mechanical behaviour of the Earth it is necessary to know something of its thermal structure, since rock rheologies depend on temperature, itself a function of depth. The temperature distribution of the Earth must reflect the inputs and outputs of heat to the Earth-system. In other words, there is a heat transfer or flow, achieved by processes of conduction, convection and radiation. The essential differences between these processes are as follows. Conduction is a diffusive process whereby kinetic energy is transferred by intermolecular collisions. Convection on the other hand requires motion of the medium to transmit heat. Electromagnetic radiation, such as that of the Sun, can also transmit heat, but it is of relatively minor importance in the Earth's heat budget.

The processes of conduction and convection are of differing importance in different zones. In the lithosphere heat is transported primarily through conduction, whereas in the mantle convection of heat from the Earth's deep interior is dominant.

The fundamental relation for conductive heat transport is given by *Fourier's law*. It states that the heat flux q is directly proportional to the temperature gradient and takes the mathematical form

$$q = -K\frac{dT}{dy} \qquad (2.30)$$

where K is the coefficient of thermal conductivity, T is the temperature at a given point in the medium and y is the coordinate in the direction of the temperature variation.

The heat flux at the Earth's surface gives a good indication of processes within the interior. Temperature measurements can be made on land in caves, mines, and better still, in deep boreholes and allow the heat flux to be calculated as long as the thermal conductivity K is known (see Section 9.4.1). K can be measured in the laboratory on rock samples by subjecting them to a known heat flux and measuring the temperature drop across the sample (Table 2.1). The large range of values for sedimentary rocks is due, to a large extent, to large variations in porosity (see also Section 9.3.1). Heat flow is in units of mW m^{-2} or cal cm^{-2} s^{-1}. Surface heat fluxes are sometimes expressed in heat flow units where 1 HFU is equivalent to 10^{-6} cal cm^{-2} s^{-1} or 41.84 mW m^{-2}.

Temperature measurements of the ocean floor can also be made by penetrating sea floor sediments with a temperature probe. The same probe contains a heater which enables the *in situ* thermal conductivity to be calculated.

The results of heat flow measurements of the oceans and continents reveal important variations (Table 2.2, Sclater, Jaupart and Galson 1980, Sclater, Parsons and Jaupart 1981). Regions of high heat flow on the continents generally correspond to active volcanic areas, such as the Andes, or to regions of extensional tectonics like the Basin and Range province of western USA. Continental collision zones typically have low to normal surface heat flows (further details in Section 9.6). In areas devoid of active tectonics and vulcanicity, the heat flow appears to be inversely correlated to age of the

Table 2.2 Regional variations in surface heat flow (data from Sclater, Jaupart and Galson 1980).

	Mean surface heat flow	
	mW m^{-2}	HFU
Continents:	56.6	1.35
Africa	49.8	1.19
North America	54.4	1.30
Australia	63.6	1.52
Oceans:	78.2	1.87
North Pacific	95.4	2.28
Indian Ocean	83.3	1.99
South Atlantic	59.0	1.41
Worldwide	69.9	1.67

rocks. This can be explained by the decreasing abundance with age of the radioactive heat-producing isotopes of uranium, thorium and potassium.

In the oceans the surface heat flows are related not to the concentration of radioisotopes but to the age of the sea floor. Newly created oceanic crust cools by conduction as it travels away from the mid-ocean ridge, thereby explaining this relationship. Mean oceanic surface heat flows are somewhat larger than their continental counterparts (Table 2.2).

The results of Table 2.2 can be expressed somewhat differently. The Earth's average surface heat flow corresponds roughly to one household light bulb (100W) over an area of a tennis court. Approximately 60 per cent of the heat loss of the Earth takes place through the ocean floor (Parsons 1982).

Since radioactive isotopes decay to stable daughter products, there must be a steadily decreasing heat production with time from radioactive decay. The rate at which heat is being transferred to the Earth's surface is therefore also decreasing with time, in turn slowing down the mantle convection system. Analysis of the abundances of the heat-producing radioisotopes and their stable products suggests that heat production was twice the present value 3000 million years ago, with ^{238}U and ^{232}Th taking over from ^{235}U and ^{40}K as the main heat producers because of the latter's relatively short half-lives.

2.2.2 Geotherms

For heat conduction in one direction only and at a rate that does not vary with time, the temperature T at any point in a plate can be related to the surface temperature T_s, depth y, heat production rate per unit mass H, density of the plate ρ and the surface heat flow q_s (Fig. 2.16):

$$T = T_s + \frac{q_s}{K}y - \frac{\rho H}{2K}y^2 \qquad (2.31)$$

This result can be used to calculate the variation of temperature with depth, the *geotherm*, assuming heat transfer by conduction only. However, this conduction model predicts a fully molten interior below about 150 km, yet the mantle is able to propagate shear waves. If surface heat flows were being contributed to substantially by crustal radioactivity rather than solely by mantle heat sources the problem would still remain, and the influence of mantle convection on heat flows appears to be the reason why the mantle conduction geotherm is invalid.

However, a conductive temperature profile appears to match *continental geotherms* where radiogenic heat production and conductive heat transport apply. Assuming that in the continental lithosphere the heat production due to radioactive elements decreases exponentially with depth (Fig. 2.17a), the geotherm is as shown in Fig. 2.17b.

In some cases, it is possible that heat is conducted in more than one direction, an application which is common in regions of large lateral varia-

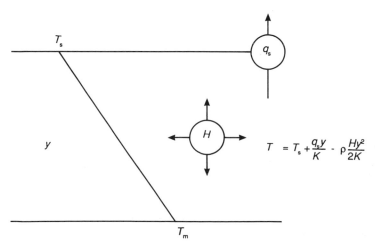

$$T = T_s + \frac{q_s y}{K} - \rho\frac{Hy^2}{2K}$$

Fig. 2.16. Heat conduction notation and the conduction geotherm. H is the rate of internal heat generation, q_s is the surface heat flow, T_m and T_s the mantle and surface temperatures, and K the thermal conductivity (see Turcotte and Schubert 1982, pp. 142–143, for derivation).

tions in surface temperatures. Such lateral variations might be due to topographic effects or to sea–land boundaries. The problem also arises where the lithosphere is stretched over a relatively narrow zone, so that there is both upward and

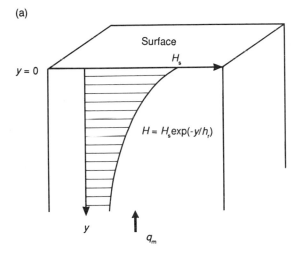

(a)

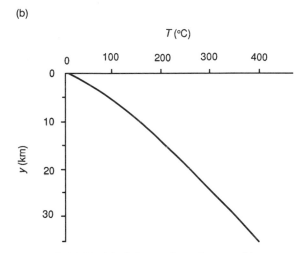

(b)

Fig. 2.17. (a) Model of the continental crust with an exponential radiogenic heat production with depth. H is the surface (y = 0) radiogenic heat production rate per unit mass, and h_r is the length scale for the exponential decrease in radiogenic heat production, that is, at a depth of y = h_r, the value of H has decreased to $1/e$ of its surface value H_s. (b) A continental geotherm resulting from the exponential distribution of radiogenic heat production shown in (a).

lateral loss of heat by conduction (see Section 3.1.6).

2.2.3 Time-dependent heat conduction: the case of cooling oceanic lithosphere

Many problems involve heat flows which vary in time. An obvious example is the heat flow associated with the intrusion of an igneous body, but the example used here is the cooling of the oceanic lithosphere and its consequent subsidence. This process finds great application in sedimentary basins experiencing a period of cooling following rifting. Heat sources *within* the medium are unimportant in the time-dependent problem of the cooling of oceanic lithosphere (that is, H = 0).

At the crest of an ocean ridge hot mantle rock is suddenly subjected to a cold surface temperature and then continues to lose heat to the cold seawater as the sea floor spreads away from the ridge. The initial cooling can, however, be treated as instantaneous and there will be a characteristic distance over which the sudden localized temperature change will be felt, the *thermal diffusion distance*. This distance must be a function of time since cooling or heating, and is equal to κt where κ is the *thermal diffusivity*, with units of length^2 time^{-1}.

Let the surface plates move away from the ridge with a velocity u (Fig. 2.18). The cooling rocks form the oceanic lithosphere and the boundary between this relatively rigid upper layer and the easily deformed mantle is an isotherm with a value

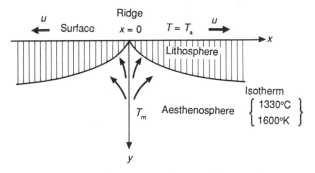

Fig. 2.18. Schematic diagram of the cooling oceanic lithosphere at a mid-ocean ridge. The oceanic plate moves away from the ridge at a velocity u. Its age is therefore determined by x/u, where x is the horizontal distance from the ridge crest.

of about 1600 °K (1300 °C). The thickness of the oceanic lithosphere is clearly a function of its age, where age can be expressed in terms of x/u.

Using a time-dependent instantaneous cooling model, the isotherms below the sea floor as a function of age are parabolic and illustrated in Fig. 2.19. The surface heat flow as a function of age is based on Fourier's law, the general form of which is found in equation (2.30). It is compared with actual heat flow measurements of the ocean floor in Fig. 2.20 (Sclater, Jaupart and Galson 1980). The regions of poor correspondence between the predicted and observed surface heat flow measurements is probably due to cooling by hydrothermal circulation of seawater. These effects become less important with age as impermeable sediments blanket the ocean floor. For large ages (> 80 Ma) however, an additional heat source appears to be recognizable, which may be mantle convection beneath the lithospheric plates.

In Section 2.1.1 the concept of vertical balancing of forces to a depth of compensation (isostasy) was introduced. As the lithosphere cools during its movement away from the ocean ridge it increases in density and therefore exerts a larger lithostatic pressure on the underlying mantle. This extra weight of cold lithosphere causes it to subside into the mantle and ocean water fills in the vacated space. Because the mantle is able to behave essentially as a fluid, it accommodates the subsidence of the overlying ocean lithosphere by flowage. Any vertical columns through the oceanic lithosphere can now be balanced (Fig. 2.21).

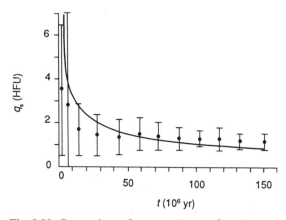

Fig. 2.20. Comparison of measured ocean floor heat flows (mean and standard deviation) and those predicted using the instantaneous cooling model, as a function of age. Data are from Sclater, Jaupart and Galson (1980).

The depth of the ocean floor w as a function of age (or distance from the ridge crest x) can be obtained using the geometrical arrangement of Fig. 2.21. The mass per unit area in a column of any age is made up of two components, the contribution of the rock column and that of the water column. At the ridge crest, however, there is just the effect of the overlying column of mantle, $w + y_L$ thick. For isostasy, the two columns must

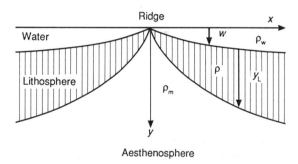

At any point, mass per unit area = $\int_o^{y_L} \rho \, dy + w \rho_w$
At ridge crest, mass per unit area at depth
$(w + y_L) = \rho_m(w + y_L)$
For equilibrium, $w(\rho_w - \rho_m) + \int_o^{y_L}(\rho - \rho_m)dy = 0$

Fig. 2.21. The principle of isostasy requires the ocean to deepen with age to offset the effects of thermal contraction of the ocean lithosphere. The water depth below the level of the ridge crest is w, the thickness of the oceanic lithosphere is y_L and ρ_m, ρ_w and ρ are the mantle, water and lithospheric densities respectively.

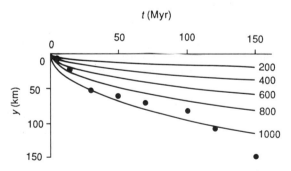

Fig. 2.19. Calculated isotherms for an oceanic lithosphere that is instantaneously cooled. The values of the isotherms are $T - T_s$ °K. The dots are estimated thicknesses of the oceanic lithosphere in the Pacific from Leeds, Knopoff and Kausel (1974).

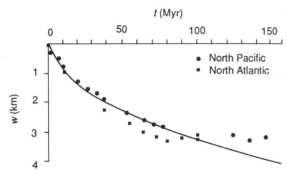

Fig. 2.22. Depth of the ocean floor below the level of the ridge crest as a function of age of the sea floor (after Parsons and Sclater 1977). The solid line shows the theoretical result for an instantaneous cooling model. It is in close agreement with observations from the North Pacific and North Atlantic.

balance. Introducing the volume coefficient of thermal expansion (following section) and an oceanic cooling curve (for instantaneous cooling of a semi-infinite half space), the water depth w as a function of the distance from the ridge crest can be obtained (Fig. 2.22). The theoretical result (using $\rho_m = 3300$ kg m^{-3}, $\rho_w = 1000$ kg m^{-3}, $\kappa = 1$ mm^2 s^{-1}, $T_m - T_0 = 1300$ °K and $\alpha_v = 3 \times 10^{-5}$ °K^{-1}) agrees reasonably well with measurements in the Atlantic and Pacific Oceans (Parsons and Sclater 1977).

2.2.4 Thermal stresses

The laws of thermodynamics say that the equilibrium state of a material is governed by any two state variables, examples being temperature T, pressure p and density ρ.

If a material is subjected to a change in pressure with temperature held constant, its volume will change. The change in volume for a certain pressure change is determined by the *isothermal compressibility* β. If pressure is held constant and temperature is varied, there will also be a fractional change in volume. This time, the factor determining the volume change is known as the *volumetric coefficient of thermal expansion* α_v. In both of these cases we consider volume to be specific volume, i.e. volume per unit mass, which relates to density by specific volume $V = 1/\rho$.

β and α_v are material properties that can be found from laboratory experiments, typical values for rock being $\beta = 10^{-11}$ Pa^{-1} and $\alpha_v = 3 \times 10^{-5}$ °K^{-1}. β and α_v can be related since any changes in specific volume are due to changes in both temperature and pressure. The total volume change is therefore the net change resulting from the pressure change ($- V\beta dp$, where dp is the pressure change) and from the temperature change ($+ V\alpha_v dT$, where dT is the temperature change). Note the opposing signs. With the material properties β and α_v given above, a 100 °K change in temperature will result in a pressure change of 3 kbar (300 MPa), implying large changes in pressure and therefore *stress* for moderately small temperature changes. This is an extremely important result since sedimentary basins are commonly associated with thermal disturbances (see especially Chapter 3).

What effects do temperature changes have on the strains expected using linear elastic theory (Section 2.1.4)? Clearly, thermal associated volume changes necessitate the modification of the laws of linear elasticity. The total strain in a body is the sum of the stress-associated strains and the temperature-associated strains. The former are given in equations (2.10) to (2.12) for an elastic material. The thermal-associated strains for an isotropic body are the same along each of the principal axes of strain, and it is known from the discussion above that the magnitude of the volume change is proportional to the temperature change by α_v (i.e. $\varepsilon_1 = \varepsilon_2 = \varepsilon_3 = -1/3 \; \alpha_v dT$). The minus sign implies an extensional strain by convention. However, it is convenient to consider strains as linear quantities, so it is necessary to introduce a new coefficient, the *linear coefficient of thermal expansion* α_1. This coefficient relates the thermally induced strains to the temperature change, where $\alpha_1 = 1/3 \; \alpha_v$ (i.e. $\varepsilon_1 = \varepsilon_2 = \varepsilon_3 = - \alpha_1 dT$ for an isotropic material). The linear coefficient of thermal expansion is therefore the change in strain per degree of change in temperature.

Thermal and stress induced strains can act in opposite senses. The removal (erosion) or addition (sedimentation) of overburden causes deviatoric stresses. In short, sedimentation causes extensional elastic horizontal stresses, whereas the thermal effect is to produce compressional stresses. The thermal effect appears to be predominant at geothermal gradients of greater than 23 °K km^{-1} (Turcotte and Schubert 1982, Fig. 4.43). Erosion causes the reverse to happen. The elastic effect causes surface compression, the thermal effect causes surface extension.

2.2.5 Thermal structure of the upper mantle: effects of convection

Whereas the thermal structure of the lithosphere is dominated by conduction, that of the mantle is determined primarily by convection. The lithosphere simply serves as a thermal boundary layer exhibiting high temperature gradients. Extensional basins, other than thin-skinned varieties, involve upwelling of convecting aesthenosphere. We therefore present here a very brief account of the thermal structure of the upper mantle.

In the interior of a vigorously convecting fluid, the mean temperature increases with depth along an *adiabat*, so that the *adiabatic temperature gradient* is the rate of temperature increase with depth caused by compression due to the overlying rock column. The compressional pressure forces cause a decrease in volume and therefore an increase in density. The relationship between the density and pressure changes is given by the *adiabatic compressibility*, β_a. For a solid it is somewhat smaller than the isothermal compressibility β given in equation (2.14). The purely adiabatic expressions for the variations of density and pressure with depth in the mantle do not perfectly match observed values based on seismic velocities (Fig. 2.23). In particular, at about 400 km there is a density discontinuity thought to be due to the phase change of olivine. This change to the denser spinel structure is thought to occur at a pressure of 14 GPa or 140 kbar and a temperature of 1900 °K. The phase change is *exothermic*, causing heating of the rock by *c.* 160 °K. The olivine–spinel phase change probably enhances mantle convection rather than blocking it. There is a second discontinuity in density at about 650 km, but its origin is less clear – it is probably also due to a change in mantle composition.

If a substance is *incompressible*, its volume is incapable of contracting, so adiabatic heating cannot take place. Rocks, however, are sufficiently compressible that adiabatic temperature changes are extremely important under the large pressure changes in the mantle. The adiabatic geotherms in the upper mantle underlying oceanic and continental lithospheres are different since continental lithosphere has its own near-surface radioactive heat

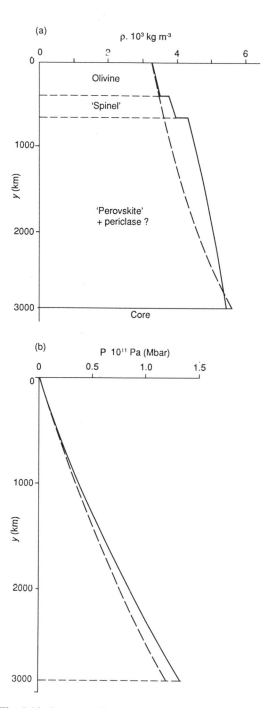

Fig. 2.23. Depth profiles of density (a) and pressure (b) in the mantle. Observed values are shown in solid lines; values calculated for a purely adiabatic behaviour are shown in dashed lines.

source. Below this near-surface layer the heat flow is assumed to be constant at about 30 mW m^{-2} (Fig. 2.24). There is an important feature underlying Fig. 2.24, namely that the continental lithosphere as a thermal entity extends to about 200 km below the surface (see Section 1.2).

It has been suggested that mantle convection takes place across the 400 km temperature and density discontinuity, essentially because subducted lithosphere appears to descend smoothly through this discontinuity. The influence of the 650 km discontinuity is less certain. Deep focus earthquakes provide evidence that convection takes place to 650 km but there is no direct evidence below this depth. Two contrasting models of mantle convection are therefore possible, one involving convection of the whole mantle, and one involving two major convecting systems, one shallow system linked to plate tectonics and a deeper system extending to the outer margin of the core. Smaller convecting cells may be superimposed on these major systems.

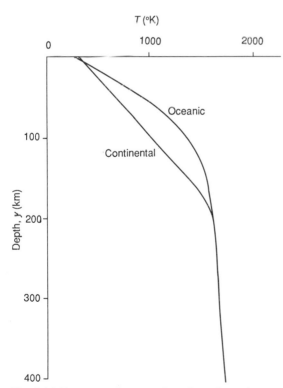

Fig. 2.24. Representative oceanic and continental geotherms in the shallow upper mantle.

The *onset of convection* is a threshold above which the influence of the temperature difference in the fluid layer exceeds that of the viscous resistance to flow (see Section 2.2.6). The temperature difference is conventionally evaluated as the fluid temperature at a point as a departure from the temperature expected from a purely conduction profile. The onset of convection is marked by a critical dimensionless number known as the *Rayleigh Number*, Ra. A Rayleigh Number analysis of Earth's mantle suggests that it must be fully convecting.

There are also large *lateral* temperature heterogeneities in the mantle, such as where cold lithospheric plates are subducted at ocean trenches. These lateral temperature variations are extremely important in providing a driving force for mantle convection. In the case of a descending cold lithospheric slab, the low temperatures of the plate cause it to be more dense than surrounding mantle, providing a gravitational body force tending to make the plate sink. A second factor is the distortion of the olivine–spinel phase change since the pressure at which this phase change occurs depends on temperature. The upward displacement of the phase change in the cold descending plate provides an additional downward-acting body force, helping to drive further plate motion. These two processes are often referred to as *trench pull*. If trench pull forces are transmitted to the oceanic plate as a tensional stress in an elastic lithosphere with a thickness of 50 km, the resultant tensional stress would be as high as 1 GPa.

The forces arising through the elevation of the ridge crest relative to the ocean floor constitute a *ridge push* force, also helping to drive plate motion (see also Section 3.1.3). Ridge push is an order of magnitude smaller than trench pull, but since the latter is countered by enormous frictional resistances, the two agents of net trench pull and ridge push are probably comparable.

2.2.6 Mantle viscosity

One of the fundamental differences between fluids and solids is their response to an applied force. Fluids deform continuously under the action of an applied stress whereas solids acquire a finite strain. Stress can be directly related to strain in a solid, but in fluids applied stresses are related to *rates* of strain, or alternatively velocity gradients. In Newtonian fluids there is a direct proportionality between applied stress and velocity gradient, the coefficient of proportionality being known as *viscosity*. (In non-Newtonian fluids there may be complex relationships between applied stresses and resultant deformations.) The problem of direct relevance to sedimentary basin analysis is the viscous flow in the mantle. One way in which the viscosity of the mantle can be estimated is by studying its response to loading and unloading.

Mountain building is an example where the crust–mantle boundary is depressed through loading, but orogeny is so slow a process that the mantle manages to constantly maintain hydrostatic equilibrium with the changing near-surface events. In contrast, the growth and melting of ice sheets is very rapid, so that the mantle adjusts itself dynamically to the changing surface load and the way in which it does so provides important information on mantle viscosity. The displacement of the surface leads to horizontal pressure gradients in the mantle which in turn cause flow. In the case of positive loading, fluid is driven away from the higher pressures under the load, the reverse being true on unloading. The surface displacement decreases exponentially with time as fluid flows from regions of elevated topography to regions of depressed topography. If w is the displacement at any time and w_m is the initial displacement of the surface, the form of the exponential decrease in surface topography with time is

$$w = w_m \exp(-t/\tau_r) \qquad (2.32)$$

where τ_r is the characteristic time for the exponential relaxation of the initial displacement. It is given by

$$\tau_r = (4\pi\mu)/(\rho g\lambda) \qquad (2.33)$$

where μ is the viscosity and λ is the wavelength of the initial displacement. Mantle viscosity can therefore be estimated from post-glacial rebound if this relaxation time can be found.

Elevated beach terraces which have been dated (usually by ^{14}C) provide the basis for quantitative estimates of the rate of post-glacial rebound. Large lateral variations in rebound rates are also found. For example, the northern shore of Lake Superior is rising relative to the southern by as much as 0.46 m per century following the Canadian glaciation (Kite 1972). The Baltic region of Scandinavia is also renowned for its raised beaches. After correcting the uplift of Swedish beaches for absolute

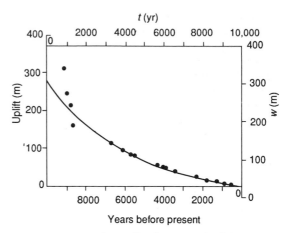

Fig. 2.25. Post-glacial uplift of the mouth of the Angerman River, Sweden over the last 10 000 years, compared with the exponential relation in equation (2.32). Such data can be used to estimate mantle viscosity. After Turcotte and Schubert (1982, p. 248).

(eustatic) changes in sea level, a relaxation time of $\tau_r = 4400$ years is found. Assuming a reasonable wavelength of the displacement for the Scandinavian glaciation to be $\lambda = 3000$ km, the viscosity of the mantle is estimated to be $\mu = 1.1 \times 10^{21}$ Pa s (10^{22} poises) (Fig. 2.25). However, this approximate analytical solution does not take into account (1) the flexural rigidity of the elastic lithosphere and (2) the depth-dependency of mantle viscosity. A summary of the possible distribution of viscosity in the mantle (Cathles 1975) is given in Table 2.3.

2.3 ROCK RHEOLOGY

2.3.1 Fundamentals

Joints and faults are evidence that rocks can behave as *brittle* materials, that is, they behave elastically

Table 2.3 Viscosity of the mantle from glacial rebound studies (Cathles 1975)

Region	Depth (km)	Dynamic or absolute viscosity	
		(Pa s)	(Poise)
Lithosphere	0–100	Elastic	
Aesthenosphere	100–175	4×10^{19}	4×10^{20}
Mesosphere	175–2848	10^{21}	10^{22}

up to a limit, beyond which they fail by fracturing. On the other hand, the widespread occurrence of folds suggests that rocks can also behave in a *ductile* manner. Three parameters are important in determining the brittle to ductile transition (Paterson 1958, Heard 1960):

1 Pressure.
2 Temperature.
3 Strain rate.

In order to model the behaviour of mantle and crustal rocks it is often necessary to assume an ideal rheology of initially linear elastic behaviour, followed after the yield strength has been attained, by perfectly plastic behaviour.

Mantle convection is thought to be attributed to two process that allow rocks to deform by thermally activated creep.

(a) *Diffusion* processes operate at very low stress levels, in which the crystalline solid behaves as a Newtonian fluid with a viscosity which depends exponentially on pressure and the inverse absolute temperature.

(b) Motion of *dislocations* is more effective at higher stress levels. It is a non-Newtonian fluid behaviour with the same exponential pressure and inverse temperature dependence.

Creep also takes place in the lower lithosphere where it can relax elastic stresses. In this case the rheology is a combination of an elastic and viscous behaviour – a *viscoelastic* rheology (Section 2.3.4).

Within the crust rocks may exhibit ductile behaviour under stress by *pressure solution creep* (Rutter 1976, 1983). Dissolution of minerals in zones of high pressure and their precipitation in areas of low pressure causes creep even at low temperatures and pressures. Otherwise at low temperatures and pressures rocks may deform in a brittle manner by fracturing.

The viscosities of both diffusion creep and dislocation creep are temperature dependent. If dislocation creep is the dominant mechanism of flow in the mantle, then the effective viscosity of the mantle will be *stress-dependent* as well as temperature dependent. This dependence of effective viscosity on stress is well illustrated by the flow of a power-law rheology fluid in a channel (Fig. 2.26).

For a power-law fluid the strain rate or velocity gradient is proportional to the power n of the stress. Velocity gradients and therefore strain rates are greater near the walls of the channel where

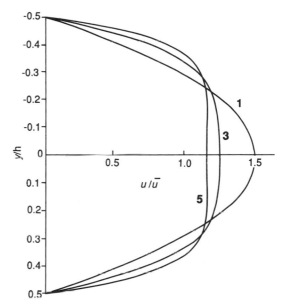

Fig. 2.26. Velocity profiles in a channel of thickness h for power-law fluid rheologies with $n = 1$ (newtonian), $n = 3$ and $n = 5$. Distance from the channel wall is expressed by y/h. The velocity u is scaled by the average velocity, $\bar{u}$.

shear stress is a maximum, whereas a core region experiences small strain rates. This plug-flow appearance of the velocity profiles for large n is a direct consequence of the stress dependency of the effective viscosity μ_{eff} which changes from low at the walls to high at the centre of the flow. This must be true because

$$\mu_{eff} = \tau \Big/ \frac{du}{dy} \qquad (2.34)$$

i.e. the viscosity is the coefficient linking the shearing stresses to the resultant velocity gradient or strain rate.

The idea can be applied to shear flow in the aesthenosphere which can be treated as having a heated lower boundary and an upper cooler boundary at the base of the rigid lithosphere. Shear in the aesthenopheric velocity profile is concentrated in zones close to the lower boundary of the aesthenosphere where the fluid is hottest and viscosity the smallest. The upper part of the aesthenosphere on the other hand tends to behave like a rigid extension of the overlying lithosphere as a result of this temperature and stress dependency of viscosity. Frictional heating can also have important consequences for shear flow of a fluid with a strongly temperature-dependent viscosity.

2.3.2 Rheology of the mantle

It was initially believed that at the high temperatures and low strain rates of the upper mantle, Newtonian (linear) flow would take place (e.g. Orowan 1967). Alternatively, the mantle may act as a power-law fluid, in which case the mantle's viscosity would be the stress-dependent effective viscosity associated with dislocation creep.

Most views of mantle rheology, however, come from laboratory studies. Since the mantle is composed primarily of olivine, laboratory studies of the creep of olivine at high temperatures are particularly relevant to mantle rheology. The strain rate in dry olivine $\dot{\varepsilon}_{xx}$ (or $-\dot{\varepsilon}_{yy}$) at 1400 °C as a function of stress appears to follow a cubic power-law rheology reasonably well (Ashby and Verall 1977) (Fig. 2.27). Other rocks deform in a non-linear way at high temperatures in the laboratory, with slightly different coefficients in the power law (ice $n = 3$, halite 5.5, limestone 2.1, dry quartzite 6.5, wet quartzite 2.6, diabase 3). The theoretical formula connecting strain rate and stress in dislocation creep closely matches the empirical relation for dry olivine in the laboratory, suggesting that dislocation creep is the dominant deformation mechanism. Although the laboratory experiments were at

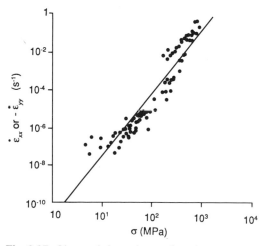

Fig. 2.27. Observed dependence of strain rate on stress for olivine at a temperture of 1400 °C (after Ashby and Verall 1977). Dry olivine obeys an approximate cubic power law rheology in these laboratory experiments.

strain rates several orders of magnitude greater than mantle strain rates, the dislocation creep process is also thought to be dominant in the mantle. Only at exceptionally low strain rates, considerably smaller than those expected in the mantle, would diffusion creep become dominant.

It is reasonable therefore to assume that the mantle has a power-law rheology. This power-law effect is likely to be small, however, compared with the temperature dependence of mantle rheology.

2.3.3 Rheology of the crust

Faulting of the brittle upper crust is very familiar. However, there is also observational evidence that near surface rocks deform in both a plastic and fluid-like manner.

The texture of many folded rocks suggests that the deformation responsible for the folding was the result of diffusive mass transfer. But the lower temperatures involved preclude the thermally-activated diffusion processes discussed in Section 2.3.1. Instead, the diffusive processes take place through *pressure solution* whereby material is transported in solution from areas of high intergranular pressure and stress and precipitated in regions of low pressure and stress, leading to creep. Point and line contacts between grains in a sand would represent such high pressure/stress zones and the pressure solution process would lead to compaction. The presence of water around the grains acts as the solvent and facilitates the transport of dissolved material.

The solubility of a mineral increases under a compressive stress, for example a compressional deviatoric or tectonic stress, causing solution, and decreases under a tensional stress, leading to precipitation.

Strain rate is linearly proportional to applied stress in pressure solution creep, so the deformation is equivalent to that of a Newtonian fluid. It explains viscous folding of rocks at quite low temperatures.

Although pressure-solution probably dominates as the main ductile mechanism at low temperatures in the upper crust, the rheology of the crust is likely to be complex as a result of the many compositional changes taking place within it. Seismic refraction velocities have been used, for example, to interpret an amphibolite facies upper crust,

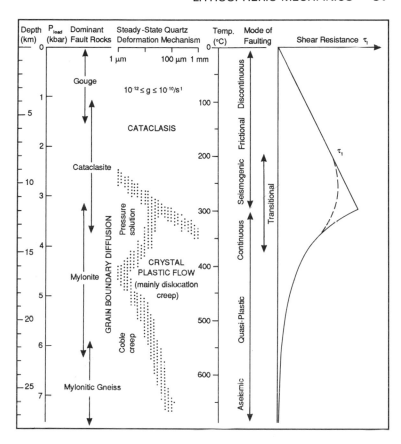

Fig. 2.28. Sibson's (1983) conceptual model for a major fault zone in the continental crust showing the dominant quartz deformation mechanism.

granulite facies middle crust and ultramafic lower crust in NW Britain (Bamford 1979, Hall and Al-Haddad 1976). Other authors have suggested much higher quartz contents, with significant proportions of acid and intermediate rocks metamorphosed to amphibolite and granulite facies in the lower crust (Fountain and Salisbury 1981). A layered continental crust should show low stress-low strength regions localizing strain during extension (Kusznir and Park 1987). In particular, a low stress-low strength region in the middle crust is thought to be the location of detachment of major extensional faults (Kusznir, Karner and Egan 1987). This topic of decoupling in mid-crustal regions is dealt with in greater depth in considering extensional basins in Chapter 3.

The consideration of rock deformation textures, laboratory studies of quartz-bearing rock and continental seismicity, led Sibson (1983) to conclude that a seismogenic upper crust characterized by discontinuous-frictional (brittle) faulting passed below 10 km (*c.* 300 °C) depth into an aseismic continuous quazi-plastic region where dislocation creep takes place in mylonitic fault zones. In the deep crust at temperatures greater than 450°C fully ductile continuous deformation is thought to dominate in gneissose shear zones (Grocott and Watterson 1980). The vertical zonation of deformation mechanism envisaged by Sibson (1983, p. 744) is shown in Fig. 2.28.

2.3.4 Viscoelasticity

The fact that seismic shear waves can be propagated through the mantle suggests that it is an elastic solid, yet it also appears to flow like a viscous fluid, enabling, for example, postglacial rebound to take place. A material that behaves as an elastic solid on short time scales but viscously on long time scales is known as a *viscoelastic* or *Maxwell* material.

In a Maxwell material, the rate of strain $\dot{\varepsilon}$ is the sum of the linear elastic strain rate $\dot{\varepsilon}_e$ and a linear viscous strain rate $\dot{\varepsilon}_f$. The elastic strain of a material under a uniaxial stress σ is

$$\varepsilon_e = \sigma/E \qquad (2.35)$$

as will be recalled from Hooke's law (equation (2.13)), where E is Young's modulus. The rate of strain is therefore the time derivative of ε_e, $d\varepsilon_e/dt$. The rate of strain in a viscous Newtonian fluid subjected to a deviatoric normal stress σ is a velocity gradient. The effective viscosity is taken as the ratio between stress and twice the strain rate. Therefore

$$\frac{d\varepsilon_f}{dt} = \frac{-\partial u}{\partial x} = \frac{\sigma}{2\mu} \qquad (2.36)$$

(where the minus sign simply indicates a tensional strain by convention). Since the total strain is the sum of the elastic and viscous fluid strains, the total strain rate is

$$\frac{d\varepsilon}{dt} = \frac{1}{2\mu}\sigma + \frac{1}{E}\frac{d\sigma}{dt} \qquad (2.37)$$

This is the fundamental rheological equation relating strain rate, stress and rate of change of stress for a Maxwell viscoelastic material. When a strain is initially rapidly applied to the viscoelastic medium, the time-derivative terms in (2.37) dominate and the material behaves elastically. Subsequently, if there is no change in the strain, the stress relaxes to l/e of its original value in a time $2\mu/E$ which is known as the *viscoelastic relaxation time*. For the aesthenosphere this relaxation time is of the order of 30 to 40 years, that is, longer than the period of seismic waves but shorter than the duration of postglacial rebound. The stress relaxation time is a strong function of temperature, rheological parameters and initial stress and its range of values for the lithosphere is controversial. Some workers believe that the relaxation time is very short ($< 10^6$ yr) while others believe it to be sufficiently long (20–30 Myr) to have a major impact on sedimentary basin geometry and subsidence. A viscoelastic rheology with a long relaxation time has been used in models of foreland basin evolution in particular (Beaumont 1978, 1981, Quinlan and Beaumont 1984). The application is treated in more detail in Chapter 4.

2.3.5 Elastic-perfectly plastic rheology

When confining pressures approach a rock's brittle strength a transition takes place from brittle (elastic) behaviour to plastic behaviour (Fig. 2.29). This transition is at the rock's yield stress σ_0. The elastic strain can be regarded as recoverable, but the deformation in the plastic field is not recoverable when the stress or load is removed (Fig. 2.30). If the deformation continues indefinitely without any

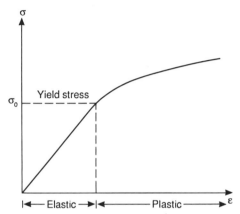

Fig. 2.29. Deformation of a solid showing a transformation from elastic to plastic behaviour.

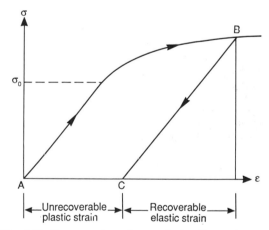

Fig. 2.30. Stress–strain trajectory for loading and unloading of an elastic-plastic material. Unloading of the material once it has passed into the plastic field results in an unrecoverable deformation, or plastic strain.

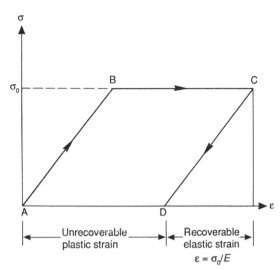

Fig. 2.31. The stress–strain trajectory for a material with an elastic-perfectly plastic rheology. Plastic strain continues indefinitely without any addition of stress above the yield stress.

addition of stress above σ_0, the material is said to exhibit an *elastic-perfectly plastic* behaviour (Fig. 2.31). Laboratory studies of the mantle rock dunite show that it conforms closely to the elastic-

perfectly plastic rheology (Griggs, Turner and Heard 1960). A typical depth at which the brittle-plastic transition would occur for dunite is about 17 km; below this the rock should yield plastically under large deviatoric stresses.

The above discussion can be applied to the bending of a plate. If the plate is flexed, a yield stress in the plate may be attained, marking the onset of plasticity. This onset corresponds to a characteristic plate curvature and has associated with it a characteristic bending moment. Consider the situation where the bending moment exceeds the critical value for the onset of plasticity. We have previously seen that the longitudinal strains in a flexed plate are large in its outer parts but small in the interior. As a result, the outer regions may deform plastically while the interior deforms elastically (an elastic core). After the bending moment for the onset of plasticity is attained there is a rapid increase in plate curvature due to a process known as *plastic hinging*. The plate then bends further, up to a maximum bending moment which is 1.5 times the critical moment for the onset of plasticity (Turcotte and Schubert 1982, p.344). The observed profile across the Tonga Trench, which shows a curvature far in excess of that predicted of a purely elastic plate may be due to this process of plastic hinging (Fig. 2.32). A similar process is discussed in Section 4.3.3 in relation to the bending of continental plates in mountain belts.

Fig. 2.32. Observed profile across the Tonga trench (solid) compared with an elastic plate profile assuming x_b = 60 km and w_b = 0.2 km. The excessive bending of the oceanic plate may be due to plastic hinging (see also Turcotte and Schubert 1982, pp. 342–345).

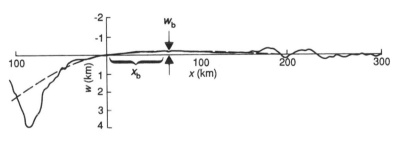

SECTION 2
THE MECHANICS OF SEDIMENTARY BASIN FORMATION

3 Basins due to lithospheric stretching

'When Earth breaks up and heaven expands
How will the change strike me and you
In the house not made with hands?'

(Browning, *By the Fire Side*)

SUMMARY

Basins due to lithospheric stretching fall within an evolutionary sequence from intracontinental rifts to passive margins.

Rifts are sites of crustal extension and present-day examples are characterized by negative Bouguer gravity anomalies, high heat flows and volcanic activity, all suggesting that a thermal anomaly exists at depth. Many contemporary rifts are associated with crustal doming, as in East Africa. It has been suggested that the rifting process falls into two classes

• *active rifting*, in which the impingement of a thermal plume or sheet on the base of the lithosphere causes convective thinning, domal uplift and crustal extension

• *passive rifting*, in which tensional stresses in the continental lithosphere cause thinning and the passive upwelling of hot aesthenosphere.

The heating of the overlying lithosphere by the introduction of a mantle hot plume may be due to (1) *conduction*, but this appears to be too slow to produce observed uplift rates in rifts, (2) *magma generation* and migration, but the absence of widespread plutonic activity suggests that this too is unimportant, and (3) *convection*; convective heating of the lithosphere may be rapid enough to explain rift uplifts when the basal heat flow from the aesthenosphere is several times the normal rate.

It is necessary to explain the source of the extensional deviatoric stresses responsible for rifting. Such stresses may result from uplift, imparting excess potential energy to the elevated region, or from disturbance of the isostatic equilibrium by thickness changes of the crust and mantle lithosphere. The thermal events, the thinning of crust/lithosphere and the elevation history of the surface are therefore all closely related, and it is not easy to discriminate the driving mechanism from the resultant process.

Salveson (1976, 1978) proposed a qualitative model of passive, mechanical extension of both the crust and subcrustal lithosphere, assuming the crust to fail by brittle fracture and the subcrustal lithosphere to flow. Such a process of extension leads to upwelling of aesthenosphere to maintain isostatic equilibrium. McKenzie (1978a) considered the quantitative implications of the passive rifting or mechanical stretching model, assuming the crust and mantle lithosphere to extend by the same amounts (i.e. *uniform stretching*). The subsidence is dependent primarily on the amount of stretching, β and the ratio of initial crustal to lithospheric thickness. For initial conditions of 'normal' lithosphere (i.e. crust *c.* 35 km, lithosphere *c.* 125 km) there is no initial uplift associated with the extension. Following mechanical stretching, the upwelled aesthenosphere should gradually cool, causing thermal contraction and regional subsidence. This thermal subsidence depends uniquely on β and characteristically takes the form of a negative exponential.

The lithosphere may extend non-uniformly with depth. Such non-uniform or *depth-dependent* extension is based on the view that the lithosphere is layered into different rheologies. The distribution of extension with depth may either be discontinuous or continuous. In the former case, deformation in the upper layer is required to be decoupled from deformation in the lower layers. In the latter case no such decoupling is required, but the subcrustal region stretches over a wider region than the crust. The elevation history of a particular point initially on the Earth's surface at sea level will vary according to its position in the developing basin. For example, if crustal thinning is concentrated in a central axial zone and subcrustal thinning is more widely distributed, locations near the basin centre should experience large amounts of early fault-controlled subsidence followed by moderate amounts of thermal subsidence. Near the basin-edge, however, the point should experience early uplift followed by thermal subsidence. Secondary convection in the upwelled aesthenosphere provides an alternative to depth-dependent stretching to explain rift flank uplift.

Extension may be asymmetrical, whereby faults propagate outwards in essentially one direction from the initial rift in the brittle upper crust, so that the upper crustal extension is spatially separated from the lower crustal/mantle lithosphere extension. Wernicke's shear zone model is an example of such simple shear, with a transcrustal shear zone transferring or relaying extension in the upper crust in one region to the lower crust and mantle lithosphere in another region. Both the depth-dependent models and the asymmetrical models substantially modify the uniform stretching predictions in terms of (1) the contributions of fault-controlled and thermal subsidence to the total subsidence, (2) the magnitude of the total subsidence. In particular, estimates of β from the phase of thermal subsidence are likely to be poor estimates of the amount of upper crustal extension, and (3) the spatial distribution of syn-rift uplift.

Uplift causes deviatoric tension, rather like the 'push' force at mid-ocean ridges. In active rifting, uplift may be due in part to thermal expansion above an aesthenospheric hot spot, but it is more likely to be due to mechanical uplift caused by lateral temperature gradients at the base of the lithosphere inducing horizontal variations in the normal stresses on the base of the plate. Uplift would then occur more or less instantaneously rather than be delayed by the time necessary for thermal conduction (say, 50 Myr). Both the aesthenosphere/lithosphere and crust/atmosphere boundaries are physically elevated by such a process, so each rheological layer will extend according to its own deformation mechanism. Basins formed by this process should possess some evidence of early uplift, perhaps recognized by regional unconformities.

Lateral transmission of mechanical energy through the lithosphere as a result of plate collision has been proposed as a cause of continental rifting. Such rifts, at high angles to the orogenic belt, have been termed *impactogens*. Numerical models using simple plane stress suggest that collision events are not responsible for the development of continental rifts. However, continental deformation can be regarded as a response of a continuous viscous medium to two kinds of forces (a) those applied at its edges by plate collisions and (b) those in continental interiors arising from crustal thickness contrasts. The relative importance of these two forces is described by the *Argand Number*, and an analysis of this type has been successful in explaining the deformation in Asia related to the collision of the Indian indenter. However, this viscous plate model applied to Asia does not satisfactorily explain the existence of the Baikal Rift.

If rifting continues to critical values of stretching, passive continental margins and spreading centres are developed. Passive margins are characterized by thick seaward-thickening prisms of sediments deposited in the post-rift phase. This massive post-rift subsidence may be due to sediment loading, phase changes (gabbro to eclogite), creep of ductile lower crust toward the ocean or thermal contraction following lithospheric thinning.

The subsidence history of thickly sedimented passive margins, such as off the eastern seaboard of the USA, can be studied by removing the effects of sediment loading. With the subsidence due to sediment loading removed, the residual subsidence is thought to reflect the true tectonic driving force. The tectonic subsidence appears to obey a $\sqrt{time}$ relationship. For highly stretched margins the thermal subsidence approximates that of the cooling half-space model of mid-ocean ridges. However, for younger margins or regions of reduced stretching, the thermal subsidence is of the form $e^{-t/\tau}$ (where t is time since the end of rifting and τ is the thermal time constant of the lithosphere), as in the uniform extension model of McKenzie (1978a).

The uniform extension model has been applied to the sediment-starved Bay of Biscay and Galicia continental margin. The observations from seismic reflection profiling fit the uniform extension model extremely well, with Airy isostasy prevailing throughout on this rapidly stretched margin. However, uniform extension may be associated with *melt segregation*. Such a model is more in harmony with the requirement for the formation of new oceanic crust. The basaltic melt may either be extruded, be intruded into the stretched lithosphere or underplate the lithosphere, forming an 'antiroot' and compensating to some extent for its thinning. As in the case of continental rifts, passive margins may be a response to *depth-dependent* extension. If the upper zone of brittle deformation is decoupled from the lower zone of ductile deformation, they may extend by different amounts. If the lower zone stretches more than the upper zone, uplift should occur if the depth of decoupling approximates the crustal thickness and should occur at the same time as extension. This would help to explain the high elevations interpreted prior to the rift-drift transition. On the thickly sedimented, mature, northeastern USA and Scotian Shelves, the depth-dependent model was particularly successful in explaining the initial uplift landward of the shelf edge, whilst all three models (uniform stretching, melt segregation and depth-dependent) satisfactorily accounted for the main features of the basin configuration.

Extensional faults may be rotational or nonrotational, listric or planar. Faults responsible for large magnitude earthquakes appear to be planar and inclined at 30° to 60° for as far as data are available. Seismogenic faults therefore do not appear to be listric. However, originally steep planar faults may be rotated, after slip has been transferred elsewhere, into subhorizontal or arched positions. This post-slip flattening of fault dip may explain many of the low-angle, apparently listric, extensional faults, with strata intersecting the fault plane at high angles in provinces such as the Basin and Range.

Fault patterns in rift systems were originally thought to delimit a symmetrical arrangement of horsts and grabens. Seismic reflection results suggest, however, that rifts are characterized by series of half-grabens, with the polarity of downthrow reversing over transform faults which segment the rift system.

Estimates of the amount of extension that has taken place can be obtained from (1) subsidence history curves (2) deep seismic results on crustal thickness changes (3) dip of planar rotational (domino) faults and less commonly (4) extensional structures in sediments and (5) offset patterns on transform linkages. Deep seismic reflection data from the northern North Sea indicate a crustal thinning of β = 2, which is in close agreement with subsidence history curves for boreholes in the same region.

3.1 RIFTS AND SAGS ON CONTINENTAL LITHOSPHERE

3.1.1 Introduction to observations and models

Rifts are areas of crustal extension and seismic studies show them to overlie thinned crust. Regions of rifting at the present day are characterized by negative Bouguer gravity anomalies, high heat flow (90 to 115 mW m^{-2}, i.e. over 2 HFU) and volcanic activity (Table 3.1), all of which suggest that a thermal anomaly exists at depth. The East African Rift may be regarded as a classical continental rift system. (Baker, Mohr and Williams 1972, King and Williams 1976). It is characterized by two updomed areas, the Ethiopian swell in the north and the East African swell in the south east (Fig. 3.1). Other domal uplifts are found in northern Africa such as those in the Tibesti and Hoggar regions. More often than not these swells are associated with widespread volcanic activity. Other striking rift systems are the Rio Grande Rift of the western United States, the Rhine Graben of Europe and the Baikal Rift of central Asia.

It has been suggested that rifting falls into two classes (Sengor and Burke 1978, Baker and Morgan 1981, Turcotte 1983, Morgan and Baker 1983, Keen 1985), active and passive. In *active rifting* the surface deformation is associated with the impingement on the base of the lithosphere of a thermal plume or sheet. Conductive heating from the mantle plume, heat transfer from magma generation or convective heating may cause the lithosphere to thin. If heat fluxes out of the astheno

sphere are large enough, relatively rapid thinning of the continental lithosphere would cause isostatic uplift. Tensional stresses generated by the uplift may then cause rifting. In *passive rifting* tensional stresses in the continental lithosphere cause it to fail, allowing hot mantle rocks to penetrate the lithosphere. Crustal doming and volcanic activity are only secondary processes (Turcotte and Oxburgh 1973). McKenzie's (1978a) widely accepted model for the origin of sedimentary basins belongs to this class of passive rifting. If passive rifting is occurring, rifting takes place first and doming may follow but not precede it. Rifting is therefore a passive response to a regional stress field. Sengor and Burke (1978) discuss the relative timing of rifting and volcanism. It is not easy to determine whether a given rift is either active or passive, since for small mantle heat flows the amount of uplift may be minimal. In addition, active and passive models are idealized abstractions which represent 'end-members'. Real world cases may exhibit aspects of each. The East African rift appears to be a good candidate for active rifting whereas it has been suggested that the Rio Grande rift may well be due to passive rifting.

Some rifts are at high angles to associated plate boundaries, such as orogenic belts (Burke 1976, 1977). Some of these rifts appear to be linked to arms of triple junctions associated with the early stages of ocean opening – these are termed *aulacogens*. Others are aligned at high angles to associated collision zones and are termed *impactogens* (Sengor, Burke and Dewey 1978, pp. 29–37) or collision grabens. Whereas aulacogens are formed contemporaneously with the ocean opening phase, impactogens clearly post-date this period, being related temporally with collision. The Upper Rhine Graben has been cited as an example of an impactogen (Sengor, Burke and Dewey 1978, pp. 29–37) and collision in the Grenville orogeny has

Table 3.1 Relation between Bouguer gravity anomaly, extension and volcanism in four continental rifts.

Rift	Volcanism	Crustal extension	Bouguer anomaly
Kenya Rift	Large volume; pre- and late-rift	25–35 km	– 100 to 200 mgal
Rio Grande	Smaller volume; early- and late-rift in S, late-rift in N	32 km	– 160 mgal
Baikal	Syn-rift	10 km	– 20 to 30 mgal
Rhine	Late-rift	5 km	+ 10 mgal

Fig. 3.1. The volcanic provinces
and sites of rifting in the African
continent. The black areas represent
regions with volcanic rocks with
ages of less than 26 Ma. Large
domal uplifts such as the Ethiopian
and East African Swells are
characterized by well-developed
rifts. Smaller uplifts such as the
Tibesti and Hoggar examples have
volcanic activity but lack rifts.

also been invoked as a cause of the Keweenawan
Rift in North America (Gordon and Hempton
1986). Impactogens are discussed in relation to
plate boundary forces in Section 3.1.10.

The fact that extension is taking place in rift
systems indicates that substantial tensional deviatoric stresses are present in the crust. It is the
source of these deviatoric stresses that is the
problem currently being debated.

Since active and passive rifting are idealized
end-member processes, it is useful initially to
consider the formation of continental rifts in terms
of the relation between the forces at work and their
effects, that is, between the uplift/subsidence,
crustal and lithospheric thinning, and tensional
deviatoric stresses.

In active rifting, an anomalous low-density mantle causes an isostatic uplift which generates extensional stresses or differential normal (i.e. vertical)
stresses on the base of the lithosphere resulting in
mechanical (non-isostatic) uplift. The heating of

the lithosphere may be due to:

1 Conduction from a sub-lithospheric heat source
causing thermal expansion.

2 Penetrative convection of magma into the lithosphere.

3 Convective heating at the base of the lithosphere
over a rising hot aesthenospheric plume.

Simple conduction appears to be too slow to
produce observed uplift rates (Mareschal 1983),
and magmatic heating is thought unlikely because
of the absence of evidence for widespread plutonic
activity together with a space problem caused by
the intrusion of vast volumes of magma (to heat a
volume of rock by 500 °C would require a volume
of magma equal to 20 or 30 per cent of the rock
volume). Convective thinning is widely regarded as
an important mechanism for thermal uplift but
there are differing opinions as to its efficacy. Some
believe that an increased heat flux into the lithosphere from an aesthenospheric convection system
of between 5 and 10 times the background heat

transport would be sufficient to thin the lithosphere to crustal levels in a few tens of millions of years (Spohn and Schubert 1983, Wendlandt and Morgan 1982). This is in agreement with the reported uplift rates of more than 1000 m in the last 10^7 years in the Rio Grande and East African rifts (Stewart 1977). Others believe that convective thinning would be too slow to be the primary mechanism for thinning and uplift (e.g. Turcotte and Emerman 1983).

Lateral temperature gradients in the aesthenosphere should result in lateral variations in the normal stresses exerted on the base of the lithosphere (Houseman and England 1986). The larger normal stresses over the postulated hot plume or sheet result in a mechanical uplift of the overlying lithosphere, imparting an excess potential energy in the updomed area and thereby generating extensional deviatoric stresses at both surface and subcrustal levels. Uplift should therefore essentially instantaneously accompany heating, and rifting should start as soon as the rock strength is exceeded.

The rise of an aesthenospheric diapir is the consequence of cold, dense mantle lithosphere overlying less dense, hot aesthenosphere. The development of a diapir depends on the effective viscosities of the aesthenosphere–lithosphere system, gravitational energy driving an initial thermal perturbation in this system. Numerical calculations suggest that large scale diapirism should occur in a few 10^6 to 10^7 years if the viscosity of the lower lithosphere is reduced to about 10^{20-22} Pa s (10^{21-23} poise), so significant and effective diapirism is to a large extent dependent on a 'preconditioning' of the lithosphere (by heating or stretching for example). Aesthenospheric diapirism is therefore unlikely to be the *sole* cause of rifting, but has been invoked to explain the Rhine Graben by Neugebauer (1983).

A number of models take passive rifting as their starting point. The uniform stretching of the lithosphere predicts, for 'normal' crustal and lithospheric thicknesses, that updoming will not accompany the extension. Although the uniform stretching model explains in outline the post-rifting cooling of extensional basins therefore, it has failed to explain the rift-flank uplifts so characteristic of many present-day rifts. This problem has been tackled in a number of ways:

1 Models involving non-uniform, or *depend-*ent stretching wherein the subcrustal lithosphere stretches by an amount greater than that of the crust. This introduces more heat into the rift zone encouraging early uplift, but it necessitates a decoupling of the two zones and causes a space problem in the mantle (Section 3.1.6.1).

2 If the geometry of the stretching in the subcrustal lithosphere is of an upward tapering region, so that *continuous non-uniform stretching* takes place with depth, the wider zone of diffuse mantle stretching would cause additional heating and uplift of rift shoulders (Section 3.1.6.2).

3 *Rifting over protracted periods of time* causes lateral heat losses during the rifting phase, causing rift flank uplift (Section 3.1.5.2).

4 *Secondary mantle convection* in the region of the extended lithosphere is another mechanism of transferring heat to the rift margins, causing thermal expansion and uplift (Section 3.1.7).

There is a further class of model which involves the lateral relay of the zone of subcrustal stretching away from the zone of near-surface faulting. Such a relay is achieved by means of a low angle shear zone penetrating the lithosphere, and gives rise to a characteristically asymmetrical geometry of crustal extension (Sections 3.1.8.1 and 3.1.8.2).

Table 3.2 summarizes these varied models.

3.1.2 Relationship between heat flow, thinning and uplift during convective heating

Continental surface heat flows are close to 50 mW m^{-2} in provinces older than 250 Ma, while in younger provinces they average 77 mW m^{-2} (Sclater, Jaupart and Galson 1980, Sclater, Parsons and Jaupart 1981). Assuming a radiogenic heat production confined to the continental crust, the basal heat flux from the aesthenosphere is approximately one half of the surface heat flux, or 25 mW m^{-2}. For a given thermal conductivity K and near-surface concentration of radioactive heat producing elements, the initial lithospheric thickness can be found to fit these surface heat flow data. If there is a sudden increase in the heat flux from the aesthenosphere by placing a hot plume beneath it, the thermal equilibrium is disturbed and the isotherm marking the lithosphere–aesthenosphere boundary starts to rise towards a new equilibrium position, the region below being occupied by the

Table 3.2 Models for stretching of the continental lithosphere.

Model	Selected references
Active	
Swell push	
(1) Gravity spreading	Houseman and England 1986
	Neugebauer 1978
	Bott 1981
(2) Diapirism	Woidt and Neugebauer 1981
Passive	
A: Pure shear	
(1) Uniform extension	McKenzie 1978a
(2) Uniform with induced mantle convection	Buck 1986, Steckler 1985
(3) Discontinuous non-uniform extension	Hellinger and Sclater 1983
(4) Continuous non-uniform extension	Rowley and Sahagian 1986
B: Simple shear	
(5) Asymmetrical with relay	Coward 1986, Wernicke 1985
Related to convergent plate boundaries	
(1) Impactogen	Sengor, Burke and
	Dewey 1978

plume. The uplift caused by thermal expansion can be estimated by isostatically balancing rock columns before and after thinning.

The time required to thin the lithosphere to a given fraction of its initial thickness can be illustrated (Fig. 3.2) by plotting the relative thinning $(y_L - y)/y_L$ against dimensionless time τ

$$\tau \equiv (K/\rho c y_c^2)t \qquad (3.1)$$

where

K is the thermal conductivity (W m $^\circ$K^{-1})
ρ is the uniform density of the lithosphere (kg m^{-3})
c is the specific heat (J kg^{-1} $^\circ$K^{-1})
y_L is the initial lithospheric thickness (km)
y is the thickness at time t(s)

Figure 3.2 shows this relationship between relative thinning and the square root of dimensionless time for two values of increased basal heat flux. The important result of this analysis is that the change in surface heat flow lags behind the thinning by an amount which increases with the magnitude of the thermal anomaly. Figure 3.2 in fact suggests that

by the time the surface heat flow has increased by 10 per cent of its total temperature change, the lithosphere has already been thinned by 50 per cent for $q = 2q_i$ and by 90 per cent for $q = 10q_i$ (q is the heat flow and q_i is the initial heat flow). Note also that there is a total amount of thinning possible which is dependent on initial, transitional and final subcrustal temperature gradients.

In order to determine the circumstances under which convective thinning of the continental lithosphere can be effective, it is possible to calculate the time required to thin the lithosphere by 95 per cent (t_{95}) for lithospheres of differing initial thickness and subject to different increases in basal heat flux (Fig. 3.3). Very thick lithospheres of 300 km, possible only under the oldest shields, can be thinned to crustal levels in tens of millions of years, the time scale usually associated with rifting (Siedler and Jacoby 1981), if basal heat flows increase by a factor of 10. A smaller factor of say 5 would suffice for thinner lithospheres. Surface heat flows from contemporary rift valleys (compiled by Morgan 1981) of 70 to 125 mW m^{-2} are somewhat less than expected from convective thinning models, but this can be explained by:

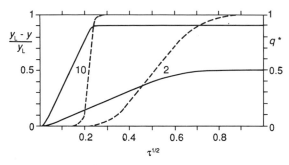

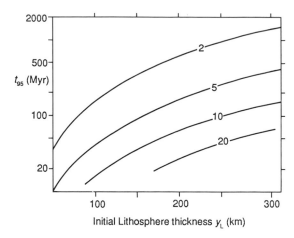

Fig. 3.2 The relative thinning of the lithosphere caused by convective heating as a function of the square root of dimensionless time for different values of increased basal heat flow. The relative thinning is the lithospheric thickness in relation to its initial thickness $(y_L - y)/y_L$ and the dimensionless time is given by equation (3.1). The relative thinning for basal heat fluxes representing an increase over the initial flux by a factor of 2 and 10 are illustrated. The increase in surface heat flow towards a final equilibrium surface heat flow (q^*) is shown by the dashed lines for the same increased basal heat fluxes. Note that following the thermal disturbance the thinning outpaces the change in surface heat flow. Young rifts may therefore have considerably more extension than can be accounted for by the observed surface heat flows. However, both the thinning and the increase in surface heat flow are completed at approximately the same time (after Spohn and Schubert 1982).

Fig. 3.3. The time taken to thin the lithosphere to 95 per cent of its initial thickness (t_{95}) as a function of its initial thickness (y_L) for different values of increase in basal heat flow. After Spohn and Schubert (1983).

1 The time lag between the increase in surface heat flow and thinning.
2 Consumption of part of the advected heat by partial melting once the lithosphere is substantially thinned – this heat will eventually reach the surface in the form of volcanism.

The magnitude of uplift of the continental surface caused by thermal expansion and isostatic equilibration varies according to two parameters in particular:
1 Uplift increases with initial lithospheric thickness (y_L).
2 It also increases with q/q_i provided that the final lithospheric thickness is greater than the thickness of the crust.

Uplift varies from 1 to 2 km for low q/q_i and small y_L. Most contemporary rift shoulders are uplifted between 1 and 4 km. Since uplifts are counteracted by subaerial erosion, the fit between calculated (model) results and observation is good.

3.1.3 Deviatoric stresses caused by uplift

It has been shown that large lithospheric swells tend to rift apart simply due to the lateral pressure gradients caused by the uplift (Artyushkov 1973, Bott and Kusznir 1979). This rifting force caused by elevation is directly analogous to the ridge-push forces helping to drive the motion of oceanic plates. In analysing the deviatoric stresses caused by swell-push forces an important observation should be borne in mind. Although most hot-spots are associated with swells (Crough 1979), very few of them are associated with significant rifting, the East African and Ethiopian uplifts being important exceptions (Morgan 1982). There may well be additional factors which need to be considered. These include:
1 Ambient state of stress in the lithosphere prior to uplift.
2 Variation of swell height with lithospheric age and hot-spot size.
3 The absolute motion of the plate overlying the hot-spot. Stationary or near-stationary plates with respect to the hot-spot are more likely to be updomed.
We shall consider the first of these three factors as follows.

Let us assume that the intraplate stress is determined solely by ridge-push forces, which is a reasonable assumption for plates such as Africa and Antarctica which are almost completely surrounded by spreading ridges. Consider the oceanic plate with a mid-ocean ridge, an adjacent basin and an oceanic swell (Fig. 3.4). The elevation of the ridge above the basin causes an excess density in the upper part of the ridge column with a compensatory density deficit in the lower part caused by the presence of hot aesthenospheric material. The hydrostatic pressure under the ridge is, however, always greater than that under the basin because of the excess density near the top of the lithosphere column under the ridge. This leads to a ridge-push force (the integral of the pressure difference over depth) which puts the oceanic lithosphere in the basin into compression. In relation to an oceanic swell, there is still a ridge-push force, although it is diminished by the effect of the relatively elevated

bathymetry over the swell. The swell should therefore experience a compressive stress, although somewhat smaller than in the ocean basin. Swells in the ocean generally rise to only 4 km depth below sea level (Crough 1978), whereas ridges rise to about 2 to 3 km below sea level. This would explain why oceanic swells do not rift.

The addition of continental crust radically changes the states of stress (Fig. 3.5). Using an identical concept of excess density and pressure in lithospheric columns, the net force from ridge to continental platform is mostly positive, giving a net compressive or ridge-push force which helps to maintain the integrity of the continental edge, stabilizing it against tensional failure. However, a negative pressure difference results from the elevated continental swell, so the swell pushes laterally harder than the ridge does, overcoming the

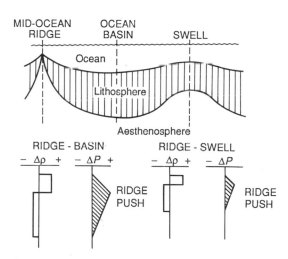

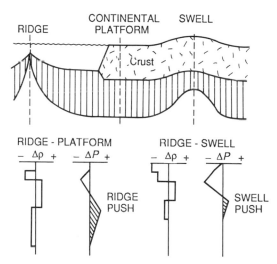

Fig. 3.4. Schematic diagram to illustrate ridge push. The cross-section shows an oceanic region with an elevated mid-ocean ridge, an oceanic swell and an intervening ocean basin. The lateral forces between the ridge, basin and swell can be considered by comparing the density and the hydrostatic pressure of the columns in the three regions. The difference in density is $\triangle\rho$, the pressure difference is $\triangle P$. The ridge crest has a positive density contrast with the ocean basin, so there is a larger hydrostatic pressure under the ridge. The net lateral force or 'push' is the integral of this excess hydrostatic pressure (shaded). The same applies in contrasting the ridge to the swell, but the lateral push is smaller. This may explain why oceanic swells seldom rift (after Crough 1983).

Fig. 3.5. Similar set-up to Fig. 3.4 but with the addition of continental crust. The pressure difference between the ridge crest and the continental platform at sea level is mostly positive, so there is a net lateral push from the ridge to the platform. A comparison of the ridge and the continental swell, however, shows a pressure deficit under the ridge, so the swell exerts a larger lateral force than the ocean ridge. If rock strength is overcome, the continental swell should extend by rifting (after Crough 1983).

ambient state of stress and promoting rifting. By quantifying this simplified configuration it can be shown (Crough 1983) that, as we have previously seen, the continental uplift is a direct function of the local lithospheric thickness.

The following conclusions can be made:

1 Continental areas at sea level are under compression due to ridge push.

2 There is a neutral height of continental crust above sea level at which a perfect force balance exists – between 500 m and 1.5 km for the model parameters.

3 Above the neutral height, the continental swells are subject to deviatoric tension.

This may explain why the Hoggar and Tibesti swells, at approximately 1 km above sea level are not rifted, whilst the Ethiopian plateau at 3 km height is extensively rifted (Fig. 3.1). Uplift-induced stresses must overcome rock strengths to cause faulting or flow (Kusznir and Park 1987). But on the other hand, the deviatoric tension in the swell can be added to by regional stresses and swells would therefore be the first places to fail. We consider mechanical stretching in a regional stress field in Section 3.1.10.

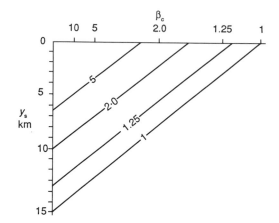

Fig. 3.6. Sediment-filled rift subsidence as a function of the lithospheric thinning (β_L) and crustal thinning (β_C). The uplift caused by lithospheric thinning offsets the subsidence caused by crustal thinning. The thickness of rift sediments predicted for a crustal and lithospheric thinning of 2 is approximately 2.5 km (after Turcotte and Emerman 1983).

3.1.4 Vertical movements and mean lithospheric stress caused by stretching

Irrespective of the active or passive mechanism for rifting, the shallow Moho under rift zones indicates that stretching of the lithosphere has taken place. Just as in Section 3.1.3, it is possible to balance isostatically two lithospheric columns, one representing the 'reference' continental lithosphere before stretching, and the other representing the stretched lithosphere along the rift axis (Turcotte and Emerman 1983). By ignoring any surface elevation changes during the stretching, and filling the central rift graben with air, sediment or water, the effects of uplift on deviatoric stresses can be ignored, and the effects of stretching can be considered.

The effects of lithospheric thinning and crustal thinning work in opposite directions in terms of vertical movements (Fig. 3.6). Large amounts of crustal thinning increase the subsidence in the rift axis, whilst large amounts of lithospheric thinning reduce it. This is a result of paramount importance which we shall return to in some detail in Sections

3.1.5 and 3.1.6, since we can now gain insights into the horizontal stresses in the rift zone resulting from crustal and lithospheric thinning but irrespective of an elevated topography in a swell.

By comparing the distribution of pressure in normal continental lithosphere and in the rift, the resultant tectonic force, F, can be estimated. If this force acts uniformly across the upper elastic lithosphere of thickness $y_{L,e}$ (stresses below this level being relaxed plastically or viscously), the mean stress in this elastic portion is

$$\bar{\sigma} = F/y_{L,e} \tag{3.2}$$

Equivalent elastic thicknesses of the continental lithosphere are poorly understood (Section 4.3), but taking $y_{L,e}$ as 50 km, the mean lithospheric compressional stress as a function of crustal thinning and lithospheric thinning is given in Fig. 3.7. By convention, a positive $\bar{\sigma}$ implies a compressive stress in the reference lithosphere or extension in the rift. Clearly this process is self-propagating, since thinning of the lithosphere causes tensional stresses which in turn promote rifting and further tensional

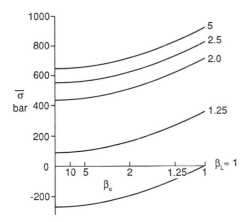

Fig. 3.7. The calculated mean lithospheric stress (averaged over an elastic lithosphere of thickness 50 km) as a function of the lithospheric and crustal thinning. There is zero stress for the undisturbed continental plate ($\beta_L = \beta_c = 1$). If the lithospheric and crustal thinning factors are both equal to 2, the predicted mean lithospheric stress in the rift is extensional and approximately 500 bar (after Turcotte and Emerman 1983).

stresses. Thinning of crust on the other hand reduces the tensional stresses, trying to equilibrate and stabilize the lithosphere–aesthenosphere system.

3.1.5 Mechanical stretching – pure shear

3.1.5.1 Kinematic models of Salveson and McKenzie

Salveson (1976, 1978) proposed that the subsidence histories of various continental rift basins and margins could be explained qualitatively by extension in both crust and subcrustal lithosphere (Fig. 3.8). He assumed the crust to fail by brittle fracture and the subcrustal lithosphere to flow plastically. The isostatic disequilibrium caused by the crustal extension leads to a compensating rise of the aesthenosphere, and consequent regional uplift. Partial melting of upwelled aesthenosphere leads to volcanism and further upward heat transfer. Eventually, after the continental lithosphere has been extended and thinned to crustal levels, new oceanic

crust is generated as the rift evolves into a continental margin.

McKenzie (1978a) considered the quantitative implications of a passive rifting or mechanical stretching model, assuming the crustal and lithospheric extension to be the same (*uniform stretching*). The stretching is symmetrical, no solid body rotation occurs, so this is the condition of *pure shear* (Section 2.2.1). He considered the instantaneous and uniform extension of the lithosphere and crust with passive upwelling of hot aesthenosphere to maintain isostatic equilibrium (Fig. 3.9). The initial surface of the continental lithosphere is taken to be at sea level and since the lithosphere is isostatically compensated throughout, the subsidence or uplift consequent upon mechanical stretching can be obtained.

The results of McKenzie's (1978a) quantitative model of uniform stretching can be summarized as follows:

1 The total subsidence in an extensional basin is made of two components; an initial fault controlled subsidence which is dependent on the initial thickness of the crust and the amount of stretching β; and a subsequent thermal subsidence caused by relaxation of lithospheric isotherms to their pre-stretching position, and which is dependent on the amount of stretching alone.

2 Whereas the fault-controlled subsidence is modelled as instantaneous, the rate of thermal subsidence decreases exponentially with time. This is the result of a decrease in heat flow with time. The heat flow reaches 1/e of its original value after about 50 Myr for a 'standard' lithosphere, so at this point after the cessation of rifting, the dependency of the heat flow on β is insignificant.

The assumptions, boundary conditions and development of the model are elaborated below. Whilst it is not necessary to understand the quantitative aspects of the model in order to appreciate its significance, a grasp of the methodology is extremely helpful in explaining the physical requirements of the system.

The lithostatic column before rifting is made up of two components (Fig. 3.10) (see Section 2.1, equation 2.1).

$$y_c \rho_c g + (y_L - y_c) \rho_{sc} g \qquad (3.3)$$

where y_c and y_L are the thicknesses of the crust and

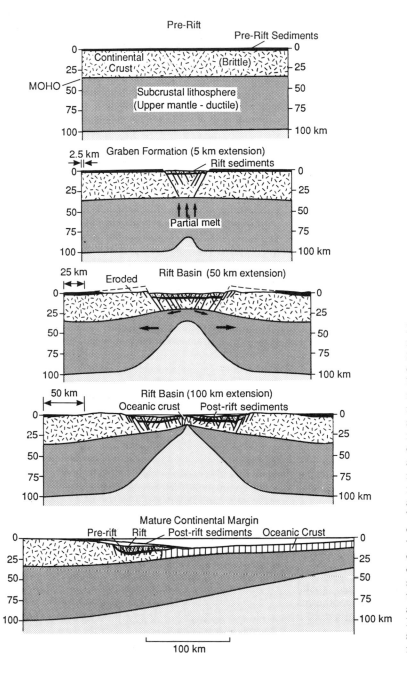

Fig. 3.8. Diagrammatic model of rift basin evolution following Salveson (1978). Tensional stresses causes the continental crust to fail by brittle fracture, whereas the mantle lithosphere fails by ductile necking. The formation of a sediment-filled graben causes isostatic disequilibrium and the compensating rise of the aesthenosphere; this leads to regional uplift. Partial melting of mantle promotes surface volcanism and an upward transfer of heat. The uplifted rift shoulders become eroded and the rift continues to fill with sediment. Eventually, as crustal extension continues, oceanic crust is created and the continent starts to cool as extension is transferred to the oceanic realm and a passive margin develops. Post-rift sediments drape the syn-rift fill and spread onto the newly created ocean floor.

lithosphere respectively, ρ_c and ρ_{sc} are the average densities of the crust and subcrustal lithosphere and g is the gravitational acceleration, which, because it is common to all lithospheric columns, can be ignored in the following analysis. These densities are assumed to have a linear relationship to temperature, and the geotherm is also assumed to be linear from T_m at the base of the lithosphere to T_0 at the surface (Fig. 3.10). As a result the average densities are given by

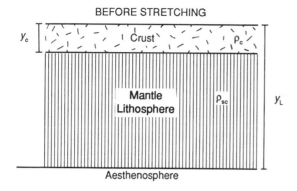

BEFORE STRETCHING

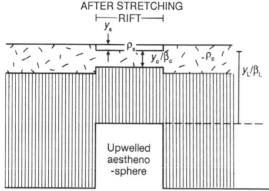

AFTER STRETCHING

Fig. 3.9. Schematic diagram to introduce the notation and fundamental precepts of the uniform stretching model. The crustal and lithospheric stretch factors are β_c and β_L. For uniform stretching $\beta_c = \beta_L$.

$$\rho_c = \rho_c^*(1 - \alpha_v T_c) \qquad (3.4)$$

and

$$\rho_{sc} = \rho_m^*(1 - \alpha_v T_{sc}) \qquad (3.5)$$

where ρ_c^* and ρ_m^* are crustal and mantle densities at 0 °C, α_v is the volumetric coefficient of thermal expansion and T_c and T_{sc} are the average crustal and subcrustal temperatures. Since the geotherms are linear, T_c and T_{sc} can be easily obtained as

$$T_c = \frac{(T_m - T_0)}{2} \cdot \frac{y_c}{y_L} \qquad (3.6)$$

and

$$T_{sc} = \frac{1}{2}\left\{T_m + \frac{y_c}{y_L}(T_m - T_0)\right\} \qquad (3.7)$$

By assuming that the surface temperature is 0 °C, these expressions simplify to

$$T_c = \frac{T_m}{2}\frac{y_c}{y_L} \qquad (3.8)$$

and

$$T_{sc} = \frac{T_m}{2}\left(1 + \frac{y_c}{y_L}\right) \qquad (3.9)$$

After rifting there are four components to the lithostatic stress at the depth of the original lithospheric thickness (Fig. 3.9):

$$y_s\rho_s + (y_c/\beta_c)\rho_c + \{(y_L/\beta_L) - (y_c/\beta_c)\}\rho_{sc}$$
$$+ \{y_L - (y_L/\beta_L) - y_s\}\rho_m \qquad (3.10)$$

where the new terms introduced are

y_s, the thickness of sediment, water or air filling the rift axis

β_c and β_L, the stretch factors for the crust and lithosphere, and for uniform stretching $\beta_c = \beta_L$

ρ_s, the average sediment, water, or air bulk density and

ρ_m, the mantle density (at a temperature of T_m), equal to $\rho_m^*(1 - \alpha_v T_m)$ where ρ_m^* is the mantle density at 0 °C.

Balancing the columns before and after uniform stretching ($\beta_L = \beta_c = \beta$)

$$y_c\rho_c + (y_L - y_c)\rho_{sc} = y_s\rho_s$$
$$+ (y_c/\beta)\rho_c + (y_L - y_c)(1/\beta)\rho_{sc}$$
$$+ (y_L - y_s - y_L/\beta)\rho_m$$

Regrouping the terms

$$y_s = \frac{(1 - 1/\beta)}{\rho_m - \rho_s}\{\rho_m y_L - y_c\rho_c - (y_L - y_c)\rho_{sc}\} \qquad (3.11)$$

Substituting the correct terms for ρ_c, ρ_{sc}, ρ_s and ρ_m into (4.11) we have after rearrangement

$$y_s =$$

$$\frac{y_L\left\{(\rho_m^* - \rho_c^*)\dfrac{y_c}{y_L}\left(1 - \alpha_v\dfrac{T_m}{2}\dfrac{y_c}{y_L}\right) - \dfrac{\alpha_v T_m \rho_m^*}{2}\right\}(1 - 1/\beta)}{\rho_m^*(1 - \alpha_v T_m) - \rho_s}$$

$$(3.12)$$

where

β is the stretch factor

y_L is the initial thickness of the lithosphere

y_c is the initial thickness of the crust

$\rho_m^* = \rho_{sc}^*$ is the density of the mantle (at 0 °C)

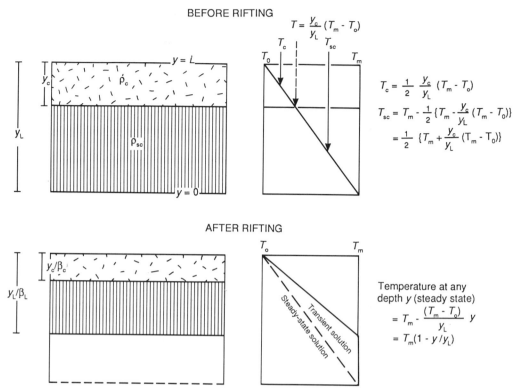

BEFORE RIFTING

$$T = \frac{y_c}{y_L}(T_m - T_o)$$

$$T_c = \frac{1}{2}\frac{y_c}{y_L}(T_m - T_o)$$

$$T_{sc} = T_m - \frac{1}{2}\{T_m - \frac{y_c}{y_L}(T_m - T_o)\}$$

$$= \frac{1}{2}\{T_m + \frac{y_c}{y_L}(T_m - T_o)\}$$

AFTER RIFTING

Temperature at any depth y (steady state)

$$= T_m - \frac{(T_m - T_o)}{y_L}y$$

$$= T_m(1 - y/y_L)$$

Fig. 3.10. Thermal consequences of uniform extension. On the right are continental geotherms resulting from the lithospheric stretching. The continental geotherms are linear with depth and the temperature at any depth y is determined by the mantle and surface temperatures and the initial thickness of the lithosphere. This can be regarded as a steady-state solution. After stretching, all points in the lithosphere are hotter and a new elevated geotherm results. Following rifting this geotherm decays to the initial geotherm in an unsteady fashion. This is the transient or unsteady temperature solution; it depends primarily on the amount of lithospheric stretching and the thermal diffusivity of the lithospheric materials. Fuller details in text.

ρ_c^* is the density of the crust (at 0 °C)
ρ_s is the average bulk density of sediment or water filling the rift
α_v is the thermal expansion coefficient of both crust and mantle
T_m is the temperature of the aesthenosphere

and y_s is positive for subsidence, negative for uplift. Using values for the above parameters, mostly derived from oceanic lithosphere studies (Parsons and Sclater 1977),

$$y_L = 125 \text{ km}$$

$$\rho_m^* = 3330 \text{ kg m}^{-3}$$

$$\rho_c^* = 2800 \text{ kg m}^{-3}$$

$$\rho_w = 1000 \text{ kg m}^{-3}$$

$$\alpha_v = 3.28 \times 10^{-5} \text{ °C}^{-1}$$

$$T_m = 1333 \text{ °C}$$

The initial subsidence S_i is positive for values of crustal thickness y_c greater than about 18 km. McKenzie's model therefore predicts an initial subsidence with no uplift where continental crusts are 'normal' in thickness (Fig. 3.11). This clearly is at variance with the generally elevated topographies of present day rift zones. The initial subsidence caused by isostatic adjustment to mechanical stretching is followed by a long-term gradual subsidence caused by cooling and thermal contraction of the lithosphere following extension. This *thermal subsidence* is dependent of β alone.

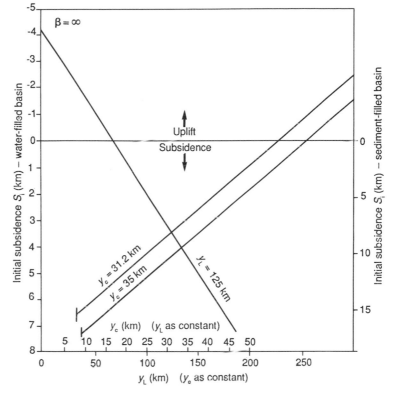

Fig. 3.11. Elevation changes resulting from uniform stretching of the lithosphere for different values of crustal/lithospheric thickness at $\beta = \infty$. For a 'standard' lithosphere of thickness 125 km and a 35 km thick crust, the predicted elevation change is between 4 and 5 km of subsidence for a water-filled basin. For a neutral surface (no elevation change) or uplift the crust must be thinner than about 17 km for a 125 km thick lithosphere (after Dewey 1982, p 386). Airy isostasy is assumed to prevail throughout.

The stretching of the lithosphere causes two responses
• the thinning of the crust and the fault-controlled subsidence is *permanent*, i.e. the brittle crust cannot regain its original thickness
• the thinning of the mantle lithosphere and any elevation changes caused by the presence of hot aesthenosphere are *transient*.

In order to predict the cooling of the lithosphere following rifting that causes the thermal subsidence, we must know the heat flow through time. Following the *instantaneous* increase in heat flow accompanying rifting in the McKenzie (1978a) model, the heat flow decreases exponentially with time.

We know two boundary conditions and we make two assumptions (Fig. 3.10):

Boundary conditions:
   (1) $T = 0$ at $y = L$
   (2) $T = T_m$ at $y = 0$

Assumptions:
   (1) lateral temperature gradients are much smaller than vertical gradients

$$\left(\frac{\delta T}{\delta x} \sim 0 \sim \frac{\delta T}{\delta z}\right)$$

   (2) internal heat production from radioisotopes is ignored

The one-dimensional unsteady (that is, time-dependent) heat flow equation is

$$\frac{\delta T}{\delta t} = K\frac{\delta^2 T}{\delta y^2} \tag{3.13}$$

where the second derivative gives the curvature of the geotherm as it relaxes to its pre-stretching gradient. The temperature at any depth and time ($T(y, t)$) is made up of a steady-state solution ($s(y) = T_m (1 - y/y_L)$) which applies to a linear geotherm in the lithosphere), and an unsteady-state component ($u(y, t)$). The general solution for the unsteady term is

$$u(y, t) = \sum_{n = 0}^{\infty} An \sin\left(\frac{n\pi y}{y_L}\right) \exp\left(- n^2\pi^2 \kappa t/y_L^2\right) \tag{3.14}$$

where A is a constant and n is an integer which expresses the order of the harmonic of the Fourier transform, and κ is the thermal diffusivity. Therefore, as n increases, the negative exponential becomes very small and ceases to contribute to the unsteady temperature field. As a gross approximation it is frequently satisfactory to consider only $n = 1$. The constant A depends on the aesthenospheric temperature (T_m) and the amount of stretching.

At $t = 0$, $An = \left\{ \dfrac{2}{\pi} (-1)^{n+1} \dfrac{\beta}{n\pi} \sin \left(\dfrac{n\pi}{\beta} \right) \right\} T_m$

so if $n = 1$, An simplifies to $\dfrac{2}{\pi} \cdot \dfrac{\beta}{\pi} \sin \left(\dfrac{\pi}{\beta} \right) T_m$

The full solution for $T(y,t)$ is therefore the sum of the steady and unsteady components,

$$T(y,t) = T_m \left(1 - \frac{y}{y_L} \right) + \left\{ \frac{2}{\pi} \cdot \frac{\beta}{\pi} \sin \left(\frac{\pi}{\beta} \right) \cdot T_m \right\}$$

$$\times \exp \left(- \frac{\pi^2 \kappa t}{y_L^2} \right) \cdot \sin \left(\frac{\pi y}{y_L} \right)$$

(3.15)

or regrouping and simplifying

$$\frac{T(y,t)}{T_m} = \left(1 - \frac{y}{y_L} \right) + \frac{2}{\pi} \cdot \frac{\beta}{\pi} \sin \left(\frac{\pi}{\beta} \right) \exp \left(\frac{-t}{\tau} \right) \sin \left(\frac{\pi y}{y_L} \right)$$

(3.16)

where $\tau = y_L^2 / \pi^2 \kappa$, and is known as the thermal time constant of the lithosphere.

The surface heat flux is given by Fourier's law (Section 2.2.1) which states that the flux is the temperature gradient times the thermal conductivity ($q_{y=L} = K \, \delta T / \delta y$). For $n = 1$, this is

$$q = \frac{K T_m}{y_L} \left\{ 1 + \frac{2\beta}{\pi} \sin \left(\frac{\pi}{\beta} \right) e^{-t/\tau} \right\}$$

(3.17)

The subsidence caused by the thermal contraction is then given by

$$S(t) \sim E_0 \frac{\beta}{\pi} \sin \frac{\pi}{\beta} (1 - e^{-t/\tau})$$

where $E_0 = 4 y_L \rho_m^* \alpha_v \, T_m / \pi^2 \, (\rho_m^* - \rho_s)$ (3.18)

and ρ_s is the average density of the water or sediment filling the basin, ρ_m^* is the mantle density at $0 \, °C$, α_v is the volumetric coefficient of thermal expansion and the remaining terms are defined above.

The result of equation (3.17) is that heat flux is strongly dependent on the amount of stretching,

the dependency becoming insignificant after about one thermal time constant (τ), or 50 Myr. At times younger than ~ 30 Myr, higher harmonics ($n = 2$, $3 \ldots$) should be used in equation (3.14). Heat flux is sketched in Fig. 3.12a. Subsidence due to thermal relaxation is shown in 3.12b for comparison. If the subsidence history of a basin is known, it is therefore possible to estimate β from the thermal subsidence curve.

Although some authors have claimed success in matching the crustal extension β, initial subsidence (eq. 3.12) and thermal subsidence (eq. 3.18) predicted by McKenzie with geological observations, others have shown serious discrepancies (Sclater *et al.* 1980, Royden and Keen 1980). [The various methods of estimating β are given in Section 3.3.3. Initial and thermal subsidence can be estimated from plots of depth versus time or 'geohistory' plots (Chapter 8)]. Specifically, the crustal extension and initial subsidence are commonly much less than the model prediction, or, equivalently, the thermal subsidence is much greater than that predicted from the observed extension β. Blackwell and Chockalingam (1981) have also investigated the thermal and elevation effects of extension using time-dependent models of the lithosphere and concluded that a mechanical source of extension is unable to generate the heat flow and uplift history actually observed in rift zones. McKenzie's (1978a) uniform stretching model is still, however, the basic starting point for the consideration of more complex variations.

1 Worked example on uniform extension

The North Sea is an intracontinental rift basin that underwent fault-controlled subsidence in the Permo-Triassic to the Early Cretaceous and thermal subsidence from about Aptian-Albian (*c.* 100 Ma) to the present day. This view is based on the observation that faults do not cut rocks younger than Early Cretaceous, suggesting that tectonic activity ceased at this time. In the post-rift phase of thermal contraction there are no major unconformities. Conoco 15/30-1 is situated in the north of the Central Graben (Sclater and Christie (1980)). The decompacted thicknesses of the stratigraphic units are derived in Chapter 8. Removing the effects of the sediment and water loads and variations in absolute sea level and bathymetry (see

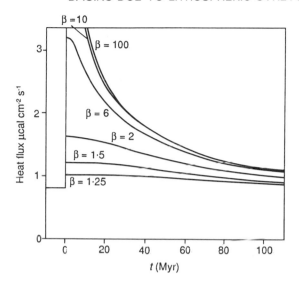

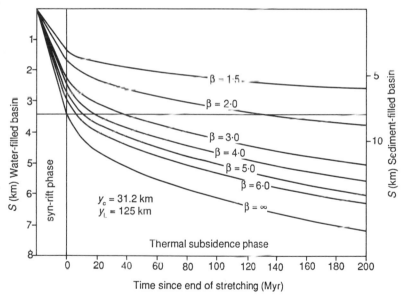

Fig. 3.12a. Heat flux as a function of time for various values of the stretch factor β, obtained using an equation of the form given in equation (3.17). The magnitude of the stretching-related heat flux is dependent on the stretch factor, but at long periods after the cessation of stretching, the heat fluxes are very similar irrespective of the initial amount of stretching (after McKenzie 1978a, p. 28).

Fig. 3.12b. The elevation change (subsidence) resulting from the heat flux patterns in Fig. 3.12a, for various values of the stretch factor, obtained from equation (3.18). The rate of thermal subsidence decreases exponentially with time. The curves shown refer to a lithosphere with an initial thickness of 125 km and an initial crustal thickness of 31.2 km (following Sclater, Jaupart and Galson 1980). After Dewey (1982, p. 387).

Chapter 8.1), the following tectonic subsidence values are obtained:

Time before present (Ma)	Thermal subsidence (km)
100	− 0.217
65	+ 1.031
55	+ 1.251
0	+ 1.854

McKenzie (1978a) shows that if this subsidence is the result of cooling of lithosphere of thickness y_L, thermal diffusivity κ, coefficient of thermal expansion $α_v$ and basal temperature T_m, following extension by a factor β, then the subsidence at time t due to cooling is given by equation 3.18.

Assume the lithosphere is 125 km thick (where undisturbed) under the North Sea area, $ρ_m$ = 3330 kg m^{-3}, $ρ_s$ = 1030 kg m^{-3} (for marine water), and κ = 10^{-6} m^2 s^{-1}.

1 Calculate the thermal time constant τ.

2 Plot thermal subsidence against $(1 - e^{-t/τ})$

where t is the time since the end of rifting. The slope of the best-fit line through the data points is $E_0 (\beta/\pi) \sin (\pi/\beta)$.

3 Calculate E_0 and thereby estimate β.

The borehole stratigraphy for Conoco 15/30-1 (Fig. 8.10) shows that the syn-rift subsidence of the basement is 2.23 km for the sediment loaded case. This value can be compared with an estimate of syn-rift subsidence from theory. McKenzie provides an expression in terms of crustal and mantle densities (ρ_c, ρ_m) and crustal thickness y_c, in addition to the parameters introduced above. The correct form of this expression is (note the error in the 1978 paper) given in equation (3.12).

Assume the following values for the parameters below:

$y_L = 125$ km

$y_c = 31.2$ km

$\alpha_v = 3.3 \times 10^{-5}{}^\circ\text{C}^{-1}$

$T_m = 1300\ {}^\circ\text{C}$

$\rho_m = 3330$ kg m^{-3}

$\rho_c = 2700$ kg m^{-3}

$\rho_s = 2066$ kg m^{-3} (bulk density of sediment column at 100 Ma).

Try to explain the discrepancy between syn-rift subsidence from theory and from the borehole stratigraphy.

Solution to uniform extension problem

In this problem the tectonic subsidence in the post-rift phase was calculated using an Airy isostatic model (Table 3.3).

The thermal time constant for a lithosphere of thickness 125 km and thermal diffusivity of 10^{-6} m^2 s^{-1} is 15.83×10^{14} s, or 50 Myr.

The slope of a best-fit line to the points on a $(1 - e^{-t/\tau})$ versus S_t graph is about 2.35. Since $E_0 = 3.15$ km, $\beta = 2.4$.

The syn-rift subsidence can be calculated from equation (3.12). ρ_s is equal to 2066 km m^{-3} because the figure of 2.23 km for the syn-rift subsidence of the basement applies to the total subsidence, not the water-loaded tectonic subsidence.

With $\beta = 2.4$ and the parameter values in the problem $S_i = 5.5$ km. This is far in excess of the observed value of 2.2 km. This discrepancy may be due to a number of factors:

1 An uplift event prior to Early Cretaceous and eroding down to Middle Jurassic levels may have removed a considerable thickness of syn-rift sediments.

2 The Conoco 15/30-1 area may have been uplifted prior to fault-controlled subsidence. The decompacted thickness of syn-rift sediment will therefore underestimate the true amount of fault-controlled subsidence of the basement.

3 The theory (which assumes essentially instantaneous stretching) may be inappropriate.

In conclusion, estimation of syn-rift subsidence is difficult because of the possibility of erosion over fault blocks, effects of regional uplifts, and the poor palaeotopographic control in predominantly continental sequences.

2 Worked example on heat flow associated with uniform extension

McKenzie (1978a) gives an expression for the surface heat flux from a cooling, stretched piece of continental lithosphere found in equation (3.17).

This expression gives the heat flux through the top of the basement, so the correct term for thermal conductivity is K_b, the thermal conductivity of the basement.

Table 3.3 Thermal subsidence calculated for uniform extension problem

Time before present (Ma)	Time since end of rifting (Myr)	Thermal subsidence (km)	$(1 - e^{-t/\tau})$
100	0	-0.217	0
65	35	1.031	0.503
55	45	1.251	0.593
0	100	1.854	0.865

Assuming that heat flux is independent of depth within the sediments, the temperature as a function of depth is given by

$$T(y) = \frac{yq(t)}{K_s}$$

which simply states Fourier's law for heat conduction (and we assume that the temperature at the surface is 0°C). K_s is the average conductivity of the sediments. The temperature at any depth is therefore

$$T(y) \approx \frac{K_b}{K_s} \frac{y}{y_L} T_m \left\{ 1 + \frac{2\beta}{\pi} \sin \left(\frac{\pi}{\beta} \right) e^{-t/\tau} \right\}$$

Calculate the temperature at the base of the Lower Cretaceous in Conoco 15/30-1 (140 Ma horizon) as a function of time since the end of rifting, with K_s = 1.25 W m^{-1} °K^{-1}, K_b = 3.0 W m^{-1} °K^{-1}.

Solution to heat flow exercise

The base of the Lower Cretaceous was at the following depths during the post-rift stage of thermal subsidence:

	Ma			
	100	65	55	0
Decompacted depth in km	0.851	2.098	2.524	3.976

The temperatures as a function of time since the end of rifting are shown in Table 3.4.

3.1.5.2 *Effects of finite rifting times and lateral heat conduction*

The quantitative model of McKenzie (1978a) assumed instantaneous rifting of the lithosphere fol-lowed by thermal subsidence as the lithosphere re-equilibrates to its pre-extension thickness. This is an attractive assumption since it gives a simple, well-defined initial condition for the thermal calculations. Using this one-dimensional model, Jarvis and McKenzie (1980) found that as long as rifting times were less than 20 Myr in duration, the uniform extension model was a very reasonable approximation. However, many sedimentary basins appear to have undergone protracted periods of rifting, considerably in excess of 20 Myr. The Paris Basin, for instance, rifted in the Mid-Permian and sedimentation was restricted to elongate rift troughs until close to the end of the Triassic. The rift phase was therefore close to 60 Myr in duration. Post-Triassic thermal subsidence allowed the basin to broaden out from the earlier rift troughs. The Triassic (Carnian–Norian, 212–200 Ma) continental red beds and evaporites of the Atlantic margin of north-eastern USA and Canada were deposited in fault-bounded rifts, but sea floor spreading did not commence until the Bajocian at about 170 Ma. The stretching in the Red Sea (Cochran 1983) appears to have occurred diffusely through a combination of extensional block faulting and dyke injection over an area of the order of 100 km wide. This phase of diffuse extension has lasted for 20 to 25 Myr in the northern Red Sea and is still occurring. The effect of a finite time of rifting is to cause a loss of heat and thus additional subsidence prior to the end of rifting or the onset of sea floor spreading. It effectively transfers a portion of the cooling from the post-rift phase to the syn-rift phase. The effects of finite rifting on subsidence should therefore be considered.

The possibility of significant lateral heat loss during prolonged periods of rifting is considerable. This process, which was not considered in the prototype uniform extension model of McKenzie (1978a), is caused by the large lateral temperature gradients set up by the lithospheric thinning. The additional heat loss away from the basin centre promotes increased

Table 3.4 Palaeotemperatures as a function of time since the end of rifting for the uniform extension heat flow problem

Time since rifting (Myr)	K_b/K_s	yT_m/y_L	$1 + (2\beta/\pi) \sin (\pi/\beta)e^{-t/\tau}$	T °C
0	2.4	8.85	2.4757	52.6
35	2.4	21.82	1.7328	90.7
45	2.4	26.25	1.6000	101.0
100	2.4	41.35	1.1997	119.1

basinal subsidence early in the basin's history, accompanied by lateral uplift (Fig. 3.13, 3.14) The longer the rifting event, the more likely lateral heat loss is to be important. Lateral heat conduction is also important in narrow basins such as pull-aparts in strike-slip zones (Section 5.6).

The one-dimensional (vertical) heat flow model considered by McKenzie (1978a) (Section 3.1.5.1) can be expanded to two dimensions by considering in addition conduction in the horizontal direction i.e.

$$T = T(x, y, t)$$

where T, the temperature, is a function of the horizontal coordinate x, the vertical coordinate y and the time t. Steckler (1981) has evaluated the horizontal heat flow term. This two-dimensional model can be adapted to estimate the effects of finite rifting by considering the extension to be composed of a large number of discrete, short, rifting events. During each event, the lithosphere is allowed to rift and then cool, so that after a large

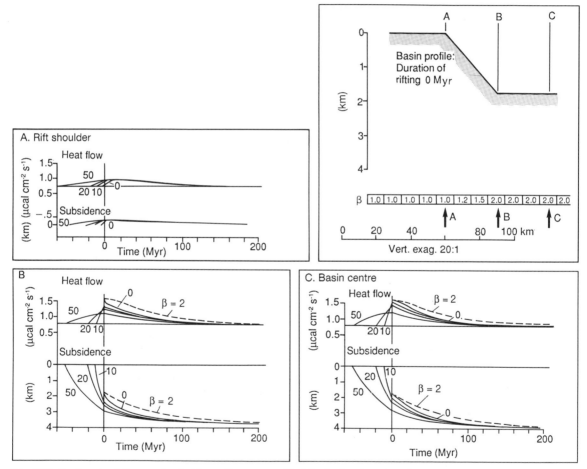

Fig. 3.13. Surface heat flow and subsidence for a water filled basin for three locations A, B and C located on the right flank, close to the hinge line and at the basin centre for the case of lateral heat conduction and finite periods of rifting (after Cochran 1983). Inset shows location of A, B and C in a basin with laterally varying stretching. Location A experiences increasing heat flows in the syn-rift phase, causing rift flank uplift, then gentle post-rift subsidence. Locations B and C experience rapid syn-rift subsidence, the total amount of syn-rift subsidence increasing with the duration of stretching. There is a commensurate reduction in the post-rift subsidence since much of the heat loss is transferred to the syn-rift phase. The dashed curves labelled $\beta = 2$ refer to the one dimensional instantaneous model of McKenzie (1978a).

number of steps a process of continuous rifting and heat loss is simulated.

The model results for a water-filled basin (Cochran 1983) show that there are important deviations between the instantaneous case and the finite rifting case for rifting times as short as 5 Myr, the differences becoming greater as the period of extension increases. The effects are greatest near the basin edge ('hinge-zone' between strongly extended lithosphere and little-extended lithosphere) where horizontal temperature gradients are largest. The effects decrease towards the centre of the basin.

The post-rift subsidence near the centre of the basin is decreased (relative to the instantaneous case) by a minimum of 10–15 per cent for a 10 Myr rifting event and by about 25 per cent for a 20 Myr rifting event. Since the syn-rift subsidence is increased at the expense of the post-rift subsidence, the effect is to flatten the post-rift subsidence curves relative to those predicted by the instantaneous one-dimensional model (McKenzie 1978a). This should lead to underestimates of the stretching β (Fig. 3.14). The model subsidence curves cross-cut the lines of equal β because of the lateral transport of heat out of the basin. The basin can be filled with sediment rather than water and the effects of the load of the sediments distributed by flexure (see Chapters 4 and 8). For a 20 Myr rifting time, about 70 per cent of the sedimentary column is represented by syn-rift sediments in Cochran's (1983) model. This is excessive compared to the thicknesses actually observed in extensional basins. The contribution of syn-rift sediments to the total sedimentary pile would be reduced if the lithospheric thinning exceeded the crustal thinning, a topic we investigate in the following section.

3.1.6 Non-uniform stretching

3.1.6.1 Discontinuous stretching with depth

Simple geometrical stretching in which the amount of stretching remains constant with depth without regard to changing rheology is thought by some to be unrealistic. It has been proposed that heat input could be increased during extension by thinning the lower crust and subcrustal lithosphere more than the upper crust, thereby raising the lithosphere–aesthenosphere boundary more than in the McKenzie model. This not only solves a heat flow problem, but it necessitates less crustal thin-

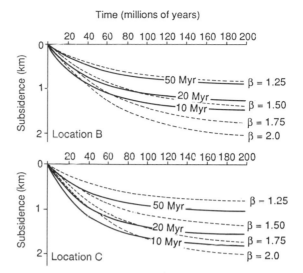

Fig. 3.14. Post-rift subsidence curves for locations B and C of the basin shown in Fig. 3.13 for durations of rifting of 10, 20 and 50 Myr (solid lines) compared to the subsidence curves resulting from the instantaneous one-dimensional model (dotted lines). Note that the two sets of curves cross-cut, making estimates of the amount of stretching from the post-rift thermal subsidence problematic unless the duration of stretching is known (after Cochran 1983).

ning for the same overall subsidence. Structural evidence (e.g. seismic studies of the Basin and Range Province, Bay of Biscay and Northwestern European Continental Shelf) suggests that some steep faults near the surface become listric into near-horizontal detachments where a transition to ductile behaviour takes place (Section 2.3). The focal depths of earthquakes in old cratons further suggest that while the upper part of the lithosphere has relatively high strength and is seismically active, the lower part is aseismic, possibly due to the operation of ductile deformation mechanisms (see useful discussion in Sibson 1983). Clearly, in some instances at least there may be decoupling of the upper and lower zones at about mid-crustal levels and it is therefore possible that the two rheological layers extend by different amounts, giving a non-uniform discontinuous stretching.

Instead of uniformly extending the lithosphere (Fig. 3.15a), consider a lithosphere of thickness y_L and crustal thickness y_c which undergoes an instantaneous stretching event. The crust is extended by

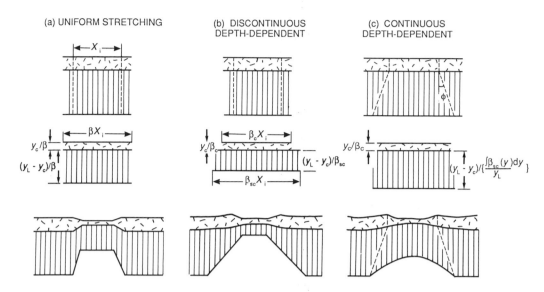

Fig. 3.15. Cartoons to illustrate three types of lithospheric extension (a) *uniform extension* in which the crust and mantle (subcrustal) lithosphere extend by identical amounts (McKenzie 1978a), (b) *discontinuous depth-dependent or non-uniform extension* in which the crust extends by a different amount to the subcrustal lithosphere; the crustal and subcrustal extensions are independent but are uniform throughout the crust and subcrustal lithosphere respectively (Royden and Keen 1980, Beaumont *et al.* 1982, Hellinger and Sclater 1983), and (c) *continuous depth-dependent stretching* in which the stretching is a continuous function of depth in the subcrustal lithosphere and the crustal stretching is the same as in (a) and (b) (Rowley and Sahagian 1986). Continuous depth-dependent stretching removes the need for a decoupling between crust and mantle.

an amount β_c and the subcrustal lithosphere by a different amount β_{sc} (Fig. 3.15b). The entire lithosphere stretches by an amount β_L and the relation among these parameters is

$$\frac{y_L}{\beta_L} = \frac{y_c}{\beta_c} + \frac{y_L - y_c}{\beta_{sc}} \tag{3.19}$$

Simplifying the notation for convenience, let $\gamma_L = 1 - 1/\beta_L$, $\gamma_c = 1 - 1/\beta_c$ and $\gamma_{sc} = 1 - 1/\beta_{sc}$ where γ represents the reduction in thickness as a percentage of the original thickness. Equation (3.19) then becomes

$$y_L \gamma_L = y_c \gamma_c + (y_L - y_c) \gamma_{sc} \tag{3.20}$$

If passive upwelling of aethenosphere accompanies extension and local isostatic equilibrium is maintained throughout, the initial fault controlled subsidence S_i will be given by a solution of the same general form as that provided by McKenzie (1978a) (eq. 3.12). Two features stand out from this analysis:

1 S_i is a linear function of γ_c and γ_L (and therefore also of γ_{sc}).
2 Crustal thinning has a much greater effect on S_i than whole lithosphere thinning.

For a neutral land surface, crustal thinning must be approximately half of the lithospheric thinning and for uplift the crustal thinning must be even less. Cooling and thermal contraction will of course contribute a thermal subsidence to this initial fault-controlled subsidence. As long as heat is not lost during the rifting process (that is, it is effectively instantaneous) the post-rift thermal subsidence reflects the amount of subcrustal lithospheric thinning, β_{sc}. This thermal subsidence is essentially of the form expressed by McKenzie (1978a) and given in equation (3.18) where β_{sc} must be substituted for β. This offers the possibility of estimating the stretch factors of lithosphere and crust from the amount of syn-rift and post-rift subsidence (see Chapter 8 for techniques).

In the centre of the Pattani Trough, Gulf of Thailand, the stretch factors β_c and β_L were 2.35 and 1.90 derived from subsidence plots, indicating that crustal thinning was 20 per cent greater than lithospheric thinning in the graben region (Hellinger and Sclater 1983).

If pre-rift sediments are preserved in the graben but eroded from the flanks, it is a good indication that doming did not precede rifting. If the extension itself caused uplift of the flanks (Salveson 1976, 1978) then this might imply that the subcrustal lithosphere was extended over a larger region than the more confined crustal extension. At first the crustal thinning, causing subsidence, outstrips the uplift from subcrustal thinning in the graben area, but the reverse is true beyond the graben edge. Later, after extensional tectonics have ceased, both flanks and graben should subside due to cooling and thermal contraction of the upwelled aesthenosphere.

Regions which are uplifted, such as graben flanks, are subject to erosion and will therefore subside to a position below sea level after complete cooling. But a second effect is the added uplift caused by isostatic adjustment to the removed load. These two processes govern how much erosion will take place before the rift flanks subside below sea level and erosion effectively stops. Hellinger and Sclater (1983) estimate that the amount of erosion, assuming it to be instantaneous, may be six times the initial uplift, and the water-loaded depth to which the surface will finally subside is roughly 1.5 times the initial uplift. This implies an increase in the area of subsidence, so non-uniform extension models incorporating erosional effects predict larger subsiding basins than uniform extensional models. However, erosion is clearly not instantaneous, so the amounts of erosion and subsidence given above are overestimates, probably by a factor of at least 2.

3.1.6.2 Continuous stretching with depth

The implication of different amounts of stretching in the crust and mantle lithosphere is that there must be a surface or zone of discontinuity separating the regions with the different values of β. Models involving such intralithospheric discontinuities can be labelled *discontinuous non-uniform stretching*. Although such models, as we have seen,

are successful in explaining the common uplift which accompanies extension (e.g. Nova Scotian and Labrador continental margin, East African Rift, Red Sea), they have a number of requirements

• the existence of an intralithospheric discontinuity, which although evident in some documentations (e.g. Biscay Margin), is by no means universally 'proven'
• a mechanism by which the mantle is detached and stretched by a different amount to the overlying crust, and a means of solving the attendant space problem in the mantle.

If the stretching is non-uniform but *continuous* with depth, these objections are removed (Rowley and Sahagian 1986). The mantle is likely to respond to extension as a function of depth, the strain rate decreasing as the extension is diffused over a wider region. This can be modelled by considering a geometry of an upward tapering region of stretching (Fig. 3.15c). If ϕ is the angle between the vertical and the boundary of the stretched region in the mantle lithosphere, the amount of stretching depends on the depth beneath the crust and angle ϕ. The variation of β_{sc} with depth can then be integrated from the base of the crust to the base of the lithosphere to obtain estimates of initial and total subsidence in a similar fashion to the uniform stretching case. The effect of ϕ is to increase the initial subsidence for larger angles (wider zones of diffuse mantle stretching), but to reduce the amount of post-extension thermal subsidence with increasing ϕ (Fig. 3.16). The wider zone of mantle stretching and therefore heating results in an uplift of the rift shoulders (Fig. 3.17), and the horizontal length scale of the uplift is an indication of the value of ϕ, since from geometry.

$$\tan \phi = X/y_{sc} \qquad (3.21)$$

where X is the horizontal width over which the uplift occurs (~250 km in the southern Red Sea, Cochran 1983), and y_{sc} is the thickness of the subcrustal lithosphere.

A point in the rift shoulder region should therefore initially experience uplift, followed by subsidence to approximately its initial elevation, or, if erosion has occurred, to a level below its initial height (Sleep 1971). The same general pattern is observed where the crustal stretching varies from a minimum at the rift margin to a maximum at the rift centre (see also White and

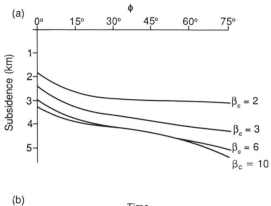

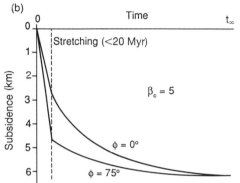

Fig. 3.16. (a) Dependence of the initial crustal subsidence on φ for various values of stretching using the continuous depth-dependent stretching model illustrated in Fig. 3.15c. The initial width of the rift is 10 km. Wider tapers in the region of subcrustal stretching cause larger values of initial subsidence for the same crustal stretch factor. (b) Initial (syn-rift) and thermal (post-rift) subsidence for two values of φ (0 and 75 degrees) and the same crustal stretch factor of β_c = 5 (after Rowley and Sahagian 1986). Rifting lasts 20 Myr.

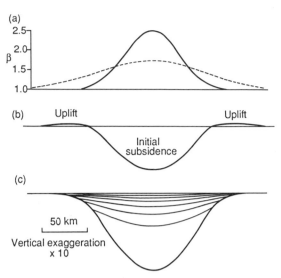

Fig. 3.17. Basin-filling patterns resulting from continuous depth-dependent stretching: (a) stretch factors in the crust and the subcrustal lithosphere as a function of horizontal distance. Crustal stretching is concentrated in an axial zone, (b) initial subsidence and uplift immediately after stretching, showing rift flank uplift, (c) total subsidence 150 Myr after rifting, showing the progressive onlap of the basin margin during the thermal subsidence phase, giving a steer's head geometry (Rowley and Sahagian 1986; White and McKenzie 1988).

McKenzie 1988). The implications of stretching the mantle over a wider region than the crust (but with equal total amounts of extension) is that stratigraphic onlap should occur over previous rift shoulders (Chapter 6) during the post-rift phase, a feature commonly found in rift-sag or 'steers-head' type basins (Fig. 3.17c).

3.1.7 Induced mantle convection as a cause of rift flank uplift

Uniform stretching models fail to explain the presence of major rift flank uplifts such as those of the East African Rift System (e.g. Morgan 1983). We have previously seen how modifications of the passive, uniform stretching model have sought to explain this phenomenon. Some authors invoke active convective heating of the lithosphere from an aesthenospheric heat source causing surface uplift (Spohn and Schubert 1983, Wendlandt and Morgan 1982), others the mechanical uplift of the lithosphere by differential normal stresses at the base of the lithosphere caused by temperature variations in the convecting aesthenosphere (Section 3.1.9) (Houseman and England 1986), others the rise of an aesthenospheric diapir in a thermally 'pre-conditioned' lithosphere (Woidt and Neugebauer 1981), and finally a number of authors have resorted to models of depth-dependent stretching wherein the subcrustal lithosphere stretches over a wider area than the crust, causing uplift on the rift margins (references in Section 3.1.6).

Models involving an active aesthenospheric heat source should predict *uplift before rifting*. At present there is no consensus on the temporal relationships between uplift and extension. However, in the Gulf of Suez (Fig. 3.18), there is good evidence that the rift flank uplifts were *not* formed as a precursor doming event prior to rifting, but rather formed *during* the main phase of extension (Steckler 1985). The rift appears to have initiated (by Miocene times) at near sea level, since the tilted fault blocks associated with early rifting suffered both subaerial erosion and marine deposition. The early Miocene topography of the Gulf of Suez region was subdued and stratigraphic thicknesses are uniform over the area (Garfunkel and Bartov 1977). However, at the end of the early Miocene, 8–10 Myr after the onset of rifting, a dramatic change took place – there is a widespread unconformity, and conglomerates appear at the rift margins suggesting major uplift and unroofing at this time.

The two main causes of vertical movements during rifting are of course
• uplift caused by heating of the lithosphere
• subsidence caused by thinning (or densification) of the crust.

Figure 3.19 shows the contribution of these two processes to the rift topography, where the effects of erosion and sedimentation have been removed (see Steckler 1985, pp. 136–137). The lithospheric and crustal extensions needed to produce the observed uplift and rift subsidence can be calculated using Airy isostasy. For the Gulf of Suez, the lithosphere must have extended by 2.5 times as much as the crust to explain the uplift and rift subsidence. This brings the uniform extension model seriously into doubt in terms of its ability to predict the lithospheric heating. How then does one explain the additional amount of heating that the lithosphere under the Gulf of Suez has undergone? Steckler believes that this extra heat results from convective flow induced by the large temperature gradients set up by rifting. Numerical experiments (Buck 1984) confirm that secondary small-scale convection should take place beneath rifts (Fig. 3.20). Convective transport should heat the lithosphere bordering the rift, causing uplift of rift shoulders concurrent with extension within the rift itself.

If this mechanism is correct, it removes the need for active, sub-lithospheric heat sources. It also removes the requirement for two-layer extension to explain high heat flows at rift margins.

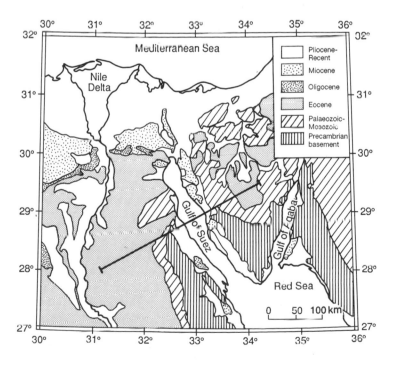

Fig. 3.18. Geological map of the Gulf of Suez region. The heavy line is the location of the topographic profile shown in Fig. 3.19.

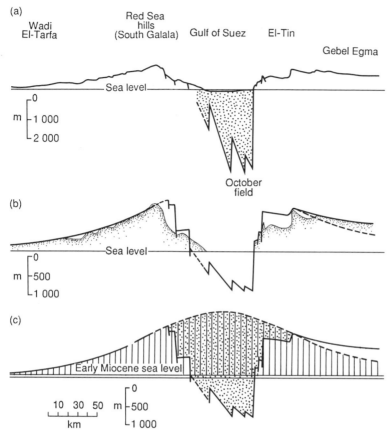

Fig. 3.19. (a) Topographic profile across the central Gulf of Suez, vertical exaggeration of 20:1. The stippled area represents the sedimentary fill of the Gulf. (b) Reconstructed profile with the effects of erosion and sedimentation removed using Airy isostasy, vertical exaggeration of 40:1. (c) Interpretation of profile in (b) involving tectonic uplift due to thermal expansion of the lithosphere (vertical lines) and downfaulting of the rift axis caused by crustal stretching (stippled pattern) (after Steckler 1985).

3.1.8 Relaying of zones of crustal and subcrustal extension – simple shear

3.1.8.1 Asymmetrical stretching

Coward (1986) believes that the lithosphere may stretch by a combination of widespread upper crustal fault controlled extension and a more localized region of concentrated lower crustal/ mantle lithosphere extension. Such lower crustal/ mantle lithosphere extension might be favoured by strain softening due to increased geothermal gradients, fabric development and metamorphic processes. In contrast, brittle faults in the upper crust tend to lock after finite amounts of extension and deformation is transferred to new faults. As a result, faults may spread outwards, but essentially in one direction away from the initial rift, widening the zone of upper crustal extension. This process is clearly analogous to the development of thrust wedges during compressional tectonics.

The asymmetrical extension represents the situation of simple shear (Fig. 3.21) (Section 2.1.2, Fig. 2.7). In this model of asymmetrical stretching patterns of uplift and subsidence should vary across the basin (Fig. 3.22). A typical basin may have an outer zone where only the upper crust is stretched and an inner zone where the upper crust is stretched by β whilst the lower crust and mantle lithosphere are stretched by a greater amount β + β′. The model predictions are therefore:

1 *Outer zone*: initial subsidence due to thinning of upper crust; no thermal subsidence.
2 *Inner zone*: both initial tectonic subsidence and later thermal subsidence, but the additional stretching β′ in the lower crust and lithospheric mantle may cause early uplift. With extreme crustal density variations, the stretched upper crust may even be raised above sea level, causing unconformities (Fig. 3.22). The end-Jurassic to Early Cretaceous so-called late Cimmerian

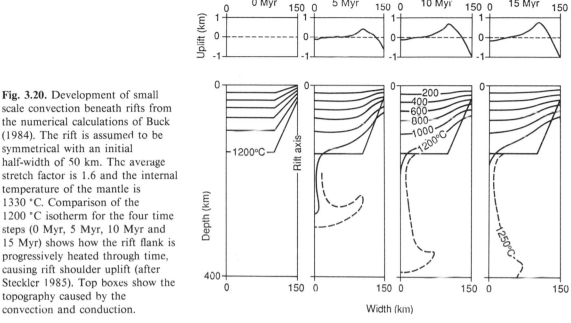

Fig. 3.20. Development of small scale convection beneath rifts from the numerical calculations of Buck (1984). The rift is assumed to be symmetrical with an initial half-width of 50 km. The average stretch factor is 1.6 and the internal temperature of the mantle is 1330 °C. Comparison of the 1200 °C isotherm for the four time steps (0 Myr, 5 Myr, 10 Myr and 15 Myr) shows how the rift flank is progressively heated through time, causing rift shoulder uplift (after Steckler 1985). Top boxes show the topography caused by the convection and conduction.

Fig. 3.21. End-member models of strain geometry in rifts (Buck, Steckler and Cochran 1988): (a) *pure shear* geometry with an upper brittle layer overlying a ductile lower layer, producing a symmetrical lithospheric cross-section. The ductile stretching may be accompanied by dilation due to intrusion of melts (cf. Royden *et al.* 1980). (b) *simple shear* geometry with a through-going low-angle detachment dividing the lithosphere into an upper 'plate' or hangingwall and a lower 'plate' or footwall. Thinning of the lower lithosphere is relayed along the detachment plane, producing a highly asymmetrical lithospheric cross-section (after Wernicke 1981, 1985).

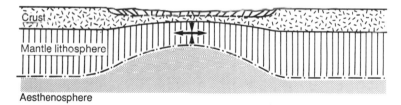

uncomformity in the northern North Sea may be due to this kind of process (Rawson and Riley 1982, Coward 1986).

3.1.8.2 Wernicke's shear zone model

Wernicke (1981) proposed, based on studies of Basin and Range tectonics, that lithospheric extension may be accomplished by displacement on a large scale, gently-dipping shear zone which traverses the entire lithosphere. Such a shear zone transfers or relays extension from the upper crust in one region to the lower crust and mantle lithosphere in another region. It necessarily results in a physical separation of the zone of fault-controlled extension from the zone of upwelled aesthenosphere.

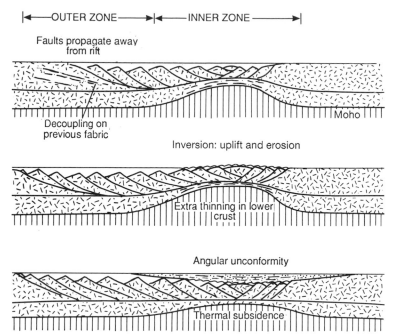

|←——OUTER ZONE——→|←—— INNER ZONE ——→|

Faults propagate away from rift

Decoupling on previous fabric

Inversion: uplift and erosion

Moho

Extra thinning in lower crust

Angular unconformity

Thermal subsidence

Fig. 3.22. Heterogeneous thinning of the lithosphere (after Coward 1986). The upper crustal extension spreads outwards asymmetrically over a wide region, possibly reactivating previous tectonic fabrics. The lower crust and subcrustal lithosphere, however, are shown extending over a much smaller region. This lower crustal/subcrustal thinning may produce thermal domes and erosional unconformities and older extensional faults may be inverted.

Wernicke (1985) suggests that there are three main zones associated with crustal shear zones (Fig. 3.23):

1 A zone where upper crust has thinned and there are abundant faults above the detachment zone.

2 A 'discrepant' zone where the lower crust has thinned but there is negligible thinning in the upper crust.

3 A zone where the shear zone extends through the subcrustal (mantle) lithosphere.

We can study the effects of low angle shear zones on basin formation and evolution. An example from the Basin and Range is given in Fig. 3.24. Since, as we have seen previously, crustal thinning by fault-controlled extension causes subsidence but subcrustal thinning produces uplift, we should expect subsidence in the region of thin-skinned extensional tectonics but tectonic *uplift* in the region overlying the lower crust and mantle thinning (the discrepant zone of Wernicke 1985). Subsequent aesthenospheric cooling may result in one of two things:

1 Thermal subsidence of the region above the discrepant zone may simply restore the crust to its initial level.

2 If subaerial erosion has taken place in the meantime, thermal subsidence will lead to the formation of a shallow basin above the discrepant zone. The basement of the basin should be un-faulted. However, beneath the zone of thin-skinned extensional tectonics there should be absent or minimal thermal subsidence.

The Wernicke shear zone model does not explain, however, basins which have a thermal subsidence spatially superimposed on a fault-controlled subsidence.

3.1.9 Houseman and England's dynamical model of lithospheric extension

An effect of convection beneath the lithosphere is to vary the normal (i.e. vertical) stress on the base of the plate because of lateral temperature gradients (McKenzie, Roberts and Weiss 1974, McKenzie 1977). Such excess normal stresses have been held responsible for swells in the oceans and for elevation of the continents (Houseman and England 1986). Elevated above regional, the lithosphere (both crust and subcrustal lithosphere) extends according to its layered rheologies.

In Houseman and England's (1986) dynamical model, the continental lithosphere is assumed to

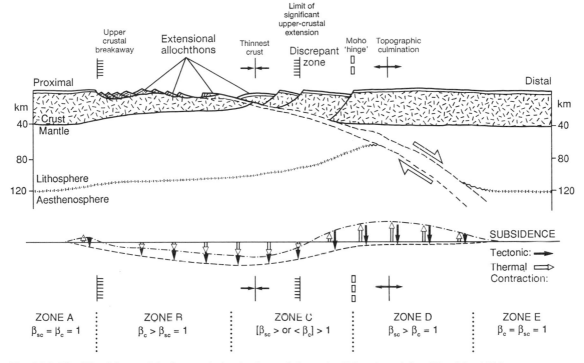

Fig. 3.23. The Wernicke model of normal simple shear of the entire lithosphere (after Wernicke 1985).

consist of two layers, crust and mantle lithosphere, each with a density as a linear function of temperature. Differences in depth-averaged horizontal deviatoric stresses between the undisturbed lithosphere and those over the mantle plume are proportional to differences between their gravitational potential energies. These depth-averaged deviatoric stresses are related to average horizontal strain rates $\dot{\varepsilon}_{xx}$ according to the deformation mechanism of the lithosphere (see Section 2.3):

1 Upper crust, dominated by brittle failure in tension.

2 Lower crust, dominated by ductile power law flow ($n = 3$).

3 Upper mantle in which a number of rheologies are possible: brittle failure, plastic deformation at high deviatoric stresses (> 100 MPa) and ductile power law flow ($n = 3$) for smaller deviatoric stresses.

Following impingement on the base of the lithosphere of the hot plume, the entire column of lithosphere is elevated by a height e_0, and $\rho_a g e_0$ is the increment in the vertical stress (ρ_a is the density of the aesthenosphere). If the stretch factor β

reflects the resulting horizontal strain, the potential energy of the extending column at any subsequent time t depends on the time history of β ($= \beta(t)$) and the time and depth history of temperature ($= T(y, t)$). Half of the surface heat flux is assumed to be generated internally within the crust.

The lithosphere responds to convection-induced uplift in one of three ways in these experiments:

1 *Accelerating extension* when the driving stress caused by uplift exceeds a threshold value and resulting strain rates are high ($> 3 \times 10^{-16}$ s^{-1}).

2 *Negligible extension* when the driving stress is due to less than about half of the uplift required for the threshold level.

3 *Self-limiting extension* at intermediate values of driving stress when cooling of the ductile portion of the lithosphere causes an increase in strength and a decrease in extensional strain rate.

The anomalous normal stress induced by convection is assumed to be applied instantaneously (at $t = 0$) to the base of the lithosphere and to persist throughout the time scale of the basin. Because the lithospheric rheology on extension is

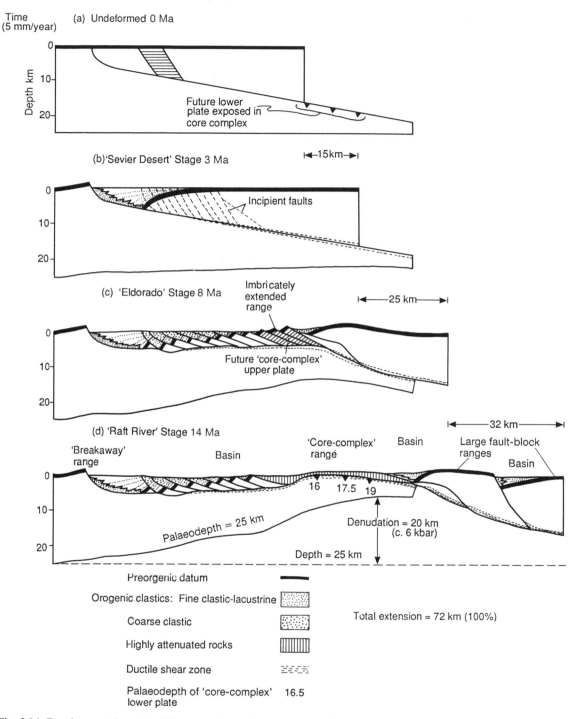

Time
(5 mm/year)

(a) Undeformed 0 Ma

Future lower
plate exposed in
core complex

|←—15km—→|

(b)'Sevier Desert' Stage 3 Ma

Incipient faults

(c) 'Eldorado' Stage 8 Ma

Imbricately
extended
range

|←—25 km—→|

Future 'core-complex'
upper plate

(d) 'Raft River' Stage 14 Ma

|←—32 km—→|

'Breakaway'
range

Basin

'Core-complex'
range

Basin

Large fault-block
ranges

Basin

16   17.5   19

Palaeodepth = 25 km

Denudation = 20 km
(c. 6 kbar)

Depth = 25 km

Preorgenic datum

Orogenic clastics: Fine clastic-lacustrine

Coarse clastic

Total extension = 72 km (100%)

Highly attenuated rocks

Ductile shear zone

Palaeodepth of 'core-complex'   16.5
lower plate

Fig. 3.24. Developmental model of the extensional shear system of the Basin and Range region, USA (after Wernicke 1985). Mid-crustal rocks in the hangingwall may initially pass through greenschist or amphibolite metamorphic conditions in the ductile shear zone, followed by uplift, cooling and deformation in the brittle field.

temperature-dependent (Sonder and England 1989), the extensional strain rate depends on the time scales of uplift (generating driving stresses) and thermal diffusion (controlling effective strength). For an initial uplift of 1.2 km, extension begins immediately (Fig. 3.25), and the heating of the crust and lithosphere causes its effective strength to decrease and the extension rate to increase. However, subsequent cooling at the Moho causes the effective strength to increase, slowing down the strain rate. Continued cooling of the upper mantle causes a return of the effective strength to its pre-uplift level, so that extension is finished by about 35 Myr (Fig. 3.25a). This can be understood by considering the way the rheologically layered lithosphere responds to the changing temperature field. For the 1.2 km uplift experiment, stress is initially concentrated in two relatively strong but rather thin horizontal layers (Fig. 3.26), the temperature-activated ductile rhe-

ologies not being able to store stresses (see Sections 2.3.3, 1.2.1 and Fig. 1.4). After a geologically short time ($t = 15$ Myr) the stress-bearing layer in the crust is thinned so that the upper mantle supports a greater proportion of the stress and takes up a much increased strain rate in the form of plastic deformation. With increased lithospheric thinning ($t = 33.5$ Myr) the upper mantle takes even more of the horizontal stresses, causing the onset of the brittle-ductile transition at this level. Significant cooling however, particularly in the ductile power-law part of the mantle causes a major drop in strain rate. Finally, as cooling continues ($t = 109$ Myr), the temperature-activated ductile rheologies slow down the strain rate to negligible values, implying that no extension is taking place. The total amount of extension (β_{max}) is therefore limited by the rheological properties of the continental lithosphere. We term this 'self-limiting' above.

The maximum amount of extension can also be controlled by the original geotherm and the elevation e_o. For a given geotherm, if e_o is larger than a threshold value extension will 'run away' (accelerating extension) and lead to the formation of ocean floor. For a value less than the threshold, extension will be self-limiting, as in the previous discussion. For e_o less than a lower critical value, extension will be negligible. The critical value for e_o is usually about half of the threshold value for run-away extension. The hotter the mantle temperature T_m, the smaller the stress difference required to produce the same amount of extension. For extension to be self-limiting the mantle lithosphere needs to cool on at least as short a time scale as the extension.

What is the typical time scale of extension? If the duration of extension is defined as t_{max}, which is the time at which 95 per cent of the extension has taken place, the relation of maximum extension to duration of extension can be shown for a number of geotherms (Fig. 3.27). Basins for terrains which plot to the left and under the curve for a particular lithospheric model have extended very quickly and the extension is not likely to be self-limiting in these cases.

The convection-related uplift may not remain constant with time. If it decays, there will be a thermal contraction phase of subsidence superimposed on the trend in Fig. 3.25c. Removal of the driving stress may limit subsidence (rather than cooling) if it takes place early in the basin history,

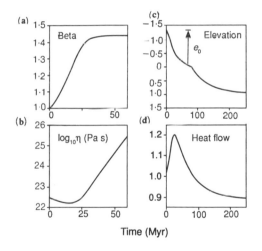

Fig. 3.25. Extension of a rheologically layered lithosphere experiencing an enhanced basal heat flux from the aesthenosphere causing an initial mechanical uplift of 1.2 km (Houseman and England 1986). Stretch factor (a), effective strength (b), elevation relative to sea level (c) and surface heat flux (d) as a function of time since instantaneous uplift. The heat flow is normalized by the initial surface heat flow value. Lithospheric model has thickness 110 km, basal temperature of 1450 °C, and internal heat generation of 0.892 $\mu W\ m^{-3}$ (other parameter values in Houseman and England 1986, Table 2b, p. 722). See text for discussion.

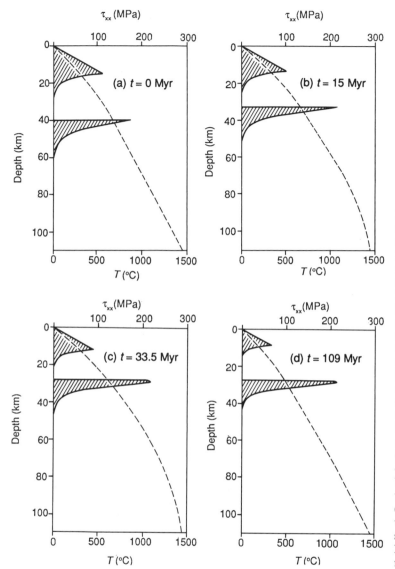

Fig. 3.26. Distribution of deviatoric stress (solid line enclosing hatched areas) and temperature (dashed line) for four stages in the experiment shown in Fig. 3.35. (a) initial condition at 0 Myr, β = 1. The two peaks in deviatoric stress mark the high strength brittle portions of the lithosphere; the crustal brittle-ductile transition occurs at about 15 km, the base of the mantle upper power-law layer being at about 40 km. (b) 15 Myr, with β = 1.21; both the high strength regions have become elevated because of the temperature increase, the mantle layer taking up relatively more of the stress. (c) 33.5 Myr, with β = 1.43; the crust has thinned to about 12 km, but cooling causes a drop in the temperature-dependent strain rate. (d) 109 Myr, with β largely unchanged at 1.44 and a negligible strain rate (after Houseman and England 1986). Same lithospheric model as in Fig. 3.25.

but if removed late, the extension will be limited by the thermally activated rheologies as described above. The various possibilities are sketched out in Fig. 3.28; clearly a great deal of variability can be introduced to the system by a process upon which there are few constraints – the time scale of the plume.

3.1.10 Effects of convergent plate boundary forces

Lateral transmission of mechanical energy through the lithosphere as a result of plate collision has been proposed as a cause of continental rifting by Molnar and Tapponnier (1975). In the case of the Rio Grande Rift, the convergent plate boundary of the western United States may be responsible. Similarly, the Baikal Rift may be influenced by plate collision in the south and a further convergent plate boundary exists in the Pacific on the east, but both boundaries are roughly 3000 km from the rift itself. The Rhine Rift has the collision boundary of the Alps in close proximity in the south and the opening Atlantic (1000 km distant) and subsiding North Sea (500 km distant) in the west and north. However, the frequency of colli-

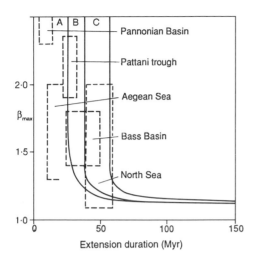

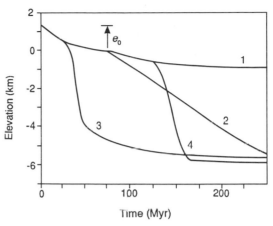

Fig. 3.27. The concept of self-limiting extension; the maximum extension versus the duration of extension for three lithospheric models A, B and C. A represents a 108 km thick lithosphere (y_L) with a basal temperature (T_1) of 1500 °C and an internal heat generation (H) of 0.938 μW m^{-3}. In B, y_L = 112 km, T_2 = 1395 °C and II = 0.845 μW m^{-3}. C is a lithospheric model where y_L = 115 km, T_1 = 1300 °C and H = 0.770 μW m^{-3} (see Houseman and England 1986, p. 722 for further parameter values). The duration of extension (t_{max}) is defined as the time at which 95 per cent of the extension has been accomplished, that is, at which β = 1 + 0.95 (β_{max} – 1). The curves for the three lithospheric models are superimposed on estimates from the Pannonian Basin (Sclater *et al.* 1980), the North Sea (lower estimates from Christie and Sclater 1980, Wood and Barton 1983, Barton and Wood 1984; higher estimates from Badley *et al.* 1988 and Klemperer 1988), the Bass Basin (Etheridge *et al.* 1984), the Aegean Sea (McKenzie 1978b, Le Pichon and Angelier 1979), and the Pattani Trough, Thailand (Hellinger and Sclater 1983). After Houseman and England (1986).

Fig. 3.28. Basement elevation as a function of time for the same set of parameter values as in Fig. 3.25 with an initial mechanical uplift of 1.2 km. Line (1); elevation change as in Fig. 3.25c, with the stress perturbation on the base of the lithosphere constant over time. The subsidence history is made up of two segments. The rapid initial subsidence (first 25 Myr) is during a period of negligible diffusion of heat and the basin is actively faulted. This phase is therefore analogous to the initial fault-controlled subsidence of McKenzie (1978a). The second segment of reduced subsidence rates is controlled by the thermal evolution of the stretched lithosphere after the extension has ceased, and is produced by the gradual decay of the thermal perturbation induced by the extension. The total elevation change will be determined therefore by the maximum amount of extension. Lines (2), (3) and (4) show the effects of removing the convection-related driving force (the plume) at various stages in the history of the basin. Line (2); same as line (1) for first 50 Myr, then the convection-related uplift (e_0 = 1.2 km) is removed gradually and becomes zero at 300 Myr. Line (3); same as line (2) but with plume decay starting at 25 Myr and finishing at 50 Myr. Line (4); as for line (2), but with a delayed plume decay beginning at 125 Myr and ending at 165 Myr. After Houseman and England (1986, p. 727).

sion boundaries compared to the large number of modern and ancient rifts is negligible. It therefore seems inconceivable that collision events have a *primary* role to play in the rifting of the continents. Some numerical models support this belief.

A quantitative estimation of the tectonic effects of continental collision can be gained by adopting a simple plane stress plate model where a plate of constant thickness is loaded horizontally by the collision of an indenter (Neugebauer 1983). The plate is assumed to exhibit a combination of a linear flow and a cubic power law creep. The modelled stress regime within the plate is strongly affected by the width of the plate compared to that of the indenter, or aspect ratio. The distribution of tensile and compressive regimes of stress can then be evaluated as a function of distance from the indented or loaded boundary.

The aspect ratios between indenter width, plate extent and rift position are approximately 1:4:2

and 1:4:<1 for the Himalaya–Baikal Rift and Alps–Rhine Rift respectively. Using these ratios and the plate model outlined above, the minimum principal stresses at the sites of the rifts are compressional (especially in the case of the Rhine Rift) or slightly extensional. The effects of a given stress regime are of course dependent on lithospheric strength. Using limits of lithospheric stress under compression and extension (Brace and Kohlstedt 1980) the anticipated stress regime falls considerably short of that required of a self-sustaining mechanism of rifting.

There are two immediate responses to this kind of numerical analysis. One is that the present orogenic arc, representing the edge of the indenter needs to be palinspastically restored to its position at the onset of rifting. Since, in the case of the Upper Rhine Graben, this onset was Lutetian (Eocene), almost the entire shortening of the meso- and neo-Alpine phases needs to be accounted for. This may amount to a figure of the order of 100 km, severely affecting the aspect ratio of indenter width : plate extent : rift position. Secondly, rifts at high angles to orogenic belts which contain a syn- to post-collision stratigraphy and which are therefore strongly correlated in time to the collision 'event' most likely have a dynamical as well as temporal link to the plate boundary processes. The exact nature of the link may be too complex to be revealed in simple indentation models.

The fact that the continental interiors are deforming and are experiencing earthquakes at very large distances from plate boundaries (Fig. 3.29) suggests that such deformation might be regarded in terms of the response of a continuous *viscous* medium to two kinds of forces:

1 Those applied on its edges by plate collisions.
2 Those in its interior arising from crustal thickness contrasts (England and McKenzie 1983, England 1983).

As we have seen previously in this chapter, the elevation of the continental lithosphere causes pressure differences to exist between it and neutrally elevated lithosphere. The relative importance of this driving force compared to that caused by applied forces at the plate boundaries can be gauged by an *Argand Number* Ar. If the Argand Number is small, that is, the effective viscosity is large for the ambient rate of strain, the deformation of the continental lithosphere will be entirely

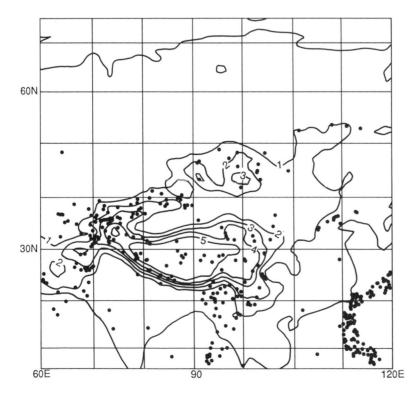

Fig. 3.29. Average topographic elevation of Asia (after Lee 1966 in England 1983) with locations of earthquakes recorded by at least 50 stations between 1961 and 1977. Note that earthquakes occur diffusely over a very wide region on the Asian continent.

due to boundary forces. If Ar is large, however, the effective viscosity will appear to be small and the medium will not be strong enough to support the elevation contrasts, so the forces due to crustal thickness changes will dominate deformation. Numerical modelling and comparison with the deformation of Asia related to the collision of the Indian indenter, suggests that the plate must have a non-Newtonian rheology ($n > 1$ in the power law equation relating stress to strain rate), and furthermore, the tectonic styles of major crustal shortening in the Himalaya, extension in Tibet (England and Houseman 1988, 1989) and a large area of predominantly strike slip motion to the north (Tapponnier and Molnar 1976) (Fig. 3.30) are adequately predicted. What the viscous plate model does *not* predict is the extension in north-central Asia, including the Baikal Rift, and it is therefore unlikely that the regionally transmitted stresses from the India-Asia collision zone can be directly responsible for Asia's intracontinental rifting.

3.1.11 Relation between rifts, sags and passive margins

The main difference between rift basins and basins experiencing regional subsidence but lacking major extensional faulting (sags) can be simply explained as follows (Fig. 3.31). In true rifts, tensile deviatoric stresses due to uplift, thinning or regional stress fields are sufficient to overcome rock strength, causing faulting. If the deviatoric stresses are insufficient to cause brittle fracture, then uplift or subsidence may occur without fracturing. If the thermal 'supply' is switched off and not renewed, the thermal contraction effect will load the lithosphere and cause subsidence, so that sags can be regarded as immature continental basins which have failed to develop into either rifts or spreading centres, whilst rift basins which have experienced a subsequent sag phase can likewise be regarded as having failed to develop fully into oceanic spreading centres. Houseman and England's concept of

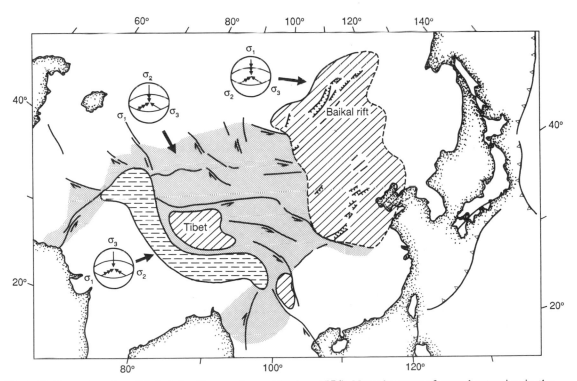

Fig. 3.30. Tectonic styles of Asia (after Tapponnier and Molnar 1976). Note the areas of crustal extension in the elevated region of Tibet and in north-central Asia including the Baikal Rift. Horizontal ornament, compression; stipple, strike slip; diagonal hatch, extension.

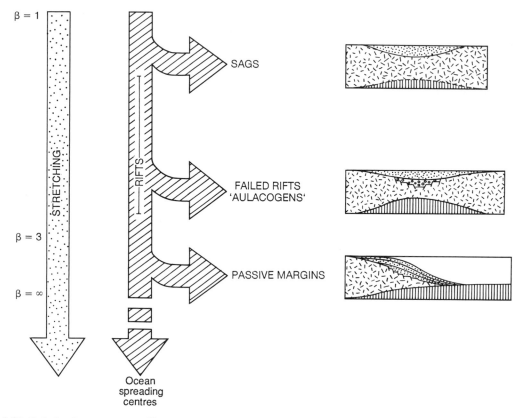

Fig. 3.31. Relation between sags, rifts, aulacogens and passive margins.

self-limiting thinning is particularly useful here. The evolution of a rift system into a passive continental margin takes place when new oceanic lithosphere is created at a spreading centre, the rift-drift transition. In terms of mechanisms for basin formation, continental margin wedges record the response of the lithosphere to its continued cooling and to the considerable loading of the sediment itself.

The evolutionary sequence from continental rifting to passive margin as a kinematic model has been postulated by many workers (at first within the framework of the geosyncline, Dietz 1963, Dewey and Bird 1970 and subsequently by Kinsman 1975, Falvey 1974 and Veevers 1981). Although the physical manifestations of continental break-up are well described, the underlying mechanism is poorly understood at present.

3.2 BASINS AT PASSIVE CONTINENTAL MARGINS

3.2.1 Introduction to observations and models

Passive continental margins (Atlantic-type margins) are characterized by seaward thickening prisms of marine sediments overlying a faulted basement with syn-rift sedimentary sequences, often of continental origin. The post-rift seaward-thickening prisms of sediments consist predominantly of shallow water deposits.

Several mechanisms have been postulated to cause the massive subsidence characterizing the post-rift (or syn-drift) phase of passive margin development (over 14 km of sediments have accumulated off the east coast of New Jersey).

• *Subsidence due to sediment loading*: sediment loading alone can only produce subsidence roughly two to three times the initial water depth (Bott 1980). Sedimentary loads *enhance* tectonically driven subsidence and acting in isolation are therefore inadequate to explain the formation of thick sequences of shallow water deposits.

• *Subsidence due to phase changes* (gabbro to eclogite) in lower crustal or mantle-lithosphere rocks: although postulated, it is not known whether this process can be widespread enough to be responsible for the subsidence of passive margins.

• *Subsidence due to creep of ductile lower crust toward the ocean*: this is thought to be caused by unequal topographic loading across the margin (Bott 1980)

• *Subsidence due to cooling following lithospheric thinning*: the upwelling of aesthenosphere (McKenzie 1978a), possibly accompanied by the intrusion of ultrabasic dykes and/or diapirs (Royden et al. 1980), is followed by thermal contraction. Support for the growing sediment prism may be by lithospheric flexure. Some workers utilize elastic plate models (Watts, Karner and Steckler 1982) whilst others prefer a viscoelastic rheology (Beaumont, Keen and Boutilier 1982).

It is possible that more than one mechanism operates in passive margin evolution. However, a basic model of lithospheric extension followed by cooling is the starting point for any analysis of passive margin subsidence.

The sediments accumulating at passive margins represent a load on the lithosphere. The lithosphere should respond by flexure as long as the wavelength of the load is sufficiently short. The two most widely applied flexural models involve either an elastic plate (Section 2.1.4) or a viscoelastic (Maxwell) plate (Section 2.3.4) overlying a weak substratum. As is outlined elsewhere (Chapter 4), the essential feature of the elastic model is that the elastic thickness of the lithosphere depends only on its thermal age at the time of loading, whereas for the viscoelastic model, the elastic thickness decreases with time since loading. As the time since loading approaches infinity, the viscoelastic model is identical to Airy isostasy. These two flexural models can only be tested in regions which have escaped major basin-wide erosional breaks. The stratigraphy of passive margins is therefore an ideal data base for evaluating the applicability of flexural models to the drift phase of subsidence (Watts et al. 1982).

Since active faulting and high heat flows accompany the early stages of rifting it is assumed that an Airy isostatic model is most applicable during this period. However, post-rift sediments are gently dipping and of wide extent, suggesting that flexure takes over at some stage after the end of rifting. Watts (1982) believed that the characteristic pattern of stratigraphic onlap on the eastern Atlantic and other margins suggests an increasing rigidity of the lithosphere with time – the expected result of an elastic lithosphere heated during the rifting stage and subsequently cooling. Other workers, for example, Beaumont et al. (1982) have used a viscoelastic model to explain subsidence on the Nova Scotian shelf (Section 3.2.3.1)

The basic geological observations of passive margins can be summarized as follows:

• they overlie earlier rift systems which are generally sub-parallel to the ocean margins, or less commonly at high angles to the ocean margin (as in the case of failed arms of triple junctions such as the Benue trough, Nigeria), or along transform fault zones (e.g. Grand Banks and Gulf of Guinea);

• an early syn-rift phase of sedimentation can be recognized from a later drifting phase, and the two are often separated by an unconformity (the 'break-up' unconformity of Falvey 1974);

• some passive margins exhibited considerable subaerial relief at the end of rifting (leading to major unconformities), as in the case of the Rockall Bank, whereas in others, the end of rifting may have occurred when the sediment surface was in deep water as in the Bay of Biscay and Galicia margin (Montadert et al. 1979a, b);

• multichannel seismic reflection sections show some passive margins to be underlain by linked listric extensional fault systems which merge into low-angle sole faults, although this is not always the case. The drifting phase in contrast is typically dominated by gravity-controlled deformation (salt-tectonics, mud diapirism, slumps, slides, listric growth faults in soft sediments);

• two kinds of margin can be identified based on the thickness of sediments (1) starved margins (2–4 km thick) and (2) nourished margins (generally 5–12 km thick). In the central Atlantic the American margin is nourished whilst the European margin is starved;

• the subsidence history of passive margins can be analysed using deep borehole information and the technique of geohistory analysis or 'backstripping' (see Chapter 8: 'Subsidence History'). If the effect of the sediment load on driving subsidence is removed, the 'tectonic subsidence' can be isolated. An analysis has been carried out on thickly sedimented passive margins such as the Baltimore Canyon region, off the eastern seaboard of the USA (Steckler and Watts 1978, 1981), using the deep boreholes COST B-2 and B-3. The backstripped thermal subsidence (i.e. the driving tectonic subsidence after the effect of the sediment loading has been removed) was plotted against $\sqrt{time}$ using an Airy model and a flexural model. A flexural model with the onset of thermal subsidence at 170 Ma fits the data very well with a $\sqrt{t}$ relationship. However, on the young (< 24 Myr) Gulf of Lions margin in the western Mediterranean, a local Airy model fitted the subsidence data better. It appears therefore that for sufficiently high values of stretching the thermal subsidence closely approximates the half-space cooling model of the mid-ocean ridges (Section 2.2.3). Young margins (< 25 Myr) or regions of reduced stretching may have exponential thermal subsidence of the form $e^{-t/\tau}$ where τ is the thermal time constant of the lithosphere (~ 50 Myr), as in the uniform extension model of McKenzie (1978a).

3.2.2 Formation of passive margins by uniform stretching

The uniform stretching model of McKenzie has been applied to the formation of passive margins (e.g. Le Pichon and Sibuet 1981). For the situation where stretching of the whole lithosphere occurs instantaneously at time $t = 0$ (or within a period of 20 Myr according to Jarvis and McKenzie (1980)), heat production from radioactivity is ignored, local (Airy) isostatic compensation is maintained throughout, and the continental surface is initially at sea level, the initial subsidence is given by equation (3.12). But since we are now dealing with passive margins, it is important to investigate the subsidence at time $t = \infty$, S_∞. The initial subsidence S_i (eq. 3.12) is a linear function of $(1 - 1/\beta)$ and S_∞ can also be expressed as a linear function of $(1 - 1/\beta)$. By introducing new parameters ρ'_m and ρ'_c as the average densities of the mantle part of the lithosphere and crust at time t infinite, the final subsidence S_∞ can be expressed

$$S_\infty = \frac{y_L(\rho'_L - \rho_L) + y_c\{\rho_L - (\rho'_L/\beta) + (\rho'_c/\beta) - \rho_c\}}{\rho_a - \rho_w}$$

(3.22)

The average densities of the mantle part of the lithosphere and the crust at time t infinite can be expressed

$$\rho'_L = \rho_m^*\left(1 - \frac{\alpha_v}{2}T_a - \frac{\alpha_v}{2}T_a\frac{y_c}{\beta y_L}\right)$$

(3.23)

and

$$\rho'_c = \rho_c^*\left(1 - \frac{\alpha_v}{2}T_a\frac{y_c}{\beta y_L}\right)$$

(3.24)

where, as a reminder y_L is the initial lithospheric thickness, y_c the initial thickness of the continental crust (at $t = 0$), the temperature at the base of the lithosphere is equal to the temperature of the aesthenosphere T_a, ρ_m^* and ρ_c^* are the densities of the mantle material and continental crust at $0\ °C$, ρ_w is the density of the seawater, α_v is the coefficient of thermal expansion. The average densities of the mantle part of the lithosphere, the crust before stretching and the aesthenosphere are ρ_L, ρ_c and ρ_a.

From (3.22) to (3.24),

$$S_\infty = y_c(1 - 1/\beta)\frac{\rho_L - \rho_c + \rho_m^*(\alpha_v/2)T_a + \varepsilon}{\rho_a - \rho_w}$$

(3.25)

where $\varepsilon = \frac{\rho_m^* - \rho_c^*}{\beta}\left(\frac{\alpha_v}{2}T_a\frac{y_c}{y_L}\right)$

(3.26)

or by slight rearrangement,

$$S_\infty = y_c(1 - 1/\beta)\frac{(\rho_m^* - \rho_c^*)\{1 - (\alpha_v/2)T_a(y_c/y_L)\}\varepsilon}{\rho_a - \rho_w}$$

(3.27)

In fact, neglect of the additional term ε introduces an error of less than 0.5 per cent.

The difference between the initial subsidence S_i and the final subsidence S_∞ is the subsidence caused by the progressive return to thermal equilibrium, i.e. that due to thermal contraction on cooling. The latter is termed thermal subsidence S_t and of course $S_t = S_\infty - S_i$.

3.2.2.1 Application to the sediment-starved Bay of Biscay and Galicia margins

As an example of this kind of analysis, it is possible to examine seismic reflection profiles across the northern Bay of Biscay and Galicia continental margins (Montadert *et al.* 1979a, b) (Fig. 3.32) and to test whether the uniform stretching model accurately predicts the observed subsidence. Active extensional tectonics started in the Late Jurassic – Early Cretaceous (*c.* 140 Ma). The extensional graben extended outwards, creating new fault blocks which were progressively tilted. By the time oceanic crust was emplaced (120 Ma) the subsiding trough had reached a depth of about 2.4 km, and simultaneously active tectonics ceased, giving way to thermal subsidence. The Bay of Biscay is

(a)

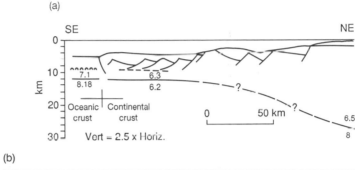

(b)

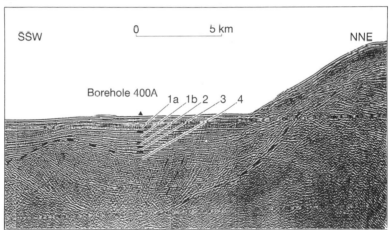

Fig. 3.32. (a) Profile across the Armorican continental margin showing the prominent tilted fault blocks, based on the multichannel seismic reflection results of Montadert *et al.* (1979). The ocean–continent transition occurs at a present-day water depth of about 5.2 km. (b) Detail of tilted fault block geometry from seismic reflection profile OC 412 (Montadert *et al.* 1977, Le Pichon and Sibuet 1981) across the lower part of the Armorican (northern Bay of Biscay) continental margin. After Montadert *et al.* (1979). (c) Line drawing interpretation of (b). 1a is Quaternary to late Pliocene, 1b is Pliocene to early Miocene, 2 is Miocene to late Palaeocene, 3 is Maastrichtian to Campanian, 4 is late Albian–late Aptian. After Montadert *et al.* (1979).

(c)

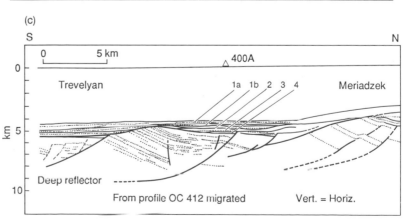

relatively starved of sediment, minimizing the effects of sediment loading compared to, say, the US Atlantic continental margin.

The following constants were chosen to fit the Biscay and Galicia data (Le Pichon and Subuet 1981).

y_L = 125 km (from Parsons and Sclater 1977)
y_c = 30 km (from refraction data)
ρ_m^* = 3350 kg m^{-3}
ρ_c^* = 2780 kg m^{-3}
ρ_w = 1030 kg m^{-3}
α_v = 3.28 × 10^{-5} °C^{-1} (from Parsons and Sclater 1977)
T_a = 1333 °C (from Parsons and Sclater 1977).

Using these constants, the initial fault controlled subsidence simplifies to

$$S_i = 3.61 \ (1 - 1/\beta) \ \text{km} \qquad (3.28)$$

and the final subsidence becomes

$$S_\infty \sim 7.83 \ (1 - 1/\beta) \ \text{km} \qquad (3.29)$$

so the thermal subsidence is the difference between S_i and S_∞,

$$S_t \sim 4.22 \ (1 - 1/\beta) \ \text{km} \qquad (3.30)$$

However, since the Bay of Biscay margin is 120 Myr old rather than being infinitely old, S_{120} is somewhat smaller than S_∞. As a result,

$$S_{120} \sim 7.23 \ (1 - 1/\beta) \qquad (3.31)$$

and

$$S_{t120} \sim 3.64 \ (1 - 1/\beta) \qquad (3.32)$$

Mid-ocean ridge crests are generally at about 2.5 km water depth, suggesting that zero-age oceanic lithosphere under 2.5 km of water is in equilibrium with a 'standard' continental lithospheric column. Therefore during rifting, the aesthenosphere should theoretically not be able to break through the thinned continental lithosphere as long as S_i is less than 2.5 km. Using equation (3.28) for the initial subsidence, the stretch factor required to produce 2.5 km of subsidence is 3.24; the crust by this time will be reduced in thickness to 9.25 km and will most likely be highly fractured – it is likely therefore that the aesthenosphere would break through when this depth was reached. This represents the continent–ocean transition.

In the Bay of Biscay the estimated total subsidence at 120 Myr since rifting (S_{120}) for β = 3.24 is

5.2 km and the final subsidence (S_∞) for an infinitely large β is 7.8 km (eq. 3.29). One should therefore expect to find continental crust in the Bay of Biscay at depths of 5.2 km or shallower in the absence of sedimentation. [With the addition of a sediment load driving further subsidence, the entire sedimentary column could be as much as 15 to 20 km thick.] Oceanic crust should not be found at shallower depths.

The extension estimated from fault block geometries in the upper crust is about β = 2.6 based on migrated seismic reflection profiling. The high value of β = 2.6 from the Biscay seismic reflection profile indicates that the crust is substantially thinned and close to the value at which the aesthenosphere could break through. Along the seismic profile the depth to the surface of the continental block is 5.2 km, which suggests that the model very satisfactorily explains the main features of the Biscay margin.

A regional synthesis suggests that water depth varies linearly with the thinning of the continental crust, following a relation close to

$$S = 7.5 \ (1 - 1/\beta) \qquad (3.33)$$

which is almost identical to the model prediction (eq. 3.31) (Fig. 3.32).

It is perhaps surprising that the Biscay and Galicia data fit the simple uniform stretching model so well. This is probably because during phases of rapid stretching, and in the absence of large sediment loads, the lithosphere is compensated on a local scale rather than responding by flexure.

3.2.3 Comparison with other models of passive continental margin development

3.2.3.1 The sediment-nourished eastern US and Canadian passive margin

It has been known for well over a decade (Sheridan 1974) that the US and Canadian Atlantic margin is very thickly sedimented, with over 10 km along most of the margin, and considerably more in areas such as the Baltimore Canyon where a deep offshore well (COST B-2) was drilled in March 1976 (Poag 1980). The subsidence history of thickly sedimented margins such as this is profoundly affected by the *sediment load*. The sedi-

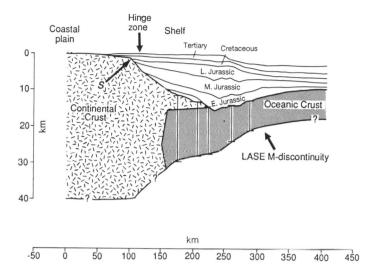

Fig. 3.33. Deep crustal structure of North American Atlantic margin interpreted from seismic reflection profiling (Poag 1985) and from the large aperture seismic experiment LASE (1986), and overlying sedimentary prism (after Watts 1988). The transect is offshore New York in the region of the Baltimore Canyon. S_J indicates the approximate position of the Jurassic pinch-out.

ment load is supported by the rigidity of the plate, and the borehole records (and seismic reflection profile data) need to be *backstripped* to obtain the tectonic subsidence (see Chapter 8: Subsidence History).

The main structural regimes of the US continental margin are shown in Fig. 3.33. From landward to seaward, the *coastal plain* contains a seaward-thickening wedge of sediments. Basement depth increases rapidly in the *hinge zone*, thought to represent the transition between relatively undeformed continental crust and the highly attenuated and heated crust that has been modified by rifting. Seaward of the hinge zone lies the locus of the thickest sediment accumulations and still further seaward lies the true *oceanic crust*.

The backstripped COST B-2 well can be compared with theoretical curves to estimate the amount of extension. Using a simple uniform extension model, β varies from 3–3.5 for Airy isostasy, to 5 (up to mid-ocean ridge values) using a flexural isostatic model. The Airy results and the flexural results probably bracket a realistic estimate of about β = 4, which would suggest that the continental crust has been thinned to almost the oceanic thickness. This is supported by gravity modelling of crustal thickness (Grow, Mattick and Schlee 1979).

The magnitude of the tectonic subsidence (and extension) varies systematically across the margin:
1 Extension seaward of the hinge zone reaches

values as great as β = 4.
2 The hinge zone has suffered much less extension (β = 1.5–2.0).
3 There is no syn-rift extension landward of the hinge zone, post-rift subsidence being accomplished by flexure.

The passive margin off Nova Scotia is one of the best studied mature examples available. The margin underwent rifting between 200 and 180 Ma. Its thick stratigraphy is well known from deep exploratory boreholes and seismic reflection data, and the deepest parts of the basin together with the variation in crustal thickness in the ocean–continental transition are known from seismic refraction. The composite cross-section is shown in Fig. 3.34.

Three models of passive margin development have been tested on the northeastern USA and-Scotian Shelves in terms of subsidence, thermal history, stratigraphy and free air gravity anomaly (Beaumont, *et al.* 1982, Steckler and Watts 1981, Watts 1982, Watts and Thorne 1984). These three models are as follows:

1 Uniform extension model

This model considers extension β to be uniform throughout the lithosphere and not to vary horizontally. This is a reasonable approximation if the gradient in β across a margin (from β = 1 on the unstretched continent to β = ∞ in the ocean) is

Boreholes

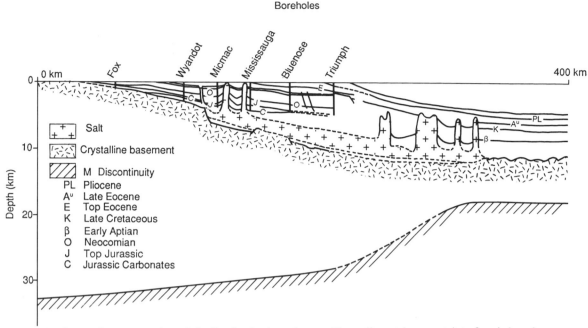

Fig. 3.34. Composite cross-section of the Scotian basin and crust. The sediment–basement interface is based on wells that penetrate it in addition to seismic reflection and refraction observations (Jansa and Wade 1975). After Beaumont *et al.* (1982).

small enough that there is insignificant lateral heat transport during cooling.

2 Uniform extension and melt segregation model

The simple uniform extension model does not address the problems of the formation of oceanic crust and the nature of the continental to oceanic crustal transition, since as $\beta \to \infty$ continental lithosphere is thinned to zero and replaced by aesthenosphere at the solidus temperature. This would imply enormously large initial subsidence for the oceanic region. This cannot be the case, since newly formed oceanic crust occurs at about 2.5 km depth at mid-ocean ridges. The melt segregation model involves greater heating (Royden and Keen 1980) than the simple uniform model of McKenzie (1978a).

The melt segregation model assumes that a basaltic melt segregates from aesthenosphere which has upwelled to replace stretched lithosphere. The segregated melt can be considered to either constitute new oceanic crust, or to be intruded into or to

underplate the stretched lithosphere. The stretching β controls the amount of segregated melt available so that in the case of a laterally varying stretching (with β increasing towards the rift centre), the thickness of segregated melt increases towards the site of ocean floor creation, compensating to some extent for the rapidly thinning continental crust.

3 Depth-dependent extension model

As we have previously seen, extension may not be uniform with depth because of the changing rheological properties of the lithosphere. If the lithosphere extends inhomogeneously and discontinuously, there must be a depth d at which the upper and lower parts of the lithosphere are decoupled. This zone of detachment or shear is where listric faults in the overlying brittle zone sole out (Montadert *et al.* 1979, Kusznir, Karner and Egan 1987).

The initial subsidence and thermal subsidence for the case of depth-dependent extension where

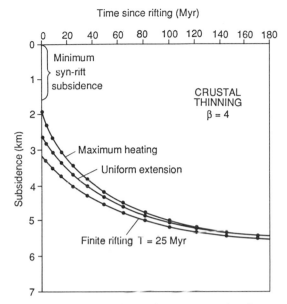

Fig. 3.35. Plot of tectonic subsidence versus time for three theoretical cases: uniform extension, rifting over a finite period of 25 Myr, and uniform extension with additional heating caused by melt segregation. In all cases the crust is thinned by a factor of 4, as indicated by the deep COST B-2 borehole. Note that there is a minimal difference between the three curves at long periods after rifting, suggesting that the 'memory' of the initial rifting event fades with time (after Steckler, Watts and Thorne 1988).

the zone above d extends by β_c and the zone below d extends by ductile deformation by a different amount β_{sc}, is given by Royden and Keen (1980).

If the lower zone stretches by ductile deformation more than the brittle upper zone, uplift should occur if the depth to decoupling approximates the crustal thickness ($d \approx y_c$). This uplift occurs at the same time as extension, and is an attractive feature of the model in view of the updoming characteristic of many present-day rift systems (e.g. East Africa, King and Williams 1976).

Figure 3.35 compares the uniform extension model with $\beta = 4$, the effects of a finite rifting period of 25 Myr (Cochran 1983), and the effects of greater heating by melt segregation (Royden and Keen 1980) on the eastern USA margin. All three curves fit the tectonic subsidence of the post-rift phase of the COST B-2 well, although syn-rift subsidences differ by a factor of 2 between the

finite case and the maximum (mid-ocean ridge) heating case.

A comparison of the stratigraphies generated by the uniform (one-layer) stretching model and the two layer (depth-dependent) stretching model is shown in Fig. 3.36. Both models show a well-developed coastal plain, hinge zone and inner shelf region underlain by a thick sequence of seaward dipping strata. The main difference is in the stratigraphy of the coastal plain, the one-layer model over-predicting syn-rift sediment thickness. The two-layer model, however, explains the lack of syn-rift (Jurassic) stratigraphy by the lateral loss of heat to the flanks of the rift, causing uplift and emergence. This result is similar to the recent studies of the Gulf of Suez by Steckler (1985) where the strong lateral temperature gradients are thought to induce secondary convection (Buck 1984, 1986) (Section 3.1.7).

The results from the US Atlantic margin suggest that a model of uniform extension coupled with melt segregation from the aesthenosphere and migration into the crust, or a depth-dependent extension model which takes into account the changing rheological properties of the lithosphere with depth makes little difference to the thermal subsidence seaward of the hinge zone, but the effects are larger on the rift flanks.

Similar conclusions were reached on the Scotian Shelf (Beaumont *et al.* 1982). With the uniform extension model and the temperature corresponding to the base of the lithosphere supporting stresses elastically (elastic or rheological lithosphere) of $T = 250\,°C$, there is a good agreement between the observed basin configuration and stratigraphy and that predicted by the model, although there are also discrepancies in detail. The differences between the three models result mainly from differences in the water-loaded *initial* subsidence; the differences between the models are greatest in the oceanic region.

Model	S_i (m)
Uniform	3305
Melt segregation	2628
Depth-dependent extension	2923

Landward of the shelf edge, the uniform extension and melt segregation models both predict too much early subsidence, whereas the depth-

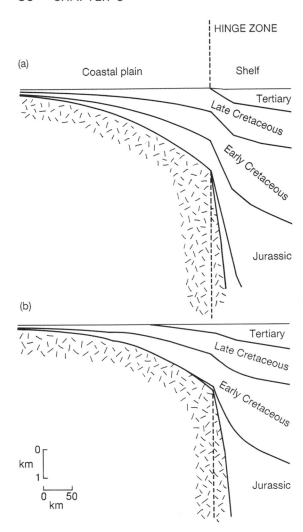

Fig. 3.36. Two synthetic stratigraphic profiles along a transect crossing the coastal plain and shelf off New Jersey constructed using the flexural loading model for passive margins given by Watts and Thorne (1984). (a) One-layer lithospheric model with uniform extension. (b) Two-layer model in which the lithosphere and crust are thinned by equal amounts seaward of the hinge zone, but only the lithosphere is thinned landward of the hinge zone. The lithospheric thinning promotes early uplift of the zone landward of the hingeline, and helps to explain the absence of Jurassic strata from this region (after Steckler, Watts and Thorne 1988).

dependent extension model predicts initial uplift (*c.* 400 m) for this inner shelf region because of the large value of subcrustal lithosphere stretching β_{sc}. This is in closer agreement with the observed stratigraphy. However, these features are subtle and it is unlikely that different rifting mechanisms can be detected unless the stratigraphy is very well known.

The different rifting mechanisms have little impact on subsidence history and thermal maturation, and all that is needed is an estimate of β_c and its lateral variations. In other words, the initial thermal disturbance causing rifting fades with time, so that after one lithospheric thermal time constant, roughly 50 to 60 Myr, the evolution of the margin proceeds essentially independently of the precise rifting mechanism.

In summary, studies of the thickly-sedimented eastern USA and eastern Canadian passive margins show that models involving melt segregation or depth-dependent stretching best explain the early uplift history of the region landward of the hinge zone or shelf edge. However, all three models including simple uniform extension fit reasonably well the post-rift thermal subsidence of the margin, suggesting a loss of 'memory' of the initial rifting mechanism with time.

3.3 STRUCTURAL STYLES OF EXTENSIONAL BASINS

3.3.1 Fault shapes – listric and planar

There has been, and continues to be, debate as to the shape of extensional faults in rift zones. The geometries which are possible (shown schematically in Fig. 3.37) can be subdivided according to whether the faults are listric or planar, and rotational or non-rotational.

• Planar normal faults cutting an originally flat layer cannot produce rotations of strata into the fault plane. Rotations are possible if the original layer is arched, but to produce large rotations implies unrealistically large uplifts. Both geometries produce a 'keystone-effect' of a deep keel under the graben system. Such keystones have never been observed, the Moho instead being typically raised under rifts.

• Planar normal faults accompanied by backward tilting in a 'domino' manner can produce

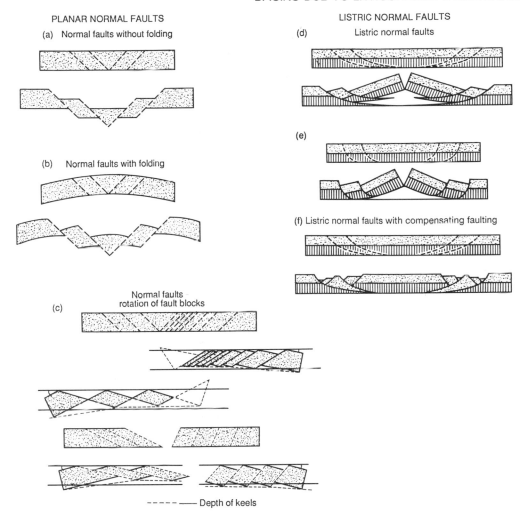

Fig. 3.37. Schematic illustrations of various fault geometries. (a) Planar normal faults without folding do not show antithetic rotation of beds into the fault plane. Such rotation is achieved in (b) by folding or uplift prior to or at the same time as faulting. However, unrealistically large uplifts would be required to cause large angles of rotation. Both (a) and (b) involve the formation of a keel (the 'keystone effect'). On a crustal scale the Moho appears to be elevated below rifts, so there is no direct evidence of the existence of deep keels. (c) Domino model in which steep planar faults 'tip over' during extension (like a stack of dominoes) until the normal component of stress is too great to allow additional displacement on the flattened fault plane. At this stage new faults may form, which causes additional tilting of the earlier fault blocks (see Wernicke and Burchfiel 1982). The bases of the fault blocks again show keels, but these may be removed by ductile flow in the lowest zones if the blocks are not assumed to be rigid and brittle throughout. The dashed fault block outline is geometrically necessary but is unrealistic. (d) Listric normal faults, that is, faults which decrease in dip downwards, becoming parallel or subparallel to a detachment zone close to the ductile–brittle transition. The problem of deep keels becomes unimportant but the 'toes' of the fault blocks are uplifted by rotation to form a central uplift. In (e) the faults are tightly curved and have the effect of almost removing the pointed, uplifted keels of the individual fault blocks. In both (d) and (c) an unrealistic central uplift is generated with an axial wedge-like opening in the central block. (f) Small compensatory (antithetic) faults are added to the listric geometries in (d) and (e). This removes the excessive central uplift, the rotation of the pointed keels and the wedge-like gap in the central block. An overview of listric normal faulting is given in Bally *et al.* (1981). The results of model experiments generating listric fault geometries are given by Ellis and McClay (1988).

considerable rotations (see Wernicke and Burchfiel 1982). Although the bases of the blocks show keels which are not observed on a crustal scale, ductile flow may eliminate them at these levels.

• Listric fault planes solve many geometrical problems by removing the need for deep keels, and where accompanied by small compensatory faults dipping toward the listric master fault, avoiding the need for major uplift in the centre of the extended zone.

Normal fault systems that exhibit rotation of beds into the fault plane are likely to be listric in shape. They provide an easy mechanism of stretching the brittle upper crust over a ductile lower crust and mantle. Because listric faults produce differential tilt between hangingwall and footwall, a series of imbricate listric fault blocks should display successively steeper tilts in the main downthrow direction (Fig. 3.37). A set of planar but rotational fault blocks would tilt strata by an equal amount.

Listric and planar normal faults may link to low angle masterfaults which traverse the entire lithosphere. If this occurs, the type of shear on the fault must change from brittle at near-surface levels to ductile at deeper levels. The prominent reflector at the base of a set of imbricate fault blocks in the Bay of Biscay (Fig. 3.32b, c) may be a crustal-scale low angle normal fault. The ductile shear on these 'detachments' may be restricted to the shear zone and not distributed uniformly. As a result, the amount of stretching on linked extensional faults in the brittle crust may be completely different from the extension in the lower crust.

Faults which show strong updip convergence of beds (Fig. 3.38) indicate that the displacement was taking place slowly on the fault at the same time as sedimentation. However, if sub-horizontal beds showing no major thickening or thinning fill the graben or half-graben, the displacement may be effectively instantaneous compared to the rate of sedimentation.

It is important to note that seismically active faults appear to be planar and inclined at roughly 45°, and no convincing listric faults have been identified from earthquake data (Jackson and McKenzie 1983, Jackson 1987). Some near-horizontal faults with steeply dipping strata in the hanging wall may be rotated from steep to flat-lying after seismic activity has ceased on them, deformation being transferred elsewhere in the fault system.

3.3.2 Polarity of fault systems

Early views of the fault pattern in rift systems were of a symmetrical distribution of horsts and grabens (e.g. Liggett and Ehrenspeck 1974, Liggett and Childs 1974) (Fig. 3.39). If the zone of stretching is to be continuous on a plate boundary length scale, it must be offset by transform faults.

Seismic reflection results suggest, however, that rifts are characterized by half-grabens over rotated

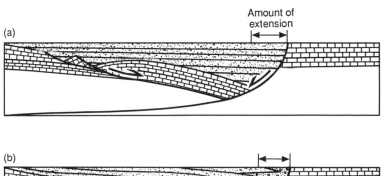

Amount of extension

(a)

(b)

Fig. 3.38. Extension in a half-graben with and without syntectonic sedimentation or 'growth'. In (a) the strata in the half-graben are *parallel* and do not show growth into the fault plane. In (b) the strata are *divergent* and thicken into the fault plane, indicating tectonic displacement at the same time as sedimentation.

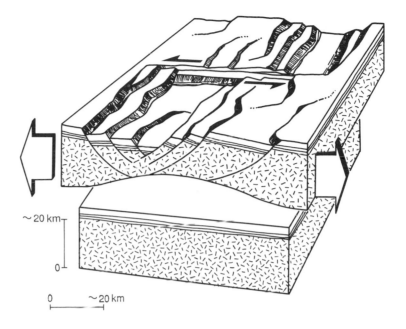

Fig. 3.39. Symmetrical rift system with listric normal faults accommodating crustal extension, offset along the rift axis by transforms (after Liggett and Ehrenspeck 1974 and Liggett and Childs 1974).

fault blocks. The polarity (or downthrow direction) commonly reverses over transform faults which segment the rift system (Fig. 3.40). The axis of the graben system may be offset to varying degrees by the transforms. The offsets along transforms are particularly important because they offer the opportunity of placing in close juxtaposition two segments or compartments with very different histories and magnitudes of stretching. Some segments may therefore experience considerably greater subsidence than others. This may explain, for example, the thick sedimentary fill of the Baltimore Canyon region compared to the Georges Bank on the Atlantic margin of North America.

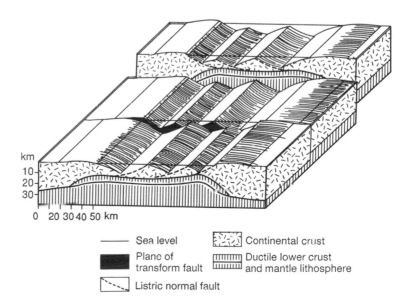

Fig. 3.40. Reversing polarity of extensional fault systems across transforms. The sedimentary fill of the half-grabens is omitted (after Bally and Oldow 1984).

The reasons for a change in polarity of faults across transforms is not fully understood. It is possible that the pre-existing heterogeneity of the basement has a major control. In this way Alleghenian fault systems in the Appalachians may have exerted some control on later Mesozoic Atlantic-related half-graben systems.

In contrast to the rifting sequence which is characterized by block-faulting, the drifting sequence at passive margins is dominated by gravity-driven tectonics.

3.3.3 Methods of estimating the amount of extension

An estimate of the amount of extension that has taken place can be obtained from a number of methods (Fig. 3.41).

1 Subsidence history

It has previously been shown (Section 3.1) that the lithospheric stretching model of the type formulated by McKenzie predicts initial uplift or subsidence depending on the ratio of crustal to lithospheric thickness y_c/y_L. If y_c/y_L is known for a particular basin, the fault-controlled initial subsidence, or more importantly, the thermal subsidence can be used to estimate the amount of stretching. Of course this assumes uniform extension and an accurate backstripping process (see Chapter 8).

2 Crustal thickness changes

In some circumstances the attenuation of the crust can be estimated from deep seismic (refraction and wide angle reflection) results. For example, the

METHODS OF ESTIMATING THE AMOUNT OF EXTENSION

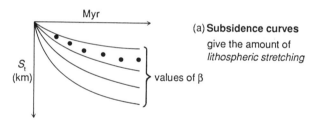

(a) **Subsidence curves** give the amount of *lithospheric stretching*

(b) **Crustal thickness changes from seismic experiments** give the amount of *crustal stretching*

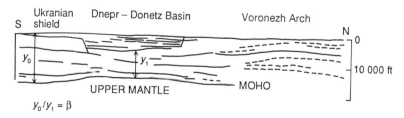

$y_0/y_1 = \beta$

(c) **Rotation of fault blocks** gives the amount of stretching of the *brittle upper crust*

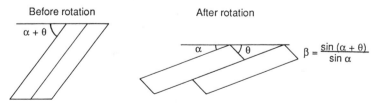

$$\beta = \frac{\sin(\alpha + \theta)}{\sin \alpha}$$

Fig. 3.41. Different methods of estimating the amount of extension.

Moho rises by 5 km in the southern part of the Rhine Rift and 3 km of sediment has accumulated. The crust has therefore thinned by 8 km (Mueller *et al.* 1973, Emter 1971). Similarly, the nearby Limagne Graben (France) contains 2 km of sediments and the Moho rises to within 24 km of the surface (Hirn and Perrier 1974). Since the crust is about 30 km thick in the Massif Central, which separates the Rhine–Bresse rift system from the Limagne Graben, the amount of crustal attenuation in these two cases can be estimated to be between 1.2 and 1.3.

3 Dip of planar rotational faults

The amount of stretching can be estimated if the dips of faults and bedding are known. In the case of domino faults, where successive fault blocks have been tilted back at a constant angle (α) and the bedding or basement-fill surface dips into the fault plane at (θ), the stretch factor is given by

$$\beta = \frac{\sin(\alpha + \theta)}{\sin \alpha} \qquad (3.34)$$

For example, if $\alpha = 25°$ and $\theta = 20°$, $\beta = 1.7$. The strain as a percentage is then

$$e\% = \left\{ \frac{\sin(\alpha + \theta)}{\sin \alpha} - 1 \right\} 100 \qquad (3.35)$$

which for the above example is 70 per cent. Where the fault has moved during sedimentation, the strain rate can be estimated by a sequential evaluation of (3.34) and (3.35).

4 Extensional structures in sediments

These estimates, based on analysis of the strain ellipsoid (see Ramsay and Huber 1983 for examples from the Swiss Helvetics), gives local amounts of strain which are extremely unreliable as a basis for extrapolation to the entire basin.

5 Offset patterns on transform linkages

In rare circumstances early structures will show a larger amount of offset on linking transform structures than later structures, including compressional/inversion structures. If offsets can be determined

and restored, it may be possible to estimate amounts of stretching.

Klemperer (1988) has recently assessed the values of stretching from deep seismic reflection lines and has compared them with estimates from independent subsidence data in the North Sea (Fig. 3.42). The North Sea is an aulacogen that underwent rifting in the Jurassic–Early Cretaceous. Regional deep seismic reflection profiles in the Viking Graben–Shetland platform area have been used to estimate the depth to the Moho. The reflector thought to represent the Moho shallows from about 32 km beneath the unrifted Shetland platform to about 20 km below parts of the Viking Graben, implying a stretching of about 2 in the Mesozoic (Fig. 3.42). This figure is in close agreement with that estimated from a refraction line (Solli 1976). The region of greatest thinning is directly below the region of greatest sedimentary thickness in the Viking graben. Major steeply-dipping faults are imaged by prominent reflectors in the (?brittle) upper crust but there is no evidence of faults cutting the entire crust down to the Moho. This may be due to deformation taking the form of a distributed shear in the (?ductile) lower crust. Dipping reflectors are once again imaged in the upper mantle; these are interpreted by Klemperer (1988) to represent localized shear zones accommodating a roughly symmetrical extension.

The stretching estimated from the slope of the subsidence curves (geohistory analysis – Chapter 8) in the same area as Klemperer's reflection lines by Badley *et al.* (1988) ($\beta = 1.5$) and Giltner (in Klemperer 1988) ($\beta = 1.8$) is close to that suggested from Klemperer's deep seismic reflection results. Beach, Bird and Gibbs' (1987) estimate of $\beta = 3.3$ seems, however, far too high and close to a value at which oceanic crust would form (see Section 3.2.2.1).

There is therefore good agreement between estimates of stretching obtained from crustal thickness changes and from subsidence history. There is, however, an important caveat. If partial melting takes place (Foucher, Le Pichon and Sibuet 1982) synextensional intrusion is likely to take place (Keen, Beaumont and Boutilier 1983). This may thicken the crust and give a false impression of the thinning. In the North Sea there is no evidence of large scale Mesozoic volcanic activity, suggesting that igneous additions to the crust were minor. Intrusion may have taken place without surface

(a) TIME SECTION   NSDP 84 LINE 1

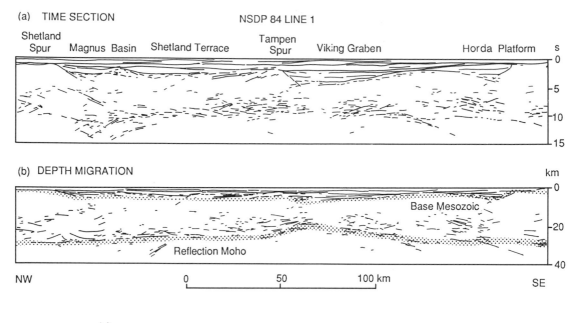

(b) DEPTH MIGRATION   km

NW   0   50   100 km   SE

(c)

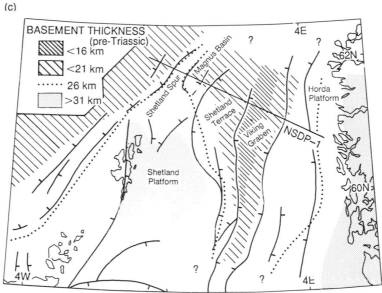

Fig. 3.42. Crustal thickness changes in the North Sea area (after Klemperer 1988). (a) Unmigrated line drawing in two-way travel time of NSDP line 1 from the Shetland Spur to the Norwegian coast (location in (c)). (b) Depth-migrated version showing the depth of the reflection Moho. (c) Contour map of the interpreted thickness of the pre-rifting basement (pre-Triassic) showing that the Viking Graben has been stretched by a factor of about two compared to the Shetland Platform.

igneous activity – if this were the case, isostatic uplift should have taken place as a result of the crustal thickening. Since the North Sea basin has subsided throughout the post-rift phase, the intrusion of igneous rocks is thought to have been minor.

4 Basins due to flexure

'Every valley shall be exalted, and every mountain and
hill shall be made low: and the crooked shall be made straight,
and the rough places plain.'

(Isaiah 40:4)

SUMMARY

Foreland basins and ocean trenches are examples
of basins caused primarily by the deflection of the
lithosphere by applied force systems. A general
flexural equation can be established for the case of
a thin elastic plate overlying a weak fluid subjected
to vertical applied forces, horizontal forces and
torques or bending moments. The general flexural
equation can then be used in different geodynamical situations by applying different boundary conditions.

The deflection of the oceanic lithosphere along
seamount chains such as the Hawaiian Islands can
be explained by either the flexure of a continuous
plate loaded by a vertical applied force (represented by the excess mass of the seamount chain),
or by the flexure of a plate broken beneath the
vertical applied force. Some useful and simple
expressions for the geometry of the deflection can
be derived which involve the maximum amplitude
of the deflection, the width of the basin, the
location of the forebulge and the height of the
forebulge. In particular, the wavelength of the
deflection is dependent on the *flexural rigidity* of
the plate, or *flexural parameter*. The maximum
deflection is dependent on the flexural rigidity and
the magnitude of the applied load.

The bending of the ocean lithosphere at trench-
arc systems can be explained by the flexure of an
elastic plate by a combination of vertical and horizontal forces and bending moments. These forces
cannot, however, be directly determined. Forebulges (or outer rises) are very well developed seaward of deep trenches in the northwestern Pacific.

The oceanic lithosphere appears to become stronger as it ages, but it does not weaken in its ability to
support loads with time since loading. This suggests
that any viscous relaxation that takes place, does so
very quickly (less than 10^5 or 10^6 years), after which
flexural rigidity approaches an asymptotic value.
The factors determining the flexural rigidity of the
continental lithosphere are less clear.

Many studies of continental flexure have been
made in collisional zones. Bouguer gravity anomaly profiles across mountain belts and foreland
basins suggest a great variability in the make-up of
the force systems deflecting the continental plate.
In the Ganga Basin the topographic load of the
Himalaya appears to be too large to account for the
deflection of the Indian plate beneath the Ganga
Basin, suggesting that an additional buoyancy force
must exist. On the other hand, the topographic
load of the Zagros mountains is insufficient to

93

cause the observed flexure of the Arabian plate in the Mesopotamian foreland basin, suggesting that an additional force with a downward-acting component must be present.

A viscoelastic rheological model of the lithosphere predicts that following loading the flexural rigidity should decrease with time. The time scale of the relaxation is debated. If the relaxation time is long (say > 20 Myr), the weakening of the plate will have a major impact on the geometry of the deflection and should profoundly influence the stratigraphic patterns in the basin-fill. Specifically, the depression should narrow and deepen with time and the forebulge should become higher.

Some authors believe that the continental lithosphere can be weakened by being subjected to large bending stresses, rather than having a flexural rigidity determined by thermal age. It has been suggested that lithosphere approaching collision zones may be segmented. An example from the Italian Apennines suggests that tears in the subducting plate may cause a stepped, compartmentalized deflection. This complex geometry of deflection at depth appears to be reflected in a similar geometry of near-surface deformation.

Orogenic wedges act as vertical loads on the deflected plate in collision zones. These wedges are viewed as dynamic units with critical surface slopes in which the gravitational and deviatoric forces strive to achieve a steady state. Disturbances of equilibrium by externally imposed forces (such as changes in convergence rate) may result in major changes in the rates of shortening or extension of the wedge. These processes in turn influence the erosion of the mountain belt to provide detritus for the foreland basin and the configuration of the load causing the deflection of the overriden plate.

Foreland basins are elongate or arcuate, highly asymmetrical basins intimately associated with continental collision zones. The alternative term 'foredeep' was introduced by Price (1973). Dickinson (1974) proposed two genetic classes of foreland basin:

1 Peripheral foreland basins situated against the outer arc of the orogen during continent–continent collision (e.g. Indo-Gangetic plain, north Alpine foreland basin).

2 Retro-arc foreland basins situated behind a magmatic arc linked with subduction of oceanic lithosphere (e.g. Andean examples, Late Mesozoic–Cenozoic Rocky Mountain basins, North America).

Both classes of foreland basin overlie cratonic lithosphere and are associated with crustal shortening in tectonically active zones.

A mechanically similar situation exists at ocean–ocean and ocean–continental collision zones where oceanic lithosphere is subducted. The complex collision zone contains many structurally controlled basins, but the most dramatic bathymetric expression is the ocean trench, and the mechanical equivalent of the thrust belt of the foreland basins is the accretionary prism.

The reader is referred to Section 2.1 for some of the fundamentals of flexure of linear elastic materials. Section 2.3 contains background to more complex rheologies. In this chapter we concentrate on the application of ideas on flexure to the geometry of ocean trenches and particularly foreland basins. Chapter 6 discusses the stratigraphic patterns in basins of this type.

We have seen in Section 2.1 that the strength of the lithosphere is capable of redistributing the support of a load by a horizontal transmission of stresses. A general flexural equation (eq. 2.29) involving vertical applied loads, horizontal pressure forces and torques or bending moments can be established for the case of a thin linear elastic plate in a state of plane strain, a reasonable approximation for the lithosphere. In the case of foreland basins, continental lithosphere is bent under the sum total of loads imposed by orogenesis. In the case of ocean trenches (forearc basins) the oceanic lithosphere is bent prior to subduction. We shall now examine these cases in more detail in order to investigate the precise nature of the basin-forming forces.

4.1 FLEXURE OF OCEANIC LITHOSPHERE

4.1.1 Geometry of the deflection

Some general concepts of flexure can be appreciated by considering the bending of oceanic lithosphere under loads acting a long way from the edge of the plate. A geological situation of this type is the flexure of the lithosphere under the load of mid-plate oceanic islands such as the Hawaiian archipelago (Vening Meinesz 1941, 1948, Watts and Cochran 1974). The Hawaiian ridge is a long (thousands of km) line of volcanic islands of

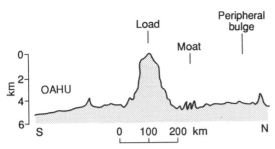

Fig. 4.1. Bathymetric profile across the Hawaiian archipelago based on Chase *et al.* (1970), showing the main morphological elements, a line load represented by the Hawaiian ridge, an adjacent moat and an outer rise or peripheral forebulge.

about 150 km width. The ridge therefore approximates a line load. The ridge is flanked by a depression, the Hawaiian Deep and then an outer rise (Fig. 4.1). These three morphological elements (load, basin and outer rise or forebulge) are a constant theme in flexural problems.

Taking the boundary conditions for the flexural force balance (eq. 2.29) as follows:
1 The applied vertical load, $q_a(x) = 0$, since the applied load is zero except at the location of the Hawaiian ridge ($x = 0$).
2 $P = 0$ since no horizontal forces are applied, then

$$D\frac{d^4w}{dx^4} + \Delta\rho g w = 0 \tag{4.1}$$

where w is the deflection, x is the horizontal scale, D is the flexural rigidity and $\Delta\rho$ in this case is the difference in density between mantle and infilling ocean water.

The general solution for a fourth order differential equation such as this is accomplished by breaking it into exponential, sine and cosine components (it is in the form of damped sinusoids)

$$w = e^{x/\alpha}(C_1 \cos x/\alpha + C_2\sin x/\alpha) + e^{-x/\alpha}(C_3 \cos x/\alpha + C_4 \sin x/\alpha) \tag{4.2}$$

where the constants C_1, C_2, C_3 and C_4 are determined by the boundary conditions and α is the *flexural parameter* (Walcott 1970), given by

$$\alpha = \left\{\frac{4D}{\Delta\rho g}\right\}^{1/4} \tag{4.3}$$

Following from the determination of the constants for the case of a line load bending the plate symmetrically (see Turcotte and Schubert 1982,

pp. 125–126), some useful and simple expressions emerge for the geometry of the deflection.

If the maximum deflection (w_0 at $x = 0$) is known, the profile of the deflection of the plate obeys

$$w = w_0 e^{-x/\alpha} (\cos x/\alpha + \sin x/\alpha) \tag{4.4}$$

The half-width of the depression (x_0) can be found since it is defined by the horizontal distance from the maximum deflection ($x = 0$) to the point where the deflection is zero ($w = 0$). Setting $w = 0$, equation (4.4) gives

$$x_0 = \frac{3\pi\alpha}{4} \tag{4.5}$$

The distance from the line load ($x = 0$) to the highest part of the forebulge (x_b) can be found since where the deflection is maximum at the forebulge crest, the slope of the deflection is zero. That is, $dw/dx = 0$ at $x = x_b$. Equation (4.4) then gives

$$x_b = \pi\alpha \tag{4.6}$$

The height of the forebulge w_b above the datum of zero deflection can also be found using the condition that $x = \pi\alpha$ at the point where $w = w_b$. As a result equation (4.4) reduces to

$$w_b = -w_0 e^{-\pi} = -0.0432\, w_0 \tag{4.7}$$

The theoretical profile is given in Fig. 4.2. By comparison of the observed (Fig. 4.1) with the theoretical (Fig. 4.2) profiles something can be said about the thickness of the equivalent elastic lithosphere under the Hawaiian islands, always assuming the theory to be an adequate representation of the geodynamics (Watts and Cochran 1974). This equivalent elastic thickness should not be thought of as the depth to a physical interface. It is merely the numerical result following from numerous assumptions about the force system and the mechanical properties of the plate. Flexural rigidity D is an alternative and perhaps better way of expressing the strength of the plate during flexure.

For example, the bathymetric profile suggests that the crest of the outer rise is about 250 km from the line load of the Hawaiian Islands, i.e. $x_b = 250$ km. Assuming the moat to be water-filled, $\Delta\rho = 2300$ kg m^{-3}. With a gravitational acceleration of 10 m s^{-2}, the flexural parameter from (4.6) is 80 km which gives from (4.3) a flexural rigidity of 2.4×10^{23} Nm. If Young's modulus E is 70 GPa and Poisson's ratio $\nu = 0.25$

(a)

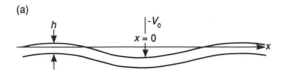

(b)

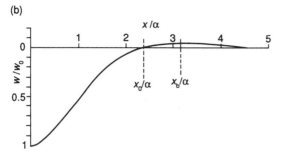

(a)

(b)

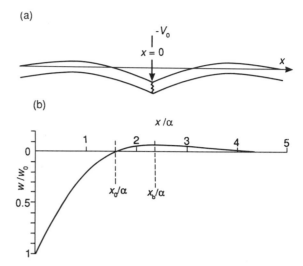

Fig. 4.2. (a) Deflection of the elastic lithosphere under a line load. (b) Theoretical deflection of the elastic lithosphere under a line load applied at the centre of the infinitely extensive plate. Parameters are defined in the text. The deflection w is scaled against the maximum deflection w_0. The horizontal distances are scaled against the flexural parameter α. Bending moments and horizontal in-plane forces are zero.

Fig. 4.3. (a) Deflection of a broken, elastic lithosphere by a line load applied at its end. Horizontal in-plane forces and applied bending moments are again zero. (b) Theoretical deflection of the broken plate under a line load applied at its end. Note that the half-width of the basin is narrower than for the unbroken plate, and that the elevation of the forebulge is greater than for the unbroken plate. Vertical and horizontal axes are scaled by w_0 and flexural parameter respectively.

$$x_b = \frac{3\pi\alpha}{4} \qquad (4.10)$$

showing that narrower forebulges characterize broken plates. Finally, the height of the forebulge (where $x_b = 3\pi\alpha/4$) is given by

$$w_b = w_0 e^{-3\pi/4} \cos 3\pi/4 = -0.0670 \, w_0 \qquad (4.11)$$

indicating a considerably larger forebulge amplitude for a broken plate.

Returning to the case of the Hawaiian Islands with $x_b = 250$ km (see above), equation (4.10) gives 106 km for the flexural parameter and flexural rigidity is therefore 7.26×10^{23} Nm. The equivalent elastic thickness with $E = 70$ GPa and $v = 0.25$ is 49 km.

Seismic refraction studies (Shor and Pollard 1964) suggest that the Moho is deflected downwards by approximately 10 km under the centre of the Hawaiian Islands. If 10 km is the maximum deflection w_0, the height of the outer rise above the undeflected sea floor can be found from equation (4.7) for the unbroken plate and equation (4.11) for the broken plate. The results are 432 m and 670 m

this gives an equivalent elastic thickness of 34 km.

If the Hawaiian plate were broken under the Hawaiian Islands, the boundary conditions of the model of the lithosphere would need to be modified (Walcott 1970). In this case, we would consider the deflection of a semi-infinite elastic plate subjected to a line load $V_0/2$ *applied at its end* (Fig. 4.3). Assuming that no external torque is applied at $x = 0$, simple expressions describing the geometry of the deflection can be obtained as follows.

The maximum deflection for a broken plate of the same flexural rigidity and under the same vertical load is twice that of an unbroken plate (Turcotte and Schubert 1982). If the maximum deflection is known, the deflection as a function of x for a broken plate loaded at its end is given by (Turcotte and Schubert 1982, p. 127)

$$w = w_0 e^{-x/\alpha} \cos x/\alpha \qquad (4.8)$$

The half width of the basin (at $w = 0$) is given by

$$x_0 = \frac{\pi\alpha}{2} \qquad (4.9)$$

showing that the basin is narrower for the case of a broken plate. The distance to the crest of the forebulge (where $dw/dx = 0$) is given by

respectively. Measurements of the bathymetry of the sea floor surrounding the Hawaiian Islands indicate that the outer rise is elevated above regional by about 500 m (Chase, Menard and Mammerickx 1970) but this does not permit us to tell whether the continuous or broken plate best fits the observational data.

Submarine processes such as erosion or draping with sediment, volcanism and tectonics may cause significant changes to the bathymetric profile. Gravity profiles have therefore been used to investigate the deflection curve across the Hawaiian Islands (Watts and Cochran 1974). A good fit was found between the observed free air gravity anomalies and those calculated for the continuous elastic plate if $D = 5 \times 10^{22}$ Nm and for the broken elastic plate if $D = 2 \times 10^{23}$ Nm. These values are

close to the flexural rigidities estimated from the wavelength of the deflection given above.

Another geological situation is of the bending of oceanic lithosphere at arc-trenches. This configuration is similar to that of an end load on a broken plate, but in this case we cannot ignore the bending moments acting on the subducting plate. Turcotte and Schubert (1982, p. 128) give the solution for the deflection by applying the boundary conditions for an ocean trench to the general flexural equation. This solution incorporates the contribution of bending moments to the deflection. The physical significance of the vertical force and bending moment is less clear in the case of oceanic trenches. They may result from a combination of gravitational body forces acting on the descending plate and the forces imposed by the magmatic arc–

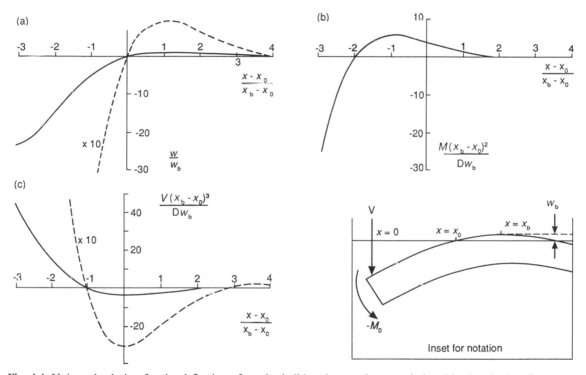

Fig. 4.4. Universal solution for the deflection of an elastic lithosphere under a vertical end load and a bending moment, approximating to the situation at oceanic trenches. (a) Dependence of the non-dimensional displacement w/w_b on the non-dimensional position $(x - x_0)/(x_b - x_0)$. Dashed line shows vertical exaggeration of 10:1 to show geometry of forebulge. (b) The non-dimensional bending moment $(M(x_b - x_0)^2/Dw_b)$ versus non-dimensional position. Note that the maximum bending moment is found roughly one third of the distance from $x = x_0$ to the maximum deflection. (c) The non-dimensional vertical shear force $(V(x_b - x_0)^3/Dw_b)$ as a function of non-dimensional position, showing that it reaches a maximum at the point of maximum deflection. Dashed line is at vertical exaggeration of 10:1. *Inset* shows notation.

accretionary prism in the overriding plate. Whatever their origin, they cannot be directly determined. However, it is possible to obtain some estimates of the geometrical aspects of the flexural depression by expressing the trench profile in terms of the height of the forebulge (w_b) and the half width of the forebulge ($x_b - x_0$) (Fig. 4.4). Turcotte and Schubert (1982, p. 128) give the solution for the latter as

$$x_b - x_0 = \pi\alpha/4 \tag{4.12}$$

In other words, the half width of the forebulge is a direct measure of the flexural parameter. A forebulge is well developed seaward of deep sea trenches in the northwestern Pacific (Watts and Talwani 1974). The observed bathymetric profile across the Mariana Trench (Fig. 4.5) suggests that the half-width of the forebulge (measured on the flank facing the trench) is about 55 km. Equation (4.12) then gives 70 km for the flexural parameter.

Both the deflection w and the height of the forebulge w_b are a function of the flexural parameter and bending moment. Dividing the expression for w by the expression for w_b (i.e. w/w_b) eliminates the unknown parameters and gives a universal flexure profile of the type shown in Fig. 4.4. This universal profile applies to any two-dimensional elastic flexure under end loading, so can be applied to a variety of geological contexts.

As an example of its application, let us return to the case of the Mariana Trench. Taking horizontal distances from the point $x = 0$ at the oceanward intersection of the plate surface with sea level (Fig.

4.5), the distance to the crest of the forebulge is approximately 55 km and its height is taken as 0.5 km. The fit between the theoretical deflection and the observed bathymetry is shown in Fig. 4.5. Although the fit is not perfect, particularly where volcanoes have built cones in the forebulge region, the correspondence is generally good, suggesting that the lithosphere at ocean trenches can indeed be modelled in this way.

The flexure of a semi-infinite elastic oceanic plate at trenches was studied by Watts and Talwani (1974). They found that the bathymetry of the Mariana trench could be explained by purely a vertically applied force at the end of the plate. The Aleutian and Kuril trenches, however, were better explained by a model involving both a vertical applied force (V) and a horizontal applied force (P where $P = 5V$). However, a bending moment at the end of the plate would reduce the necessity of a horizontal force to explain the deflection (Hanks 1971).

4.1.2 Flexural rigidity of the oceanic lithosphere

The lithosphere can be regarded as a thermal boundary layer losing heat to the atmosphere and oceans by conduction. Such a thermal entity does not, however, have a homogeneous rheology (Section 2.3), being separated into an elastic upper layer which stores and transmits elastic stresses, and a lower layer characterized by a power-law creep which serves to dissipate stresses. The oceanic lithosphere thickens as a function of age, being ~ 6 km at ridge

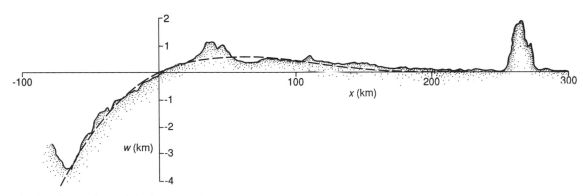

Fig. 4.5. Comparison of the bathymetric profile across the Mariana Trench (solid line) (Watts and Talwani 1974, p. 65) with that predicted from the universal flexural equation (dashed line), with x_b taken as 55 km and w_b as 0.5 km (after Turcotte and Schubert 1982, p. 130). The fit is good, except for newly constructed volcanoes in the forebulge region.

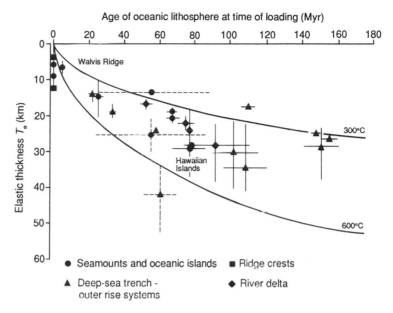

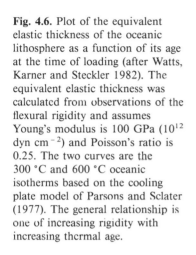

Fig. 4.6. Plot of the equivalent elastic thickness of the oceanic lithosphere as a function of its age at the time of loading (after Watts, Karner and Steckler 1982). The equivalent elastic thickness was calculated from observations of the flexural rigidity and assumes Young's modulus is 100 GPa (10^{12} dyn cm^{-2}) and Poisson's ratio is 0.25. The two curves are the 300 °C and 600 °C oceanic isotherms based on the cooling plate model of Parsons and Sclater (1977). The general relationship is one of increasing rigidity with increasing thermal age.

axes and thickening to about 100 km under the oldest (Jurassic) ocean floor (Section 2.2). Clearly the ability of the oceanic lithosphere to support loads will also therefore be a function of its age. The plot of effective elastic thickness of the oceanic lithosphere versus age (Fig. 4.6 from Watts, Karner and Steckler 1982) shows that the lithosphere becomes stronger with time. This relationship is exponential with time, similar to that of the oceanic bathymetry using a cooling plate model such as that shown in Fig. 2.22 (Parsons and Sclater 1977).

The timing of the formation of the Hawaiian–Emperor Seamount chain varies in age from ~ 3 Ma near Hawaii to ~ 70 Ma near the northernmost Emperor Seamount. Yet gravity data indicate that there is no significant change in flexural rigidity along the chain (Watts and Cochran 1974). This evidence suggests that the flexural rigidity of the oceanic lithosphere does not change with age since loading. The oceanic lithosphere can therefore be treated as being elastic.

4.2 FLEXURE OF THE CONTINENTAL LITHOSPHERE

The continental elastic lithosphere responds differently from the oceanic elastic lithosphere on all scales of deformation. The continents appear to accumulate strains over long periods of geological

time whereas the oceanic lithosphere remains relatively intact over its short lifetime of up to about 180 Myr. This is illustrated by the presence of narrow and well-defined plate boundaries in the oceans as opposed to the wide and diffuse zones in the continents such as the Himalayas and Tibet (review in Molnar 1988). The continents are therefore exceedingly complex, particularly in their inherited fabric. Even without the complications of fundamental lithospheric heterogeneities, flexure of continental lithosphere is controversial, and there are no simple parameters that can be correlated with the observed flexural rigidity of the continental lithosphere (McNutt and Kogan 1988). The changes through time ('secular evolution') of the flexural rigidity of continental lithosphere has in particular been widely discussed.

4.2.1 Theoretical deflection

We can take as a starting-off point a model of the continental lithosphere as a linear elastic solid overlying an inviscid fluid mantle and subject to a linear load at its edge. The load is assumed to be of infinite length in the direction parallel to the edge of the plate. The universal flexural profile of Figure 4.4 should therefore be applicable. Since regions of continental collision are characterized by mountains acting as sourcelands for large volumes of detritus, basins occupying the flexural depressions tend to be rapidly filled with sediment. As a result $\Delta\rho$ in the flexural parameter (eq. 4.3) is the difference between mantle and sediment densities, $\rho_m - \rho_s$.

As in the case of the Mariana Trench, it is possible to match a theoretical deflection to the observed depth of basement in a region of continental flexure such as the Appalachians (Fig. 4.7). The best fit of the theoretical curve to the basement shape is by choosing x_b = 122 km (half width of forebulge) and w_b = 0.29 km (height of forebulge). As must be obvious when considering a Palaeozoic foreland basin, much information has been lost through erosion and the choice of parameters above is therefore neither unique nor verifiable. The close match between theory and observation in Fig. 4.7 merely gives some confidence in treating the geodynamical problem as essentially one of flexure.

It is important to note that the Appalachian continental flexure has been used as an example of a viscoelastic response to flexure, and the reader is referred to Section 6.2.2 for further details. The flexural rigidity implied by the Appalachian profile in Fig. 4.7 is 10^{24} Nm (equivalent elastic thickness of 54 km). This suggests that the flexural rigidity of continental lithosphere is generally greater than that of oceanic lithosphere.

4.3 FLEXURAL RIGIDITY OF CONTINENTAL LITHOSPHERE AND ITS SECULAR EVOLUTION

It was noted in Sections 4.1 and 4.2 how gravity data can be used to constrain flexural models of the oceanic lithosphere at mid-plate seamount chains and at deep sea trenches. Gravity anomalies also result from the flexure of continental lithosphere. Gravity (Bouguer) anomalies have been widely used to study the flexure resulting from continental collision (Karner and Watts 1983, Lyon-Caen and Molnar 1989).

Bouguer gravity anomalies along profiles in the western Himalaya and Ganga Basin indicate a mass deficit over the basin and a mass excess in the Lesser Himalaya (Lyon-Caen and Molnar 1985) (Fig. 4.8). The Bouguer anomaly over the Ganga Basin is, however, smaller than expected from the topographic load of the Himalaya, suggesting that some additional force, acting upward on the de-

(a)

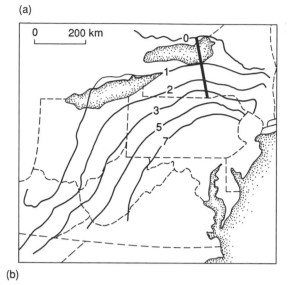

(b)

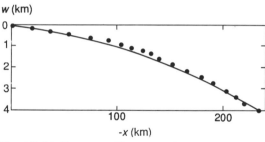

Fig. 4.7. (a) Contours of basement (in km) in the Appalachian basin of the eastern United States, based on borehole records and seismic reflection studies (Turcotte and Schubert 1982, p. 131). (b) Depths of basement below sea level along the profile given in (a) as a function of distance from the point at which the basement rocks crop out at the surface. The heavy line is the theoretical deflection given by the universal flexural equation with x_b equal to 122 km and w_b equal to 0.29 km. It should be noted that the stratigraphy of the Appalachian Basin has been analysed by Quinlan and Beaumont (1984) and Beaumont, Quinlan and Hamilton (1988) in support of a *viscoelastic* lithospheric model (fuller discussion in Chapter 6).

flected lithosphere, is necessary. In contrast, the topographic load of the Zagros mountains is insufficient to cause the observed flexure of the Arabian plate with the formation of the Mesopotamian foredeep and Persian Gulf as the foreland basin (Snyder and Barazangi, 1986). This leads to the requirement for an additional force flexing the

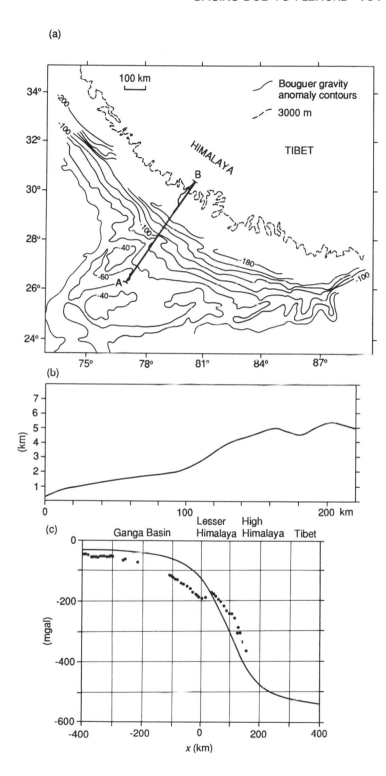

Fig. 4.8. The Himalayas and Ganga Basin. (a) Contours of Bouguer anomalies over the Ganga Basin, showing an increasingly negative anomaly as the Himalayan front is approached. A–B is location of profile shown in (b) and (c). 3 km topographic contour shown for reference. (b) Topography along the northern part of section A–B, with the northern edge of the Ganga Basin at $x = 0$. (c) Observed Bouguer gravity anomalies along section A–B, compared with the anomalies computed assuming that the topography is locally compensated (Airy model) by thickening of the crust beneath central India. Note that there is an apparent mass excess in the Himalaya and a mass deficit over the Ganga Basin (after Lyon-Caen and Molnar 1985).

Arabian plate downwards. Snyder and Barazangi believe that this additional force may be horizontal compression. It is less likely to be subsurface density variations caused by overriding an initially thinned Arabian continental margin, since there is no associated positive Bouguer anomaly, unlike the Alps (Fig. 4.9) and the Appalachians (Karner and Watts 1983).

Since the wavelength of the deflection is determined by the flexural rigidity (Sections 2.1.5, 4.1) it is possible to calculate D from the gravity profile.

Based on Bouguer gravity profiles, the flexural rigidity of the Indian plate being overridden by the western Himalayas varies considerably over even relatively small horizontal distances (Lyon-Caen and Molnar 1985). Between profiles 200 km apart, estimated flexural rigidities changed by a factor of 5 to 10 (between c. 10^{24} and 7×10^{24} Nm). The rapidity of the horizontal changes suggests that this is not a result purely of differences in convective heat supply or loss in the mantle, but may instead be a result of small temperature variations effecting large changes in strength in a highly temperature dependent lithospheric rheology, or of compositional or thickness changes of the deflected lithosphere. Secondly, it is necessary to infer a segment of the Indian plate, now deeply buried beneath the lesser Himalaya, which is steeply dipping and considerably weaker than the segment under the Ganga Basin (c. $2-5 \times 10^{22}$ Nm). This turns out to be a common feature of collision zones, since a similar abrupt steepening of the Moho is found in the Carpathian, Apennine, Andean and Zagros mountain belts.

4.3.1 Thermo-rheological models

The equivalent elastic plate thicknesses (or flexural rigidities) for continental lithosphere have been calculated in a wide variety of tectonic settings such as the Basin and Range province of continental extension in western USA using isostatic rebound of Pleistocene Lake Bonneville (Crittenden, 1963, Nakiboglu and Lambeck 1983), the Kenya dome of the East African Rift System (Banks and Swain 1978, Bechtel, Forsyth and Swain 1987) and passive continental margins, particularly that of the eastern seaboard of the USA (Watts and Ryan 1976, Watts *et al.* 1982). However, all of these examples come from regions which have undergone some form of thermal disturbance.

Whereas, to a first approximation, the loading of oceanic and continental lithosphere can be modelled as a flexed elastic plate overlying a fluid substratum, laboratory-derived rock deformation data (Goetze 1978, Goetze and Evans 1979, Kirby 1983) show that the true rheology of the lithosphere is far more complex. The lithosphere is thought to be divided into an upper region which deforms by brittle failure, a middle region which stores elastic stresses and a lower region which is subject to a nonlinear temperature dependent flow approximated by a plastic failure criterion (see also Section 2.3). Karner, Steckler and Thorne (1983) applied this thermoelastic rheological model to the flexure of continental lithosphere and concluded that:

1 in terms of flexural rigidity, the results from oceanic and continental lithosphere are compatible and can both be explained by a simple thermoelastic model;
2 the long-term thermal behaviour of the continental lithosphere is governed by a simple cooling plate model with a plate thickness of 200–250 km;
3 the effective elastic thickness for continental lithosphere corresponds with the depth to the 450 °C isotherm, and does not change following loading.

These conclusions have been contested by others (Willett, Chapman and Neugebauer 1985) who claim that there is no clear relation between thermal state and age of continental lithosphere, nor does the apparent elastic thickness correspond to the 450°C isotherm, but rather spans a wide temperature range of 300 °C to > 900 °C (Fig. 4.10). Karner *et al.*'s (1983) thermoelastic rheological model does not therefore universally explain the flexure of the continental lithosphere. An alternative is of a Maxwell viscoelastic heterogeneous lithosphere, i.e. a nonlinear thermally activated rheology (see Section 2.3.4), with viscous strain rates in the lithosphere determined by a power law ($n = 3$).

Using a numerical viscoelastic model, Willett *et al.* (1985) demonstrated that the surface displacement resulting from a supralithospheric load shows a depression which narrows and deepens with time, dragging a peripheral bulge towards the centre of the load (Fig. 4.11). This can be understood by considering the distribution of stress in the lithosphere resulting from the load. Initially, the deviatoric stress increases toward the base of the mechanical lithosphere, but immediately after

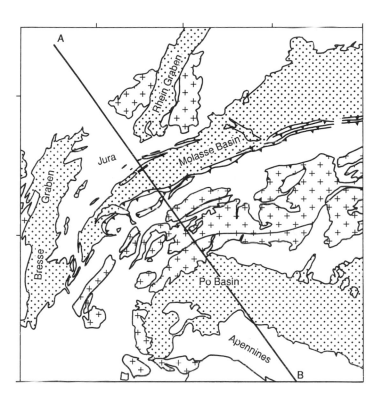

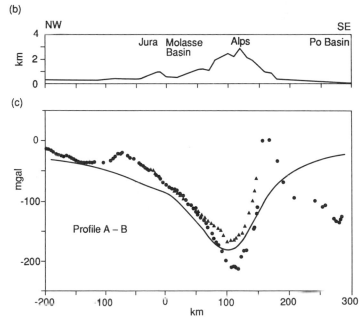

Fig. 4.9. The Alps and Molasse Basin (North Alpine Foreland Basin). (a) Simplified geological map showing the location of the Molasse Basin north of the Alps and the location of the section shown in (b) and (c). The tick on the line of the section indicates the origin where $x = 0$ and corresponds to the northern edge of the Molasse Basin. (b) Topography along section A–B. (c) Measured Bouguer anomalies along Section A–B with (triangles) and without (circles) terrain corrections. The curve is the computed Bouguer anomaly assuming local (Airy) isostatic equilibrium. The density of the load is taken as 2670 kg m^{-3}. Note the Bouguer anomaly peak under the Alps, but displaced southwards from the highest topographic elevations (after Lyon-Caen and Molnar 1989).

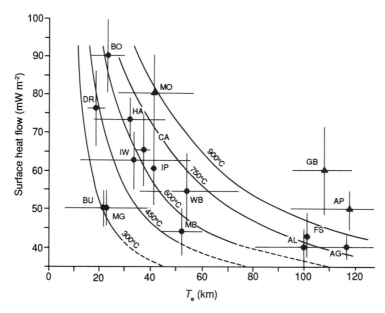

Fig. 4.10. Plot of the equivalent elastic thickness versus surface heat flow for a number of sites of flexure. The solid curves indicate depths to isotherms at a given surface heat flow for a steady-state thermal model. The plot demonstrates the very wide range in temperature (300 °C to >900 °C) for the base of the elastic lithosphere. Flexure sites are as follows: AG Lake Agassis, AL Lake Algonquin, FS Fennoscandia, CA Caribou Mountains, IP Interior Plains, GB Ganges Basin, IW Idaho-Wyoming thrust belt, MB Michigan Basin, BU Boothnia Uplift, MG Midcontinent Gravity High, HA Lake Hamilton, BO Lake Bonneville, DR North Great Dividing Range, AP Appalachian foreland basin, MO Molasse Basin, WB Williston Basin (full details in Willett, Chapman and Neugebauer 1985, p. 521).

loading, flow redistributes material and allows the stress to relax. A zone of negligible deviatoric stress develops and grows upward with time, so that the stresses are progressively concentrated in the colder upper lithosphere. The surface displacements can be used to calculate the apparent elastic thickness, which decreases in time. The model results suggest that for long times after loading, the stress field and surface displacements do not change significantly, suggesting that there is a long-term asymptotic elastic thickness (see also Kusznir and Karner 1985). The viscoelastic model satisfies the field data on ages of load and effective elastic thicknesses from a variety of flexural situations ranging from glacial rebound (e.g. Lake Agassis) to intra-continental basins (e.g. Michigan Basin). However, it does not account for those examples with complex thermal or mechanical histories such as foreland basins (Molasse, Appalachian, Ganga) (Fig. 4.12).

In summary, an elastic model modified to take account of the thermal structure of the lithosphere at the time of loading (i.e. thermoelastic model) predicts that flexural rigidity and equivalent elastic thickness do not significantly change with time since loading. Their values will depend only on the thermal age of the lithosphere. A viscoelastic model, however, suggests that flexural rigidities and elastic thicknesses should decrease immediately after loading because of viscous relaxation of stresses. Their values are therefore dependent on the time since loading. It is possible that the thermoelastic model satisfies the general dependency of flexure on thermal age, but the detailed evolution of flexural rigidity with time since loading may be better expressed by a rheological model which accounts for the temperature and stress dependency of lithospheric material of viscoelastic type. The time scale of the viscous relaxation of stresses (eq. 2.34, Section 2.3.4) is crucial to this problem. Unfortunately, there are as yet few sufficiently well-documented case studies of basin stratigraphy to adequately assess the value of this relaxation time.

4.3.2 Dependence of elastic thickness on weakening during plate bending

McNutt and Kogan (1987) repeated the study of the equivalent elastic thickness of the continental lithosphere carried out by Karner *et al.* (1983) but with an expanded data base from northern Eurasia. Contrary to the view that continental plates increase in elastic thickness with thermal age, they

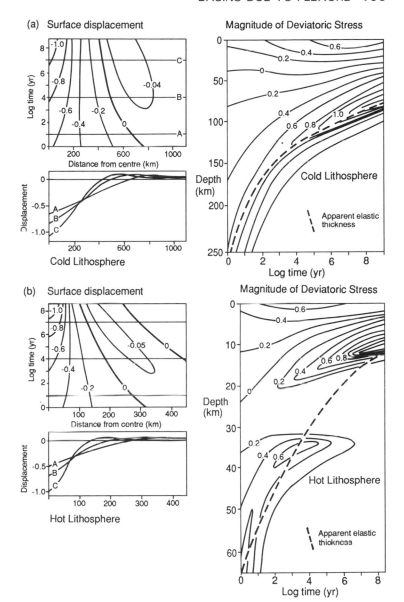

Fig. 4.11. Surface displacement and horizontal deviatoric stresses for two end-member thermal states of viscoelastic continental lithospheres. In (a) and (b) surface displacement is mapped continuously over time. The basement profile at three selected times A, B and C is also shown. The deviatoric stresses are shown for a plane directly below the centre of the load; the evolution of this stress field is shown as a function of log time since instantaneous loading. The elastic thickness calculated from the wavelength of the surface displacements is shown by the dashed line. In both (a) and (b) the equivalent elastic thickness decreases immediately after loading. (a) represents a 'cold' lithosphere with a surface heat flow of 40 mW m^{-2}, surface displacements are normalized to 1200 m and stresses are normalized to 50 MPa. (b) refers to a 'hot' lithosphere with a surface heat flow of 90 mW m^{-2}, surface displacements are normalized to 1600 m and stresses are normalized to 100 MPa. After Willett *et al.* 1985, p. 522.

found that old plates are not always stiff (Table 4.1). Instead a good correlation was found between the elastic plate thickness and the surface curvature in map view of the thrust belts of the mountain ranges. Highly arcuate mountain belts are associated with low elastic thicknesses (Alps, Carpathians) whereas long linear belts tend to be supported by a strong plate (Urals, Appalachians). McNutt and Kogan (1987) suggest that the plan view shape is related to the cross-sectional curva-

ture of the flexed plate, steeply dipping highly curved plates appearing weaker, and shallowly dipping plates appearing stronger. This would imply that the equivalent elastic thickness is not so much related to a characteristic isotherm, but is more due to the extent to which the elastic lithosphere has flexed under high bending stresses.

McNutt, Diament and Kogan (1988) analysed fifteen mountain belts in order to better establish the relationships between (1) elastic plate thickness

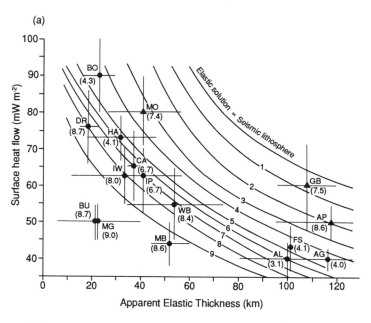

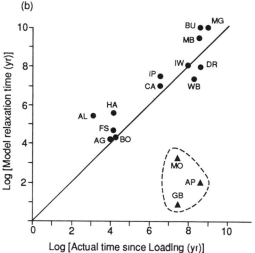

Fig. 4.12. (a) The equivalent elastic thickness of a viscoelastic continental lithosphere related to its thermal state and duration of loading. The solid curves show the predicted weakening of the lithosphere by relaxation of viscous stresses in the lower and hotter part of the lithosphere following loading. Contour values are base-10 logarithms of the load duration in Myr. A weakening trend is therefore towards the lower left of the diagram. Values in brackets are the base-10 logarithms of the actual load durations for comparison. Note the large error bars. (b) A comparison of the actual time since loading and the viscoelastic relaxation time necessary to produce the observed equivalent elastic thickness. The three foreland basins of the Appalachians, Ganga Basin and Molasse Basin show significant departures from the model prediction and exhibit values close to those expected of an elastic lithosphere (after Willett *et al.* 1985, p. 523). Abbreviations are as in Fig. 4.10.

(2) age of flexed lithosphere at the time of loading (3) radius of curvature of the thrust belts (4) total length of the thrust segment and (5) dip of the underthrust slab. The elastic plate thickness was derived from
• subsidence analysis (Carpathians, Apennines)
• gravity observations, supplemented by seismic data (Californian Transverse Ranges, Appalachians, Andes, Alps, Zagros, Caucasus, Urals, Pamir, Tienshan, Kunlun, Himalaya, Verhoyansk).

The age of the flexed lithosphere at the time of loading is the present age, derived for example from radiometric dating of basement rocks, minus the age of the orogeny. The radius of curvature and length of thrust segments were estimated from geological maps. The dip of the underthrust slab varies along its length. However, its dip was taken along traverses which best constrained the equivalent elastic thickness, and measured from the straight line which joins the top of the plate at the

Table 4.1 Compilation of equivalent elastic thickness and age of basement for continental lithosphere

Orogen [name of flexed plate]	Equivalent elastic thickness (km)		Age of basement at time of loading (Myr)	
Alps (East) [European Platform]	50	(26–56)	250	(230–320)
Alps (West) [European Platform]	25		250	
Andes [Brazilian Shield]	45 ± 20		1000	
Apennines [Adriatic]	20 ± 5		100	
Appalachians [North American Platform]	105 ± 25	(95–140)	630	(560–1000)
Carpathians [East European Platform]	30 ± 10		1600	
Caucasus [East European Platform]	85 ± 10		1600	
Himalaya [Indian Shield]	90 ± 10	(95–120)	1000	(1000–2000)
Kunlun [Tarim Basin]	40		800	
Pamir [Tadjik Depression]	15 ± 5		320	
Tien Shan [Tarim Basin]	40 ± 20		800	
Tranverse Ranges [California]	5–15		100	
Urals [East European Platform]	75 ± 25		1600	
Verkhoyansk [Siberian Platform]	50 ± 10		1600	
Zagros [Arabian Shield]	50 ± 25		630	
Idaho–Wyoming Thrust Belt [North American Platform]		(12–55)		(75–275)
Other sites (intracontinental basins and post-glacial rebound)				
Lake Agassis		(101–132)		(1600–1800)
Lake Algonquin		(80–120)		(1200–1500)
Fennoscandia		(101)		(2000–2800)
Caribou Mountains, USA		(34–40)		(1600–1800)
Interior Plains, USA		(41)		(1600–2500)
Michigan Basin		(44–60)		(800–1100)
Boothnia Uplift		(6–40)		(1000–1400)
Midcontinent Gravity High		(15–30)		(100–400)
Lake Hamilton		(18–46)		(200)
Lake Bonneville		(16)		(200)
North Great Dividing Range (Australia)		(15–22)		(0–100)
Williston Basin		(34–75)		(2000–2400)

Numbers in parentheses – estimates from Watts, Karner and Steckler (1982) and Willett, Chapman and Neugebauer (1985). Other data from McNutt, Diament and Kogan (1988, p. 8827).

outer edge of the foredeep to the top of the plate just beneath the mountain front. The estimate is likely to have large errors. This angle is clearly shallower than the dip of the plate further under the orogenic wedge; near-vertically dipping slabs have been imaged from earthquake hypocentres under the Transverse Ranges and Pamir. The main conclusions of this study were:
• there is no indication that older mountain belts have suffered viscoelastic relaxation so that their present flexural rigidities are low
• the age of the continental lithosphere at the time of loading appears to set an upper bound to the maximum possible equivalent elastic thickness that could be observed. However, many T_e values fall

well below this level, suggesting that some weakening has taken place, but regardless of plate age. This mechanism of weakening may be related to decoupling of an elastic plate at some depth (say $T_e/2$) allowing the upper half and lower half to flex independently. This reduces the equivalent elastic thickness to 63 per cent of the simple elastic plate for the same curvature (McNutt *et al.* p. 8830). The more realistic situation is of a plate with different rheologies with depth; Byerlee's law (brittle) in the upper crust and upper mantle, temperature-activated ductile creep in the lower crust and at the base of the mechanical plate (Fig. 4.13). Bent to a high curvature, the bending stresses cause failure in the ductile lower crust, leading to decoupling, and the equivalent elastic thickness drops by a factor of about two. The effect of a ductile zone in the lower crust, activated at high bending stresses, may therefore be responsible for the weakening of continental plates.

• there is a good correlation between plate dip estimated by McNutt *et al.* (1988) and equivalent elastic thickness. Since the wavelength of the flexure (which constrains T_e) and the surface and buried loads (which control the dip of the slab) are independent parameters, the authors conclude that the steep dip is responsible for weakening an otherwise strong plate. There may, however, be large errors in the estimate of plate dip.

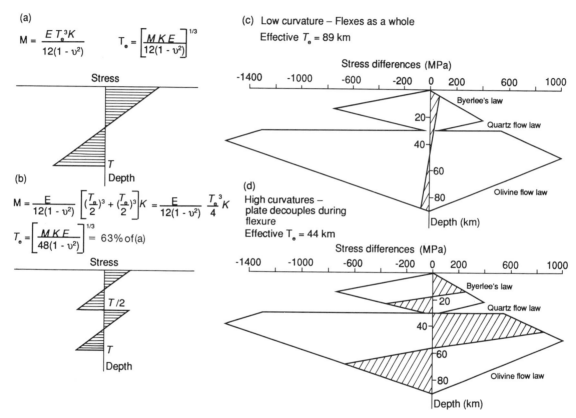

Fig. 4.13. The effects of decoupling at high curvatures illustrated by cross-sections of stress in the flexed lithosphere (after McNutt, Diament and Kogan 1988). (a) Distribution of fibre stresses in an elastic plate of thickness T showing extension in the upper half of the plate and compression in the lower half. (b) A purely elastic plate that is decoupled at $T/2$ so that the upper and lower portions flex independently with the same radius of curvature as in (a). The equivalent elastic thickness is 63 per cent of that in (a). M is the bending moment, K the plate curvature, E Young's modulus, v Poisson's ratio. (c) and (d) Rheologically layered continental lithosphere showing the failure envelope under extension (positive stress) and compression (negative stress). The strength is limited by frictional sliding (Byerlee's law) in the upper crust and uppermost mantle and by ductile flow in the lower crust and lower lithosphere. The hatched areas represent the stress resulting from flexure superimposed on the failure envelope. In (c) the plate is flexed at a very low curvature. The fibre stresses rarely exceed the failure criterion of the plate (there is a small amount of yielding at the top and bottom) and the equivalent elastic thickness (89 km) is almost exactly the same as the thickness of the elastic plate. In (d) the plate is flexed to a high curvature. This leads to the failure criterion being exceeded and a decoupling zone forms in the lower crust. The equivalent elastic thickness calculated for the total bending moment sustained in both the crust and the mantle is just 44 km. Decoupling therefore has strong potential mechanical implications for flexure.

• there is a good correlation between the radius of curvature of the thrust belt and the equivalent elastic plate thickness, suggesting that plates which are highly curved in the vertical plane are also likely to be curved in the plane of the Earth's surface, producing arcuate mountain belts. These arcuate belts are commonly associated with back arc extension (e.g. Pannonian Basin), itself related to the dip of the subducting slab (Uyeda and Kanamori 1979).

• the weaker, highly curved plates create short thrust segments, lacking the lateral integrity to form long thrusts.

4.4 SEGMENTATION OF SUBDUCTED LITHOSPHERE – EFFECTS ON GEOMETRY OF FLEXURAL BASINS

We have so far considered the deflection of the lithosphere at collision zones to be perfectly cylindrical. That is, any two-dimensional transect would be representative of the flexure as a whole. However, in some instances, there are offsets in the positions of the depocentres marking distinct fore-deep segments. The Apennines of Italy are a well-documented example (Royden, Patacca and Scandone 1987). Four forebulge segments, each characterized by Bouguer gravity highs can be distinguished (Fig. 4.14). They are thought to be linked to four slab segments which are delimited by tears or lines of weakness in the subducted slab. Along transects within individual segments, the gravity data can be used to calculate the equivalent elastic thickness of the plate. (The topographic loads on the plate are thought to be insufficient to explain the deflection (Royden and Karner 1984) and substantial subsurface loads are invoked which are several times the magnitude of the surface load.) In all four segments, the elastic thickness is in the range 15 to 20 km (Royden *et al.* 1987), suggesting a uniform but stepped response of the deflected lithosphere. The stepped deflection is thought to be due to segmentation of the lithosphere within the subduction zone. The model deflection using a semi-infinite elastic sheet is illustrated in Fig. 4.15. Applied loads and bending moments applied to the segments of the subducted lithosphere produced a series of offset forebulges and foredeep basin compartments which match well with the Apenninic geology. The segmentation

of the plate at depth is masked by a continuous zone underlying the foreland basin, the only evidence for the segmentation lying in the distribution of the swells and basins on the deflected plate.

Furthermore, the thrust belt appears to mimic the deeper segmentation. In the segment with a forebulge far to the northeast, the thrust belt has also advanced relatively to the northeast. The same is true of the lesser-travelled thrust units correlating with forebulges relatively to the southwest. Since subsurface loads (i.e. forces within a subduction zone) are thought to be much more important than surface loads in the Apennines, this implies that near surface tectonics are also intimately related to the lithospheric heterogeneity at depth.

4.5 DYNAMICS OF OROGENIC WEDGES

Foreland basins are dynamically linked to associated orogenic belts. The evolution of the orogenic wedge is important for basin development in a number of ways

• the wedge represents a supracrustal load on the foreland plate: its geometry and structure therefore influence the deflection of the foreland plate;

• the shortening and thickening of the wedge, or its extension and forward propagation change the *configuration* of the load with respect to the deflected plate, i.e. its magnitude and distribution over the deflected plate;

• the unroofing by tectonic uplift and erosion of the orogenic wedge provides the detritus for deposition in the basin.

Here we concentrate on some of the broad processes of orogenic wedge evolution that have some bearing on the deflection of the foreland plate. Further information on foreland basin stratigraphy can be found in Section 6.2.

Three possible mechanisms for the driving forces responsible for crustal shortening have been proposed; (1) *gravity sliding*, (2) *gravity spreading*, (3) *horizontal deviatoric push*. Gravity sliding requires a gradient for sliding under gravity alone. The gravitational force acting on the gradient must overcome the resistance to movement for gravity sliding to occur. Mathematical models suggest that the angle need only be small (a few degrees) where the effective normal stress on the inclined surface is reduced by the presence of high pore fluid pressures (Hubbert and Rubey 1959). However, field

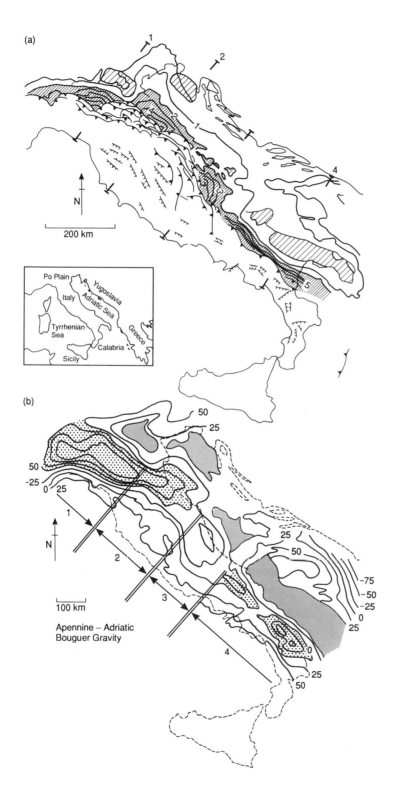

Fig. 4.14. (a) General map of the Apennine system showing depth to the base of the Pliocene in the Adriatic and Po Basins at a 1 km contour interval. There are four distinct outer-rise segments recognized in the basal Pliocene surface (shown in wide diagonal hatch). The narrow diagonal hatch shows the Apennine foreland basin and parts of the basal Pliocene surface that is below 2 km depth. (b) Simplified map of the Bouguer anomaly gravity field (milligal). Contour interval is 25 mgal. The four morphological outer rises correspond to Bouguer gravity highs, suggesting that they are maintained by regional flexure. The shaded areas represent anomalies of greater than 0 mgal in sectors 1 to 3 and 50 mgal in sector 4; stipple shows gravity anomalies less than − 50 mgal in sectors 1 to 3 and 0 mgal in sector 4. These sectors are believed to be bounded by major tears in the subducted plate, segmenting it at depth (after Royden, Patacca and Scandone 1987, compiled from Ogniben *et al.* 1985 and Morelli *et al.* 1975).

Fig. 4.15. Three-dimensional model used by Royden *et al.* (1987) to simulate the flexure of segmented lithosphere. The lithosphere is assumed to be continuous in the foreland region but to be segmented by free boundaries or faults in the subduction zone. Vertical loads and/or bending moments are applied to the ends of the broken segments of the plate. The contours of the deflected surface, assuming the three slab segments to be initially at 1 km depth, are shown in km. A uniform equivalent elastic thickness of 20 km is assumed throughout and the vertical end load is 5.10^{12} Nm^{-1}. Note the offset of the outer rise segments and the three-dimensional pattern of the foreland basin depocentres (after Royden *et al.* 1987).

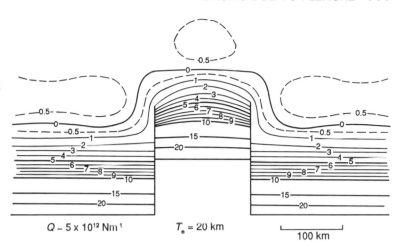

$Q - 5 \times 10^{12}$ Nm^{1} $T_e = 20$ km 100 km

studies, particularly in the Rocky Mountains (e.g. Dahlstrom 1970) suggest that many thrusts dip in the opposite sense to the mass transport direction. Gravity sliding is therefore not likely to be the sole or dominant process in mountain belt tectonics.

Price (1973) believed that orogenic wedges behave as plastic entities as a result of the interfingering of a large number of thrust sheets into a mechanically interdependent system. Gravity acting on an orogenic 'high' would provide the driving force for a spreading along thrust faults. The horizontal shear stress produced by gravity is given by (Elliott 1976)

$$\tau = \rho g h \alpha \qquad (4.13)$$

where ρ = density, g = gravitational acceleration, h = depth below surface, α = surface slope of wedge. However, although this strain system is capable of producing local shortening which must be compensated elsewhere by extension in order to maintain a constant angle of the wedge, it cannot explain the evidence of large scale net shortening in orogenic belts.

Horizontal compressional forces (deviatoric 'pushes') exerted along convergent plate boundaries may be primarily responsible for the dynamics of orogenic wedges. The convergent orogen can

be regarded as a wedge-shaped prism resting on a rigid slab with a rigid buttress at the rear (Chapple 1978). The wedge behaves as a single mechanically continuous, dynamic unit, but here the longitudinal force applied at the rear of the wedge is counterbalanced by resistance to sliding on its base. The geometry attained is one of a wedge tapering along its length with a dynamic balance between the gravitational forces arising from the slope of the wedge, the push from the rear and the basal shear force or 'traction'. This balance can be expressed as follows (Platt 1986, p. 1039)

$$\tau_b = \rho g h \alpha + 2K\theta \qquad (4.14)$$

where the additional term on the right-hand side is due to the horizontal 'push', K is the yield strength of the wedge and θ is the angle at the front of the wedge (Fig. 4.16). Davis, Suppe and Dahlen (1983) accounted for the internal strength and pore-fluid pressure of the wedge. They introduced the term '*critical taper*' to describe the surface slope which would produce a wedge in a state of yield throughout (Fig. 4.16). Davis *et al.* (1983) therefore expressed the basal shear stress in terms of the gravitational force and the material properties of the wedge as follows:

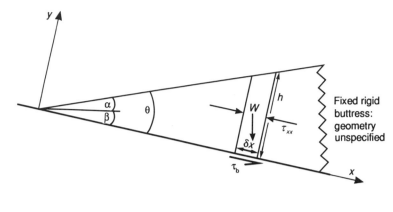

Fig. 4.16. Simplified model of an accretionary wedge (after Platt 1986). α is the surface slope, β is the basal slope and θ is the taper angle ($\alpha + \beta$). A small segment of the wedge of basal length δx and height h is subject to a body force W, to a basal traction $\tau_b \delta x$, and to 'push' forces from the rear, produced by the longitudinal deviatoric stresses τ_{xx}. The force balance is expressed in equation (4.14). Frontal accretion, underplating, erosion and changes in basal shear stress (for example by changes in the rate of convergence/subduction) place the wedge out of dynamic equilibrium. The wedge responds by shortening and thickening or by collapsing by extension.

$$\tau_b = \rho g h \alpha + (1 - \lambda)\, K \rho g h \theta \qquad (4.15)$$

where λ is the ratio of pore-fluid to lithostatic pressure and the other terms are essentially the same as in equation (4.14).

Platt (1986) predicted patterns of deformation in the orogenic wedge resulting from externally imposed changes in its geometry. The most important change results from accretion. Two types are recognizable:

1 *Frontal accretion* is the accumulation of material at the tip of the wedge, thereby lengthening the wedge. The response, if longitudinal deviatoric stresses are large enough, is internal shortening of the wedge. This may take the form of out-of-sequence thrusting or backthrusting.

2 *Underplating* of material to the underside of the wedge, causing the wedge to thicken and increase

in surface slope. The wedge may respond by extension, lengthening the wedge and lowering surface slopes.

Other factors influencing the shape of the orogenic wedge are:

3 *Erosion* at the rear of the wedge (highest terrain) encourages renewed shortening

4 *Changes in basal shear stress*: an increase in τ_b caused, for example, by an increase in the rate of subduction, leads to shortening and thickening; a decrease in τ_b may cause extension (Dahlen 1984). If subduction ceases, τ_b vanishes and the orogenic wedge should collapse by extension at the rear.

Platt (1986) explains the uplift of high pressure rocks to surface positions in the rear of orogenic belts as due to this process of extension (Fig. 4.17).

The implications of this dynamic model for

Fig. 4.17. (*Opposite*) Platt's (1986) evolutionary model of an accretionary wedge from youth to maturity. (A) Early stage with frontal accretion dominant. The gravitational effect of the surface slope (see first term on right-hand side of equation (4.15)) is too low in the frontal region, which therefore shortens and thickens internally. (B) Large scale underplating is the dominant mode of accretion, so that the rear of the wedge extends by extensional faulting and possibly by ductile flow near the base of the wedge. The deeper parts of the wedge may also undergo high pressure metamorphism. (C) Continued underplating and resultant extension has lifted the high-pressure rocks towards the surface. Extension towards the rear of the wedge promotes some shortening (late thrusting) at the front. (D) In the mature stage underplating and extension have brought the high-pressure rocks to levels accessible to erosion. The prism is now 300 km long, comparable to the Makran wedge of Pakistan (Platt *et al.* 1985). A wedge model of this type, constituting a lithospheric load, has been used to simulate the stratigraphy of the North Alpine Foreland Basin of the central Alps (Sinclair *et al.* 1989).

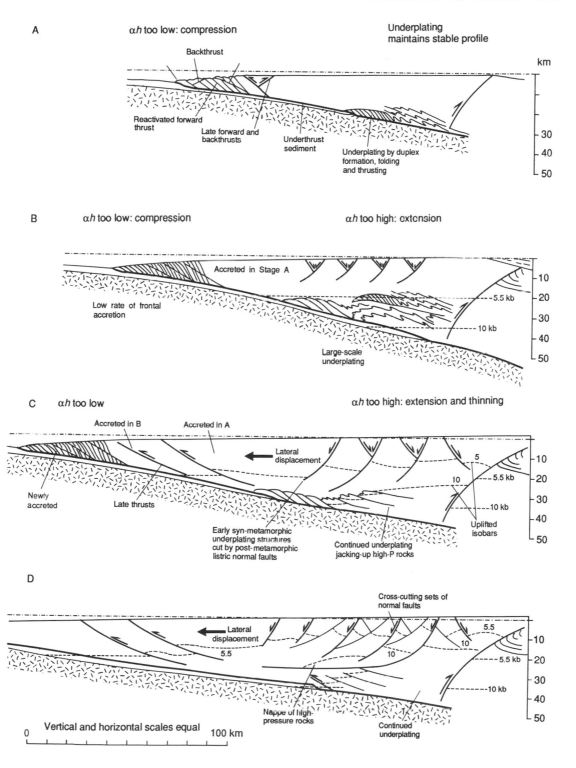

A αh too low: compression

Underplating
maintains stable profile

Backthrust

Reactivated forward
thrust

Late forward and
backthrusts

Underthrust
sediment

Underplating by duplex
formation, folding
and thrusting

B αh too low: compression   αh too high: extension

Accreted in Stage A

Low rate of frontal
accretion

Large-scale
underplating

C αh too low   αh too high: extension and thinning

Accreted in B   Accreted in A

Lateral
displacement

Newly
accreted

Late thrusts

Early syn-metamorphic
underplating structures
cut by post-metamorphic
listric normal faults

Continued underplating
jacking-up high-P rocks

Uplifted
isobars

D

Cross-cutting sets of
normal faults

Lateral
displacement

Nappe of high-
pressure rocks

Continued
underplating

Vertical and horizontal scales equal

0   100 km

basin development are clear. Variations in rate of subduction, magnitude of deviatoric compression or material properties of the wedge may cause large temporal variations in the load configuration and therefore in the deflection of the plate. In particular, lengthening and contraction of the wedge will have an impact on the position of the forebulge on the downgoing slab relative to the orogenic front. We consider the effects of erosion in modifying the geometry of the load in a later section (6.2).

5 Basins associated with strike-slip deformation

On that day his feet will stand on the Mount of Olives, east of Jerusalem,
and the Mount of Olives will be split in two from east to west,
forming a great valley, with half of the mountain moving north and
half moving south

(Zechariah 14:4)
Holy Bible New International Version

SUMMARY

Sedimentary basins form generally by localized extension along a strike-slip fault system which itself may be related to either divergent or convergent relative plate motion. Less commonly, loading resulting from local crustal thickening may result in flexure and subsidence. Although strike-slip basins form in a wide variety of geodynamical settings such as oceanic and continental transforms and arc and suture collisional boundaries, they are best known from intracontinental and continental margin environments.

In simple systems, the orientation of the strike-slip fault in relation to the plate vector is important in determining whether divergent (transtensile) or convergent (transpressive) strike-slip takes place. This guide breaks down, however, in complex regions of continental convergence such as Turkey.

The bulk of the shear strain is accommodated in a central *principal displacement zone* (PDZ) which may be linear to curvilinear in plan view and steeply inclined in section. The PDZ commonly branches upwards into a splaying system of faults producing a *flower structure*. Some fault zones, such as the Garlock fault in California, penetrate to great depths, terminating in the middle crust, whilst others link at depth with relatively shallow low-angle detachments belonging to orogenic wedges.

Strike-slip zones are characterized by en echelon arrangements of faults and folds which are orientated in a consistent pattern with respect to the strain ellipse. The most important fractures are termed *Riedel Shears*, but extension fractures are also formed. En echelon folds may form, with axes roughly at right angles to the extension fractures. The exact pattern of faults and folds produced in any particular fault zone depends on the local geological fabric and the youth or maturity of the fault system.

The precise structural pattern is controlled by a number of factors including (1) the kinematics (convergent, divergent, parallel) of the fault system, (2) the magnitude of the displacement, (3) the material properties of the rocks and sedimentary infills in the deforming zone and (4) the configuration of pre-existing structures.

The PDZ is characteristically segmented. The individual segments may be linked, both in plan view and cross-sectional view, by *oversteps*. If the sense of an along-strike overstep is the same as the sense of fault slip, a *pull-apart basin* is formed: if the sense of the overstep is opposite to that of fault slip, a *push-up range* is formed. Pull-apart basins appear to develop in a continuous evolutionary sequence at releasing bends with increasing offset, from narrow 'spindle-shaped' basins to 'lazy S' and 'lazy Z' basins to 'rhomboidal' basins and eventually into ocean-floored basins.

Thermal and subsidence modelling is poorly developed in strike-slip basins, largely on account of their complex structural history. In basins involving lithospheric thinning, the uniform extension model has been applied with modifications for the lateral loss of heat through the basin walls during the extension. Other basins appear to form over zones of thin-skinned extension, with no mantle involvement. These basins, such as the Vienna Basin in the compressive Alpine–Carpathian system, are cool and lack a well-developed post-extension thermal subsidence.

5.1 OVERVIEW

Basins associated with strike-slip deformation are generally small and complex compared to rifts, passive margins and foreland basins. They are intimately linked to the detailed structural evolution of an area and mechanical models have been slow to appear because of the extreme complexity of this history of deformation. Since the theory is poorly developed at present, we choose in the main to illustrate general principles of strike-slip basin development by summarizing observations on their tectonic and stratigraphic histories (Sections 5.3 and 5.4). It is suggested that two broad types of strike-slip basin exist in terms of thermal and subsidence history (Sections 5.5 and 5.6):
• strike-slip basins with mantle involvement; these can be thought of as 'hot' basins
• strike-slip basins that are relatively thin-skinned; they can be regarded as 'cold' basins.

Strike-slip deformation occurs where principally lateral movement takes place between adjacent crustal or lithospheric blocks. In reality, such movement is rarely purely lateral, and displacements are commonly *oblique*, that is, involving a certain amount of normal or reverse dip-slip move-

ment. Oblique slip may therefore characterize any strike-slip zone, but particular zones may experience a net contraction whilst others may suffer net extension. The stress regimes responsible for these two variants on pure strike-slip deformation are known as *transpressive* and *transtensile*.

Major-strike slip deformation and associated basin formation takes place in a wide range of geodynamical situations, including at oceanic and continental transforms, at convergent plate boundaries such as orogenic belts, and in regions of continental extension undergoing rifting. Sylvester (1988, p. 1667) has recently proposed a classification of strike-slip faults into *interplate* and *intraplate* varieties (see also Section 5.2). He recommends use of the term 'transform' fault for deepseated interplate types and 'transcurrent' fault for intraplate strike-slip faults confined to the crust. Strike-slip zones are characterized by extreme structural complexity. Individual strike-slip faults are generally linear or curvilinear in plan view, steep (subvertical) in section and penetrate to considerable depths, perhaps decoupling crustal blocks at the base of the seismogenic crust (i.e. 10–15 km). In contrast to regions of pure extension or contraction, strike-slip zones possess prominent *en echelon* faults and folds, and faults with normal and reverse slip commonly coexist. The vergence direction of folds and the mass transport indicators from thrusts associated with transpression are distinctively poorly clustered or apparently random.

Basins in strike-slip zones are correspondingly complex. They form in areas of localized extension caused by the geometry and kinematic history of the fault configuration, or in areas of net shortening, where flexural loading may drive subsidence. Characteristically, a basin experiences both extension and shortening during its life span (part of what Ingersoll (1988, p. 1715) calls the Reading cycle, after Reading (1980)), or one part of the basin may experience shortening whilst another part is undergoing extension. Strike-slip basins are therefore emphatically syntectonic and there is little evidence of substantial thermally-driven subsidence. This latter feature may be due at least in part to the small size of strike-slip basins. Their narrowness (usually less than 50 km wide) causes extreme heat loss to the sides as well as vertical conduction to the overlying sea or atmosphere.

The sedimentary fill of strike-slip basins reflects their varied structural history, with highly asymmetrical longitudinal and transverse distribution of facies and abundant and complex unconformities. Subsidence rates are commonly extremely high, but subsidence may be relatively shortlived.

The evolution of some orogenic belts, notably the North American Cordillera (e.g. Irving 1979), involves very large magnitude lateral displacements of terranes. Up to 1500 km of relative lateral motion is thought to have occurred in both the Mesozoic North American Cordillera and in the Tertiary India–Asia collision (Molnar and Tapponnier 1975). Clearly, strike-slip processes offer the possibility of immense lateral translations of crustal terranes.

Palaeomagnetic studies suggest that crustal blocks commonly undergo *rotations* about vertical axes. The amount and scale of rotation varies greatly. The Western Transverse Ranges of south-ern California (Fig. 5.1) have experienced net clockwise rotations of 30°–90° for example, whilst in the Cajon Pass region (Fig. 5.1) there has been no significant rotation since 9.5 Ma. Small blocks can rotate rapidly, such as the Imperial Valley area, which has rotated by 35° in the last 0.9 Ma! The existence of such rotating blocks supports the view that the crustal blocks deform like a set of dominoes (Freund 1970).

There is an obvious physical implication of the existence of rotating blocks, which is that the blocks must detach on their boundary faults at some level in the crust or upper mantle (Terres and Sylvester 1981, Dewey and Pindell 1985).

In this chapter we shall examine the relationship of strike-slip basins to relative plate motions and the stress patterns thus set up. The structural style

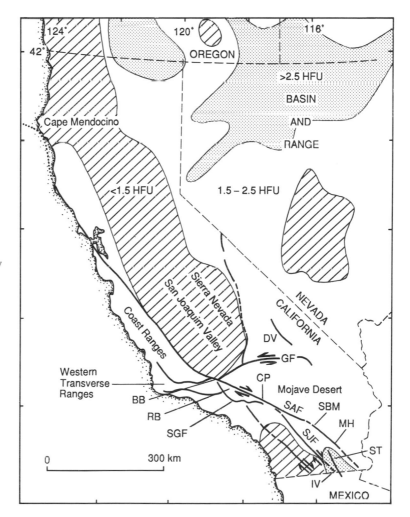

Fig. 5.1. Fault systems and heat flow provinces of California and adjacent regions. Regional heat flow provinces from Lachenbruch and Sass (1980), showing the high heat flows in the extensional Basin and Range Province and in Imperial Valley near the Mexican border, and the low heat flows in most of southern California and the coastal strip in central and northern California. Faults and localities mentioned in text: BB, Big Bend; CP, Cajon Pass; DV, Death Valley; GF, Garlock Fault; IV, Imperial Valley; MH, Mecca Hills; RB, Ridge Basin; SAF, San Andreas Fault; SBM, San Bernardino Mountains; SGF, San Gabriel Fault; SJF, San Jacinto Fault; ST, Salton Trough.

Table 5.1 Miall's (1984) classification of strike-slip faults

1 *Plate boundary transform faults*	
Intracontinental	San Andreas (California), Alpine (New Zealand), Dead Sea (Middle East)
Intraoceanic	El Pilar–Oca, Caribbean, Greater Antillean–Cayman, Magellan–North Scotia
Oceanic with continental margin or fragment	Bismarck (New Guinea), Fairweather–Queen Charlotte (western Canada)
2 *Divergent margin transform faults* (Spreading ridge offsets and fracture zones)	Spitzbergen (Hornsund Fault) (Svalbard), Romanche (Atlantic Ocean), Falkland–Agulhas (Atlantic Ocean), Mendocino (Pacific)
3 *Convergent margin transcurrent faults* (Arc parallel)	Sumatra (Sunda Arc), Atacama (Chile), Median Tectonic Line (Japan)
4 *Suture zone transcurrent faults* (Oblique collision)	Hornelen Basin faults (Norway), Altyn Tagh and Kunlun (Tibet), North Anatolian (Turkey), Vienna Basin faults (Austria–Czechoslovakia)

of strike-slip basins will also be discussed insofar as it illuminates their formative mechanisms. Although models of strike-slip basin subsidence and heat flow are as yet poorly developed, we shall study briefly the model histories of basins formed by firstly whole-lithosphere uniform extension and secondly thin-skinned upper crustal extension.

5.2 RELATION TO PLATE MOTION

Strike-slip zones may be associated with entire plate boundaries such as the San Andreas fault system of California and the Alpine fault system of New Zealand, microplate boundaries, intraplate deformations or small fractures of limited displacement. As we have seen, movement may be convergent-oblique (transpressive) or divergent-oblique (transtensile). Miall (1984) summarized strike-slip faults in terms of plate tectonic context (Table 5.1). Sylvester's (1988) classification, drawing considerably on Woodcock's (1986) genetic scheme (Fig. 5.2), is based on an interplate or intraplate setting (Table 5.2). Transform faults that delimit plates may be (1) *ridge transforms*, offsetting divergent ridges (2) *boundary transforms* that

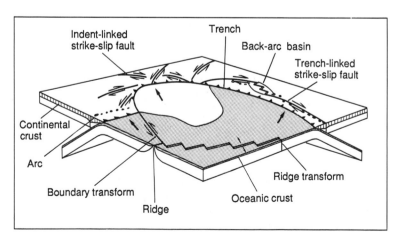

Fig. 5.2. Genetic classification of major classes of strike-slip fault according to plate tectonic setting (after Woodcock 1986). See also Table 5.2.

Table 5.2 Sylvester's (1988) classification of strike-slip faults

1 **Interplate 'transforms'** (deep-seated, delimiting plate)	2 **Intraplate 'transcurrent' faults** (confined to crust)
1.1 *Ridge transform faults* Displace segments of oceanic crust with similar spreading vectors E.g. Romanche fracture zone (Atlantic Ocean)	2.1 *Indent-linked strike-slip faults* Bound continental blocks in collision zones E.g. North Anatolian (Turkey), Altyn Tagh and Kunlun (Tibet)
1.2 *Boundary transform faults* Separate different plates parallel to the plate boundary E.g. San Andreas (California), Alpine fault (New Zealand)	2.2 *Intracontinental strike-slip faults* Separate allochthons of different tectonic styles E.g. Garlock fault (California)
1.3 *Trench-linked strike-slip faults* Accommodate horizontal component of oblique subduction E.g. Atacama fault (Chile), Median Tectonic Line (Japan)	2.3 *Tear faults* Accommodate different displacement within a given allochthon or between the allochthon and adjacent structural units E.g. Asiak fold thrust belt (Canada)
	2.4 *Transfer faults* Linking overstepping or en echelon strike-slip faults E.g. Southern and Northern Diagonal faults (eastern Sinai, Israel)

accommodate the horizontal displacement between different plates, and (3) *trench-linked strike-slip faults* (transforms) parallel to the trench in convergent settings. Intraplate transcurrent faults do not penetrate below the crust. They include (1) *indent-linked strike-slip faults* in zones of convergence and tectonic escape, (2) *intracontinental strike-slip faults* unrelated to indentation and typically separating distinct regional tectonic domains, (3) *tear faults* that accommodate the differential displacement within or at the margin of an allochthon, and (4) *transfer faults* connecting overstepping segments of parallel or en echelon strike-slip faults.

The development of regions of extension and shortening along strike-slip systems has been related to the relative orientation of the plate slip vectors and the major faults (Mann *et al.* 1983). In the case of the San Andreas and Dead Sea strike-slip systems, basins develop where the principal displacement zone is divergent with respect to the plate vector. In contrast, uplifts or push-up blocks such as the Transverse Ranges, California and Lebanon Ranges, occur where the principal displacement zone is convergent with respect to the plate vector. This simple relationship is unlikely to

apply where deformation takes place on many faults enclosing rotating crustal blocks. It is a very poor guide to patterns of uplift and subsidence in complex regions of continental convergence such as Turkey (Sengor, Gorur and Saroglu 1985).

5.3 THE STRUCTURAL PATTERN OF STRIKE-SLIP FAULT SYSTEMS

5.3.1 Structural features of the principal displacement zone (PDZ)

Strike-slip faults are linear to curvilinear in plan view and generally possess a principal displacement zone along which the bulk of the shear strain is accommodated. However, changes in the orientation of the fault and/or the influences of the local geological fabric may cause deformation to extend beyond the PDZ into the juxtaposed crustal blocks (Fig. 5.3). In cross-section, the principal displacement zones of large strike-slip faults are steeply inclined and commonly grade upwards from narrow well-defined zones cutting igneous and metamorphic basement rocks at depth, to a braided,

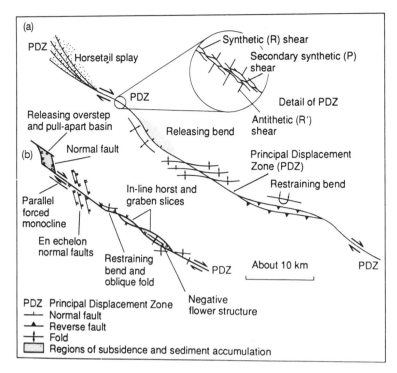

Fig. 5.3. (a) The plan view arrangement of structures associated with an idealized right-lateral (dextral) strike-slip fault. (b) Adaptation to a slightly divergent setting with the predominance of pull-aparts, en echelon normal faults and graben slices within the PDZ.

more diffuse deformation in the overlying sedimentary cover. This upward branching effect has led to the fault splays being christened '*flower structures*' (Fig. 5.4). Some strike-slip faults link at depth with low-angle detachments, as occurs in foreland fold and thrust belts, e.g. Vienna Basin (Royden 1985) and in regions of regional extension such as the Basin and Range, USA (Cheadle *et al.* 1985). In the latter case, the Garlock fault, a major strike-slip fault at a high angle to the San Andreas trend in California (Fig. 5.1), appears to terminate downwards on a low angle surface situated in the middle crust (9–21 km depth) according to deep seismic reflection profiling (COCORP).

Figure 5.3 shows *en echelon* arrangements of faults and folds associated with strike-slip displacements. These structures are consistently arranged both in orientation and sense of strain with respect to the PDZ. These structures are distinct from the *oversteps* between different segments of the principal displacement zone (Section 5.3.2). En echelon arrangements have been produced in model studies involving the deformation of clay, loose sand or artificial materials (e.g. Riedel 1929, Cloos 1955 and more recently Harris and Cobbold 1984). Similar patterns have been observed in 'natural'

environments where alluvium has been affected by seismic disturbance, as in the 1975 earthquake in Imperial Valley, California documented by Sharp (1976). These experimental results and limited natural occurrences suggest that five sets of fractures are associated with a shear displacement (Fig. 5.5):

1 Synthetic strike-slip faults orientated at small angles to the regional shear couple. These are frequently termed *Riedel (R) shears*. The sense of offset is the same as that of the PDZ.

2 Antithetic strike-slip faults orientated at high angles to the regional shear couple. These are termed *conjugate Riedel shears* or *R'*. The sense of offset is opposite to that of the PDZ.

3 Secondary synthetic faults or *P-shears* with a sense of offset similar to that of the PDZ.

4 *Tension fractures* related to extension in the strain ellipse.

5 Faults parallel to the principal displacement zone and shear couple, or *Y-shears* of Bartlett, Friedman and Logan (1981).

En echelon folds develop with their axial traces parallel to the long axis of the strain ellipse, indicating shortening perpendicular to the extension demonstrated by the tension fractures (Fig.

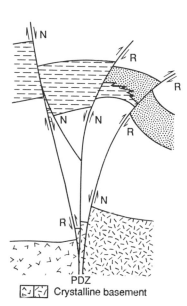

MAJOR CHARACTERISTICS

- Basement-involved
- PDZ Sub-vertical at depth
- Upwards diverging and rejoining splays

JUXTAPOSED ROCKS

- Contrasting basement type
- Abrupt variations in thickness and facies in a single stratigraphic unit

SEPARATION IN ONE PROFILE

- Normal and reverse-separation faults in same profile
- Variable magnitude and sense of separation for different horizons offset by the same fault

SUCCESSIVE PROFILES

- Inconsistent dip direction on a single fault
- Variable magnitude and sense of separation for a given horizon on a single fault
- Variable proportions of normal- and reverse-separation faults

PDZ

⬚ Crystalline basement

⬚ Time-stratigraphic unit with variable sedimentary facies

Fig. 5.4. The major characteristics in cross-sectional view of an idealized strike-slip fault. The upward branching from a near-vertical fault is termed a flower structure (after Christie-Blick and Biddle 1985).

5.5). Geological examples are invariably more complicated than the situation shown in Fig. 5.5, because the geological fabric influences fault orientations and also because faults and folds become rotated during progressive deformation so that the present fault configuration represents a cumulative picture over time. However, the idealized pattern is useful as a predictor of fault and fold occurrences if the regional shear direction is known, or, alternatively as a predictor of the latter where observations are possible on the resultant fracture and fold pattern.

The offset pattern on strike-slip faults in cross-section can be exceedingly complex. The sense of displacement may vary along one fault from horizon to horizon, and within a flower structure faults of opposite displacement occur together (Fig. 5.4). Additionally, a given fault in one cross-section may commonly switch in dip in another cross-section. These characteristics make strike-slip fault zones distinctive compared to regional extensional and contractional fault systems.

The precise structural pattern is controlled by a number of factors including (Christie-Blick and

Fig. 5.5. The angular relations between structures that form in an idealized right-lateral simple shear, compiled from clay models and from geological examples (after Christie-Blick and Biddle 1985). (a) Fractures and folds superimposed on a strain ellipse for the overall deformation. Terminology of structures from Wilcox, Harding and Seeley (1973). (b) Riedel shear terminology modified from Tchalenko and Ambraseys (1970) and Bartlett, Friedman and Logan (1981).

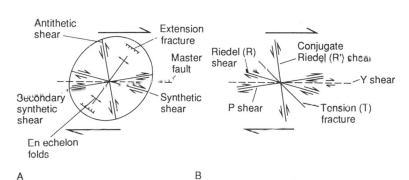

Antithetic shear

Extension fracture

Master fault

Secondary synthetic shear

Synthetic shear

En echelon folds

Riedel (R) shear

Conjugate Riedel (R') shear

Y shear

P shear

Tension (T) fracture

A   B

Biddle 1985, p.10):
• convergent, divergent or simple strike-slip (parallel) kinematics
• magnitude of the displacement
• material properties of the rocks and sedimentary infills in the deforming zone
• configuration of pre-existing structures giving a geological fabric.

1 Convergent, divergent and parallel kinematics

Convergent strike-slip causes the development of many reverse faults and parallel to en echelon folds with respect to the PDZ. Depending on the obliquity of the strike-slip, this pattern may grade into a fold and thrust belt as in the Western Transverse and Coast Ranges of California (Fig. 5.1). Upward splaying flower structures have an overall antiformal structure caused by the net shortening. They are commonly known as *positive flower structures*. Folds are less well developed in divergent strike-slip settings, taking the form of flexures associated with extensional faulting. Flower structures take on an overall synformal structure, and are consequently known as *negative flower structures*.

The scale of the distribution of convergence or divergence varies from regional to local. For example, regional divergent strike-slip may take place where major fault strands are oblique to interplate slip vectors as in the Dead Sea transform. On a much more local scale, divergent strike-slip may take place at releasing fault oversteps and fault junctions, as in the Ridge Basin, California. Divergent strike-slip faults also develop on a local scale where crustal blocks rotate between bounding wrench faults.

A detailed study of the rotations and oversteps associated with a strike slip displacement following a recent earthquake was carried out by Terres and Sylvester (1981). The deformation of a recently ploughed carrot field in Imperial Valley following the earthquake of 15 October 1979 is shown in Fig. 5.6. This very small scale example shows many of the features predicted from model experiments with both extension and contraction occurring simultaneously. The topsoil broke up into elongate blocks along the furrows and these blocks moved relative to each other along conjugate Riedel Shears (R') or antithetic shears. Extension occurred along the long axis of the strain ellipse and folding along the short axis. The elongate blocks delimited

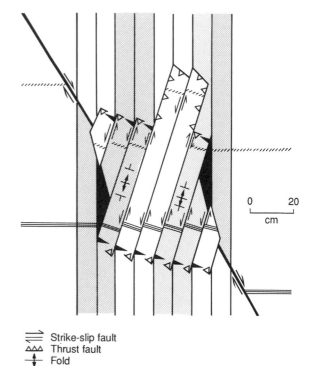

⇉ Strike-slip fault
△△△ Thrust fault
—+— Fold
◣ Area experiencing extension

Fig. 5.6. Geometrical model of the faults, folds and gaps that were associated with a surface rupture in the Imperial Valley, California following the earthquake of 15 October 1979 (Terres and Sylvester 1981). The parallel lines are furrows in a carrot field that became displaced during the rupture, individual furrows shearing against their neighbours at the same time as being rotated about a vertical axis. Note the small scale of the model.

by shears, which Dewey (1982) has termed '*Riedel flakes*', show clockwise rotations in this example of right-lateral strike-slip. A much larger rotated block with dimensions of 20 km by 70 km, the Almacik flake along the North Anatolian fault, Turkey, is also thought to be due to Riedel flaking (Sengor *et al.* 1985) (Fig. 5.7). Here, clockwise rotation of the tectonostratigraphic domains of 110° has taken place since the initiation of faulting. The right-lateral strike-slip along the North Anatolian PDZ has produced families of thrusts, Riedel and P shears.

Rotations may also take place along straight segments of adjacent strike-slip faults, causing geometrical space problems. These space problems

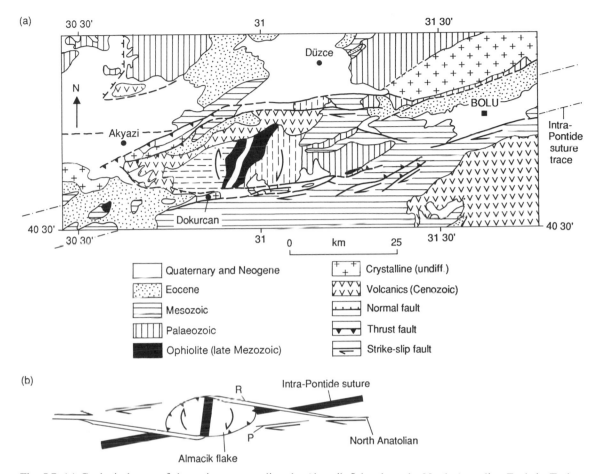

Fig. 5.7. (a) Geological map of the region surrounding the Almacik flake along the North Anatolian Fault in Turkey (Sengor, Gorur and Saroglu 1985). (b) Interpretation of the Almacik flake based on the concept of 'Riedel flaking'. R and P shears are shown (cf. Fig. 5.5).

are resolved by gaps and overlaps occurring at fault block corners. Figure 5.8 shows how regions of extension and contraction may alternate along the strike-slip fault.

2 Magnitude of the displacement

Laboratory experiments indicate that there is a sequential development of structural features in strike-slip zones (Tchalenko (1970) and Wilcox, Harding and Seely (1973) for clay models, Bartlett *et al.* (1981) for rock samples under confining pressure). In the rock sample experiments the zone of deformation first of all expands rapidly due to the development of folds and fractures. Weakening

of the deforming zone soon, however, stabilizes the spread of the zone of deformation, and later structures are concentrated in a central core zone (Odonne and Vialon 1983).

All early formed structures are rotated by later deformations, so their orientation is dependent on the magnitude of the displacement.

Some of the features observed in experiments can be found in natural occurrences, whilst other features are difficult to match. An increasing complexity from discontinuous faults and folds along low displacement boundaries to through-going PDZs along high displacement boundaries is commonly observed. Active fault systems which are accompanied by sedimentation may, however, be

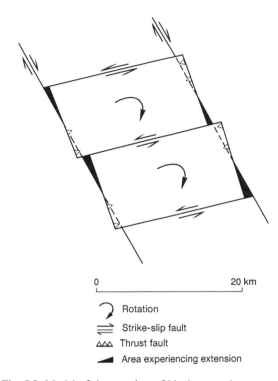

0   20 km

⤳ Rotation

⇒ Strike-slip fault

△△△ Thrust fault

◢ Area experiencing extension

Fig. 5.8. Model of the rotation of blocks near the intersection of the San Andreas and San Jacinto faults, California, based on earthquake hypocentral locations and first motion studies (Nicholson *et al.* 1985a, b, Dibblee 1977). During a large earthquake, one of the bounding faults moves by right-lateral slip. Continued aseismic motion between the bounding faults is accommodated by clockwise rotations of small blocks and by minor left-lateral movements on faults defining these rotating blocks. Rotation causes overlaps (small reverse faults) and gaps (small normal faults) to form along the major bounding right-lateral faults. Viewed on a crustal scale, these rotating 'flakes' would be detached from the underlying lower crust/mantle by thrust faults which would merge with the associated left-lateral strike-slip faults.

buried faster than they deform. In this way, the uppermost, younger sediments may record less deformation than the older stratigraphy.

3 Material properties in the deforming zone

The lithologies of rocks in the strike-slip zone and the rates and pressure–temperature conditions of deformation all control the individual structural character of the fault zone. These factors vary from one strike-slip zone to another, but they also may vary in time within an individual strike-slip fault system. Deformation of different stratigraphic units, strain weakening or strain hardening effects, uplift and erosion or sedimentation and burial, or changes in heat flow related to crustal thinning and/or volcanism can all have significant effects on the evolution of the fault zone.

4 Pre-existing geological fabric

Continental strike-slip fault systems commonly intersect or make use of older heterogeneities. For example, the Great Basin, western USA underwent Laramide shortening (NE–SW trend), then two phases of extension, one related to back-arc processes (axis of extension orientated NE–SW), the other to the development of a right lateral mega-shear (change in extension direction to NW–SE). The right-lateral faults and, locally, conjugate left-lateral faults associated with this clockwise rotation in the extension direction may well be related to the existence of older crustal structures.

The Great Basin is an example of regional extension, so the strike-slip motion is taking place in a transtensile regime. The Rhine Graben of western Europe, however, originated in an environment of continental collision north of the Alps in Eocene times. Early extensional faults in the late Eocene–Oligocene were used as strike-slip faults in the Pliocene to Holocene.

Pre-existing structures which significantly influence the location and orientation of folds and faults during strike-slip are termed 'essential' structures. On the other hand, pre-existing structures which exert no control on later deformation are termed 'incidental' (Christie-Blick and Biddle 1985, p. 13).

5.3.2 Role of oversteps

An *overstep* or *stepover* is a discontinuity between two approximately parallel overlapping or underlapping faults. Oversteps are extremely important in determining the location of regions of subsidence and uplift along a strike-slip system. There appear to be two basic types (Fig. 5.9):

• oversteps along the strike of faults, that is, observed in plan view; faults are continuous in the down-dip direction;

• oversteps along the dip of faults, that is, observed

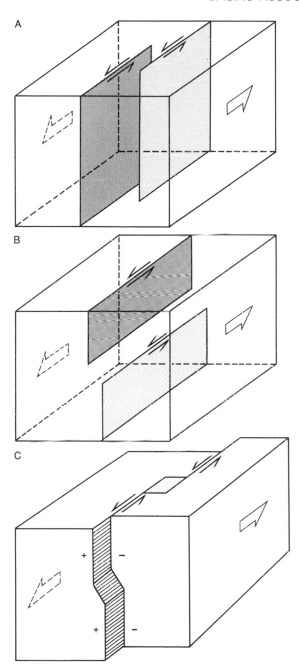

Fig. 5.9 Oversteps on strike-slip faults. (A) Along-strike oversteps in which faults are continuous in the down-dip direction, (B) Down-dip oversteps in which the faults are continuous in the plan view, and (C) Combination of along-strike and down-dip oversteps, with a pull-apart basin located at the along-strike overstep (after Aydin and Nur 1985, p. 36).

in cross-section; faults are otherwise continuous in plan view.

Both types of overstep may occur along the same fault or fault zone. If the sense of an along-strike overstep is the same as the sense of fault slip, a *pull-apart basin* is formed. If, on the other hand, the sense of the overstep is opposite to that of fault slip, a *push-up range* is formed. Down-dip oversteps are probably as important as along-strike oversteps, but are more difficult to identify and map. Seismicity studies on active strike-slip faults such as the Calaveras fault, California, show a distinct offset of the trends of hypocentres, breaking the fault into an upper segment (2–7 km depth) and a lower segment 2 km away (4–10 km depth). This suggests that a down-dip overstep is present (Reasenberg and Ellsworth 1982).

5.4 BASINS IN STRIKE-SLIP ZONES

Classification schemes have used the geometry of the basin and the nature of its bounding faults. This enables a 'static' classification to be achieved, but since the structural style and geometry change markedly through time, present-day characteristics may provide only strictly limited information on the kinematic history and controls on basin development.

In general, subsidence tends to occur where strike-slip is accompanied by a component of divergence (Fig. 5.10). This might result from a *bend* or an *overstep* in the fault trace, giving a *pull-apart basin* (Burchfiel and Stewart 1966), or through extension near a fault junction, giving a *fault-wedge basin* (synonym *wedge graben*) (Crowell 1974b). Uplift on the other hand tends to occur where there is a component of convergence (Fig. 5.10), although flexure of an underlying block by an overriding block may cause subsidence. Uplift takes place at convergences of faults into a fault junction or at bends in fault traces. These zones of uplift provide sourcelands of detritus available to fill nearby basins. Bends, oversteps and fault junctions associated predominantly with extension and subsidence are termed 'releasing', whilst those associated with shortening and uplift are termed 'restraining' (Crowell 1974b). Parallel-sided basins in which oceanic crust occurs are termed *rhombochasms* (Carey 1976).

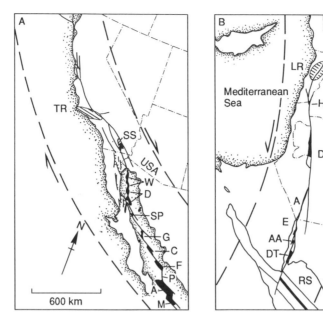

Fig. 5.10. Regions of compression and extension along strike-slip boundaries between rigid continental plates related to the orientation of the fault zone with respect to the plate slip vector (after Mann, Hempton *et al.* 1983, p. 533). (A) Pacific–North American plate boundary. The dashed lines are theoretical interplate slip lines from Minster *et al.* (1974). The major zone of compression is the push-up block of the Transverse Ranges (diagonally hatched) where the dextral San Andreas Fault Zone has a 'convergent' orientation with respect to the interplate slip line. The prominent pull-apart basins, however, are situated relatively to the south where the fault zone has a 'divergent' orientation with respect to the interplate slip line. TR, Transverse Ranges; SS, Salton Sea pull-apart at a right step between the San Andreas and Imperial Faults. Pull-aparts in the Gulf of California include: W, Wagner Basin; D, Delfin Basin; SP San Pedro Martir Basin; G, Guaymas Basin; C, Carmen Basin; F, Farallon Basin; P, Pescadero Basin Complex; A, Alarcon Basin; M Mazatlan Basin. Data sources include Crowell (1981), Henyey and Bischoff (1973), Bischoff and Henyey (1974), Niemitz and Bischoff (1981). (B) Arabia–Sinai (Levant) plate boundary zone. Theoretical interplate slip lines are from Le Pichon and Francheteau (1978). The prominent area of compression is the Lebanon Ranges (LR) push-up block where the sinistral Dead Sea Fault zone is convergent with respect to the interplate slip lines (diagonally hatched). Pull-aparts include H, Hula Basin; DS, Dead Sea Basin; A, Arava Fault trough; E, Elat Basin in the northern Gulf of Aqaba; AA, Arnona–Aragonese Basin; DT, Dakar-Tiran Basin. The Red Sea is abbreviated by RS. Data sources are Garfunkel (1981) and Ben-Avraham, Almagor and Garfunkel (1979).

The causes of fault bends, oversteps and junctions are varied. Some bends may result from pre-existing crustal heterogeneities, as in the case of strike-slip faults propagating along older extensional fractures. This appears to be the case in Jamaica (Mann *et al.* 1985, p. 212). Jamaica lies entirely within a 200 km wide seismic zone of left-lateral strike-slip deformation between the North American and Caribbean plates (Fig. 5.11a). The east-striking Plantain Garden fault zone bends abruptly as it crosses a NW-striking graben containing Palaeocene to Eocene rift sediments (Fig. 5.11b) known as the Wagwater belt. The exten-sional faults at the east of the Wagwater graben have been reactivated as reverse faults along a restraining bend in the Neogene strike-slip system. Other bends may be due to deformation of initially straight faults as a result of incompatible slip at a fault junction (e.g. the 'big-bend' of the San Andreas fault, Fig. 5.1), rotation of adjacent blocks (southern San Andreas fault), or intersection of the fault with a zone of greater extensional strain (e.g. western end of North Anatolian fault) or contractional strain (e.g. near junction of North Anatolian and East Anatolian faults in eastern Turkey, Fig. 5.12).

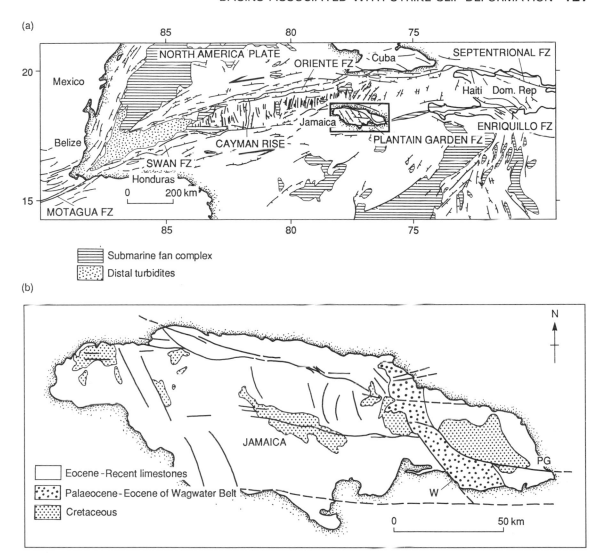

Fig. 5.11. (a) Recent faulting and sedimentation in the North American–Caribbean plate boundary zone (modified from Case and Holcombe 1980, Mann *et al.* 1983). Jamaica is situated at a compressional right step (restraining bend) between the sinistral Enriquillo–Plantain Garden fault zone in the east and the faults bordering the Cayman Trough pull-apart in the west. Box indicates area detailed in (b). (b) Schematic geological map of Jamaica with major late Neogene faults. PG is Plantain Garden Fault, W is Wagwater Fault bounding the western edge of the Palaeocene-Eocene sediments of the Wagwater Belt. After Mann *et al.* (1985).

Fault branching and oversteps may develop by a number of mechanisms including segmentation of curved fault traces, intersection of weak zones orientated obliquely to the direction of strike-slip and changes in stress fields caused by fault inter-action. The precise reasons for the formation of oversteps and branches remains obscure (Aydin and Nur 1985, p. 35).

Strike-slip basins may be *detached* and therefore thin-skinned. Examples are known from both areas of pronounced regional shortening such as the Vienna Basin and the St. George Basin in the

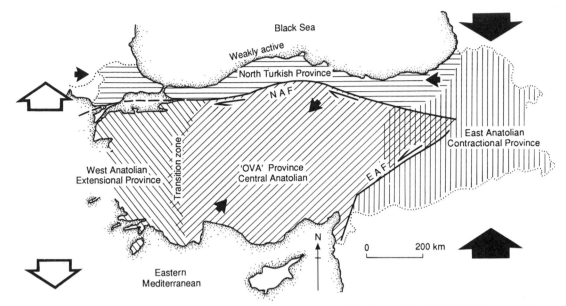

Fig. 5.12. The four main tectonic regimes of Turkey caused by the post late Serravallian (*c.* 12 Ma) westward escape of the Anatolian block from the east Anatolian convergent zone onto the oceanic lithosphere of the eastern Mediterranean Sea (after Sengor *et al.* 1985). The four regimes are (1) East Anatolian contractional province of north-south shortening, situated mostly to the east of the meeting point of the East Anatolian and North Anatolian strike-slip faults, (2) the weakly active North Turkish province characterized by limited E–W shortening, (3) the West Anatolian extensional province characterized by N–S extension, and (4) the Central Anatolian 'ova' province with NE–SW shortening and NW–SE extension, and containing large, roughly equant shaped complex basins termed 'ovas'. Black arrows indicate shortening and white arrows extension. The arrows are roughly proportional to the magnitude of the total strain.

Bering Sea, Alaska, and in areas of regional extension such as the West Anatolian extensional province of Turkey and the Basin and Range of the USA.

5.4.1 Mechanisms for pull-apart basin formation

A number of models exist for the development of pull-apart basins (Fig. 5.13):
1 Overlap of side-stepping faults.
2 Slip on divergent fault segments.
3 Nucleation on en echelon fractures or Riedel shears.
4 Coalescence of adjacent pull-aparts into larger system.

1 The simplest model is based on pioneer field studies of active pull-aparts such as the Dead Sea and the Hope Fault Zone in New Zealand. Discontinuous parallel fault segments which have a horizontal separation develop oversteps. As the master faults lengthen, producing more overlap, the basin lengthens whilst its width remains fixed by the original separation of the parallel master faults (Fig. 5.13). This model has been widely applied to the San Andreas fault system in the Gulf of California–Salton Trough area (Crowell 1974b), and the Dead Sea Fault System (Garfunkel *et al.* 1981). This configuration of lengthening, parallel master faults has been used to simulate pull-apart development using elastic dislocation theory (Rodgers 1980). This theory predicts that distinct changes occur in the pull-apart basin as the amount of overlap of the master faults increases. Specifically, when the overlap is about equal to the separation, the pull-apart develops two depocentres in regions of extension separated by a zone of secondary strike-slip faulting in the basement. As the overlap increases, the depocentres become more widely spaced and the intervening zone of strike-slip faulting broadens. Although there are

(1) Overlap of side-stepping faults

(2) Slip on divergent fault segments

(3) Nucleation on en echelon fractures

(4) Coalescence of scale-dependent basins

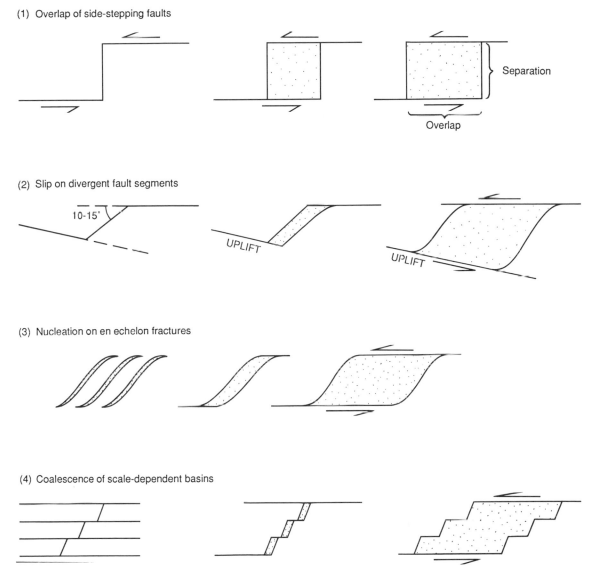

Fig. 5.13. Models of pull-apart basin development (after Mann *et al.* 1983, p. 531). (1) Pull-apart opening between left-stepping and sinistral master strike-slip faults. Fault separation and basin width remain constant through time whereas fault overlap and basin length increase with the amount of strike-slip displacement. The elastic dislocation model of Rodgers (1980) applies to this kind of pull apart mechanism. (2) Pull-apart opening across an oblique median fault and non-parallel master faults. Note that compression and uplift occurs on one side of the pull-apart while an extensional 'gap' develops on the other. (3) Pull-apart formation by nucleation from extensional fractures, based on shear box experiments (Koide and Bhattacharji 1977). (4) Pull-apart formation by coalescence of small scale-similar sub-basins, as suggested by Aydin and Nur (1982). This allows the widening of pull-aparts with increased offset, in contrast to (1).

instances where field observations fit the theoretical elastic dislocation model rather well (e.g. Cariaco Basin, Venezuelan part of southern Caribbean, Schubert 1982), basins which are filled with sediment more rapidly than the upward propagation of faults from the basement may develop fault systems at strong variance to those predicted by elastic dislocation theory.

2 Detailed field mapping of some active pull-aparts suggests that bounding strike-slip faults may be non-parallel (Freund 1971). If non-parallel, non-overlapping faults are slightly divergent and connected by a short oblique segment, continued strike slip may open up a basin along one side of the oblique fault and cause compression along the other side (Fig. 5.13).

3 Shear box experiments suggest that pull-apart basins may be structurally analogous to en echelon extensional fractures produced in clay materials. As deformation continues, the shear fractures join to form a pull-apart basin. A similar process of formation from rotated large scale tension gashes or Riedel shears has also been proposed.

4 A worldwide compilation of the dimensions of a large number of pull-aparts suggests that there is a linear relation between the basin length (fault overlap) and the basin width (fault separation). This scale dependence of pull-apart basins indicated that they are commonly three times as long as they are wide irrespective of their absolute size. This may be due to the coalescence of adjacent pull-aparts in a single larger basin with increasing offset, or to the formation of new fault strands parallel to existing ones (Aydin and Nur 1982).

5.4.2 Continuum development from a releasing bend: evolutionary sequence of a pull-apart basin

Analysis of neotectonic pull-apart basins in their embryonic stage suggests that they are associated with (A) releasing bend geometries. Master faults do not overlap but are connected by oblique median strike-slip faults, and (B) non-parallel master faults (see also (1) and (2) above). Gentle restraining bends appear to produce narrow *spindle-shaped* pull-aparts like the Clonard Basin, Haiti at the left-step in the Enriquillo–Plantain Garden Fault Zone (Fig. 5.11). In contrast, sharp restraining bends produce multiple, staggered pull-aparts as in the Salton Trough of California. Since embryonic pull-aparts appear to occur on releasing bends of through-going faults, they are unlikely to be analogous to the tension gashes or Riedel shears produced in shear box experiments (see (3) above).

Continued offset produces basin shapes known as '*lazy S*' (between sinistral faults) and '*lazy Z*' (between dextral faults) (Mann *et al.* 1983, p. 540) (Fig. 5.14), representing a transitional stage between spindle-shaped basins between master faults with no overlap and rhomboidal basins between overlapping master faults. S- and Z-shaped pull-aparts are particularly common where the master faults are widely separated (> 10 km) and their strikes are non-parallel. The Death Valley Basin, California has a pronounced Z-shape.

Lengthening of the S- or Z-shaped basins produces rhomb-shaped pull-aparts. Length to width ratios (overlap to separation) increase by this process because the separation remains fixed by the width of the releasing bend. Rhomboidal pull aparts are characterized by overlapping master faults and deep depocentres at the ends of the pull apart basin, separated by a shallow sill(s). The Cariaco Basin of the Venezuelan Borderlands is a good example, with two subcircular deeps in excess of 1400 m and a shallow sill at 900 m. Multiple deeps may be arranged diagonally across the rhomboidal pull-apart, as is the case in the Gulf of Aqaba in the northern Red Sea, the deepest segment of the sinistral Dead Sea Fault System (Ben-Avraham, Almagor and Garfunkel 1979). The rapid subsidence between the overlapping master faults generally greatly exceeds sedimentation, leading to deep marine or lacustrine environments. The presence of prominent depocentres at the ends of rhomboidal pull-aparts does not support the mechanism of simple extension between overlapped master faults (see (1) above), but is predicted by the elastic dislocation model (Rodgers 1980). An alternative view is that the depocentres may in fact be smaller pull-aparts in the basin floor of the larger pull-apart, that is, the rhomboidal basin is in a coalesced stage of development (see (4) above).

With continued offset, rhomboidal pull aparts may develop into narrow oceanic basins where the length (overlap) greatly exceeds the basin width (separation). The Cayman Trough along the boundary between the Caribbean and North American plates is a type example (Fig. 5.11a). The large

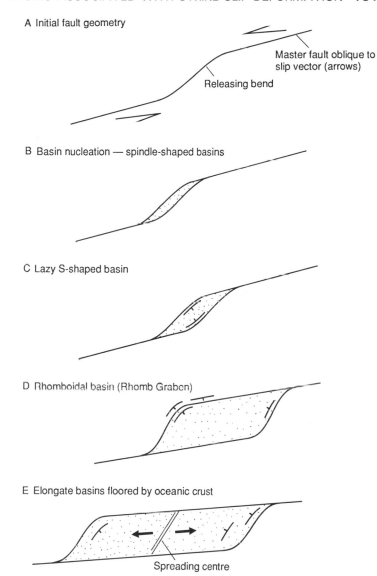

A Initial fault geometry

Master fault oblique to slip vector (arrows)

Releasing bend

B Basin nucleation — spindle-shaped basins

C Lazy S-shaped basin

D Rhomboidal basin (Rhomb Graben)

E Elongate basins floored by oceanic crust

Spreading centre

Fig. 5.14. Continuum model of pull-apart development, after Mann *et al.* (1983). (A) Pull-apart begins life on a releasing fault bend. The size of the bend controls the pull-apart basin width. (B) Spindle-shaped basin. Continued offset produces lazy S-shaped basin (C), then a rhomb-shaped basin (D), then finally a long, elongate trough floored with oceanic crust (E). Most pull-aparts do not reach the stage represented by (E). Instead they tend to be terminated after the basin length has reached about three times the original width of the releasing bend (or fault separation, Fig. 5.13).

length/width ratio suggests that coalescence is unimportant at this stage.

5.5 APPLICATION OF MODEL OF UNIFORM LITHOSPHERIC EXTENSION TO PULL-APART BASINS

The local extension in pull-apart basins causes very rapid subsidence and sediment accumulation, over 10 km of sediments accumulating in a period of less than 5 Myr in some examples such as the Miocene Ridge Basin of California (Link and Osborne 1978). The subsidence can be approximated by a model of lithospheric extension (McKenzie 1978a, Section 3.1.5) but the small size of pull-aparts implies that lateral temperature gradients must be large. Lateral heat conduction to the basin walls becomes an important source of heat loss, the narrower the basin the greater the cooling. The critical basin width below which lateral heat loss to the sides becomes important appears to be about 100 km (Steckler 1981) to 250 km (Cochran 1983). Since narrow basins cool rapidly during extension, the subsidence is greater than predicted for the rift stage by the uniform extension model (Section 3.1.5). Rapid early subsidence related to lateral heat loss may help to explain the sediment starvation and deep bathymetries characteristic of the early phases of pull-apart basin development.

Pitman and Andrews (1985) studied the subsidence history of small pull-aparts typical of the San Andreas zone of California. Here, heat flow data, seismic data and geological studies (e.g. Atwater 1970) suggest an anomalously thin lithosphere. Pitman and Andrews (1985, p. 45) constructed a model in which they successively rifted then cooled the lithosphere over small time steps of 0.05 Myr to simulate the process of syn-rift heat loss (see also Section 3.1.5.2).

The input parameters for the model were as follows:

Lithospheric thickness	y_L	= 62.5 km
Crustal thickness	y_C	= 30 km
Stretch factor	β	= 1.6 [Rate of stretching 30 mm yr^{-1}]
Initial basin widths	X_0	= 10 km, 20 km, 30 km.

For a basin filled with water only, the results of the model for a point in the centre of the basin are shown in Fig. 5.15. The rapid initial subsidence is caused by crustal thinning combined with lateral heat loss to the basin walls. This steep curve flattens abruptly as extension terminates and the remaining subsidence is attributed to cooling only. Clearly, the initial basin width has a major influence on the subsidence history, the narrowest basins losing heat most rapidly to the sides and therefore subsiding both at a greater rate and with a greater magnitude than wider basins.

If the basin is kept at all times full to the brim, the subsidence must increase by a factor related to the different densities of water and infilling sediment $((\rho_m - \rho_w)/(\rho_m - \rho_s))$. This is a requirement of isostasy discussed in Section 2.1. However, pull-apart basins are commonly starved in their youthful stages. If the rate of supply of sediment to the basin, or denudation rate, is related to the mean elevation as follows

$$\frac{d\overline{Y}}{dt} = K\overline{Y} \qquad (5.1)$$

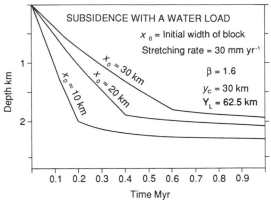

Fig. 5.15. Subsidence as a function of time for basins experiencing lateral heat loss in addition to vertical conduction. Three subsidence curves are shown for three initial widths of zones of stretching (10 km, 20 km and 30 km) for a point in the centre of the basin. The infilling medium is assumed to be water only. The blocks are stretched at 30 mm yr^{-1} up to a stretch factor of 1.6. The curves comprise two segments. The initial steep segment represents subsidence due to lateral heat loss as well as crustal stretching. The flatter later segment represents subsidence due to the remaining thermal contraction. The total subsidence and the time-history of subsidence are both influenced by the initial width of the zone of stretching between unstretched basin walls (after Pitman and Andrews 1985).

where $\overline{Y}$ is the mean elevation and K is a denudation constant (Ahnert 1970) which appears to be of the order 10^{-1}yr^{-1}, it is possible to study the lag between subsidence and basin-filling. Such a linear relationship is at best a rough approximation and an erosional model based on slope is in some cases preferable (Sinclair et al. 1991). Nevertheless, for the sake of simplicity and clarity we shall use equation (5.1).

Taking a reasonable estimate of the average elevation of the source area as a starting point, the size of the drainage basin required to continuously feed the sedimentary basin and keep pace with tectonic subsidence is extremely large (several hundred times the size of the depositional area). It is more likely therefore that sediment supply will not keep pace with subsidence in the early stages of basin development. The basin should gradually fill as the subsidence rate slackens, eventually filling completely some 5 Myr after the start of rifting

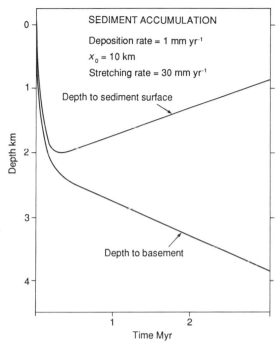

Fig. 5.16. The depth to basement and the depth to the sediment surface in a narrow (initially 10 km) basin characterized by lateral heat loss where the rate of sediment supply is restricted to 1mm yr^{-1}. Local (Airy) isostasy is assumed throughout in compensating the depth to basement for the sediment load. The early history of the basin is marked by deep water sedimentation and several million years elapse before the sediment surface approaches sea level (after Pitman and Andrews 1985).

(Fig. 5.16). The typical sedimentary fills of strike-slip basins are discussed in Section 7.2.3.

5.6 DETACHED PULL-APART BASINS IN REGIONS OF LITHOSPHERIC COMPRESSION: THE VIENNA BASIN IN THE CARPATHIAN–PANNONIAN SYSTEM

5.6.1 Geological setting

The 200 km by 60 km Vienna Basin contains up to 6 km of Miocene sedimentary rocks and formed adjacent to the coeval Carpathian thrust belt. The Vienna Basin therefore provides an excellent example of a rhombohedral pull-apart formed in a background environment of lithospheric shortening. The basin developed on top of the allochthonous thrust terranes of the Alpine–Carpathian system, which in turn overlie the autochthonous cover and basement of the overridden European plate (Fig. 5.17). Two phases of extension and subsidence have been recognized, the first occurring on top of the moving thrust sheets in early Miocene times, and the second occurring after thrusting had ceased in middle Miocene times. A period of uplift, normal faulting, tilting and erosion at the end of the early Miocene separated the two phases of subsidence. The deposits of the two phases tend to occur in separate rhombohedral sub-basins defined by a series of braided synsedimentary faults separating blocks of reduced subsidence.

The NE–SW trending major fault systems active during sedimentation in the Vienna Basin do not significantly disrupt the underlying autochthonous cover rocks of the European plate. Only a few faults intersect the autochthonous basement at a steep angle causing large displacements; many faults with large displacements at the surface instead apparently lose their displacement with depth – they are interpreted (Royden 1985, p. 326) to pass into flat detachments utilizing the older thrust planes. The thrust sheets beneath the Vienna Basin have been thinned by about 50 per cent during the extension.

The relation between thrusting, strike-slip faulting and normal faulting is summarized in Fig. 5.18. An important observation in explaining the geometry is that the thrust sheets west of the Vienna Basin became fixed at the end of the early Miocene whilst thrusting continued east of the basin. This relative displacement requires either (a) sinistral bending of the thrust belt near the Vienna Basin, or the preferred case (b) left-slip along N- or NE-striking faults within the thrust belt. Thus the Vienna Basin can be viewed as a result of diachronous thrusting in a region of oblique convergence.

If the Vienna Basin is essentially thin-skinned (< 10 km), with transtensional faults soling into a shallow-dipping decollement representing an old thrust plane, it has considerable implications for the subsidence and thermal history of this variety of pull-apart basin.

Synsedimentary faults commonly reach the surface and there is active seismicity along the southeastern edge of the basin. Consequently, since

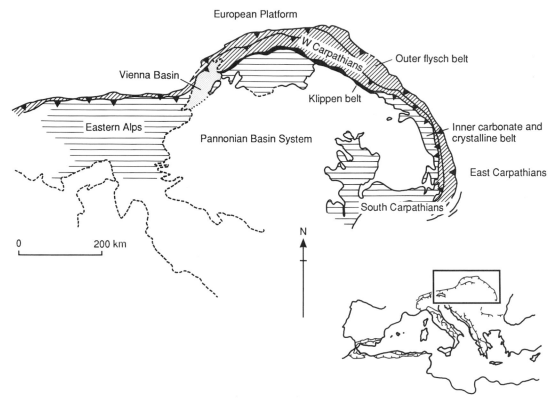

Fig. 5.17 Location of the Vienna Basin in the Alpine-Carpathian System. The outer flysch belt was deformed in the Tertiary. The inner parts of the system (carbonate and crystalline belt) was deformed mainly in the Cretaceous. The Klippen belt separates the outer from the inner West Carpathians (after Royden 1985).

extension appears to continue from Miocene to the present day, it is not possible to determine the post-extensional thermal subsidence. However, it should be possible to test whether the thin-skinned model is consistent with the observed subsidence and heat flow data.

5.6.2 Thin-skinned basin model

Some of the assumptions of the model used by Royden (1985) are as follows:
1 Since the extension is thin-skinned, the lower crust and mantle are unaffected by the extension near the surface. There is therefore no thermal anomaly beneath the basin.
2 The heat flow through the basin should be roughly equivalent to that through the adjacent Flysch Belt if the extension is confined to upper crustal levels. This is indeed the case with heat flow estimations lying between 45 and 60 mW m^{-2} across the Czechoslovakian and Austrian parts of the Vienna Basin and the Carpathian Flysch Belt (Royden 1985 and references therein).
3 Since there is no thermal disturbance at depth, there should be minimal or no post-extension thermal subsidence. Subsidence curves derived from boreholes located in the northern part of the basin where extension ceased at the end of Karpatian times supports this assumption, although other locations, such as the central and southern parts of the basin where extension continues, show different patterns.

The model used to compare the calculated thermal effects of thin-skinned extension to the observed subsidence histories and thermal data from the Vienna Basin and environs has three steps (Fig. 5.19):

Fig. 5.18. Schematic diagram to illustrate the relationship of extension in the Vienna Basin to strike-slip faulting and contemporaneous thrusting. In middle Miocene times (17.5–13.0 Ma) the nappes to the west of the Vienna Basin were fixed with respect to the European platform, whilst nappes to the east were being transported to the northeast over the European plate. This resulted in sinistral strike-slip deformation between the active an inactive nappes. Because the napp had a basal detachment, the strike-slip deformation is confined largely to the allochthon. The Vienna Basin is situated at a left step (releasing bend) in this sinistr strike-slip system. It has a stretch factor of about two, rather than extending completely down to the autochthon, as shown in the diagram (after Royden 1985).

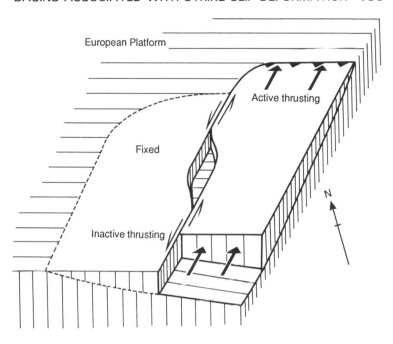

1 Initial thermal conditions are assumed to be those of a lithospheric slab of 125 km thickness at thermal equilibrium with a constant basal temperature of 1300°C. The radioactive heat generation is modelled as that produced from a plane source at 8 km giving a surface flux of 24 mW m^{-2}. This produces a total surface heat flow of 58 mW m^{-2}, identical to the present measured heat flow of the European platform near the Vienna Basin (Cermak 1979).

2 The European basement was overriden by flysch and carbonate nappes during the Miocene. This is modelled as an instantaneous emplacement of 4 km of material west of the Vienna Basin and 8 km beneath and along strike from the Vienna Basin. Using an appropriate thermal conductivity of the rocks comprising the thrust sheets, the present-day surface heat flow can be calculated. If the internal radiogenic heat production from the nappes is ignored, the surface heat flows are reduced by 10–25 per cent compared to the initial surface heat flow through the European Platform. This results from the thermal blanketing effect of the cold nappe pile. However, when the nappes are given their own radiogenic heat source (uniform at 1.0 µW m^{-2}), the estimated present-day surface heat

fluxes are very close to the observed values.

3 Extension in the basin was simulated by thinning the nappe pile by 100 per cent, accompanied by sedimentation keeping pace with subsidence. If an internal heat production for both basin sediments and nappe rocks is assumed (1.0 µW m^{-2}) there is once again an excellent match between calculated and observed values of surface heat flow. In addition, present-day surface heat flow values calculated for the Vienna Basin and adjacent flysch belt are almost identical, suggesting no heating at depth beneath the basin.

The lack of a thermal anomaly at depth results in low thermal gradients and low levels of organic maturity. This contrasts strongly with strike-slip basins with involvement of the lower crust and mantle lithosphere. Such basins, like the neighbouring Pannonian Basin (Fig. 5.18, 5.20), have much higher geothermal gradients and correspondingly high levels of organic maturity.

After nappe emplacement the thermal blanketing results in depression of the surface heat flows. However, gradually the lithosphere heats up and undergoes thermal expansion. This should result in uplift, both below the Vienna Basin and beneath

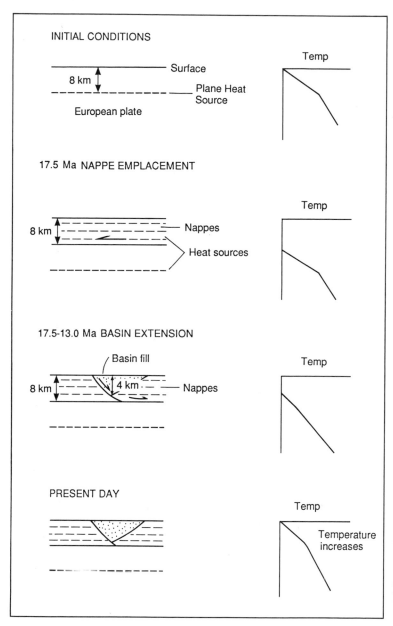

Fig. 5.19. Model used to calculate the thermal effects of nappe emplacement and thin-skinned extension of the Vienna Basin by Royden (1985). At initial conditions the European plate is at thermal equilibrium with a plane (radiogenic) heat source at 8 km depth. At 17.5 Ma (start of Karpatian stage) an 8 km thick thrust sheet with a temperature of 0 °C is emplaced over the European platform. Between 17.5 and 13.0 Ma (Karpatian–Badenian) the thrust complex is extended to form the Vienna Basin. The underlying autochthon is unaffected mechanically and thermally by the thin-skinned extension, and the temperature changes are solely due to conduction. Finally, temperatures in the basin increase by thermal conduction from the underlying European plate (after Royden 1985).

unextended parts of the thrust complex. This contrasts strongly with extensional basins involving mantle cooling (cf. Section 3.1) and therefore thermal subsidence. The post-extension uplift predicted as a result of lithospheric heating is observed in the northeast of the basin where 200–300 m of uplift has occurred since the end of

extension in middle Miocene times. This further suggests that the extension of the upper crust occurs over an undisturbed region that is undergoing slow uplift.

Basins such as the Vienna Basin with shallow detachments may be typical of zones close to the thrust front. As the distance from the thrust front

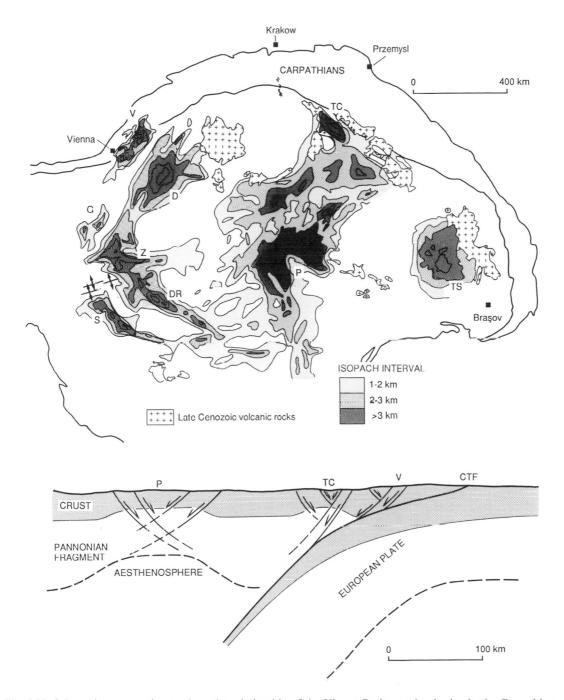

Fig. 5.20. Schematic cross-section to show the relationship of the Vienna Basin to other basins in the Carpathian–Pannonian system. The Vienna and Transcarpathian Basins are located on the leading (thin) edge of the Pannonian lithosphere and above the deflected European plate. The Pannonian Basin, however, is located entirely on the Pannonian lithosphere where it overlies aesthenosphere. Extension in the Pannonian Basin therefore involves mantle and the basin is consequently 'hot' compared to the 'cool' Vienna Basin (Royden *et al.* 1983). Inset shows locations of the basins. CTF, Carpathian Thrust Front; V, Vienna Basin; P, Pannonian Basin; TC, Transcarpathian Basin; TS, Transylvanian Basin; S, Sava Basin; DR, Drava Basin; D, Danube Basin; G, Graz Basin; Z, Zala Basin.

increases, extension may reach progressively greater depths, ultimately affecting the entire lithosphere. This may explain why the Pannonian Basin situated far to the south of the arcuate Carpathian suture involves mantle extension (Fig. 5.20) and elevated heat flows, whereas the Vienna Basin located within the Carpathian system is extremely thin-skinned and consequently cool, with no post-extension thermal subsidence.

SECTION 3
THE SEDIMENTARY BASIN-FILL

6 Controls on basin stratigraphy

Earth and Ocean seem
To sleep in one another's arms and dream
Of waves, flowers, clouds, woods, rocks and all that we
Read in their smiles, and call reality.

(Shelley, *Epipsychidion*)

SUMMARY

The stratigraphy in a sedimentary basin is the long-term response of the depositional surface to prolonged subsidence. However, the detailed patterns of subsidence and, to some extent, erosion in time and space give rise to characteristic geometries of the stratigraphic units. The fundamental meso-scale building blocks of stratigraphy are coherent and genetically related packets of strata known as *depositional sequences*. They are bounded by unconformities or lateral conformities and are thought to have a chronostratigraphic significance. The boundaries of depositional sequences are of critical importance and the relation of the internal stratigraphic horizons to the depositional sequence boundary indicates the changes taking place that gave rise to the deposition of the new depositional sequence. For example, *onlap* onto a lower depositional sequence boundary indicates a progressive overstepping of newer depositional units beyond the limits of the older depositional units. It is commonly associated with a relative rise in sea level in a marine basin. Other geometrical sequence relationships such as *downlap* onto a lower depositional boundary and *toplap* and *erosional truncation* against an upper depositional sequence boundary are also found.

The main controls on basin stratigraphy are (a) tectonic (see Chapters 3, 4 and 5), (b) related to global (that is, absolute) changes in sea level, or (c) a combination of both.

Basins such as rifts and passive margins cool following a thermal disturbance of the lithosphere. The two different end-member flexural responses of the lithosphere (elastic and viscoelastic, see Sections 2.1, 2.3 and Chapter 4) imply two contrasting stratigraphic styles; onlap in the case of an elastic lithosphere and progressive basinward shifts of the depositional limit (offlap) in the case of a viscoelastic lithosphere.

On the passive margin of North America for example, the transition from fault-controlled Airy-type subsidence to (elastic) flexurally-controlled subsidence coincides with the start of strong onlap of sediments onto the basement. Lateral heat flow causes momentary thermal uplift (cf. Sections 3.1.5.2, 3.1.7), arresting the pattern of onlap, but flexural subsidence then again outstrips uplift causing renewed onlap.

The role of flexure in controlling the stratigraphy of foreland basins is evident from the very existence of the basin, but the secular evolution of flexural rigidity is difficult to assess (Section 4.3). Movement of the orogenic load is responsible for an onlap onto the flexed plate for a time-constant flexural rigidity. Erosion of the mountain belt and *diffusion of sediment* into the basin has the same effect of onlap onto the flexed plate.

The Wilson cycle of rifting, drifting, subduction and collision implies that foreland basins should be superimposed on an inherited passive margin. As a result, the lithosphere should possess a rigidity which reflects its previous history of heating and thinning. Secondly, the first orogenic loads are emplaced on a pre-existing oceanic bathymetry. These two features allow extremely thick overthrust wedges to develop with little topographic expression. Passage of the flexural forebulge causes complex unconformities to develop and the progessive overthrusting of the plate causes migration of depocentres and pinchouts. Older stratigraphy is cannibalized by the orogenic front, sometimes being uplifted and eroded to provide new detritus.

Changes in regional stress fields caused by major plate boundary reorganizations can cause vertical movements that are large enough to have a great impact on basic stratigraphy. Application of an *in-plane stress* to an intracratonic basin, passive margin or foreland basin exploits original deformations of the lithosphere, such as the Moho, the sediment/basement interface or any other rheological boundaries. The effects of changes of in-plane stress are particularly felt on basin margins where such changes are not swamped by rapid subsidence rates. In a foreland basin, in-plane compression causes amplified basin subsidence and marginal uplift, whereas in-plane tension causes basin uplift and marginal subsidence. The effects are more subdued in intracratonic basins because the Moho and sediment/basement interfaces destructively interfere. In extensional basins in-plane compression typically causes basin margin uplift, causing an unconformity or a drop in the amount of stratigraphic onlap. Tensile in-plane stress causes basin margin subsidence, promoting stratigraphic onlap. A critical question remains as to the source and magnitude of the in-plane stress necessary to cause the vertical crustal movements necessary for the stratigraphic patterns outlined above.

Absolute changes in sea level may be caused by changes in the volume of ocean water or changes in the volume of ocean basins. Additions of juvenile water from the continuing differentiation of the lithosphere are probably compensated by losses due to hydrothermal alteration of new ocean crust. Sediment influx to ocean basins is potentially able to cause slow sea level rises but the process is counteracted by sediment removal at subduction zones, although a mass balance is not possible at present.

Volume changes in the ocean ridge system caused by variations in spreading rate are thought to be responsible for long-term changes in global sea level, involving a maximum sea level during the Late Cretaceous and falling with varying severity through the Cenozoic to the present. Glaciations and deglaciations are very rapid processes changing the volume of water in the ocean system.

The stratigraphic response to changing global sea levels has been investigated from two points of view. One method derives from *seismic stratigraphic studies* which charts *relative* changes in sea level through time, principally through the use of patterns of coastal onlap. Relative changes in sea level which can be correlated across widely separated regions are believed to be eustatic in origin. Such changes fall into long-term (first-order), medium-term (second-order) and short-term (third-order) cycles. Secondly, *synthetic stratigraphy* can be constructed quantitatively and compared with the depositional sequence stratigraphy observed from surface and subsurface (seismic stratigraphic) studies. Input of variables describing the rates of tectonic subsidence and sediment supply together with the rate of sea level variation has illuminated the complexities of the stratigraphic response, including onlap, toplap, and erosional truncation.

Depositional sequences can be subdivided into *systems tracts* which are deposited at particular intervals on the relative sea level curve. *Lowstand systems tracts* consist of basin-floor fans, slope fans and wedges accreting onto the continental slope. When relative sea level fall is gradual, systems tracts may develop at the shelf edge. During rapid sea level rises, *transgressive systems tracts* back-step onto the basin margin. After the period of maximum flooding of the basin margin, *highstand systems tracts* prograde into the basin producing marked clinoform geometries over basal downlap surfaces.

The depositional sequences of a sedimentary basin are undoubtedly a complex response to tectonic and eustatic factors. The first-order cycles of the relative sea level curves are probably due to long-term changes in the rate of crustal generation causing changes in the volume of mid-ocean ridge systems. The second-order cycles may be a result of tectonic effects at the rift-drift transition on cooling margins, changing in-plane stresses due to plate boundary reorganizations and, less likely, eustatic effects caused by glaciations and deglaciations. The third-order cycles may be due to rapidly applied in-plane stresses and/or to absolute sea level changes caused by glaciations and deglaciations. Parasequences and punctuated aggradation cycles can be explained by sea level changes of the type experienced during Late Quaternary glaciation/deglaciation in terms of amplitude and frequency of sea level change. The frequency falls within the orbital 'Milankovitch' band. The stratigraphic signatures of all of these cycles and the smaller scale paracycles can be modelled by an interaction of tectonic and eustatic factors but the causative mechanisms are not at present fully understood.

The previous chapters have provided some insights into the possible physical causes of the main classes of sedimentary basin. Rifts and continental margins originate through stretching of lithosphere and evolve primarily in response to thermal contraction and sediment and water loading (Chapter 3). Foreland basins form by flexure under the weight of supracrustal and/or subcrustal loads (Chapter 4).

Their evolution is controlled by secular (that is, through time) changes in the rigidity of the flexed lithosphere and the history (magnitude, location) of the load. The long-term response of the depositional surface is prolonged subsidence, but the detailed interplay of subsidence and sediment supply in time and place gives rise to the stratigraphy of a basin. In this chapter we shall examine how large scale factors and their interaction give rise to distinct basin stratigraphies.

6.1 PRELIMINARY REMARK ON DEPOSITIONAL SEQUENCES

The primary meso-scale units of stratigraphy are depositional sequences. They are coherent packets of strata which are genetically related and which can be traced for considerable distances across a basin. A depositional sequence is formed by an interaction of tectonics, thermal history, sea level change and sediment supply. They are recognized by bounding unconformities or laterally correlatable conformities at their tops and bases (Mitchum, Vail and Thompson 1977) (Fig. 6.1). Within a depositional sequence, individual subunits or stages can exhibit a variety of geometrical relationships to the depositional boundary (Fig. 6.2). Widespread recognition of these geometrical relationships was only possible with the availability of high-quality seismic reflection data. As well as helping to define a depositional sequence boundary, these types of discordant relationship also furnish clues as to the origin of the unconformity. Onlap, downlap and toplap indicate nondepositional hiatuses, whereas truncation indicates an erosional hiatus or it may be the result of structural disruption (Fig. 6.2) (Vail, Mitchum and Thompson in Payton 1977).

The minimum stratigraphic unit that can be called a depositional sequence must have a significance in the basin either in terms of thickness or in terms of geological time.

Sloss (1988) has traced the history of ideas on sequence stratigraphy over the last 40 years. The importance of rock-stratigraphic units traceable over wide areas of the North American continent and bounded by unconformities of 'interregional scope' has long been recognized (Sloss 1950, 1963). These sequences were of a rank higher than even supergroup and are therefore considerably larger

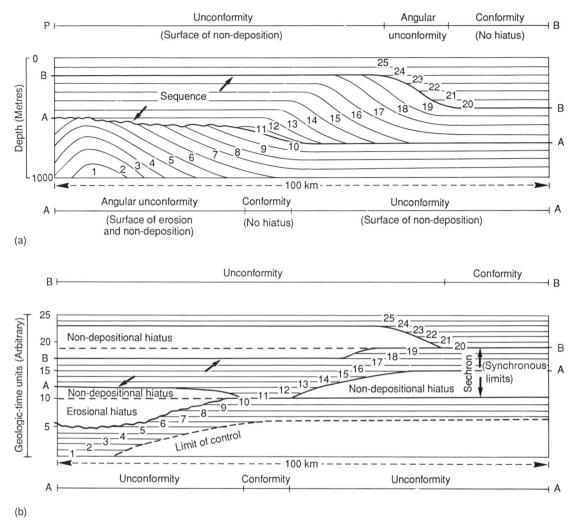

Fig. 6.1. The basic concept of the depositional sequence as outlined by Vail *et al.* in 1977a (in Payton, ed. 1977). (a) Generalized stratigraphic section of a depositional sequence. The sequence boundary A changes from an angular unconformity in the left half of the diagram to a conformity in the centre and then to a non-depositional unconformity on the right. The sequence boundary B passes from a non-depositional unconformity on the left to an angular unconformity in the centre to a lateral conformity on the right. Unconformities are dated at the points where they have laterally become conformable. Units 1 to 25 represent strata deposited during successive time intervals. (b) Generalized chronostratigraphic section of the same stratigraphic sequence as in (a). The depositional sequence between surfaces A and B ranges in age between the beginning of 11 and the end of 19. A sechron is the maximum stratigraphic age of a depositional sequence. Chronostratigraphic charts of this type are sometimes termed 'Wheeler diagrams' following Wheeler (1958).

than depositional sequences implied here (*sensu* Vail *et al.* 1977 a, b). Distinct groups of the smaller scale depositional sequences, called *supersequences*, can be recognized and appear to correspond to Sloss's original sequences.

Depositional sequences have a chronostratigraphic significance because they were deposited during a time interval limited by the ages of the sequence boundaries where they are conformities. Clearly, where unconformities mark the bound-

Fig. 6.2. The geometrical relationships of strata to a depositional sequence boundary or to any other surface within a depositional sequence. (A) Relations to upper surface, involving (1) erosional truncation, (2) toplap (commonly non-depositional rather than erosional) and (3) concordance. (B) Relations to lower surface involving (1) onlap where the overlying strata are near-horizontal and the surface is inclined, (2) downlap where the overlying strata are inclined and (3) concordance. (C) Additional geometrical patterns of (1) offlap, where there is a progressive basinward migration of the stratigraphic units and (2) basinward shifts, where the basinward movement is discrete rather than progressive.

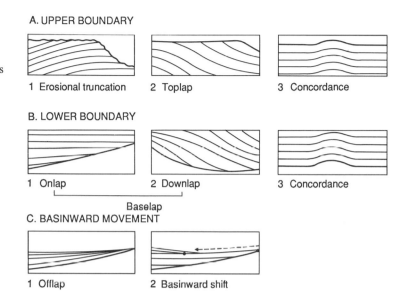

A. UPPER BOUNDARY

1 Erosional truncation   2 Toplap   3 Concordance

B. LOWER BOUNDARY

1 Onlap   2 Downlap   3 Concordance

Baselap

C. BASINWARD MOVEMENT

1 Offlap   2 Basinward shift

aries, the age range is reduced. The total time interval during which a sequence is deposited is called a *sechron*.

The boundaries of depositional sequences are crucial to their definition and correlation. Dunbar and Rogers (1957) emphasized the angularity or parallelism of the strata above and below an unconformity, and introduced the terms nonconformity, angular unconformity, disconformity and paraconformity. More important, however, in depositional sequence analysis, is the relation of strata to the unconformity itself. For example, there may be a discordant relationship recognized by *lapout* or truncation (Fig. 6.2).

Two important types of lapout at the lower boundary of a depositional sequence are recognized:
1 *Onlap* where an initially horizontal stratum laps out against an initially inclined surface, or where an initially inclined stratum laps out against a surface with a greater inclination.
2 *Downlap* where an initially inclined stratum terminates downdip against an initial horizontal, irregular or inclined surface.

In some instances, where structural movement has been complex, it may be impossible to discriminate between onlap and downlap. The general term baselap is then useful.

The lapout of strata at the upper boundary of a depositional sequence is termed *toplap*. Toplap is very common where initially inclined strata, such as delta foresets or clinoforms in general, terminate updip at the depositional sequence boundary. It is evidence of a non-depositional hiatus and may be associated with sediment bypassing or minor erosion.

Erosional truncation is the lateral termination of a stratum by erosion at the upper boundary of a sequence, and represents an erosional hiatus.

6.2 TECTONIC MECHANISMS

6.2.1 Effects of flexure on stratigraphy in basins due to stretching

Sections 3.1 and 3.2 outline the mechanisms of subsidence in stretched basins as comprising (1) a fault-controlled initial subsidence caused by mechanical stretching of the upper brittle layer of the lithosphere and (2) a thermal subsidence caused by the cooling and contraction of the upwelled asthenosphere. These mechanisms are amplified by sediment and water loading (Chapter 8). In the period of active stretching or rifting, the lithosphere is generally viewed as being in a state of purely local (Airy) isostasy. That is, the lithosphere behaves as a very weak support for any superimposed loads in the active rift. However, the presence of gently-dipping post-rift sediments suggests that a broadly-distributed subsidence characterizes the post-rift stage. The way in which the lithosphere distributes loads in the post-rift stage is by flexure. Whereas the stretching event determines the first-order depth and size of the basin, the subtleties of the flexural response of the lithosphere have a profound influence on depositional sequences in the post-rift stage.

Chapter 4 suggests that if the lithosphere behaves *elastically* over geological time periods, flexural rigidity should depend on its thermal age. Basin stratigraphy should reflect this increase in flexural rigidity with time, by a progressive overstepping of younger strata at the margin of the basin. If the lithosphere behaves *viscoelastically* (and the time constant of the viscous relaxation is long), stresses are relaxed by a viscous flow, causing the flexural rigidity to decrease with time. For the same tectonic (cooling plate) driving mechanism, this different basement response produces a pattern of stratigraphic offlap (Fig. 6.2c), with the youngest sediments restricted to the basin centre.

For a basin with an Airy compensated rift stage and a succeeding post-rift flexural stage, the widths of the basins produced on elastic and viscoelastic lithosphere are therefore markedly different (Fig.

6.3). For the elastic lithosphere, the post-rift sediments strongly overstep the syn-rift sediments giving a 'steers-head' geometry. In contrast, on a viscoelastic plate, the overstep of the post-rift sediments is minimal because of the stress relaxation with time.

If equivalent elastic thickness (corresponding to the depth to the 450 °C isotherm in Watts, Karner and Steckler's (1982) models) varies across the basin so that β increases towards the central axis, the overstep of post-rift sediments becomes even more marked, making the basin wider. A lateral transfer of heat (see Section 3.1.5.2), cooling the highly stretched basin centre but heating the flanks, causes a thermal bulge on the basin margin early in the basin history. This causes a narrowing of the basin early in its evolution. Near the edge of the unstretched lithosphere the thermal bulge and flexural subsidence compete. Viewed on a whole-basin scale the effects of a varying β and lateral heat flow appear to cancel each other out, producing a geometry almost identical to that predicted from the simple uniform stretching (McKenzie) model (Section 3.1.5).

The pattern of stratigraphic onlap has been modelled in more detail for the passive margin of eastern North America (Watts 1982) (see also

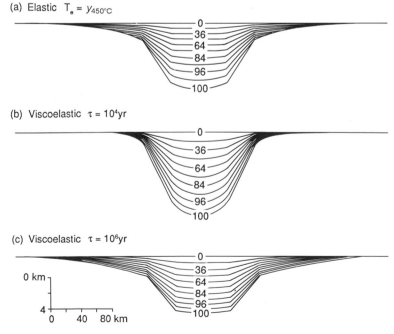

(a) Elastic $T_e = y_{450°C}$

(b) Viscoelastic $\tau = 10^4$ yr

(c) Viscoelastic $\tau = 10^6$ yr

Fig. 6.3. Calculated stratigraphy of stretched basins with $\beta = 2.0$, overlying lithospheres of contrasting rheology. (a) *Elastic* lithosphere where the equivalent elastic thickness is the depth to the 450 °C isotherm. The equivalent elastic thickness increases with thermal age, causing stratigraphic onlap and a steers-head geometry to the basin. (b) *Viscoelastic* lithosphere with a viscous relaxation time of 10^4 years. The lithosphere weakens rapidly following loading, causing stratigraphic offlap and a gradually narrowing basin. (c) *Viscoelastic* lithosphere with a viscous relaxation time of 10^6 years. The slower viscous relaxation produces a basin geometry similar to that in (a), but differs in being characterized by stratigraphic offlap. Vertical exaggeration of × 10. After Watts, Karner and Steckler (1982).

Fig. 6.4. Coastal plain and shelf stratigraphy using a thermal and mechanical model of a passive margin in which the tectonic subsidence of the margin is due to thermal contraction following heating and stretching of crust and lithosphere during rifting. Sediments rapidly infill the continental shelf, keeping a constant bathymetry with time. The sedimentary load flexes a cooling plate that increases in rigidity with time since heating. The initial lithospheric thickness is 125 km, initial crustal thickness 31.2 km, the coefficient of thermal expansion 3.4×10^{-5} °C^{-1}, the mantle temperature 1333 °C, initial densities of 2800 and 3330 kg m^{-3} for crust and lithosphere respectively, and the uniform density of infilling material 2500 kg m^{-3}. The model calculations assume the stretch factor to be 3.0 and the equivalent elastic thickness to be given by the depth to the 450 °C isotherm.

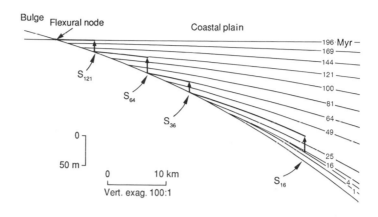

Solid lines are the boundaries of stratigraphic units with ages indicated in Myr since the end of rifting. The effects of compaction have been ignored. The heavy line shows the amounts of coastal aggradation calculated using the method of Vail *et al.* (1977). After Watts (1982, p. 471).

Section 3.3). Sedimentation is assumed to keep pace with subsidence, maintaining a constant bathymetric profile through time. Figure 6.4 shows the initial strong onlap of sediments onto the basement at the transition from fault-controlled Airy-type subsidence to flexural-controlled subsidence. However, lateral heat flow (N.B: enhanced heat flows due to secondary convection, Section 3.17) causes thermal uplift on the coastal plain, abruptly terminating onlap. By about 16 Myr after rifting, flexural subsidence outstrips thermal uplift and the sediments again progressively onlap basement. We shall see in forthcoming sections the importance of these ideas to the interpretation of relative sea level changes.

6.2.2 Role of flexure in generating foreland basin stratigraphy

The typical large scale geometry of foreland basin stratigraphy is of wedge shaped units, thick close to the orogenic load and thinning onto the foreland into a 'feather-edge'. This is a reflection of the lateral gradient in subsidence rate from the centre of the load to the peripheral bulge. Superimposed on this is commonly a spatial translation (with respect to the underlying plate) of stratigraphic units ahead of a moving load, a consequence of

continued convergence being accommodated in the orogenic belt. This effect causes a general onlap of successively younger stratigraphy onto the foreland (Fig. 6.5). Numerous case studies exist which demonstrate this simple geometrical style, such as the Magallanes Basin of southern South America (Fig. 6.6).

The details of the deflection of the foreland plate must depend on a number of factors (discussed in Chapter 4):

1 The flexural rigidity of the flexed lithosphere;
2 The nature of the distributed loads (topographic/ thrust loads, horizontal end forces, bending moments, sediment and water loads);
3 Pre-existing heterogeneities.

As a consequence, it is vital in considering foreland basin stratigraphy to consider the previous geological history of the lithosphere. The most common scenario is termed the Wilson cycle. This cycle consists of a number of stages:

1 Rifting of continental lithosphere;
2 Formation of an oceanic spreading centre and development of passive margins;
3 Ocean enlargement followed by subduction and the development of B-type subduction (active) margin(s);
4 Ocean contraction when the consumption rate exceeds the spreading rate;

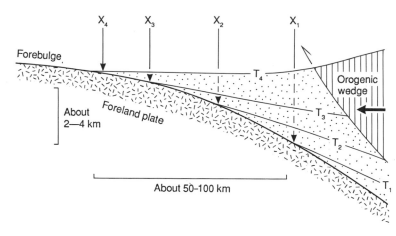

Fig. 6.5. Diagrammatic illustration of foreland basin stratigraphic pattern of onlap onto foreland plate. x_1 to x_4 show the successive positions of pinch-outs corresponding to the chronostratigraphic lines t_1 to t_4.

5 With continued and prolonged contraction, the passive margin is underthrusted during continental collision.

The collision process therefore flexes an inherited passive margin structure. Two points immediately stand out:

1 Since the lithosphere has been previously heated and stretched (Section 3.1) it will possess a flexural rigidity or equivalent elastic thickness which reflects this history, the 'memory' diminishing with time since rifting.

2 The first loads will be emplaced on a pre-existing continental margin bathymetry rather than on an imaginary flat surface representing 'regional'.

Stockmal, Beaumont and Boutilier (1986), using a depth-dependent plate rheology (see Section 2.3, 3.2.3), proposed four end-member models to account for the transition from passive margin to foreland basin (Table 6.1).

The evolution of the four end-member models is summarized in Fig. 6.7, the final stage being one of 50 Myr duration involving erosion and isostatic adjustment. Whereas at early stages the thermal age of

the passive margin is an important, even dominant control on basin development, this effect becomes less important as the overthrust wedge is progressively emplaced on stronger unstretched lithosphere. The maximum foreland basin depths prior to post-deformational erosion vary from ~ 3 km for a low topography (e.g. Zagros) to ~ 16 km for a high topography (e.g. Himalayas). Post-deformation erosion and tectonic thinning (extension) (Section 4.5) can cause massive unroofing of up to 40 km, bringing high grade metamorphic rocks to the surface. The products of erosion may entirely bypass the foreland basin, since it too will experience uplift as a result of the regional flexural response of the lithosphere. Very thick overthrust wedges can develop with little topographic expression if they are emplaced on a deep oceanic bathymetry.

Passage of the flexural forebulge can cause a complex arrangement of unconformities (Price and Hatcher 1983). Initial uplift in the forebulge region is followed by subsidence as the bulge moves onto the craton. This is thought to be the reason for the development of erosional unconformities on

Table 6.1 Four end-members of foreland basins superimposed on passive margins

	High topographic profile (e.g. Himalayas) (a) Frontal slope ~3°	Low topographic profile (e.g. Zagros) (b) Frontal slope ~0.5°
Thermally young (~10 Myr) continental margin	Young–high	Young–low
Thermally old (~120 Myr) continental margin	Old–high	Old–low

a, Seeber, Armbruster and Quittmeyer (1981); b, Chapple (1978)

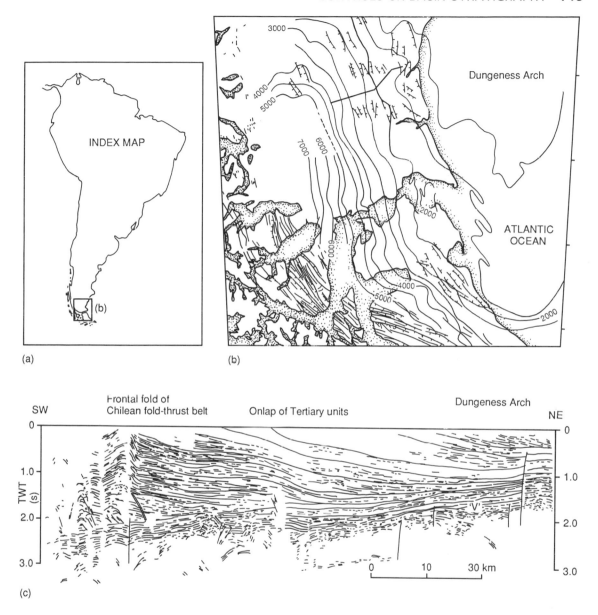

Fig. 6.6. The Magallanes Basin, southern South America (after Biddle *et al.* 1986). (a) Location map. (b) Isopachs (in m) of sedimentary fill of Magallanes Basin above the level of the Tobifera volcanics (representing the last stages of rifting prior to foreland basin flexure; Bruhn, Stern and De Wit 1978, Gust *et al.* 1985) showing thinning onto Dungeness Arch, and main structural lineaments, particularly the fold-thrust belt in the SW. The heavy solid line indicates line of section shown in (c). (c) Line drawing interpretation of seismic reflection record from the frontal fold of the Chilean fold-thrust belt to the flank of the Dungeness Arch. The Tertiary stratigraphic units show strong onlap on to the top of the Tobifera Volcanics (V). The progradational clinoforms are thought to be fan-deltas derived from the Andean mountain belt in the SW. They prograded into an environment of deep water shale deposition, which separated the fan-deltas from a zone of slow deposition of glauconitic sands close to and onlapping the basement high in the forebulge region.

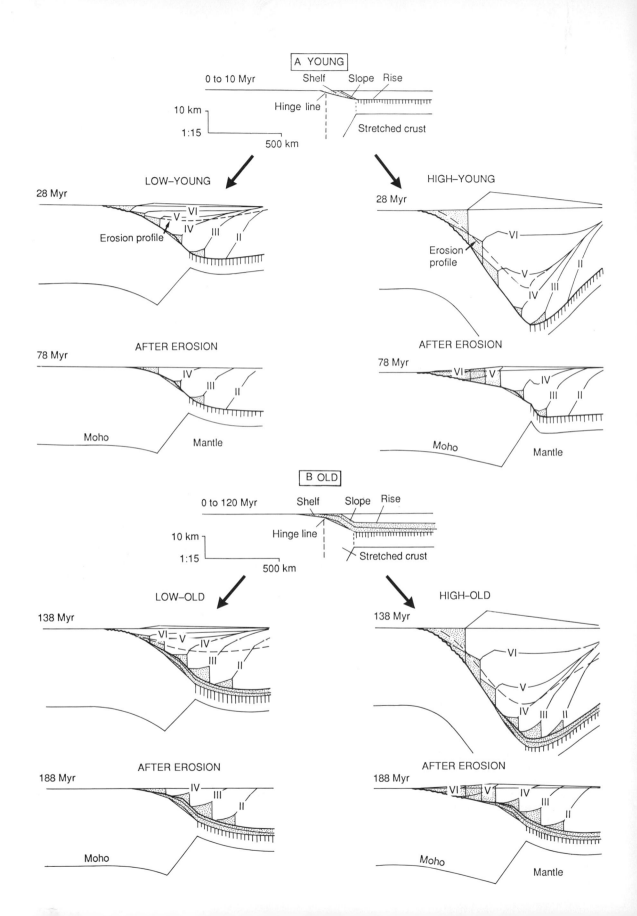

carbonate banks and their subsequent collapse. The unconformities should migrate in a time-transgressive manner onto the craton.

The Devonian (Acadian) peripheral bulge to the Appalachian orogen, for example, is characterized by a wedging system of unconformities and associated sedimentological variations. Progressive uplift caused unconformities to merge, and the intersection points of merged unconformities also overstep each other systematically toward the orogen (Fig. 6.8). This indicates contemporaneous uplift and orogenward movement of the forebulge at this time. Sedimentologically, uplift events are marked by shoaling or emergence (karst, caliche).

The migration of depocentres and of feather-edge pinch-outs of stratigraphy gives an impression of the mobility of the distributed loads and/or variations in the lithospheric response. Cretaceous foreland basin evolution in the western United States has been explained by the flexural response of an elastic lithosphere to moving thrust loads (Jordan 1981). The possible contribution of buried loads, in-plane (horizontal) forces and bending moments (see Chapter 4) were ignored. The Sevier orogenic belt has as a companion a foreland basin of retroarc type (Dickinson 1974) on its eastern side. The cross-section in the Idaho–Wyoming area allows a palinspastic restoration of the thrust belt and the foreland basin, which shows a progressive eastward shift of the depocentres through time (Fig. 6.9). The subsidence in the foreland basin was predicted using the restored thrust load configuration and the erosion of emergent thrust sheets and redistribution of sediment into the basin. Careful matching of predicted with observed stratigraphy allowed a flexural rigidity of about 10^{23} Nm to be estimated for the Cretaceous lithosphere. There was no necessity to invoke a changing flexural rigidity with time, suggesting that the lithosphere could be explained as behaving elastically over the modelled 70 Myr time span and that lateral variations in rigidity were also insignificant.

Some of the Alpine foreland basins of Europe also show a clear relationship between load

---

Fig. 6.7. (*Opposite*) End-member models of foreland basin development from an inherited passive margin (after Stockmal, Beaumont and Boutilier 1986). The change from a passive margin to an active (convergent) setting is modelled by emplacing overthrust loads onto the previously stretched passive margin. The encroachment of the allochthonous terrane flexes down the foreland plate, creating either an oceanic trench, an under-filled foreland basin or a steady-state or over-filled foreland basin. The passive margin stage is based on the model of Beaumont *et al.* (1982), (see also Section 3.2.3) involving an elastic but non-uniform lithosphere. The equivalent elastic thickness is taken as the depth to the 750 °C isotherm, which for a normal geothermal gradient, corresponds to an equivalent elastic thickness of 70 km, equal to estimates for the Appalachian foreland basin (Quinlan and Beaumont 1984) but far in excess of the value estimated for the Alps (Sinclair, *et al.* 1991). It is assumed that sediments and overthrust material do not contribute to the equivalent elastic thickness of the deflected plate. The passive margin model is modified to enable the topographic profile of the overthrust wedge at a number of time steps to be specified. That is, the mechanics of the wedge are ignored (cf. Sinclair *et al.* 1991); the overthrust wedge topography is chosen *a priori* and loads added to achieve this topography. The maximum topographic elevation of the overthrust wedge is assumed to grow sequentially, reaching a peak at the end of deformation. The toe of the overthrust wedge is advanced at 50 mm yr^{-1}, an order of magnitude faster than in some orogens such as the Alps (Hsu 1979, Homewood, Allen and Williams 1986) and the Canadian Rocky Mountains (Price 1981). This ensures that the 'young' models are end-members. At the end of overthrusting, the overthrust terranes are eroded for 50 Myr, the 'low' examples being eroded to sea level, whereas the 'high' examples are eroded to a profile similar to that of the Canadian Rockies today.

(A) Initial conditions for the YOUNG models, showing a 10 Ma passive margin with a thin sedimentary prism. (A1) is *low-young* configuration and (A2) is *high-young* configuration at the end of overthrusting (showing profiles at eight time steps indicated by roman numerals) and after 50 Myr of erosion.

(B) Initial conditions for the OLD models, showing a 120 Ma mature passive margin with a thick sedimentary cover due to thermal subsidence. (B1) and (B2) are the *low-old* and *high-old* configurations.

The influence of the age of the margin is felt primarily in the early stages of collision when the overthrust wedge loads a previously heated and therefore weaker plate. As the toe of the overthrust wedge advances landward of the hinge line, the geometry of the foreland basin is largely unaffected by the age of the margin. Instead, the thickness of the overthrust load is of great importance at this stage, with foreland basins up to 16 km deep being developed under topographies similar to the Himalayas.

(a)

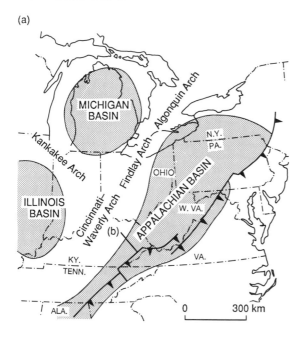

(b)

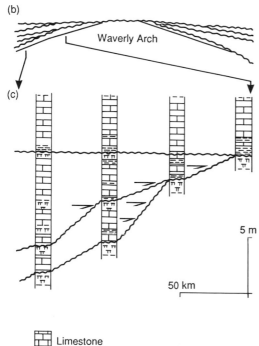

Limestone

Shale

Emergence Crust

Onlap

Unconformity

Fig. 6.8. The peripheral arches of the Appalachian foreland basin system (after Tankard 1986).
(a) Palaeozoic thrusting and flexure produced a system of swells or arches separating the Appalachian foreland basin from the Michigan and Illinois basins of the American interior.
(b) Stratigraphic relations during the late Acadian (Early Carboniferous or Mississippian) across the Cincinatti–Waverly arch in Kentucky (after Ettensohn 1981), showing wedges separated by unconformities fanning out from the crest of the arch.
(c) Stratigraphic columns on the western edge of the arch show merging unconformities (Dever *et al.* 1977) and evidence for periodic shoaling (emergence phenomena such as karsts) interpreted to be caused by forebulge uplift.

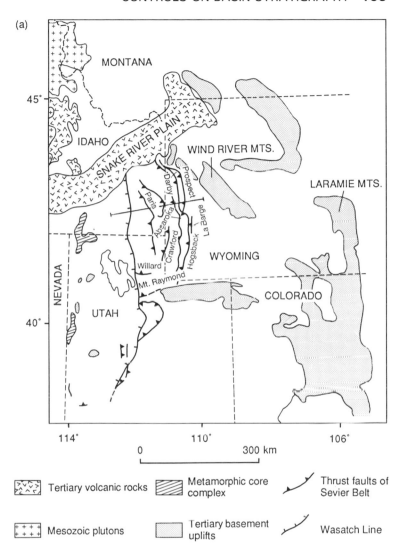

Fig. 6.9.(a) Location of Jurassic–Tertiary Sevier thrust belt in western USA. Volcanic rocks, basement uplifts and the extensional Wasatch line are all younger than thrusting and foreland basin development. Solid line is line of section in (b).

Tertiary volcanic rocks

Metamorphic core complex

Thrust faults of Sevier Belt

Mesozoic plutons

Tertiary basement uplifts

Wasatch Line

mobility and depocentre migration. In the Molasse basin of western Switzerland, a classical peripheral foreland basin, pinch-outs and depocentres of stratigraphic units migrated onto the European craton during the Oligo-Miocene, corresponding to the main collisional phase of Alpine orogenesis A palinspastic restoration of thrust units over the same time period coupled with a time-bracketing technique for dating thrust movements allowed fault tip propagation and shortening rates to be estimated (Homewood, Allen and Williams 1986). There was a very close correspondence between the rates of depocentre and pinch-out migration and of thrust-related shortening ($\sim$5 mm yr^{-1}) in western

Switzerland. This feature initially suggests, like the Cretaceous Sevier belt, that the lithospheric response is constant through time. However, the small time period analysed and the large uncertainties in the structural restoration and thrust timing all combine to make the precision low.

If the pinch-out migration rate far exceeds the rate at which the mountain belt advances, the foreland basin should progressively widen with time (Fig. 6.10). This might result from a number of processes. For example, it would be expected if the mountain belt loaded a progressively stronger elastic lithosphere as it overrode first the passive margin then the unstretched craton (Fig. 6.7). It

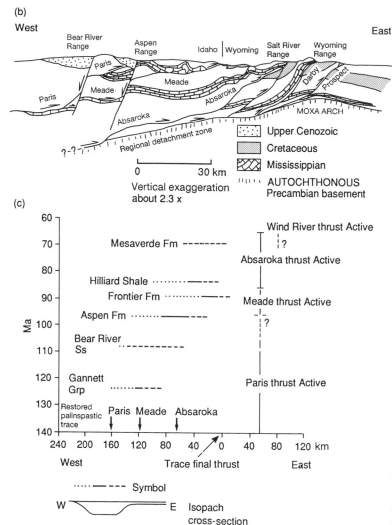

(b)

Upper Cenozoic

Cretaceous

Mississippian

AUTOCHTHONOUS
Precambrian basement

(c)

Fig. 6.9.(b) Generalized and schematic cross-section across the Idaho–Wyoming thrust belt (after Royse *et al.* 1975). (c) Graph illustrating the eastward shift of the palinspastic position of the foreland basin depocentre through time. This eastward shift is paralleled by an eastward migration of active thrusting (after Jordan 1981).

might also result from the forelandward shift of the centre of gravity of the load by erosion and deposition of sediment in the basin (see below). On the other hand, if the pinch-out migration rate slows relative to the rate of advance of the mountain belt or if the forebulge is dragged inwards toward a stationary mountain front, the foreland basin should narrow (Fig. 6.10). Some authors have interpreted this kind of offlapping relationship to result from stress relaxation in a viscoelastic lithosphere (Tankard 1986). We see below, however, how a similar effect can be produced by thickening of the adjacent orogenic

wedge loading an elastic plate (Sinclair *et al.* 1991, see also Flemings and Jordan 1989).

Both the Cordilleran (Rocky Mountains) and Appalachian foreland basins of North America originated by thrusting onto previously stretched passive margin ramps. Because of the weak lithosphere the supracrustal loads produced deep basins. Subsequent tectonic thickening with little cratonward propagation is thought to be associated with overdeepening of the shale-dominated basins together with uplift along basin-margin arches. This phase of marginal uplift and basinal deepening was cited by Quinlan and Beaumont (1984)

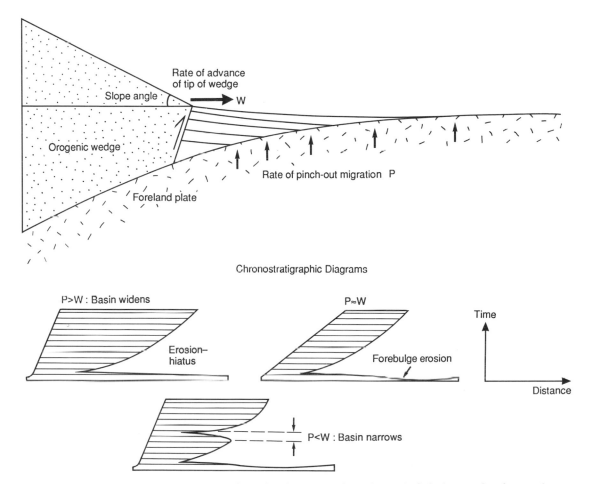

Fig. 6.10. Influence of the rates of advance of the tip of the orogenic wedge and of pinch-out migration on the foreland basin width. Three possible chronostratigraphic diagrams show basin widening, a basin with a time-constant width, and basin narrowing separating two periods of basin widening. Note that onlap onto the foreland plate occurs even in the case of the time-constant basin width. See also Fig. 6.12.

and Tankard (1986) as evidence for viscous relaxation following loading (see Sections 2.3 and 4.3). The final stage is characterized by renewed thrust propagation onto a more rigid continental lithosphere. Thrust-related uplift caused the shedding of coarse clastic wedges into the basin, and swamping the now unimportant basin-margin arches (Fig. 6.11).

The topography caused by thrusting in mountain belts is eroded to provide detritus to fill the foreland basin. Moretti and Turcotte (1985) first applied a *diffusion equation* to this process, that is, they modelled the transport of sediment as proportional to the topographic gradient. Flemings and

Jordan (1989) and Sinclair *et al.* (1991) have specifically applied diffusion modelling to the stratigraphy of foreland basins. The stratigraphy of the North Alpine Foreland Basin in the central Alps can be simulated using a diffusion model for erosion, a thrust-wedge model based on the concept of a critical taper (Section 4.5 for details), and a rheology of a linear elastic plate for the flexed European lithosphere (Sections 2.1.5, 4.1, 4.2). The amount of sediment transported to the foreland basin is proportional to the slope, the transport coefficient (K) varying over several orders of magnitude according to climatic and hydrologic setting. This relation therefore has the form:

W   E

EARLY ALLEGHENIAN (PENNSYLVANIAN)

Upper Breathitt–flexural deformation
(Overfilled basin)

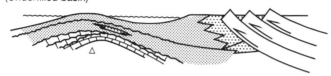

Lower-Breathitt–flexural deformation
(Underfilled basin)

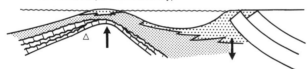

LATE ACADIAN (MISSISSIPPIAN)

Carter Caves–relaxation phase 2
(Uplift, reworking of mudsheet, barrier quartz arenites,
underfilled basin, major unconformity)

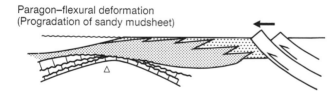

Paragon–flexural deformation
(Progradation of sandy mudsheet)

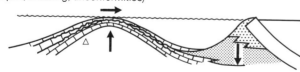

Slade–relaxation phase I
(Uplift, shoaling, unconformities)

Slade–stable platform
(Blanket limestone sedimentation)

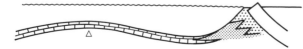

Fig. 6.11. Sequence of
diagrammatic cross-sections of the
Appalachian foreland basin to show
the alternation of periods of
flexural deformation caused by
thrusting with periods of quiescence
characterized by viscoelastic
relaxation. The first diagram shows
the Appalachian and Illinois basins
'yoked' together, the region being
dominated by blanket shale and
limestone deposition. The following
diagrams show the responses of
basin deepening and arch
uplift/erosion during relaxation, and
the shedding of clastic wedges
during periods of active thrusting.
See text for further details (after
Tankard 1986).

$$q(x) = -K(dh/dx) \qquad (6.1)$$

where K the transport coefficient is in $m^2 \, yr^{-1}$, h is the topography in m, q is the amount of material transported and x is the spatial variable. Flemings and Jordan (1989), for example, suggest that K for fluvial transport is c. $10^4 \, m^2 \, yr^{-1}$ whereas it is only $10^{-2} \, m^2 \, yr^{-1}$ for the decay of isolated slopes. The values obtained from a study of the modern subAndean foreland were of the order of $10^4 \, m^2 \, yr^{-1}$ (Flemings and Jordan 1989). The magnitude of the rate of transport of sediment into the basin clearly determines whether, for a given flexural response, it is *underfilled* or *overfilled* (see Covey 1986 for discussion of the concept of under- and over-filling).

The results for the North Alpine foreland basin in the central Alps are highly informative (Fig. 6.12). Movement of the forebulge, both vertically and horizontally, and the attendant unconformities produced in the basin-filling stratigraphy, could be simulated purely by varying the rate of thrust belt propagation and wedge thickening coupled with diffusive erosion. It was not necessary to vary the flexural rigidity of the European plate through time, nor to invoke a non-elastic rheology for the flexed plate. Subtle unconformities ('sequence boundaries') associated with abrupt basinward shifts or progressive offlap could be produced by thickening the load and slowing its propagation rate. Erosion of the load then causes onlap as the centre of gravity of the load migrates foreland-wards through redistribution of sediment from wedge to basin. Rapid advance of the thrust tip encourages underfilling of the basin, lower values of thrust belt migration rate promoting overfilling, the sedimentation keeping pace with the creation of new space in the basin.

6.2.3 Changes in states of in-plane stress

Some workers believe that variations in regional stress fields acting within inhomogeneous litho-spheric plates can cause vertical movements large enough to have a major impact on stratigraphy. The horizontal (in-plane) stress required to buckle an undeformed plate is enormous (see Section 2.1), but acting on an already deflected plate, the in-plane stresses may enhance or reduce the curva-ture of the deflection. The effect of changing horizontal (i.e. in-plane) stresses on a passive continental margin with an overlying sedimentary load has been modelled (Cloetingh, McQueen and Lambeck 1985). Effects of thermal blanketing by the sediments (De Bremaecker 1983) and of lateral heat conduction across the basin edges (Watts *et al.* 1982) are ignored.

The effect of the sediment load on a uniform elastic plate with an age-dependent equivalent elastic thickness is given in Fig. 6.13a. Both the wavelength and amplitude of the deflection in-crease with time – this is due to the combined effect of the thickening of the plate as it cools and the steadily increasing sediment load as the passive margin prism grows. Application of an in-plane force causes a modification of this deflection. For example, changing from an in-plane tensional stress of $5 \times 10^{12} \, Nm^{-1}$ (2.3 Kbar) to a compres-sional stress of the same magnitude on a litho-sphere with an equivalent elastic thickness appro-priate to an age of 30 Myr, produces a net uplift on the edge of the basin of about 100 m (Fig. 6.13b). The effectiveness of the in-plane force in producing large changes in the deflection increases with increasing thickness of sediment load, but de-creases with the increasing age of the lithosphere. As a consequence, in-plane forces are particularly important on young, rapidly loaded margins.

Changes in horizontal stresses also have effects towards the basin centre, but because the total subsidence is large in these areas, the contribution of in-plane stresses to the deflection is masked.

When in-plane stresses are superimposed on a lithosphere with a laterally varying equivalent elas-tic thickness, as expected at the continent–ocean transition on a passive margin, the differential uplift or subsidence becomes asymmetrically arranged (skewed) because of a tilting of the crust carrying the in-plane forces. Although this is more realistic geologically, the same general pattern emerges as for the uniform elastic plate (Fig. 6.13b).

Karner (1986) applied more fully the concept of in-plane stresses to a lithospheric structure where the flexural rigidity varied not only in time (cooling effect) but also in space as a result of basin formation. Using an analytical approach, Karner (1986) shows that the effect of a tensile in-plane stress is to reduce the curvature of the plate, that is, to effectively increase the plate rigidity, and that of a compressive in-plane stress to increase the cur-vature, that is to reduce the plate rigidity. These conclusions are based on an elastic rheology and

(a) Lithostratigraphy

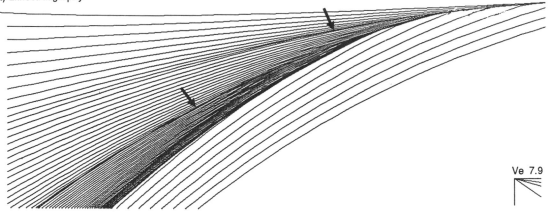

Ve 7.9

(b) Chronostratigraphy

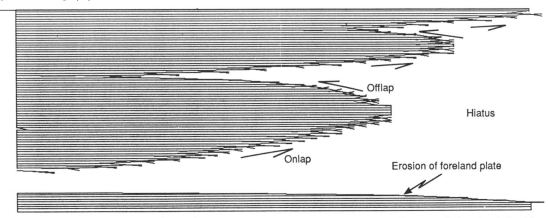

Offlap

Hiatus

Onlap

Erosion of foreland plate

Fig. 6.12. Computer-generated stratigraphy using input data from the central Alps of eastern Switzerland. In this experiment the equivalent elastic thickness was calculated to be 12 km from the curvature of the foreland plate using decompacted borehole records. (a) Two unconformities (arrows) that pass basinwards into conformities punctuate the foreland basin stratigraphy. They are overlain by strongly onlapping strata. (b) Chronostratigraphic diagram for the same stratigraphic pattern as in (a), showing the hiatus represented by each unconformity, erosion of the foreland plate caused by forebulge uplift, and patterns of offlap and onlap. The unconformities were simulated by varying the slope angle of the critically tapered orogenic wedge and its advance rate at the tip of the wedge (see also Section 4.5). Thickening in the orogenic belt, associated with an increase in the slope angle, rejuvenates the thrust load and 'pulls' the forebulge towards the basin, producing an unconformity with offlap. Erosion of the orogenic wedge and resedimentation in the foreland basin causes a 'pushing-out' of the forebulge and stratigraphic onlap. These depositional sequence boundaries are therefore produced without any eustatic change and without any weakening of the overridden plate following loading.

would not apply in other cases, such as for an elastic–plastic rheology (e.g. Chapple and Forsyth 1979, also Section 2.3). Nevertheless, changes in in-plane stress must have some effect on flexural rigidity.

The effects of changes in in-plane forces on foreland and intracratonic basins is shown schematically in Fig. 6.14, where the in-plane stress is exploiting the lithospheric deformations represented by the sediment/basement and Moho inter-

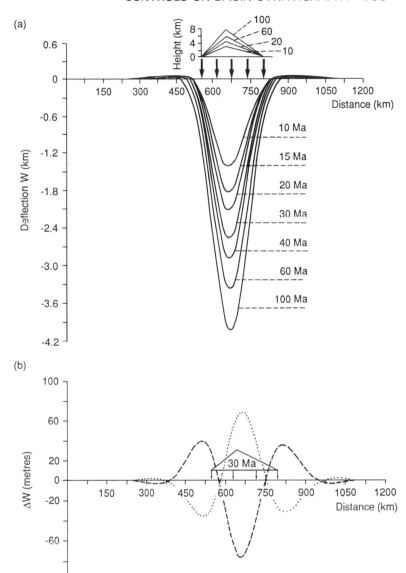

Fig. 6.13. Effects of in-plane forces following Cloetingh, McQueen and Lambeck (1985). (a) Deflection of a uniform elastic plate, whose equivalent elastic thickness depends on its thermal age, by the sediment prism shown above. This approximates the sedimentary loading of a passive margin in which sedimentation keeps pace with thermal subsidence following rifting. The plate has a laterally constant equivalent elastic thickness in this example. A laterally varying T_e does not substantially alter the results. (b) Differential subsidence or uplift resulting from the application of in-plane stresses to the plate in (a) at a thermal age of 30 Myr. Dashed curve – result of superimposing 5×10^{12} N m^{-1} axial *compression*, producing basin margin differential uplifts and basin centre differential subsidence. Dotted curve – result of superimposing 5×10^{12} N m^{-1} axial *tension*, producing basin centre differential uplift and margin differential subsidence.

faces. The effects are greater in foreland basins since the sediment/basement and Moho interfaces parallel each other, causing constructive interference of the in-plane stress effects. In-plane compression causes amplified basin subsidence and marginal uplift, whereas in-plane tension causes basin uplift and marginal subsidence. The effects are more subdued in an intracratonic rift basin since the sediment/basement and Moho deformations compete and the success of one over the other varies in time as a result of the lithospheric

thickness changes accompanying basin formation. Unconformities due to vertical movements in intracratonic basins should be rare in their early development where there is a cancelling out of the two components, but should be more common in the middle stages.

The application of a compressive (12.5×10^{12} Nm^{-1} or 4 Kbar) stress on a typical extensional basin margin causes the basin margin to uplift, producing an unconformity recognized by a sharp drop in the amount of stratigraphic onlap. In

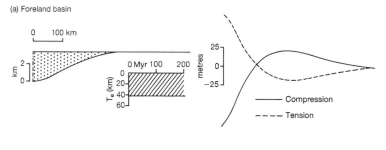

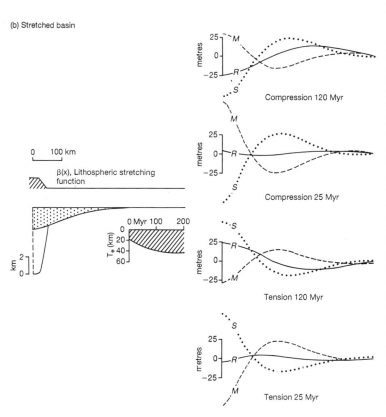

Fig. 6.14. Effects of in-plane forces on intracratonic rifts and foreland basins following Karner (1986). (a) Application of an in-plane force of $12.5 \; 10^{12}$ N m^{-1} on a foreland basin in which the Moho and sediment/basement interfaces are parallel and which therefore constructively interfere. Equivalent elastic thickness is assumed to be 42 km and not to vary laterally. In-plane compression induces basin margin uplift and basin centre subsidence, whereas in-plane tension results in basin centre uplift and marginal subsidence. (b) Application of an in-plane force of 12.5×10^{12} N m^{-1} on an aulacogen in which the Moho and sediment-basement interfaces are not parallel and which therefore destructively interfere. The equivalent elastic thickness is a function of space, time since the end of rifting and the amount of stretching. The effects of in-plane compression and in-plane tension applied at early (25 Myr after the end of rifting) and late (120 Myr after the end of rifting) stages in basin development are shown on the right. S is the contribution from the sediment-basement interface; M is the contribution from the Moho interface. The resultant is R, which becomes increasingly important (i.e. greater in amplitude) during ageing of the basin.

contrast, an identical tensile in-plane stress causes a major transgression as the basin margin is depressed (Fig. 6.15).

The critical question is therefore whether in-plane stresses of the magnitude necessary to cause vertical crustal movements of the order of 100 m exist in the real world. The deformation of oceanic lithosphere prior to subduction, broad basement undulations in the Bay of Bengal, departures from isostatic equilibrium in several Australian tectonic provinces and the rifting of continental lithosphere to form extensional basins, all suggest in-plane stresses of a few to several kilobars. Do these in-plane stresses change in sign over a sufficiently short time period to significantly influence stratigraphy? The answer to this problem may come from convergent plate margins, such as those of the Indo-Australian plate. Here, over a wide area,

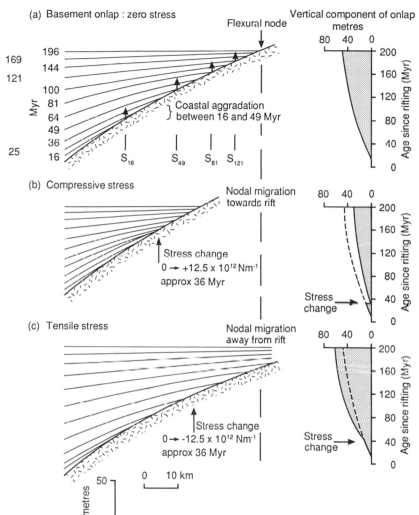

Fig. 6.15. Stratigraphic patterns in a stretched basin resulting from the experiments shown in Fig. 6.14 (after Karner 1986). Diagrams at right show the amount of coastal aggradation as a function of time, which serves as a proxy for apparent relative sea level change. (a) Stratigraphic onlap caused by the cooling of the lithosphere following rifting. (b) Effects of applying a compressive in-plane force of 12.5×10^{12} N m^{-1} at 36 Myr after the end of rifting, at which point the equivalent elastic thickness is 32 km. This produces basin margin uplift and a basinward shift in coastal onlap. (c) An increase in the rate of coastal onlap is produced by the application of a tensile in-plane force of 12.5×10^{12} N m^{-1} at 36 Myr following rifting.

major changes of stress appear to have occurred at 5 to 6 Ma. Such changes should therefore affect neighbouring passive margins, particularly those around Australia; and at this time the Otway Basin of southern Australia experienced a shallowing event leading to regression. Rapid changes in stress have also been postulated in other settings, such as the late Cenozoic evolution of Andean intermontane basins, and on a larger scale throughout the Pacific when the Farallon plate broke up about 30 Ma ago. It has been suggested that initiation of sea floor spreading in new oceans may also cause changes in stress in neighbouring continental plates. For example, the formation of the Norwegian–Greenland Sea at the beginning of the Palaeocene may be linked with the change from pelagic to thick clastic sedimentation in the intracratonic North Sea rift-sag basin.

Since it is possible that changes in stress of plate margins may be transmitted far into plate interiors, the mechanisms outlined here may be responsible for the correlations between timing of sea level changes in oceanic and continental regions observed by Sloss (1979), without resorting to global eustatic controls. A mechanism of this type is also implied by the recognition of 'megasequences' (equivalent to supersequences) by Hubbard, Pape and Roberts (1985), and Hubbard (1988).

6.3 EUSTATIC CHANGES IN SEA LEVEL: CAUSES

So far we have considered changes in basement height relative to a datum (sea level or 'regional') in terms of large scale mechanisms such as lithospheric thickness changes and flexure. The elevation changes caused by these tectonic mechanisms may be extremely widespread, being determined by mantle convection processes, flexural rigidity of the lithosphere or plate boundary forces for example, but they are not global. Global sea level changes (absolute changes relative to the centre of the Earth) result either from changes in the amount of water in the oceans or from changes in the volume of the ocean basins. We can examine global sea level change in terms of four possible causes:

1 Continuing differentiation of lithospheric material as a result of plate tectonic processes (Section 6.3.1);

2 Changes in the volumetric capacity of the ocean basins caused by sediment accumulation or by sediment abstraction (Section 6.3.2);

3 Changes in the volumetric capacity of the ocean basins caused by volume changes in the mid-ocean ridge system (Section 6.3.3);

4 Reduction of available water by locking up in polar ice caps and glaciers (Sections 6.3.4).

In considering absolute sea level change, it is important to incorporate the effects of isostasy. An increase in water depth in the ocean basins causes a subsidence of the ocean floor due to the excess weight of the water at the new sea level. A decrease in water depth causes uplift of the ocean floor due to the removal of part of the water column. This is a result of a simple isostatic balance as illustrated

in Fig. 6.16. The displacement of the oceanic basement relative to the centre of the Earth by changes in the water load would be found throughout the oceans, although the effect would only be observed at places such as oceanic islands. These regions would experience the same displacement as the oceanic basement, and would therefore act as a 'dipstick', indicating the sea level change.

Consider an ocean basin of depth h_1 which deepens to a depth h_2 (Fig. 6.16). The greater depth of the ocean is made up of two components, the subsidence of the ocean floor S, and the sea level rise Δ_{SL}. If the ocean floor has a density (ρ_m) of 3300 kg m^{-3} and the water a density (ρ_w) of 1000 kg m^{-3}, the isostatic subsidence of the ocean floor is approximately 0.4 of the sea level change. Expressed slightly differently, the sea level change is 0.7 of the increase in the water depth of the ocean ($h_2 - h_1$). These figures represent approximate upper bounds on the water loading of the oceanic lithosphere since they result from a purely local (vertical) isostasy. A flexural response of the lithosphere would reduce the water-loaded subsidence. However, the approximation is reasonable, since the wavelength of the water load (the ocean) is very large compared to lithospheric thickness (see Sections 2.1 and 8.3.4).

6.3.1 Continuing lithospheric differentiation

The volume of water in the oceans may be added to by contributions from mid-ocean ridge and island arc volcanism. Counteracting this, water may be removed by hydrothermal alteration of new crust and at subduction zones. These processes are probably roughly balanced. There is no evi-

$$\rho_w h_1 + \rho_m C_1 = h_2 \rho_w + \rho_m (C_1 - S)$$
$$\rho_m S = \rho_w (h_2 - h_1)$$
$$\rho_m S = \rho_w (S + \Delta_{SL})$$
$$S \left(\frac{\rho_m - \rho_w}{\rho_w} \right) = \Delta_{SL}$$

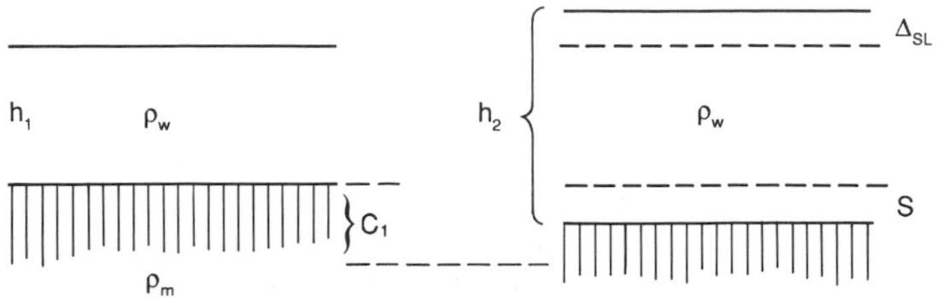

Fig. 6.16. Isostatic effect of a change of water depth in the ocean. Since the wavelength of the water load is very large, the compensation can be regarded as local (Airy).

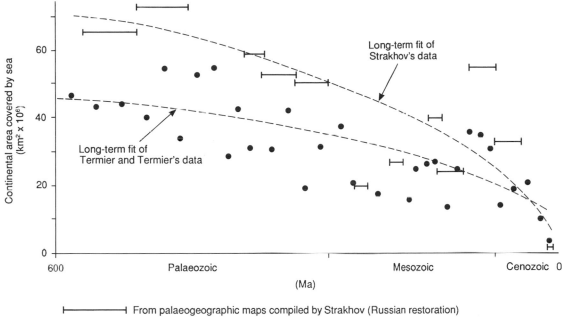

Fig. 6.17. An early attempt to plot relative sea level change using the area of the continent covered by sea as a function of geological time (after Holmes 1965, p. 969). The estimates based on palaeogeographical reconstructions of Russian and French workers were compiled by Egyed (1956). There is a general tendency of decreasing continental areas covered by seas through time; this suggested to Carey (1956) and Holmes (1965) that the Earth was expanding. Alternative mechanisms to explain this trend include long-term continental differentiation. The rate of relative sea level change indicated by the plot is very slow indeed, involving approximately 150–250 m of change over 600 Myr!

dence for appreciable, long-term changes in the chemical composition of the oceans during the past 1500 Ma, suggesting that the volume of sea water has not greatly increased as a result of Phanerozoic volcanism.

Continents may also accrete new material by a differentiation from the heavier oceanic lithosphere, causing changes in the volumes of the ocean basins. It is suspected that there is now very little or no net accretion taking place.

If there have been any net additions or losses of ocean water from volcanic or hydrothermal processes, or by differentiation of lithospheric material, over say, the past 600 million years, we should see a very gradual 'drift' in the sea level as a function of time. Holmes (1965 p. 969) presented a sea level curve for the Phanerozoic (Fig. 6.17) which can be interpreted as indicating a very gradual emergence of the continents, at a rate of 0.2–0.4 mm per 1000 years.

6.3.2 Changes in the volumetric capacity of the ocean basins caused by sediment influx or removal

The subaerial surface of the Earth is continually being eroded, but at markedly different rates in different climatic and tectonic regimes. For example, the circum-Pacific region between South Korea and Pakistan, characterized by young mountain belts and intensive chemical weathering, provides an enormous amount of weathering products, whereas a region of comparable size such as the Eurasian Arctic with lower topography and cooler temperatures provides very little. Eroded sediment makes its way eventually to the ocean basins. For every unit of sediment being transported, however, a large proportion is retained in continental environments, such as lakes, deserts, river floodplains and eventually coastal zones, deltas and the continental shelf (Chapter 7). A small proportion of the

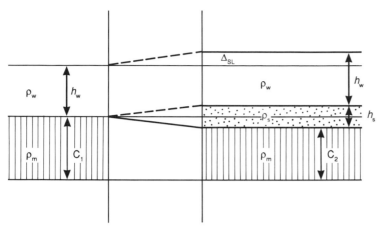

$$\rho_m C_1 + h_w \rho_w = \rho_m C_2 + \rho_s h_s + h_w \rho_w$$
$$\rho_m (C_1 - C_2) = \rho_s h_s$$
$$\text{Since } C_1 - C_2 = h_s - \Delta_{SL} \, ,$$
$$\rho_m (h_s - \Delta_{SL}) = \rho_s h_s$$
$$\therefore \Delta_{SL} = h_s (1 - \rho_s / \rho_m)$$

If ρ_m = 3300 kg m⁻³; ρ_s = 2500 kg m⁻³, then $\Delta_{SL} \approx 0.24\ h_s$
If 10 mm of sediment are deposited each 1000 yr, Δ_{SL} = 2.4 mm per 1000 yr

Fig. 6.18. Derivation of the sea level change resulting from deposition of sediment in the ocean.

total flux of sediment reaches the deep sea (Milliman and Meade 1981). If, for the sake of argument, the sediment accumulation rate in the oceans is 10 mm every 1000 years and if the average density of the oceanic sediments is 2500 kg m⁻³, the net increase in sea level is about 2.4 mm per 1000 years (derivation in Fig. 6.18) (cf. Holeman 1968). This is because the sediment depresses the oceanic lithosphere, but there is still a net displacement of water, causing the sea level rise.

Sediment is removed from the oceans at subduction zones. However, it is uncertain how much sediment is subducted and eventually melted to contribute to arc 'volcanism', and how much is tectonically accreted to the edge of the overriding plate.

If the length of today's subduction zones is 40 000 km and subduction is taking place on average at 60 mm yr⁻¹, the area of sedimentary cover being potentially consumed can be estimated. This can be compared with the volume of sediment accumulating each year on oceanic crust (10 mm thickness over an area of 3.2×10^8 km²). If the net influx per year of sediments is exactly balanced by their net removal by subduction, an average of well over 1 km thickness of sediment would need to be consumed at subduction zones

(Fig. 6.19). Since even the oldest ocean floor is generally covered by less than this, we can conclude that subduction as a process is easily *potentially* able to counterbalance the effects of sedimentation. If sediment is accreted to the overriding plate in an accretionary prism, less sedimentary material is subducted, swinging the two opposing processes of influx and removal more into balance. It is quite possible that the balance, when averaged over long periods of time, is insufficient to cause rates of sea level change of more than 1 mm per 1000 years (Pitman 1979).

6.3.3 Changes in mid-ocean ridge volume

That changes in the volume of the mid-ocean ridge systems could be responsible for major sea level changes has been argued for some time (Menard 1964, Hallam 1963, Russell 1968, Valentine and Moores 1970, Hayes and Pitman 1973). This is based on the observation that the volume of the present world ridge system (c. 1.6×10^8 km³) is a significant proportion, over 10 per cent, of the volume of ocean water (c. 1.35×10^9 km³). Because the bathymetry of the ocean is dependent on age, variations in spreading rates cause major changes in ridge volumes, the faster spreading

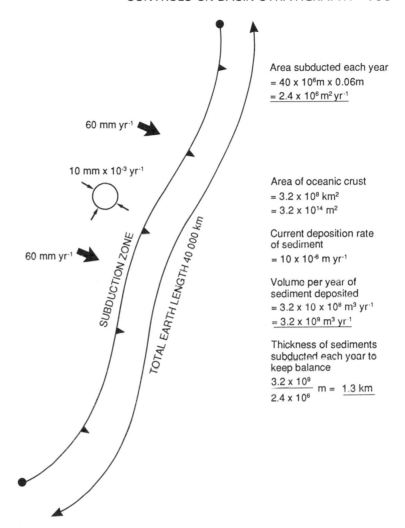

Area subducted each year
= 40 x 10⁶m x 0.06m
= 2.4 x 10⁶ m² yr⁻¹

Area of oceanic crust
= 3.2 x 10⁸ km²
= 3.2 x 10¹⁴ m²

Current deposition rate
of sediment
= 10 x 10⁻⁶ m yr⁻¹

Volume per year of
sediment deposited
= 3.2 x 10 x 10⁸ m³ yr⁻¹
= 3.2 x 10⁹ m³ yr⁻¹

Thickness of sediments
subducted each year to
keep balance
$\frac{3.2 \times 10^9}{2.4 \times 10^6}$ m = 1.3 km

Fig. 6.19. Simple calculation of amount of sediment potentially removed from the oceans along the Earth's subduction zones. For illustrative purposes only.

ridges being hotter and therefore more voluminous. A second way of modifying the world ridge volume is through changes in the total length of spreading axes. Hallam (1977) believes that at certain geological times, the latter mechanism was the more important of the two.

Bearing in mind that a change in ocean depth of magnitude Δh produces a change in sea level relative to the centre of the Earth of magnitude $\Delta_{SL} = 0.7\Delta h$, the sea level fluctuations resulting from mid-ocean ridge volume changes can be estimated if the shape of the ocean basins is known (Pitman 1979). The latter can be gauged from the *hypsometric curve.*

Figure 6.20 shows the profile of a ridge which has been spreading at 20 mm yr⁻¹ for 70 Myr. At $t = 0$ Myr the spreading rate increases to 60 mm yr⁻¹ so that at $t = 70$ Myr the ridge has a cross-sectional area which is precisely three times the original. The ridge then reverts to spreading at 20 mm yr⁻¹. The expected sea level changes to accompany this mid-ocean ridge history are a gradual increase in sea level as the ridge grows to its new equilibrium over 0–70 Myr, then a rapid fall as the ridge reverts to the slower spreading rate. In both the case of the periods of rise and fall of sea level, the maximum rate of sea level change is immediately after the spreading rate change (Fig.

Time

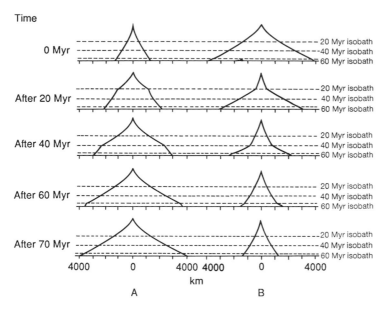

Fig. 6.20. Volume changes of spreading ridges as a function of spreading rate (after Pitman 1979). Sequence A shows a ridge that has been spreading at 20 mm yr^{-1} for 70 Myr. At 0 Myr the spreading rate changes to 60 mm yr^{-1}. Subsequent diagrams show the sequential changes in the ridge cross-section. Sequence B shows the reverse evolution; a ridge that has been spreading at 60 mm yr^{-1} for 70 Myr slows at 0 Myr to 20 mm yr^{-1}. The cross-sectional areas of equilibrium ridge profiles (after 70 Myr of spreading at the new rate) change by a factor of three in both cases.

6.21). This maximum rate is about 4 mm per 1000 years for the situation in Fig. 6.21.

Constructing an accurate sea level curve for the last 85 Myr based on ridge volumes depends on:
1 Knowledge of the spreading history of mid-ocean

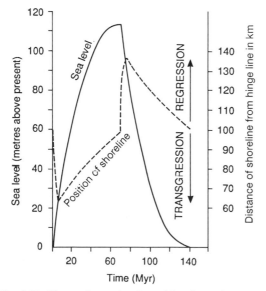

Fig. 6.21. Change in sea level resulting from the changes in spreading rates shown in Sequence A of Fig. 6.20. The maximum rate of sea level change occurs immediately after the change in spreading rate (after Pitman 1979).

ridges back to 155 Ma (since it requires 70 Myr for equilibrium). Some of the older ridges have probably been subducted, and new ones formed.
2 While some ridge segments may have expanded, others may have contracted. This lack of synchroneity must be addressed in order to estimate *net* volume changes.
3 Reliability of the magnetic polarity time scale used to estimate spreading rates. Magnetic anomalies need to be timed precisely, which commonly means dating the oldest overlying sediments palaeontologically. Palaeontological ages are then calibrated with absolute ages.

The postulated sea level changes as a result of ocean ridge volume fluctuations (Fig. 6.22, Larson and Pitman 1972) shows a maximum sea level in the latest Cretaceous (Maastrichtian) when it was elevated some 350 m above present, and a fall of varying severity through the Cenozoic to the present. The maximum rate of sea level fall based on these figures is thought to be nearly 7 mm per 1000 years, considerably greater than the other mechanisms outlined above, but still rather slow compared to glacially-forced rates (Section 6.3.4). The Late Cretaceous maximum of 350 m above present sea level is, however, in disagreement with estimates from continental margin stratigraphy (Watts and Steckler 1979), from estimates of the flooding of continents using individual hypsometric curves (Bond 1978) and from estimates of the amount of

Fig. 6.22. A comparison of estimates of sea level changes in the last 160 Myr after Watts and Steckler (1979). The heavy solid line is based on the subsidence history of boreholes from the continental margin of eastern North America (Watts and Steckler 1979). The dashed line is the estimate of Vail *et al.* (1977) based on patterns of coastal onlap recognized on seismic reflection lines, calibrated by the data of Pitman (1978) (circles) based on changes in the rates of spreading of mid-ocean ridges. The fine solid line is from estimates of the amount of flooding of the continental area of North America (Wise 1974; see also Fig. 6.17).

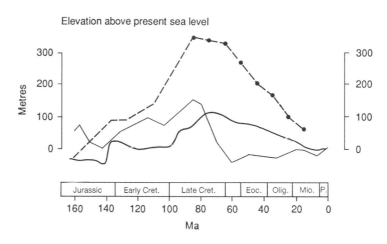

crustal generation through time (Parsons 1982), which all suggest a Late Cretaceous maximum of considerably less (< 100 to 150 m) (Fig. 6.22).

We shall later see how a synthetic stratigraphy can be constructed by comparing rates of subsidence, sedimentation and sea level change (Section 6.4.2).

6.3.4 Glaciations and desiccations

Large sea level changes can be caused by abstraction of water from the oceans into land based ice-sheets, and of course, the subsequent release on melting. Ice shelves are unimportant since the floating ice displaces its own mass of water. These changes can be called *glacio-eustatic* (Fairbridge 1961, Donovan and Jones 1979). The lithosphere also responds to changes in the water and ice loads; these adjustments are termed *glacio-isostatic* (Clark, Farrell and Peltier 1978, Peltier 1980) (see also Section 2.2.6 on glacial rates).

The volumes of ice sheets have been estimated, for example, in the Antarctic (Shumskij 1969). Total melting of the Antarctic land ice, some 2 to 3 × 10[7] km[3], would result in an increase in water depth of between 60 and 75 m. Melting of the much smaller Greenland ice cap would probably caused an additional 5 m increase in water depth. Allowing for the water-loaded depression of the ocean floor discussed above (Fig. 6.16), the actual sea level rise associated with the complete melting of today's land-based ice

sheets is expected to be in the region of 50 m. Calculations of former ice sheet volumes are speculative. Assuming that former ice sheets reached an equilibrium maximum thickness, and delineating their maximum limits from Quaternary geology, the sea level fall resulting from locking up of water in Pleistocene ice sheets is thought to have been about 100 m. The total change in sea level corresponding to the removal of Pleistocene-scale ice sheets is therefore about 150 m and liable to have major impacts on basin stratigraphy.

How fast do ice sheets form and melt? In geological terms this is a rapid process. A post glacial melting is thought to have produced an average sea level rise of 10 mm yr[−1], and sea level falls accompanying ice sheet expansion are thought to be even more rapid (Hays, Imbrie and Shackleton 1976, Imbrie 1982). The rates are about three orders of magnitude greater than those expected to result from changes in ocean ridge system volumes.

Glacio-isostatic changes in the volumes of the ocean basins are also thought to result in rapid sea level changes, from 1 to 10 mm yr[−1] (Clark *et al.* 1978) (see also Section 2.2.6). In any locality, there is therefore in the history of deglaciation a competition between rebound rates and rates of eustatic sea level rise (Belknap *et al.* 1987 for an example from the Late Quaternary of Maine).

Late Quaternary and Holocene sea level fluctuations following the Pleistocene glaciation are relatively well understood compared to more ancient sea level charges. Sea level curves derived from oxygen isotopes from benthonic foraminifera recovered in deep sea cores (Shackleton 1977, Shackleton and Opdyke 1973) and tropical coral terraces and related fauna (Fairbanks and Matthews 1978, Aharon 1983) show a high frequency fluctuation, with 8 sea level maxima (and 8 minima) in the last 120 000 years. These sea level changes vary from 20 m to 180 m in total height. They have a long primary period of about 10^5 years, and shorter secondary periods of about 40 000 and 20 000 years. The strongly periodic nature of the deep sea isotope record and its correlation with dates on elevated coral reel terraces has been put forward as supporting a control by variations in the Earth's orbit, that is, a Milankovitch control (Hays *et al.* 1976) (see

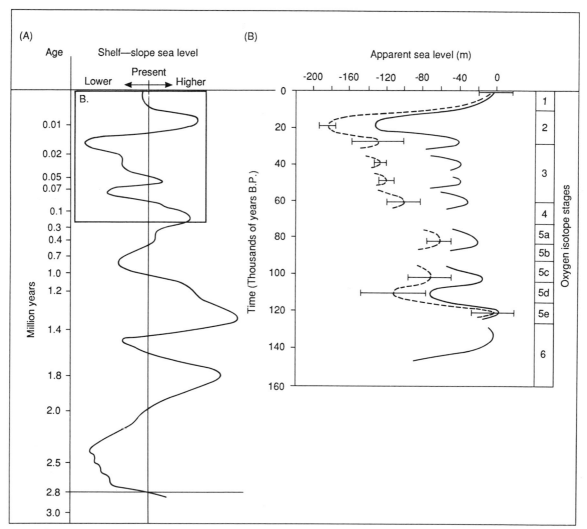

Fig. 6.23. (A) Inferred Pleistocene sea level fluctuations in the northwestern Gulf of Mexico since 3 Ma (after Morton and Price 1987, p. 183). Derived by Smith (1965) from interpreted water depth and palaeotemperatures and modified by Beard *et al.* (1982). Note nonlinear scale. (B) Detailed glacio-eustatic record (dashed line, Moore 1982) and oxygen isotope record (based on deep sea benthonic foraminifera) (solid line, Williams 1984) for the last 130 000 yr (after Suter, Berryhill and Penland 1987, p. 200).

Section 6.5.6) (Fig. 6.23). There is considerable agreement between the oxygen isotope record and the Late Quaternary stratigraphy of coastal regions, the Louisiana continental shelf being an excellent example (Suter, Berryhill and Penland 1987).

It has been shown that the Mediterranean Basin dried out during the late Miocene (Hsu, Ryan and Cita 1973 and Hsu 1979) as it became isolated from the world's oceans (see also Section 6.5.2). This would have caused a sea level rise, perhaps of 12 m, in the remaining oceans. This is considerably smaller than the variations caused by glaciations, and there is no reason to believe that the geological past was characterized by desiccation events greater than the Messinian salinity crisis of the Mediterranean region. However, rates of change of sea level would be rapid and similar to those estimated for glaciations (c. 10 mm yr^{-1}) (Donovan and Jones 1979).

6.4 EUSTATIC CHANGES IN SEA LEVEL: STRATIGRAPHIC RESPONSE

The stratigraphic response to postulated eustatic sea level changes has been investigated from two points of view:
1 Interpretations of stratigraphy, principally at passive margins, and construction of a sea level curve a posteriori.

2 Simulations based on an a priori sea level model.

6.4.1 Seismic stratigraphic methods of Vail *et al.*

Using seismic reflection results, a group of geologists from Exxon, headed by Peter Vail, have constructed a chart of relative sea level through time. This chart, colloquially known as the Vail curve, is based on the concept of depositional sequences and the kinds of baselap and toplap at sequence boundaries. It is composed of *cycles of relative change of sea level* during which relative rises or falls take place. In detail, gradual cumulative rises consist of smaller scale rises separated by stillstands. These cycles are known as *paracycles*. On a much larger time scale are *supercycles* involving successive rises to higher relative positions of sea level, followed by one or more major relative falls (Fig. 6.24).

Vail *et al.* (in Payton 1977) believed that the most reliable indicators of relative changes in sea level were the depositional limits of onlap and toplap within the coastal facies of marine sequences. Although subsequent studies (e.g. Haq, Hardenbol and Vail 1987; Section 6.6) have refined views, the 1977 study remains a fundamental basis for the study of sequence stratigraphy.
1 A relative rise of sea level is indicated by *coastal onlap* (Fig. 6.25), i.e. the progressive landward

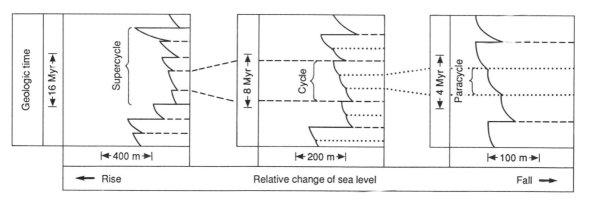

Fig. 6.24. The cycles of relative sea level change envisaged by Vail *et al.* (in Payton 1977). These cycles in fact monitor relative changes in the position of coastal onlap. The true relative sea level fluctuations would be less asymmetrical. Note the hierarchy of cycles from supercycles down to paracycles.

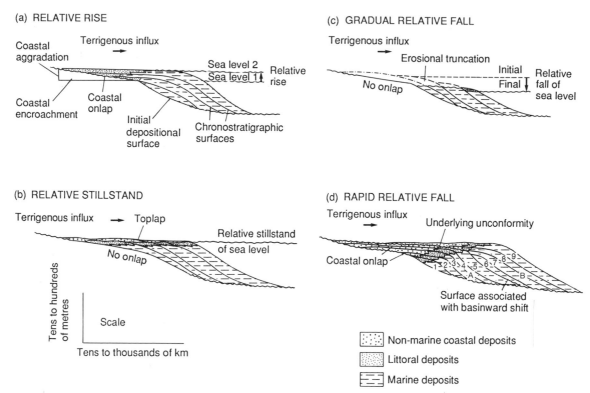

Fig. 6.25. The stratigraphic indications of relative sea level change as envisaged by Vail *et al.* (in Payton 1977). A relative rise causes onlap (a), a relative stillstand causes toplap (b), a gradual relative sea level fall causes offlap (c) and a rapid relative sea level fall causes a basinward shift in the position of coastal onlap (d). This occurs between the stratigraphic units 5 and 6 in (d), separating two depositional sequences A and B.

onlap of littoral and/or nonmarine coastal deposits. The rise in sea level, which acts as a base level, allows an increment of coastal deposition to be accommodated. Either the vertical or the horizontal component of coastal onlap can be used to estimate the relative sea level rise. These components are termed coastal aggradation and coastal encroachment respectively. Effects of compaction and effects of basinward tilting of depositional surfaces need to be removed from these measurements of coastal onlap. Depending on the magnitude of the terrigenous influx to a continental shelf, transgression, regression or a stationary shoreline may result from a relative sea level rise.

2 A *relative stillstand* of sea level is indicated by *coastal toplap* (Fig. 6.25). It may result from sea level and the underlying initial surface actually remaining stationary, or it may result if both rise or fall at the same rate. If the sediment supply is

sufficient, coastal deposits can build up to effective base level, and build out seaward. This produces coastal toplap with successive terminations of strata against the upper depositional sequence boundary lying progressively seaward.

3 A *relative fall* is indicated by a *downward shift of coastal onlap*, which is a shift downslope and seawards from the highest position of coastal onlap in a given depositional sequence to the lowest position of coastal onlap in the overlying sequence (Fig. 6.25). Seismic reflection data suggest that the downward shift is very abrupt, and where accurately dated, the relative sea level fall was thought to be correspondingly rapid, occurring within a million years or less. Measurement of the amount of downward shift in coastal onlap is hampered by erosion of the underlying depositional sequence during relative sea level fall, differential subsidence causing a basinward tilting of stratigraphy, and difficulties in dating the oldest sediments in the youngest depositional sequence, which are often fully marine rather than coastal. Consequently, relative sea level falls are difficult to evaluate quantitatively. The sea level falls in the second generation of Exxon studies (Fig. 6.26) (Haq *et al.* 1987) are noticeably less severe.

At times of lower relative sea level, the exposed shelf may be incised or bypassed, and coarse clastic bodies (e.g. fans) may accumulate at the new basin margin. Exposure of the entire shelf has been demonstrated from studies of Late Quaternary stratigraphy (e.g. Louisiana continental shelf, Suter *et al.* 1987) to have occurred during the last glacial maximum. A renewed rise in relative sea level may result in the fan being buried by finer-grained sediments and the shelf may once again receive sediment as coastal onlap proceeds. By correlating sedimentary facies across basin margins therefore, the detailed response of sedimentary systems to relative sea level changes can be investigated. We shall return to this important subject in considering sedimentary models in Chapter 7.

A sea level curve can be constructed for a particular region by determining sequence boundaries, ages, areal distributions and presence of coastal onlap or toplap. The sequence can then be placed on a chronostratigraphic chart. The identification of the cycles of relative rise and fall of sea level and the measurement of their magnitudes is by means of coastal aggradation, coastal toplap, and downward shift in coastal onlap. Sea level curves from different regions can be compared to establish a global sea level chart that includes cycles found in the same chronostratigraphic position in a number of widespread regions.

The global cycle charts presented by Vail *et al.* (1977b) show cycles of three orders of magnitude (Fig. 6.24):

First-order cycles: two such cycles are recognized (Precambrian to Early Triassic and Middle Triassic to present day) of long duration (300 Myr and 225 Myr respectively).

Second-order cycles: 14 such cycles range in age from 10 to 80 Myr.

Third-order cycles: very many cycles (> 80) of short duration (1–10 Myr) are also recognized.

Whereas the first-order cycles are nearly symmetrical, the second and third-order cycles are strongly asymmetrical, with gradual rises and abrupt and rapid falls.

The relative sea level positions are shown on a scale of 1.0 to 0.0, with the highest sea level occurring at 65 Ma (end Cretaceous) and the lowest at 30 Ma (mid-Oligocene).

How much of the global sea level curve is eustatic and how much is due to other factors? The younger of the first-order cycles is very similar to the curve suggested by Pitman (1978) and Hays and Pitman (1973) derived from changes in mid-ocean ridge volumes (Section 6.3.3). The first-order cycles may therefore be truly eustatic, although there may be some circularity in this interpretation. However, the second and third-order cycles show significant departures from the mid-ocean ridge volume curve. Vail *et al.* (1977b) explained this as follows:

a Where the Vail curve shows a higher sea level than Pitman, it is likely that subsidence was outstripping an overall sea level fall, causing a relative sea level rise on the seismic sections.

b Where the Vail curve shows a lower sea level than Pitman, abstraction of water from the oceans by glaciation, not evaluated in the Pitman curve, may have taken place. Geological evidence of glaciations in the Phanerozoic is, however, sparse, and it may be necessary to search for other explanations in many cases, particularly in the Early Tertiary and Mesozoic.

In the first generation studies (Vail *et al.* 1977a, b) the *relative* changes of sea level indicated by coastal onlap (coastal aggradation) and offlap were interpreted to represent *absolute* cycles of sea level change. This interpretation has now been relaxed, and the eustatic curve provided by Haq *et al.* (1987) (Fig. 6.26) has less severe sea level falls. The eustatic sea level curve is presented as having two components; (1) a *long-term* eustatic curve that is essentially the first-order cycle of previous reports (Vail *et al.* 1977a, b) and equivalent to the sea level change resulting from changes in mid-ocean ridge volumes, and (2) a *short-term* eustatic curve which represents the smoothing of the onlap curve, so that eustatic highstands are placed prior to major basinward shifts in coastal onlap, implying that the latest onlapping sediments were deposited at the start of the falling limb of the eustatic sea level curve.

We return to the question of the causes of the sea level changes shown by Haq *et al.* (1987) in Section 6.5. A further discussion of concepts in sequence stratigraphy is found in Section 6.6.

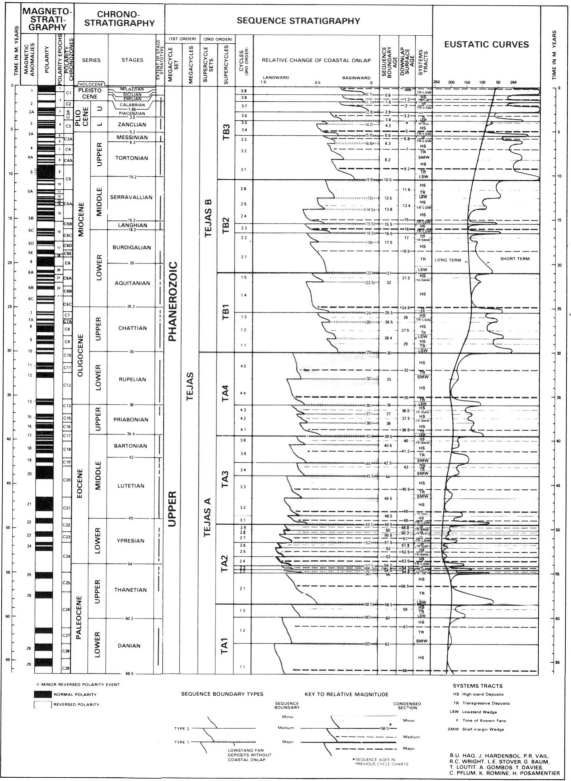

Fig. 6.26. The Vail–Haq curve. (Reproduced with permission from Haq *et al.* 1987).

6.4.2 Synthetic stratigraphy resulting from absolute variations in sea level

Sloss (1962) argued some time ago that transgressions and regressions of the shoreline were controlled (at least in part) by the relative magnitudes of the rate of sea level change and rate of subsidence. Pitman (1978) developed a model of a slowly subsiding continental margin with a more rapidly changing sea level and by treating the problem quantitatively was able to create a synthetic stratigraphy that could be compared with seismic reflection results.

The elements of the model together with the notation are shown in Fig. 6.27. A gently sloping continental margin subsides at a time-constant rate about a fixed hinge line, and the slope of the margin is maintained at a constant value by the balance of subsidence, sedimentation and erosion. Pitman's quantitative model assumes that erosion occurs landward of the shoreline and deposition occurs seaward. So that sediment could accumulate on the coastal plain, it was necessary to include an additional constant sedimentation term to the formulation. By plotting the movement of the shoreline as a function of subsidence rate and rates of sea level change, Pitman was able to simulate patterns of, for example, erosional truncation.

The factors that need to be considered in a quantitative stratigraphic model (Turcotte and Willeman 1983, Turcotte and Kenyon 1984) which accounts for a wide range of geological possibilities are:

1 *The type of sea level variation:* Turcotte and Kenyon (1984) and Turcotte and Bernthal (1984) considered two types (a) a rapid rise followed by a slow fall, and (b) slow rise in sea level followed by a rapid fall. The first sea level function involves an instantaneous rise of magnitude h_0, followed by a slow fall over a time period τ. Turcotte and Kenyon (1984) thought this to be akin to a glacially controlled variation but Quaternary sea level changes do not support this view. The second sea level function involves a slow rise of sea level of magnitude h_0 which then falls instantaneously. These two sea level functions can be regarded as idealized end-members. They are assumed to repeat themselves indefinitely in time.

2 *The rate of tectonic subsidence relative to the rate of sea level variation:* the sea level functions are superimposed on a constant velocity of basin subsidence U_0. This is equivalent to a linear long-term rise in sea level (Fig. 6.28). The amplitude of the sea level change, the rate of basin subsidence and the time period (τ) can be related as follows:

$$U_0\tau/h_0 = 1/f \qquad (6.2)$$

To put some physical significance into this expression, $U_0\tau$ is the amount of basin subsidence occurring in a single sea level cycle, so $U_0\tau/h_0$ is the *tectonic* subsidence as a fraction of the *eustatic* sea level change. Clearly, if the rate of basin subsidence is large compared to the magnitude of the sea level change, that is, $1/f > 1$, the basement surface experiences a continuously deepening water column with time and no unconformities need result.

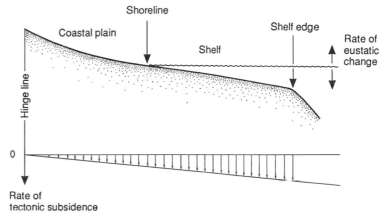

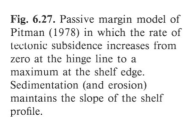

Fig. 6.27. Passive margin model of Pitman (1978) in which the rate of tectonic subsidence increases from zero at the hinge line to a maximum at the shelf edge. Sedimentation (and erosion) maintains the slope of the shelf profile.

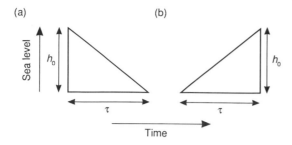

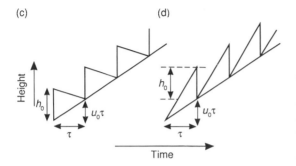

Fig. 6.28. Two types of sea level function considered to produce synthetic statigraphy (after Turcotte and Kenyon 1984, Turcotte and Willeman 1983): (a) instantaneous rise function and (b) instantaneous fall function in which the amplitude of the sea level change is h_0 and the period is τ. Height of sea level above basement is given for the two functions in (c) and (d) where the tectonic subsidence during one cycle is $u_0\tau$, and the tectonic subsidence as a function of the eustatic sea level change ($1/f$) is 3/4 in both cases.

In a sedimentary basin subjected to laterally varying subsidence rates, therefore, such as a passive margin or foreland basin, one might see unconformities bounding depositional sequences in areas of low tectonic subsidence, passing laterally into conformities where subsidence rates are higher.

3 *The rates of sedimentation and erosion relative to the rate of sea level variation:* sedimentation in the basin takes place at an aggradational velocity U_s. If sediment influx outpaces basin subsidence ($\delta = U_s/U_0 > 1$) and ignoring isostatic adjustments to the sediment load (Section 8.1), the sediment water interface rises relative to the original datum. If the reverse is the case (i.e. $\delta = U_s/U_0 < 1$), the depositional surface subsides relative to the datum. Having established the upward growing sediment surface resulting from the interaction of U_0 and U_s, its relation to the sea level change can be studied using the same philosophy as in (2). Erosion can be treated as negative sedimentation and the ratio of the erosion velocity compared to the basin subsidence velocity gives

$$E = U_e/U_0 \qquad (6.3)$$

If the magnitude of the sea level fall is greater than the rate of erosion, emergent topography will develop. If not, erosion will respond to the effects

of the falling sea level by bevelling the surface down to the base level. The interference of sea level change, tectonic subsidence, sedimentation and erosion is shown in Fig. 6.29. Products of the model are shallowing-up sequences followed by subaerial erosion for the 'instantaneous rise' function and denuded topography followed by a deepening-up sequence for the 'instantaneous fall' function (full details in captions). This clearly has major implications for the interpretation of sedimentary cycles such as those described in Chapter 7.2.

A synthetic stratigraphy can now be constructed by introducing a spatial scale to the model (Fig. 6.30).

(A) Firstly, consider the effects of a laterally varying tectonic subsidence (Fig. 6.30), from zero at the hinge line increasing linearly to a maximum $1/f = 2$ (this is so far an identical formulation to Pitman 1978), a situation closely analogous to a continental margin. Sedimentation rates are assumed to be large compared to tectonic subsidence, so that deep water sedimentation never takes place.

(B) Secondly, consider the effects of a variable sedimentation rate (Fig. 6.31). We assume a large-nearshore sedimentation rate decreasing offshore,

(a) Instantaneous rise function

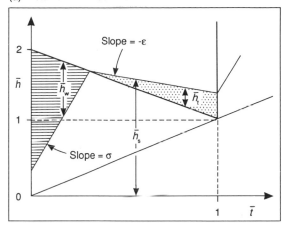

(b) Instantaneous fall function

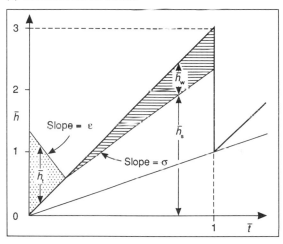

Fig. 6.29. The effects of rates of sedimentation and erosion on the instantaneous rise (a) and instantaneous fall functions (b). In (a) $1/f$ = 1/2, sedimentation outpaces subsidence by a factor (σ) of 4.5, and the erosion velocity is the same as the subsidence velocity (E = 1.0). Non-dimensional water depth $\bar{h}_w$ (= $h_w/U_0\tau$), topographic elevation $\bar{h}_t$ (= $h_t/U_0\tau$) *and sediment depth* $\bar{h}_s$ (= $h_s/U_0\tau$) as a function of dimensionless time $\bar{t}$ (= t/τ). A shallowing-up sequence results from the conditions in (a), whereas a deepening-up sequence results from the conditions in (b). After Turcotte and Kenyon (1984).

so that δ decreases linearly from 3.5 to 1.0, and a zero erosion rate E = 0. For the 'instantaneous rise' function, and $1/f$ = 2, continuous sedimentation occurs, producing prograding sigmoidal clinoforms. For smaller tectonic subsidence ($1/f$ = 1/2) coastal toplap is produced as a result of a slowly falling sea level. For the 'instantaneous fall' function, coastal onlap occurs in the region $1 < \delta < 3/2$ for the rapid subsidence rate ($1/f$ = 2), that is, relatively far offshore. But for lower tectonic subsidence rates, coastal onlap dominates the stratigraphy over a wide area of the shelf ($1 < \delta < 3$). With higher erosion rates, the geometry is modified to include a great deal more erosional truncation and the removal of coastal onlap pat-

terns (which requires low rates of erosion to enable a progressive encroachment to take place).

In summary:

• *coastal onlap* is strongly favoured by low or variable rates of erosion.

• *coastal toplap* can be produced by a slow fall in sea level at low erosion rates as well as by a stillstand.

The cyclic stratigraphy of shallow marine carbonate platforms has also been forward modelled (Read *et al.* 1986). For example, the 2200 Ma Rocknest platform of northwest Canada was modelled using a rapid rise–slow fall eustatic 100 000 years and an amplitude of less than 10 m (Grotzinger 1986a, b).

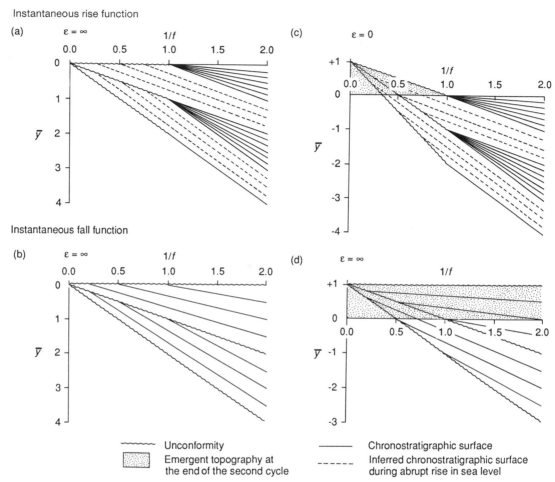

Fig. 6.30. Synthetic stratigraphic cross-sections constructed for two eustatic cycles using two sea level functions and a laterally varying subsidence rate. In (a) and (b) the erosion rate is very high so that no emergent topography is developed. In (c) and (d) erosion is assumed to be zero; the emergent topography existing at the end of the second cycle is stippled. In all cases, the rate of sedimentation is assumed to keep pace with the rate of sea level change. The horizontal scale of $1/f$ introduces the effect of a laterally varying subsidence rate from zero at the hinge line to a maximum ($1/f = 2.0$) at the right-hand edge of the sections. Note the strong onlap in (c) and (d).

6.5 SIGNIFICANCE OF THE VAIL/HAQ CURVE: TECTONIC VERSUS EUSTATIC CONTROLS

6.5.1 Reiteration of the problem

The fact that the stratigraphical packages in sedimentary basins, or depositional sequences, are characterized by a range of geometrical characters such as onlap, erosional truncation and toplap is now well established. This data base is derived largely from seismic reflection surveys of sedimen-

tary basins, particularly passive margins. What is not settled, however, is the significance of the baselap and toplap patterns in terms of mechanisms and, as a corollary, whether there is true global synchroneity to the excursions of the relative sea level curve. In this chapter we have investigated a number of processes which may lead to the preservation of distinct depositional sequences. These break into two main groups:

1 Truly eustatic, globally synchronous, sea level variations superimposed on a regime on basin subsidence, sedimentation and erosion.

Instantaneous rise function

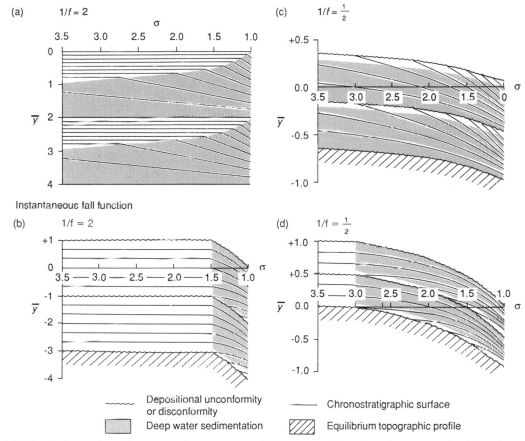

Instantaneous fall function

Fig. 6.31. Synthetic stratigraphic cross-sections constructed for two eustatic cycles using two sea level functions and a laterally varying sedimentation rate. The erosion rate is taken as zero ($E = 0$). The non-dimensional sedimentation rate (σ) varies from a maximum of 3.5 at the basin edge (left margin of section) to a minimum of 1.0 at the shelf edge on the right of the section. (a) and (b) are characterized by high rates of subsidence ($1/f = 2.0$), and (c) and (d) by low rates of basin subsidence ($1/f = 1/2$). Note the prograding clinoforms in (a), the toplap in (c), and the onlap in (b) and (d).

2 Relative changes in sea level caused by large scale tectonic factors such as changes in in-plane stress, flexural rigidity of the lithosphere or any redistribution of force systems applied to the lithosphere.

Although no consensus is currently possible, we can now briefly evaluate these varied processes. In so doing, it is informative to focus on the main criticisms of the Vail Curve as first published in Memoir 26 of the American Association of Petroleum Geologists in 1977. For our purposes, the most important have been:

1 The lack of incorporation of local and regional

(but not global) controls on sedimentation, such as tectonically driven subsidence.
2 The question of timing and global synchroneity of some of the events interpreted by Vail *et al.*

6.5.2 The synchroneity of global sequence boundaries

A revised set of sea level curves for the period from Triassic to the present has been presented by Haq *et al.* (1987) which sets them on a firmer chronostratigraphic basis. This is a necessary prerequisite for the global correlation of sequence boundaries

and therefore represents a test of their exact synchroneity. This chronostratigraphical basis is a combination of:
• radiometric dates
• magnetostratigraphy (geomagnetic polarity reversals)
• biostratigraphy.

A remaining problem is that Haq *et al.* (1987) use a combination of high pressure–high temperature radiometric dates in conjunction with low pressure–low temperature radiometric dates without clear consistency. In addition, the magnetostratigraphy differs from other well-established scales such as that of Berggren *et al.* (1985). Nevertheless, the new generation of cycle charts of sea level fluctuations is based on an improved chronology. It is also augmented by study of surface outcrops as well as subsurface (well-log and seismic) data. All of these outcrop studies are available in the public domain.

It is possible that some sequence boundaries which may be correlatable and synchronous in widely separated basins may not be perfectly and globally synchronous. Parkinson and Summerhayes (1985) point out the following possibility.

Consider three basins (A, B, C) each with its own subsidence history, but each with a decreasing rate of subsidence with time. This situation is analogous to the thermal subsidence of stretched basins (Chapter 3). If the eustatic sea level falls at a much faster rate than the subsidence rate at a particular point (such as the coastline) in the three basins, and there is a constant sediment supply, regression of the shoreline should result in each basin (Fig. 6.32). This will be a synchronous event. However, if at some later time the rate of fall of sea level changes to a lower value such that the new value is less than the subsidence rate in some of the basins (B,C) but still greater than the subsidence rate in others (A) then the shoreline should move landward in the former (B, C) but continue to be pushed seaward in the latter (A) which is subsiding more slowly. In other words there is a partial synchroneity of the sequence stratigraphy in response to the sea level change.

Since the subsidence rates are decreasing with time in all three basins, eventually the rate of sea level fall will once again exceed the rate of subsidence in all basins. A seaward shift of the shoreline should take place first in the basin B, then basin C (the fastest subsiding basin) but the onset of

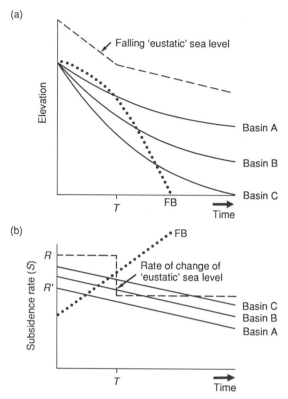

Fig. 6.32. Impact of a change in the rate of eustatic sea level change in sedimentary basins with different subsidence histories. (a) Subsidence history of three extensional basins showing exponential decrease in subsidence rate with time, and a foreland basin (FB) with an increasing subsidence rate with time. The falling eustatic curve has an instantaneous change in slope at time *T*. (b) Rates of subsidence and eustatic fall versus time. Where the rate of subsidence is greater than the rate of eustatic fall, transgression occurs (assuming no sediment influx); where the rate of subsidence is less than the rate of eustatic fall, regression occurs. The stratigraphic response therefore critically depends on the magnitudes of the rates of subsidence and eustatic change. After Parkinson and Summerhayes (1985).

regression does not take place at the same time. Not only is there a lack of synchroneity, but the seaward movement of the shoreline, which should produce coastal toplap or a downward shift in the position of coastal onlap, is accomplished at a time of constant rate of sea level fall.

The same concept can now be applied to basins with different driving mechanisms for subsidence;

for example, passive margins (Chapter 3) versus foreland basins (Chapter 4). Foreland basins do not experience the steady decline in the rate of subsidence so typical of basins on stretched lithosphere. Instead they commonly possess convex-up subsidence curves, indicating an accelerating subsidence (Allen, Homewood and Williams 1986). Placing this kind of subsidence signature on Fig. 6.32, it is clear that whilst the change from regression to transgression under the sharply falling rate of eustatic sea level change takes place synchronously with basins C and B, the accelerating subsidence rate means that the foreland basin experiences long-continued transgression. Only a massive increase in sediment supply will damp out this tendency, and this may only be accomplished after the associated orogenic wedge has developed a significant topographic expression. Once again, a change in the rate of sea level change may be responsible for *partially* synchronous stratigraphic responses in widely separated basins with different formative mechanisms.

The interaction of rates of tectonic subsidence and sea level change suggested here may explain some of the diversity of the coastal onlap curves from distinct basins, whilst maintaining the fundamental concepts upon which the Vail curve is built.

Having established this critique of the use of sequence stratigraphy in evaluating global sea level changes, it is now possible to summarize some of the different explanations for the cycles in the Haq/Vail curve.

6.5.3 First-order cycles

There seems little doubt that the first-order cycles, of long duration (300–225 Myr), are the result of long-term changes in the volume of mid-ocean ridges, itself a result of variations in spreading rates (Hays and Pitman 1973, Pitman 1978). Although the operation of tectonic processes in continental interiors (Bond 1978) and on continental margins (Watts and Steckler 1979) have been postulated, the primary signal most probably comes from variations in ocean ridge spreading rates.

Pitman's sea level curve from the late Cretaceous to the late Miocene shows a cumulative fall from 350 m to 60 m above the present position of sea level. It matches very closely the Tertiary segment of Vail *et al.*'s first order curve and Haq. *et al.*'s

'long-term' curve, both derived ostensibly from seismic stratigraphic data only. Others have argued, however, that the sea level change of about 300 m postulated by Pitman (1978) is excessive, preferring the figure of < 150 m for the sea level maximum in the Late Cretaceous above present day levels (Sections 6.3.3).

6.5.4 Second-order cycles

The second-order cycles (or 'supercycles') are of intermediate duration (10–80 Myr) and in general consist of slow rises in sea level followed by rapid sea level falls. There is some debate as to their origin.

The supercycles ('megasequences' of Hubbard *et al.* 1985) were delineated by Vail *et al.* (1977a) on the basis of patterns of coastal onlap. As we have seen, coastal onlap can result from a large number of processes, many of which should not be truly global. These include:

1 Coastal onlap may be initiated at the *rift-drift transition* during continental break up (Watts *et al.* 1982, Watts 1982). Cooling of a continental margin causes an increase in the flexural rigidity of the lithosphere. The thermal contraction and sediment loads therefore cause an increasingly large wavelength of flexural response, inundating the basin margin. Watts (1982) has pointed to the close correspondence of the timing of the rift-drift transition associated with the break-up of the Pangean supercontinent and the supercycles of Vail *et al.* (Fig. 6.33). Such tectonic factors cannot explain, however, the abrupt sea level falls terminating second-order cycles. For example, widespread Oligocene regressive sequences in several continents cannot be explained by the model of flexure of an elastic lithosphere.

2 *Plate boundary reorganizations* have been postulated as the causes of regional unconformities (Bally 1982) of the type that define depositional sequence boundaries. Plate boundary reorganizations are also responsible, it is thought, for changes in the state of in-plane stress. We have previously seen (Section 6.2.3) the effects of in-plane stress on basin stratigraphy, particularly on patterns of coastal onlap produced during in-plane compression. These are, however, short-term effects. It is possible that changes in in-plane stress are responsible for the termination of the second-order cycles

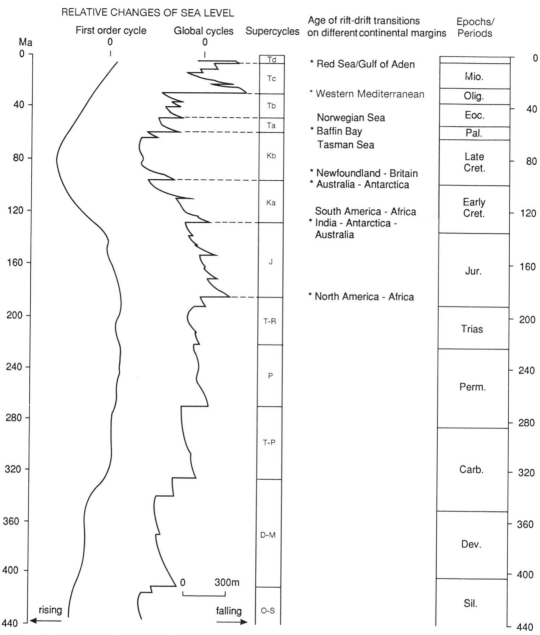

Fig. 6.33. Comparison of the second order cycles of Vail *et al.* (1977) and the age of the rift-drift transition at different continental margins (after Watts 1982).

since the expected result of a compressive in-plane stress on a stretched basin is to cause basin edge uplift.

3 *Glaciation and deglaciation* may account for some second-order cycles, especially those in the late Neogene. Sea level changes related to glaciation are rapid. This mechanism may explain the major Oligocene regressive event, but it is more problematic at geological times that lack evidence of widespread glaciation.

6.5.5 Third-order cycles

The third-order cycles of short duration (1–10 Myr) are also strongly asymmetrical, with slow rises in sea level followed by rapid falls. Any mechanism used to explain the occurrence of third-order cycles must account for their rapid repeat times.

Glaciation and deglaciation are associated with sea level changes as fast as 10 mm yr^{-1}, and major desiccations appear to occur at similar rates. Both are therefore good candidates for the causal mechanisms of third-order cycles. Glaciations have been documented in the Pleistocene, late and early Miocene and late Oligocene; these can be correlated with sea level falls at the end of third-order cycles. The Messinian desiccation event also appears at first sight to correlate with a sea level fall at 6.6 Ma. However, detailed work on ODP and DSDP cores and onshore studies in Sicily suggest that the desiccation 'event' postdates the Tortonian–Messinian boundary (now placed at 6.5 Ma) by 0.5 Myr, so that the correlation is inexact. The second point is more general, viz. that desiccating the Mediterranean by land-locking is likely to have caused a sea level *rise* in the remaining oceans by reducing the volume of the ocean basins (Section 6.3.4). In summary, it is most unlikely that every third-order cycle is related to climatic factors, especially where geological evidence of glaciations or desiccations is lacking.

In-plane stress modification of basin stratigraphy, particularly during compression, may result in an onlap pattern which is very similar to the third order cycles of Vail *et al.* (1977a, b). The rate at which the in-plane stress can change is a critical observation, since this rate should control the acceleration or deceleration of coastal aggradation. As yet, too little information is available on the rates at which such states of stress change, but indications are that they are accomplished on a time scale of a few million years. The combination of the likely magnitude of the sea level change (~100 m for a regional change of stress of a few kilobars), and the rapidity of the change (of the order of 10^6 years), makes this a suitable mechanism to cause the third-order cycles. This implies sea level changes of about 10 mm yr^{-1}, roughly comparable to those of glaciations and deglaciations.

6.5.6 A note on paracycles and PACs

Cycles of stratigraphy on an even finer scale have been identified from shallowing-upwards sequences separated by sharply defined surfaces marking an abrupt change to deeper water (Section 7.1.3.2). Such cycles are synonymous with the paracycles of Vail *et al.* (1977a), and give rise to *parasequences*, the smallest stratigraphic packages identifiable in a basin. The boundaries of parasequences, as in depositional sequence boundaries, are thought to be synchronous surfaces.

Parasequences have long been recognized in a variety of palaeoenvironmental settings ranging from fluvial to deep marine, and in both siliciclastic- and carbonate-dominated systems. Parasequences are small scale but fundamental building blocks of stratigraphy, generally up to 15 m in thickness.

In the Proterozoic Rocknest platform of northwest Canada (Grotzinger 1986a) the parasequences contain marker beds that can be traced for enormous distances (> 100 km) both perpendicular and parallel to strike. Various cycle types exist involving (1) packstone lags that form sheets covering the erosional tops of cycles, representing initial transgressive sediments over flooded tidal flat lithologies, (2) mixed carbonate-siliciclastic muds and silts representing maximum submergence of the platform, (3) thickly laminated dolosiltites formed close to fairweather wave-base, (4) stromatolitic dolomites of the upper subtidal to intertidal zone, and (5) cryptalgalaminites and tufas formed in upper intertidal and supratidal environments. 140 to 160 of these *progradational* shallowing-up cycles have been identified in the Rocknest. Recent studies of the Purbeckian of the Swiss and French Jura (Strasser 1988) reveals the same stratigraphical signal of small-scale shallowing-upwards cycles separated by surfaces due to rapid deepening.

Some workers believe, however, that shallowing-up cycles are produced primarily by *aggradation* rather than progradation (Goodwin and Anderson 1985). These cycles, termed *punctuated aggradation cycles* are 1–5 m thick in the carbonate-dominated U. Silurian–L. Devonian Keyser Formation of Pennsylvania.

Although there have been many recent documentations, the significance of this kind of episodic stratigraphic hypothesis to carbonate and clastic sedimentary sequences has been appreciated for some time, and is the basis for an understanding of, for example, the deepening-upwards cycles in the Triassic Lofer Limestone of the Alps (Fischer 1964), and the shallowing-upwards Carboniferous cyclothems of Kansas (Duff, Hallam and Walton, 1967).

The driving mechanisms of the paracycles must be of high frequency. Karner (1986), for example, has postulated changes in states of in-plane stress to explain these cycles. The synthetic stratigraphic model outlined in Section 6.4.2 using an 'instantaneous rise' function is, however, of special interest since it simulates the gradual shallowing and rapid deepening of many parasequences. A perfectly sinusoidal sea level fluctuation superimposed on a constant rate of tectonic subsidence is equally able to simulate cyclothemic sedimentation (Turcotte and Willeman 1983). The glacially-forced sea level changes of the late Quaternary (Section 6.3.4) have a similar frequency to those required to explain some parasequences. The time scale of the fluctuation is also reminiscent of the frequency of climatic change attributed to eccentricities in the orbital motion of the Earth (Hays, Imbrie and Shackelton 1976, Imbrie 1982). These variations are commonly termed *Milankovitch* cycles, and were originally proposed by Fischer (1964) to explain the Lofer cyclothems. Milankovitch climatic fluctuations therefore appear to be a powerful mechanism for producing parasequences (Grotzinger 1986a, b). Subsequent time-series analyses of the periodicity (e.g. Goldhammer, Dunn and Hardie 1987) confirm that the fluctuation fits with predicted Milankovitch frequencies. A number of periodicities occur, including that due to precession of the equinoxes (19 000 and 23 000 years), obliquity (41 000 years) and ellipticity (100 000 years). Ruddiman and MacIntyre (1981) believe the 23 000 year frequency to be dominant at latitudes south of 45°N. This figure is in close agreement with the 20 000 year frequency estimated from coral reef terraces (Turcotte and Bernthal 1984), but often the precision of dating ancient sequences is too poor to allow comparison of paracycle durations with Milankovitch frequencies.

6.6 FURTHER CONCEPTS IN SEQUENCE STRATIGRAPHY

We have seen (Section 6.1) that the fundamental unit of stratigraphy that has chronostratigraphic significance is the *depositional sequence*. These units, which are bounded top and bottom by unconformities or their correlative conformities, can be subdivided into smaller units of stratigraphy which have distinct stacking patterns of chronostratigraphic increments. These small units are termed *systems tracts* (Van Wagoner *et al.* 1988), and they are themselves composed of *parasequences* (Section 6.5.6). Parasequences are relatively conformable successions of genetically related beds which are bounded by surfaces which reflect abrupt changes in water depth.

The location and nature of depositional systems has been related to relative sea level fluctuations by the Exxon group (Van Wagoner *et al.* 1988, Posamentier, Jervey and Vail 1988, Posamentier and Vail, 1988). The significant departure of these studies from the pioneering works of Vail *et al.* (1977a, b) is that the *rate* of sea level change is a critical parameter in determining the stratigraphic response, particularly in relation to the *rate* of subsidence. A new model, which in terms of its assumptions is very similar to that of Pitman (1978) (Fig. 6.27, Section 6.4.2), can be set up as follows.

Let us assume that there is a true eustatic (i.e. absolute) sea level fluctuation as sketched in Fig. 6.34. The relative sea level change at the sea floor, assuming no sedimentation, will be the sum of this eustatic component and the subsidence. That is, eustatic sea level rises will be amplified by the subsidence, eustatic sea level falls will be damped by the subsidence. For small time periods, the subsidence can be treated as linear with time.

Now consider the rates at which these processes take place. The rate of eustatic change is zero at the high points and low points of the eustatic curve, and the rate of change is greatest at the inflection points of the curve. For those familiar with calculus, we were merely making the observation that the slope or rate of change is the first differential of the eustatic curve. The rate of change of the subsidence is constant over time because the subsidence is linear. These two rates can now be summed to obtain the curve for the *rate of relative sea level change*. For a situation where no sediment influx is taking place, and ignoring the effects of the isostatic response to the water load, this latter curve gives us the rate at which new space is available to be filled by sediment.

Bearing this in mind, it is possible to introduce a further parameter, the *sediment supply*, in order

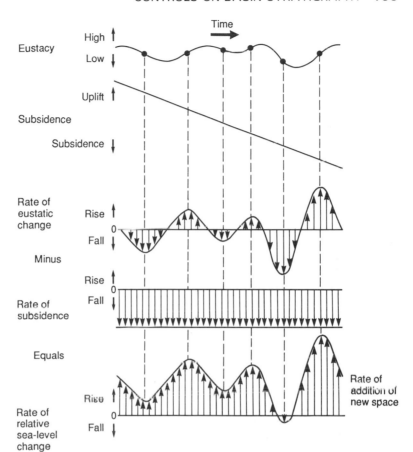

Fig. 6.34. Relative sea level as a function of eustatic change and tectonic subsidence. The stratigraphic pattern is strongly influenced by the resultant rate effects of eustatic change, subsidence (shown in this figure) and sedimentation (not shown). After Posamentier *et al.* (1988).

to model the stratigraphic response. Such a response must be viewed of course in terms of the three-dimensional assemblages of lithofacies comprising *depositional systems* (see also Chapter 7), and the linkage of contemporaneous depositional systems into *systems tracts* (Brown and Fisher 1977) which correspond to particular segments of the relative sea level curve (Posamentier *et al.* 1988). The geodynamic setting which is customarily used for this kind of approach is a passive margin with a well-developed shelf, slope and basin, a seaward increasing subsidence rate, and a time-constant sediment supply. Whilst a constant sediment supply is clearly a poor assumption, the effects of variation in sediment supply are unlikely to be felt strongly at the landward termination of stratal packets, since here the pattern of onlap and aggradation is more likely controlled by the base level (sea level or graded stream profile). However, variations in sediment supply should exert themselves more strongly on the seaward extent of deposition.

The sediment supply interacts with the rate at which new space is made available in the following manner. If the rate of addition of new space slows down, the sediment supply will easily fill the available space. Aggradation will, however, slow down and surplus sediment will begin to be 'exported' or 'bypassed' offshore, leading to progradation. If the reverse happens and the rate of new space available increases, sediment deposition will be restricted close to the supply point, terminating progradation. The period of greatest increase in new space available is at the *inflection point* of the *rising limb* of the curve of rate of relative sea level change, causing increased aggradation but no progradation. The period of greatest decrease in the new space available is at the inflection point of the *falling limb* of the curve of rate of relative sea level (Fig. 6.34), leading to progradation. Progradation and aggradation are therefore inversely related for a constant sediment supply.

Since the inflection points on the falling limb of the rate of relative sea level curve mark times of least available new space, sedimentation commonly shifts to the basin plain and the shelf is subjected to erosion. An unconformity is produced and is accompanied by a basinward shift in coastal onlap and a cessation of fluvial deposition. If the rate of relative sea level fall is rapid, streams are rejuvenated and incise into the former shelf giving *Type 1 unconformities* (Fig. 6.35). However, if the rate of relative sea level lowering is slow, there is a gradual but widespread denudation without substantial river incision, giving *Type 2 unconformities.*

6.6.1 Relation of systems tracts to a relative sea level cycle

Systems tracts can be broadly divided into the following classes according to their relationship to specific segments of the relative sea level change curve (Fig. 6.36):

1 *Lowstand systems tract*, comprising (a) a *lowstand fan* deposited during a time of rapid relative sea level fall, (b) a *slope fan* deposited during the late relative sea level fall and early relative sea level rise, and (c) a *lowstand wedge* deposited during the late relative sea level fall or early rise. Where sedimentation is transferred not as far as the basin floor or slope, a *shelf margin* systems tract may develop at times of falling but decelerating relative sea level.

2 *Transgressive systems tract* formed during a rapid relative sea level rise.

3 *Highstand systems tract* deposited during the late part of a relative sea level rise, a relative sea level stillstand and the early part of a relative sea level fall.

6.6.1.1 Lowstand systems tract

When the relative sea level fall is rapid, no space is available for further sedimentation, the former shelf is incised by streams and sedimentation is transferred to the basin floor and slope. Base-of-slope fans (*lowstand fans*) are nourished by sediment bypassed through the shelf and slope by valleys and canyons. The base of the lowstand fan is therefore a Type 1 sequence boundary. *Slope*

fans result from deposition on the middle or base of the continental slope and may be coeval with the basin-floor fan. *Lowstand wedges* are characterized by onlap onto the slope and at the same time by progradation and downlap onto the previous basin-floor or slope fans. They are deposited at times of low but very slowly changing sea level, particularly during a slow relative rise. Lowstand fans, slope fans and lowstand wedges are all associated with underlying Type 1 sequence boundaries.

When the relative sea level fall is gradual, sedimentation may be progressively shifted basinwards to the shelf edge where both onlap in a landward direction and downlap in a basinward direction takes place. The base of the *shelf margin* systems tract is therefore a Type 2 sequence boundary.

6.6.1.2 Transgressive systems tract

During a rapid relative sea level rise, the underlying lowstand or shelf margin system tracts are transgressed (the transgressive surface). Sets of parasequences making up the transgressive system tract are commonly retrogradational, that is, they back-step onto the basin margin, with strong onlap in a landward direction and downlap onto the transgressive surface in a basinward direction. As the rate of relative sea level change slows down, the sets of parasequences change from being retrogradational to being aggradational, the surface at which this occurs being that of the *maximum flooding* (Fig. 6.36). *Condensed sections* occur in the basin during times of transgression.

6.6.1.3 Highstand systems tract

After the maximum flooding, the relative sea level rise slows and sets of aggradational parasequences are succeeded by progradational parasequences with clinoform geometries. These highstand systems tract parasequences onlap onto the underlying sequence boundary in a landward direction and downlap onto the top of the transgressive systems tract or lowstand systems tracts in a basinward direction, so that there is a prominent *downlap surface* below a highstand system tract. Condensed sequences also occur during the early stages of a highstand systems tract before major progradation takes place (Loutit *et al.* 1988). Highstands offer the possibility of thick subaerial deposition of

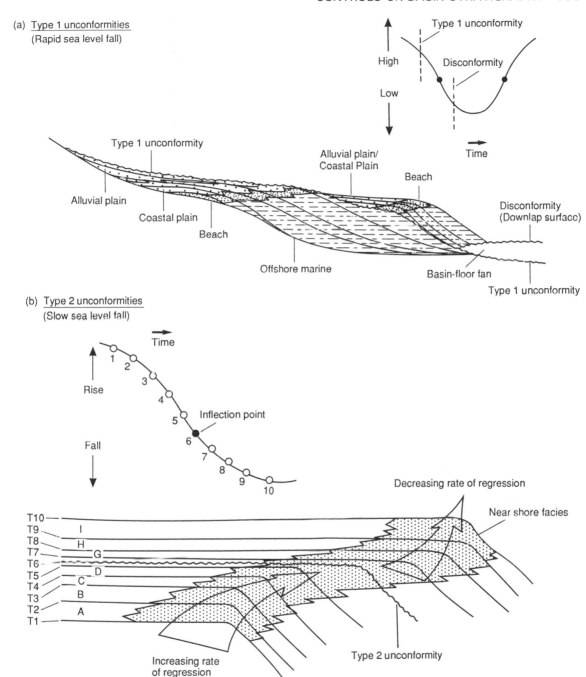

Fig. 6.35. Type 1 and Type 2 unconformities (after Posamentier *et al.* 1988). Type 1 unconformities (a) erode into the older shelf sediments and may underlie basin-floor fans. They are believed to be formed during a rapid sea level lowering. The sediments of the second depositional sequence downlap onto a disconformity or onlap the Type 1 unconformity. Type 2 unconformities (b) do not strongly erode the underlying depositional sequence, but signal a change from prominent progradation to aggradation as the rate of regression slows. Overlying sediments may onlap the Type 2 boundary landward of the shelf edge, although this is not illustrated above (see Fig. 6.36). Type 2 unconformities are interpreted to form at the inflection point of a slow relative sea level fall.

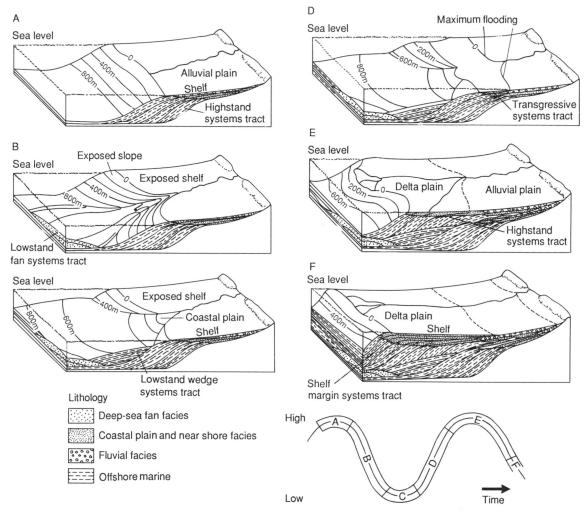

Fig. 6.36. Characteristic systems tracts and their relation to the eustatic curve, according to the Exxon group (Posamentier *et al.* 1988). These block diagrams are extremely idealized and have a strong vertical exaggeration.

fluviatile sediments. Initially, as much new space is added, rivers will aggrade with a vertical stacking pattern. However, as a final equilibrium or 'graded' profile is reached, the rate of new space added reduces, encouraging lateral stacking patterns. Widespread fluviatile deposition should cease at the inflection point of the falling limb of the relative sea level curve.

There is an urgent need for these concepts to be rigorously tested by observational studies using either high resolution seismic reflection data or detailed surface geology. There is also a need to

consider other basinal settings other than the siliciclastic shelf-basin floor type. There is a growing number of cases where these concepts have been applied to seismic reflection data, but the resolution of the data makes the checking of details impossible. One example of a detailed surface study is that of the Eocene Hecho Group of the Spanish Pyrenees, where excellent biostratigraphical control allows fluviatile to shelf to basin floor variations in depositional systems to be evaluated in the context of relative sea level change (Mutti *et al.* 1985).

6.6.2 Sequence stratigraphy of the Eocene Montañana and Hecho Groups, Spanish Pyrenees

The Eocene foreland basin of the south-central Pyrenees developed in two main phases, separated by a period of tectonic deformation in the Bartonian (late middle Eocene) which shifted depocentres southwards. The older, *inner basin* has a late-Palaeocene to late middle Eocene fill of fluvio-deltaic sediments in the east (Montañana Group, Nijman and Nio 1975) and turbiditic marine sediments in the west (Hecho Group, Mutti *et al.* 1972). This basin has three main palaeogeographical sectors (Fig. 6.37):

1 An eastern sector contains thick sequences of fluvial, deltaic, nearshore and shelfal deposits (Montañana Group). These sediments accumulated on top of a southward-transported thrust sheet known as the Cotiella unit. The western margin of this sector was delineated by the lateral

ramp of the Cotiella Unit.

2 A central sector roughly divided in two by a north-trending anticline (Boltaña High). East of the Boltaña High the sector is characterized by channel-fill turbidite bodies and thick intervening mudstones. West of the anticline the sediments consist of unchannellized sandstone lobes. However, the Boltaña only existed as a palaeohigh during the deposition of the youngest two depositional sequences.

3 A western sector contains a thick succession of basin plain turbidites with intercalations of resedimented carbonate units.

The Eocene strata can be divided into seven main depositional sequences within which it is convenient to recognize three main facies associations (1) alluvial, nearshore and shelfal, (2) marine mudstones and (3) turbidites. The most informative relationships from a sequence-stratigraphic viewpoint occur in the Hecho Group. In the

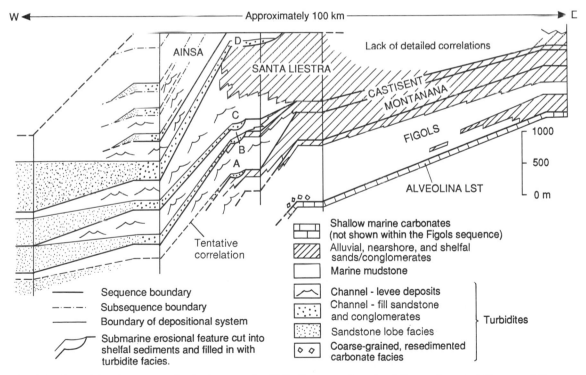

Fig. 6.37. Depositional sequences in the lower and middle Eocene sediments of the south-central Pyrenees (after Mutti *et al.* 1985). The sequence boundaries incise the shelf in the east and underlie basin-floor sandy turbidites in the west. The stratigraphic cross-section represents the transition from the eastern (Montañana) and western (Ainsa and Hecho) sectors of the south Pyrenean foreland basin. This hinge zone was tectonically controlled.

eastern part of the central sector, four main standstone-rich turbidite depositional systems are separated vertically by thick wedges of locally slumped, fine grained and thin-bedded turbidites that thin out to the west. These fine grained units are thought to be channel-levee complexes developed near the margins of actively prograding deltaic systems.

Within the sand-rich systems there are marked changes from east to west. In the east the sediments are dominated by channelled sandstones and conglomerates filling large scale submarine erosional features up to 4–5 km wide and over 200 m deep. In the west, this facies association is replaced by unchannellized sandstone lobes that build laterally extensive (up to several tens of km) tabular units.

The sandstone-rich systems of the Hecho Group in the west are correlative with the basal unconformities of the depositional sequences in the east.

In addition, channel-levee complexes that separate vertically the different sandstone-rich systems are correlative with the fluvio-deltaic deposits of the Montañana Group in the eastern sector. Turbidite deposition therefore appears to have taken place at both highstands and lowstands of sea level (Mutti 1985):

• at lowstands; turbidite systems were sand-rich owing to subaerial or submarine erosion of the deltaic shelves

• at highstands; fine grained sediments accumulating along the margins of actively prograding deltaic systems were resedimented as muddy turbidites down the palaeoslope to the west, forming channel-levee complexes.

The Eocene of the south-central Pyrenees therefore provides an excellent example of the sedimentological and stratigraphic response to relative sea level change.

7 The basin-fill: depositional style

Earth felt the wound, and Nature from her seat
Sighing through all her works gave signs of woe
That all was lost.

(Milton, *Paradise Lost*)

SUMMARY

Sediment dispersal links depositional environments into depositional systems, and changes in one part of the system are generally felt in other parts of the system. Depositional systems and their sedimentary products reflect the sum total of autogenic (internal to the system) and allogenic (external to the system) controls.

Depositional systems are highly variable. Terrestrial systems in basins with through-drainage are dominated by well-established river systems and perennial lakes. Basins with internal drainage are characterized by ephemeral river systems, generally shallow and strongly fluctuating lakes, continental sabkhas and deserts.

Fluvial depositional systems are extremely sensitive to autogenic and, especially, allogenic controls such as climate and tectonics. These controls determine run-off (discharge), sediment load, slope and therefore fluvial style. Braided systems are favoured by high and variable discharges, steep slopes and large amounts of bedload. High-sinuosity, meandering systems are common in regions of low slope, equable discharge and relatively more suspended load. Ephemeral systems are rarely able to establish an 'equilibrium' fluvial style. Terminal fans and 'goose-foot' patterns are found in ephemeral systems. The effect of extensional neotectonics on fluvial systems is principally to influence the position of avulsed channels, and field studies support the computer-simulated alluvial stratigraphic patterns of Bridge and Leeder (1979). Compressional neotectonics commonly causes barriers to regional downslope fluvial transport, deflecting rivers for considerable distances along tectonic strike. Glacio-eustatic changes also have a profound influence on the entire fluvial depositional system by varying base levels. Fluvial facies models have reached a relatively high level of sophistication. Meandering and braided rivers have well-established facies models, but there is a limited data base on anastomosed rivers.

Aeolian and especially lacustrine depositional systems are very sensitive to climatic change. Much interest is currently focused on aeolian systems and the first models of aeolian–fluviatile and aeolian–lacustrine interaction are being developed. The fluctuating water levels and water chemistries of lakes emphasize the close dependence of facies on climate. A number of distinctive facies typify the central lake and lake margin in hydrologically open and hydrologically closed systems.

Coastal depositional systems are a response to the competing effects of fluvial input, and wave power and tidal energy of the receiving basin. Deltas are accordingly traditionally classified as fluvial-dominated, wave-dominated or tide-dominated. The power of the river, profile of the nearshore region and the basinal wave and tidal energies control the dynamics of the dispersion of fluvial outflows and hence river mouth processes and depositional patterns. Deltas have well-established facies models centred on the coarsening-upward sequence caused by delta progradation. Non-deltaic coastlines are dominated by beaches, barrier islands and cheniers. In areas of moderate tidal range, barrier islands become dissected by tidal channels and in macrotidal environments estuaries are common. Coasts with barriers produce distinctive sequences depending on whether preservation results from transgression, regression or barrier-inlet migration.

Arid shorelines are typified by the coastal carbonate–evaporite sabkhas of the Persian Gulf, although siliciclastic sabkhas are also found. Conspicuous elements of the Persian Gulf-type sabkha system are fringing coastal barriers and inlets, back barrier lagoons, and algal flats in the intertidal-supratidal zone. Evaporites grow within and as encrustations on supratidal sediments, and also in small ponds.

The shallowing-upward sequence is typical of coastal to nearshore carbonates and evaporites. The sequence usually has four parts: (1) a basal high-energy transgressive deposit, (2) subtidal carbonates, (3) intertidal low-energy (stromatolitic) or high-energy (beach) deposits, (4) supratidal to terrestrial sediments which are evaporitic in arid climatic zones. Both an allocyclic eustatic mechanism and an autocyclic mechanism based on the size of the subtidal source area of sediment have been proposed to explain these shallowing-upward carbonate cycles.

Siliciclastic shelf depositional systems contain modern, relict and reworked (palimpsest) components. They may be storm-dominated, tide-dominated, or oceanic-current dominated in order of present day frequency of occurrence. Facies models for siliciclastic shelves suffer from a poor dovetailing of geological observations and those of physical oceanography. Models of tidal sandwave deposits are now sophisticated but tidal sand ridges are poorly known. Storm shelf models are plagued by the paradigm of hummocky cross-stratification and the delayed consideration of the effects of sea level change. The upward coarsening sequence from turbidites to hummocky cross-stratified sandstones to beach and fluviatile units has been presumed to be due to delta or coastal progradation over a storm dominated shelf and shoreface. Alternative facies sequences due to the migration of storm-driven linear sand ridges have been proposed.

The continental shelf in subtropical latitudes is the site of widespread carbonate sedimentation. Rimmed shelves shelter protected lagoons, and their margins commonly drop precipitously into the abyssal depths. Open shelves, however, lack a guarding rim and have the form of a ramp. Relative changes of sea level cause profound changes in the distribution of facies on the carbonate shelf.

Reef models represent an integration of sedimentological and palaeontological observations. Four growth stages from a pioneer stage to colonization to diversification and finally to domination, at which point the reef has built up to an elevation to produce a surf zone, are distinguished. There are substantial variations in development and in facies between high-energy reefs, low-energy patch reefs, mud mounds and stromatolite reefs. The primary reef builders have also changed through geological time.

The deep sea is characterized by pelagic fall-out over vast areas of the abyssal plain. Clastic sediments supplied from neighbouring continents accumulate in slope aprons, submarine fans and on the flat basin plain. Deep sea clastic facies models have been dominated by the concept of turbidity currents and the *Bouma sequence*. Submarine fan models contain as key elements a leveed, channellized upper fan, a mid fan with aggradational lobes, and a flat lower fan lacking channels. Thinning and fining up sequences are thought to be due to channel filling and abandonment, while thickening and coarsening upward sequences are interpreted as the result of the progradation and aggradation of mid fan lobes. Allocyclic (sea level) controls are now recognized to be extremely important, with most sand being emplaced in the deep sea during lowstands. Traditional fan models do not explain the deep narrow troughs filled by longitudinal turbidites, nor the contourites of the slope and basin plain.

The sedimentary systems of *intracratonic* sags are commonly continental and endorheic. Whereas the Chad Basin, Africa is predominantly composed of siliciclastic fluviatile, lacustrine and aeolian deposits, the Michigan Basin, USA is dominated by shallow marine carbonates and evaporites.

Rift basins have tectonically controlled syn-rift fills that are commonly lacustrine and fluviatile in the early stages and may become shallow marine in later stages. Vulcanism accompanies rift sedimentation. The nature of the sedimentary fill depends on its climatic zone – the East African rift system illustrates this phenomenon, with alkaline shallow lakes in the semi-arid north and deep, freshwater lakes in the south. Supply of clastics to rift lakes is determined by the geometry of tilted fault blocks producing half-graben. *Failed rifts* such as the Benue Trough, Africa or North Sea pass from the syn-rift stage into a period of thermal subsidence characterized by marginal and open marine, commonly deltaic, sedimentation.

Passive margins are commonly sediment-starved in their early stages and accumulate evaporites, organic-rich shales and pelagic carbonates. This is generally followed by an increase in sediment derived from the continent, building thick seaward-prograding, principally shallow marine clastic wedges. Other margins are typified by thick carbonate banks.

In convergent settings, sediment accumulates in foredeep and thrust sheet top basins in the flexural downwarp (*foreland basin*) ahead of orogenic wedges. Early deposits are commonly turbiditic and the basin underfilled. Later deposits are shallow marine or continental as the basin reaches steady-state or overfilled status. Retroarc foreland basins differ in having a composition of the sedimentary fill that reflects the large amounts of plutonic and volcanic rocks in the orogenic belt.

Ocean trenches, accretionary basins and *forearc basins* are found at convergent ocean–ocean or ocean–continent boundaries. The sedimentary fills are typically strongly tectonized and their compositions are dominated by detritus from adjacent ocean crust and arc. Sediments are generally deep water and commonly turbiditic, although some accretionary complexes are above sea level, as in the Makran of Pakistan. Backarc basins on oceanic crust are dominated by deep marine sedimentation of pelagic oozes with marginal shallow water or alluvial fringes. There are some examples of backarc spreading on continental crust, such as the Pannonian Basin of central Europe.

Strike-slip basins have complex sedimentary fills indicating rapid subsidence, and major lateral facies changes from active fault scarps with breccias and conglomerates to centrally located finergrained zones. The Ridge Basin California and the Dead Sea are classic examples on land – the California Borderland basins represent deep marine equivalents.

7.1 DEPOSITIONAL SYSTEMS AND FACIES MODELS

7.1.1 Introduction

Depositional systems are sets of depositional environments linked by the process of sediment dispersal. They are responsible for large stratigraphic thicknesses, and environmental changes in one part of the system can generally be recognized in the stratigraphy of another part of the system, however distant. Some of the large scale stratigraphic responses to tectonic and eustatic factors have been described previously (Chapter 6). Here we concentrate on (1) the variability and controls on large scale depositional systems, (2) the depositional facies models which have been applied to these systems and their component parts, and (3) the relation between depositional systems and the background tectonic environment of the basin. It is not our purpose to enter into a discussion of the complexities of individual depositional environments and facies – the reader is referred to the texts by Reading (2nd edition 1986), Brenchley and Williams (1985, pp. 31–172), Walker (2nd edition 1984a), Miall (1984) and Reineck and Singh (1975) for this purpose. However, in the context of the basin, an examination of the large scale correlations within depositional systems and the smaller scale lateral and vertical variations in facies allow us to better understand the relationship between the *allocyclic* controls and *autocyclic* controls on the basin-fill (Beerbower 1964). A hierarchy of depositional products based on scale exists from facies/beds to assemblages of facies in parasequences, to depositional systems and finally depositional sequences. As the scale of the product increases, the cause of its deposition also changes from short-term, local and often autogenic causes at one extreme, to long-term, large scale causes involving lithospheric and climatic allogenic processes at the other extreme.

Depositional systems and their sedimentary products therefore reflect the integration of autogenic and allogenic controls. Sedimentary basins with different driving mechanisms (Chapters 3 to 5) consequently have distinctive assemblages of depositional systems and facies. We give an overview of this linkage of tectonic environment and sedimentary character in Section 7.2.

The processes acting in particular depositional environments give rise to a suite of relatively small scale characteristics of the resultant sedimentary deposits. The sum total of these small scale attributes that make a particular rock unit distinctive are embodied in the term *facies*. The term was introduced by Steno in 1669. Its modern usage follows Gressly (1938) who used the term to imply the full assemblage of lithological and palaeontological features of a rock unit. The term has been used subsequently in a number of ways, both purely descriptively and in terms of interpreted process of deposition. Discussion of these problems can be found in Middleton (1978) and Reading (1986).

A *facies model* is an interpretive device used to explain the association of facies. Facies models generally achieve this explanatory function by linking observations on modern processes and ancient deposits into a coherent synthesis. The summary which the facies model provides should be usable in a number of different ways. It should incorporate large volumes of data into a form which generalizes sedimentary processes. It should

be a stimulant for further investigations and should act as a predictor in new geological situations. Finally, it should help to give insights into the dynamic interpretation of the sediment unit (Walker 1984a, p. 6).

The art of facies modelling is, however, subject to its own problems. Most facies models depend on studies of *modern environments*: it is therefore necessary to assume that the present is an accurate reflection of the past. The time scale of the observations is critical, and we are often left asking the question of how typical are the last 100 years (a time of markedly elevated sea levels) of the geological past, and what is the effect of the infrequent or catastrophic 'event' on the geological record? (see Dott 1983, Reading 1986). Secondly, it is difficult to assess the true preservation potential of modern sedimentary sequences, and studies of present-day and sub-Recent deposits may be strongly biased compared to the ancient record.

The *vertical sequence* has played a very important role in the development of facies models. Cycles in clastic sediments are of two basic types:
• upward increase in transport energy (generally coarsening and thickening of beds upwards)
• upward decrease in transport energy (generally fining and thinning upwards).

Cycles may be the result of the natural processes within a depositional system (*autocyclic*), or they may be caused by external controls (*allocyclic*). Typical autocyclic mechanisms are the meandering or avulsion of a river channel. Allocyclic mechanisms include tectonic movements, eustatic sea level changes and climatic variations. The significance of allocyclic versus autocyclic mechanisms in building stratigraphic packages (such as parasequences – Chapter 6) is a subject currently receiving a great deal of attention.

Although vertical profiles in carbonate sediments are now widely used, a great deal of emphasis has been placed on grain type, structure and microstructure, fauna and chemical signature of carbonates. *Shoaling-upward sequences* formed in shallow subtidal to supratidal settings are common (e.g. Ginsburg 1975, James 1984a), although *deepening-upward* cycles have also been described, as for example in the so-called Lofer cycles of the Alpine Triassic (Wilson 1975).

Cycles are extremely well developed in evaporitic sediments because of their sensitivity to variations in water chemistry. The best-known vertical sequence is that of the coastal sabkha, where coastal progradation results in bioturbated subtidal sediments being overlain by intertidal and then supratidal sediments with displacive gypsum. Somewhat different vertical sequences have been described from other evaporitic basins such as the Messinian of the Mediterranean. Vai and Ricci Lucchi (1977) attributed the repetition of sequences of algal mats with selenite overlain by gypsum sands and reworked selenite passing up into shales with terrestrial vegetation and an insect fauna to shoaling-upward intertidal to supratidal cycles.

Vertical sequences can be interpreted from geophysical logs, particularly where there are pronounced vertical variations in porosity. An early contribution was that of Fisher *et al.* (1969) who distinguished a number of typical profiles in coastal and marginal marine clastic sediments in the Gulf Coast region (Fig. 7.1). These vertical profiles appear as distinctive shapes when the gamma ray or spontaneous potential logs are placed alongside a resistivity log:

1 Upward-fining cycles appear as 'bell-shaped' log patterns.
2 Upward-coarsening cycles appear as 'funnel-shaped' log patterns.
3 Amalgamated coarsening-up and fining-up units may be 'symmetrical'.
4 Units with no vertical trend in porosity or clay content appear as 'cylindrical' log patterns.

These patterns, though useful, are broad approximations only and do not permit unambiguous interpretations.

The vertical profile is clearly an inadequate and sometimes misleading representation of the sedimentary evolution of a sequence. Compared to established texts (Reading 1986, Walker 1984) therefore, we have chosen to give less emphasis to the one-dimensional view of vertical sequences. Two-dimensional panels, constructed from well-exposed surface outcrops or from high quality seismic reflection lines, are now a standard tool in stratigraphic and facies analysis.

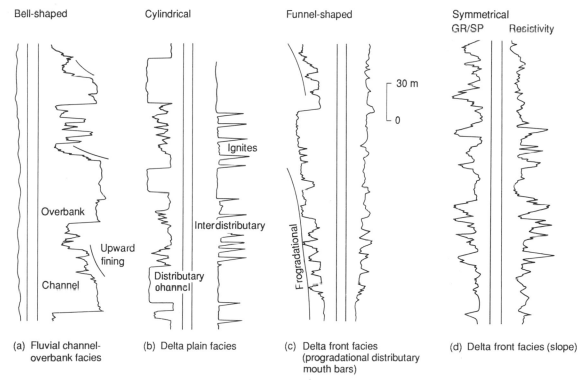

Fig. 7.1. Typical electrical log patterns from lobate river-dominated delta facies in the Gulf Coast region (modified from Fisher *et al.* 1969).

Three-dimensional geometries can often be interpreted from a series of differently orientated two-dimensional sections. The variability of depositional units, both vertically (in time) and laterally, contribute to its *heterogeneity* (see also Section 10.4.2). For example, in a sand-dominated sequence, the primary heterogeneity results from the intercalation of shales. Weber (1982) has suggested that the lengths of shale intercalations are largest in fully marine environments such as the shelf. The lengths of shale breaks decrease in deltaic and barrier environments and are shortest in fluviatile environments where stacking of channels results in erosion of fine members.

The two- and three-dimensional study of sediment geometry has given rise to the concept of '*architecture*', the way in which individual sediment bodies are stacked in time and space. There are now a very large number of studies of architecture, most progress having been made in alluvial sedimentary systems. One of the most comprehensive architectural field studies known to us has been carried out in the linked basin compartments of the southern Pyrenees on the Eocene Castissent Sandstone of the Montañana Group (Nijman and Puigdefabregas 1978, Marzo, Nijman and Puigdefabregas 1988). Several types of channel filling bodies are found (Atkinson 1983, Marzo, *et al.* 1988) representing sheet-like braided channel networks, point-bar deposition in mixed-load meandering streams and stable bedload channels giving lenticular ribbons (see Section 7.1.2). These different elements, together with overbank and brackish-water fine-grained components, amalgamate to make up the architectural characteristics of the Castissent.

7.1.2 Continental systems and facies

Terrestrial depositional systems may include the deposits of rivers, deserts, lakes, slope wastage and ice sheets and glaciers. We concentrate here on fluvial, desert and lacustrine systems. Continental basins have a basic two-fold subdivision:

1 Basins with *through drainage* are dominated by well-established river systems and perennial lakes.

2 Basins with *internal drainage* are characterized by ephemeral river systems, generally shallow and short-lived lakes, continental sabkhas and deserts.

7.1.2.1 Fluvial systems

Fluvial style is a complex response to a number of autocyclic and allocyclic controls. The primary allocyclic controls include (1) climate, which controls run-off (discharge) and the weathering of parent rocks, and (2) tectonics, which controls basin slopes and relief of hinterlands in the drainage basin. This interplay results in a distinctive set of characters of the river system, e.g. sediment load, discharge, slope, vegetation which in turn determine channel pattern. The general attributes of fluvial depositional systems are emphasized below. The resultant facies models are then discussed.

Alluvial basin types can be classified according to a number of simple criteria (Miall 1981):

• existence of transverse or longitudinal drainage systems (Fig. 7.2). The Atlantic coastal plain of North America, for example, consists almost entirely of transverse rivers. Transverse rivers may alternatively join a major longitudinal trunk river such as the Po system in northern Italy.

• nature of proximal, medial and distal elements. Tectonically active basin margins commonly have alluvial fans as proximal elements. Medial elements include braidplains and high-sinuosity alluvial systems of transverse or longitudinal type. Distal elements may be lake margins, terminal fans and sabkhas, deltas and estuaries.

River channel patterns are a sensitive response to type of load, discharge, slope and vegetation. They can be divided into a number of types based on:

• suspended-, mixed- or bed-load transport (Schumm 1963)

• channel sinuosity (Leopold and Wolman 1957, Rust 1978)

• single or multiple thalweg (cf. Rust's (1978) braiding parameter).

The channel pattern, lateral migration and vertical accretion of the river and floodplain all respond to a delicate balance in a wide range of variables. Four principal river types exist:

1 Meandering.

2 Braided.

3 Straight.

4 Anastomosed.

The *lateral stability* of the river channel has a profound influence on the sediment body geometry

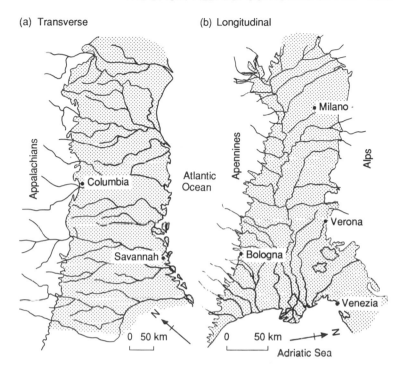

(a) Transverse   (b) Longitudinal

Fig. 7.2. Transverse and longitudinal drainage patterns. (a) The Atlantic coastal plain of North and South Carolina and Georgia. (b) The Po Basin of northern Italy. After Miall (1984).

and has been emphasized in the classification of Friend (1983). *Fixed channels* produce channel deposits of very limited across-stream extent, or ribbons, whereas *mobile-channel* belts are associated with lateral migration and the formation of sediment sheets. *Unchannelized* transport in sheetfloods also produces laterally extensive sheets.

Alluvial fans form where rivers emerge from valleys onto an unconfined plain or major trunk valley, building semi-conical depositional landforms. Fans in semi-arid regions are relatively small and sediment transport takes place on them in mass flows (muddy debris flows) as well as in high-discharge, short-duration floods. Humid fans in tropical regions are less common because intense chemical weathering liberates very fine-grained detritus and also because dense vegetation protects slopes. Where they occur, they are large, of low gradient and are dominated by perennial streamflow (Schumm 1977). Steep, paraglacial fans form where valley glaciers are retreating, high levels of spring run-off causing most of the sediment transport.

High and low sinuosity river systems may occupy small intermontane basins, linear rifts or vast intracratonic sags (Section 7.2). Braided systems are favoured by high and variable discharges, steep slopes and large amounts of bedload. High sinuosity, meandering systems are common in regions of low slope, equable discharge and relatively more suspended load.

Ephemeral fluvial systems in regions of internal drainage may contain a wide variety of river types in close proximity. This results from the large down-stream variations in discharge caused by distributive branching and losses by evaporation and soak-in to the floodplain. The Oligocene fluvial systems along the Catalanides mountains in northeastern Spain are an ancient example. Large, low sinuosity gravel feeder channels close to the mountainous hinterland are arranged in a '*goose-foot*' pattern, while intervening areas contain small, highly sinuous river channels (Fig. 7.3) (Allen *et al.* 1983).

Fluvial systems respond rapidly to environmental change, and the effects of tectonic activity can be assessed on-land in neotectonically active regions. They therefore serve as sensitive 'barometers' of allocyclic processes.

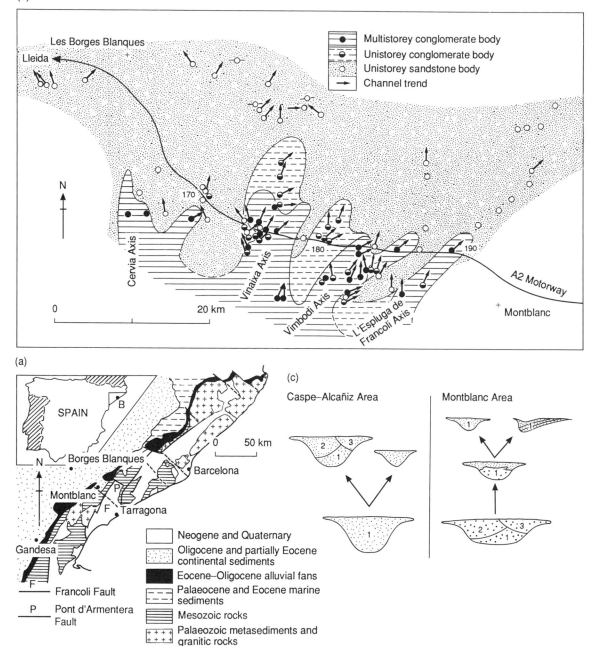

Fig. 7.3. An ancient ephemeral fluvial system from the Eocene–Oligocene Scala Dei Group of the Ebro Basin, NE Spain (after Allen *et al.* 1983). (a) Location of fans abutting the basin edge represented by the Catalanides chain. (b) Detail of one of these fluvial systems (Scala Dei Group) in the region of Montblanc and Les Borges Blanques. Gravels in multistorey channels are concentrated in distinct axes which spread into the basin like goose feet. Between the axes are finer-grained sediments with composite palaeosols and single storey sand and gravel filled channels. Basinward, the channels are small in scale, single storey and composed of sandstone. (c) Lateral variations in the geometry of channel bodies in the Scala Dei Group of the Montblanc area (Eocene–Oligocene) and the Miocene deposits of the Caspe–Alcañiz area of the southern Ebro Basin, some 70 km to the west. The Caspe fluvial sediments are also thought to have been deposited in a strongly ephemeral system (Williams 1975, Friend, Slater and Williams 1979).

The effects of tectonic tilting on alluvial processes was pioneered by Mike (1975) and has been pursued by Alexander and Leeder (1987) and Leeder and Alexander (1987) by studies of Holocene extensional tilt-block basins in Montana (Fig. 7.4), part of the wider Basin and Range province. Present-day stream patterns in the Hebgen lake area of southwestern Montana are strongly influenced by a magnitude 7.5 earthquake in 1959 which caused displacements of up to 6.7 m. The modern Madison River hugs the extensional faults forming the northeastern margin of the valley and has clearly left behind abandoned meander scars to the SW of the present river course (Fig. 7.5). These abandoned former meanders have a pronounced asymmetry with a highly preferential preservation of meander loops that are convex towards the W or SW. This suggested to Leeder and Alexander

(1987) that the river had gradually translated towards the present active fault scarp as a result of surface tilting, causing an asymmetrical meander belt to develop. This is opposed to the view that the river abruptly avulsed to its present position along the northeastern margin. In ancient alluvial sequences a process as described above should produce many lateral accretion surfaces, representing the migration of the inner bank on meanders (point bars), that dip up the regional or local tectonic slope.

River channel patterns and depositional systems may change rapidly as a result of base level changes. A good example is the Mississippi Valley, USA (Fisk 1947). The valley was strongly incised by rivers during the last glacio-eustatic lowstand (see Sections 6.3, 6.4). Rising base levels following melting of ice caps allowed a coarse braided system

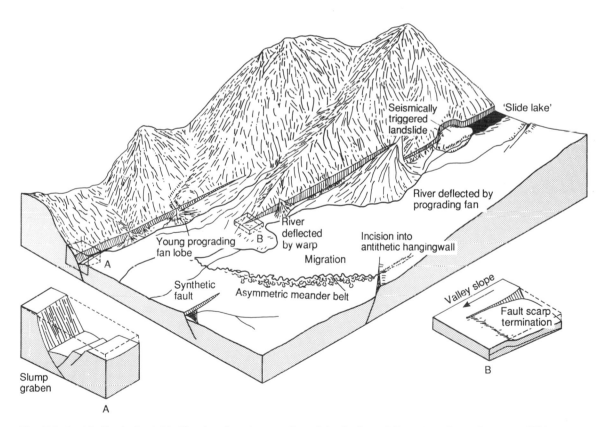

Fig. 7.4. An idealized alluvial half-graben based on studies of the Basin and Range province of western USA, showing some of the common geomorphological features resulting from extensional neotectonics (after Alexander and Leeder 1987).

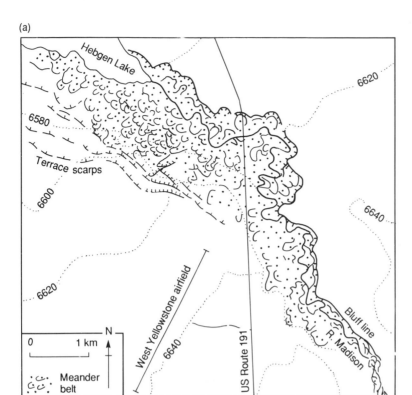

(a)

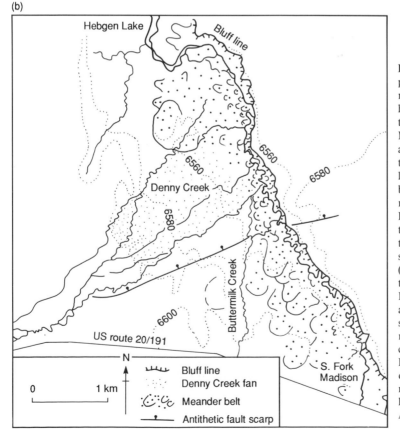

(b)

Fig. 7.5. Maps drawn from aerial photographs to show the nature of meanderbelts associated with a half-graben. (a) River Madison in the Hebgen Lake region of Montana. Note the highly asymmetrical meanderbelt bounded to the NE by a prominent bluff line. Terrace scarps were mapped by Nash (1984). The abandoned meander loops of the Madison River are predominantly convex towards the SW, that is, away from the bluff line along which most subsidence has taken place. (b) South Fork Madison River in the Hebgen Lake region of Montana. The meanderbelt is again asymmetrical. An antithetic fault scarp intersects the modern meanderbelt and influences the course of the Buttermilk Creek. The Denny Creek alluvial fan flows down the low gradient of the roll-over in the hangingwall of the half-graben. After Leeder and Alexander (1987).

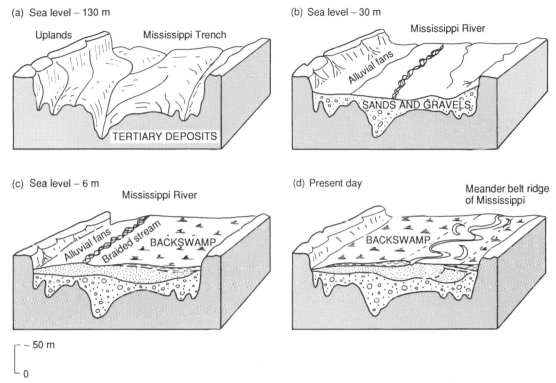

(a) Sea level – 130 m

Uplands   Mississippi Trench

TERTIARY DEPOSITS

(b) Sea level – 30 m

Mississippi River

Alluvial fans

SANDS AND GRAVELS

(c) Sea level – 6 m

Mississippi River

Alluvial fans   Braided stream   BACKSWAMP

(d) Present day

Meander belt ridge
of Mississippi

BACKSWAMP

~ 50 m

0

Fig. 7.6. Schematic illustration of the late Pleistocene to Recent history of the Mississippi River Valley (after Fisk 1944). (a) Late Pleistocene entrenchment during a sea level fall/lowstand at 130 m below present sea level. A low sinuosity river system carried gravel to the Gulf of Mexico. (b) Aggradation of gravels and sands during the sea level rise to 30 m below present. (c) Development of fine-grained floodplains (backswamps) adjacent to a braided alluvial ridge, at 6 m below present sea level. (d) The modern high-sinuosity Mississippi River meanders in a well-developed alluvial ridge with flanking extensive floodplains. Modified from Leeder (1982).

to be developed. Continuing sea level rise resulted in this low sinuosity system evolving to the present highly sinuous, fine grained system (Fig. 7.6).

Bridge and Leeder (1979) simulated alluvial stratigraphy using a computer model which accounted for the floodplain geometry and its rate of aggradation, the compaction of the floodplain sediments during burial, the magnitude and frequency of tectonic events causing subsidence, and the rate of avulsion. Figure 7.7 shows an example of such computer-generated stratigraphy, for the case of an alluvial valley with varying amounts of tectonic tilt. The effect of the tectonic tilting is to focus the channel belts close to the active down-faulted margin of the half-graben. The outer hinge zone of the half-graben is noticeably starved of coarse-members in these experiments.

Straight rivers are rather uncommon and there is a limited data base on anastomosing rivers, but meandering and braided rivers have well-developed facies models.

Meandering river facies model

The morphological elements of a meandering river system, such as the modern Mississippi and Brazos Rivers, are shown in Fig. 7.8. Sand deposition is normally restricted to the main channel with its point-bars and chute bars. The floodplain and levees receive finer grades of sediment during river floods. The channel floor commonly has a coarse lag of pebbles which are only moved at maximum flood velocities. During 'average' discharges, sand is moved through the system in dunes on the

(a)

(b)

(c)

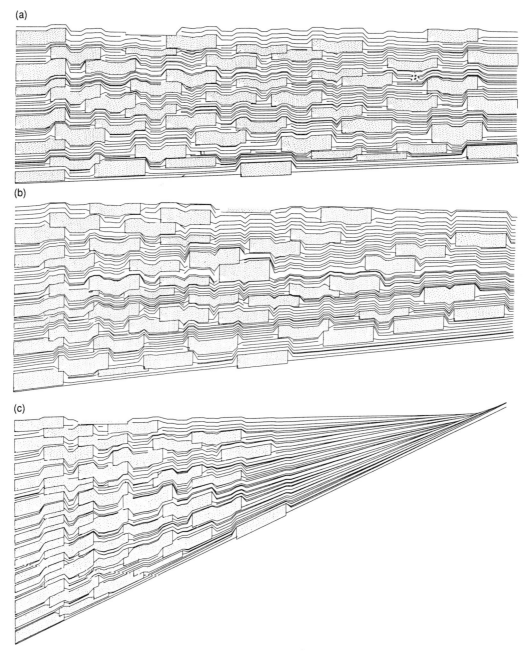

Fig. 7.7. Computer-generated alluvial stratigraphy for half-graben. (a) to (c) show increasing asymmetry of subsidence in the half-graben. In (a) the frequency of tectonic downfaulting (in 1.0 m steps) of the basin margin is 890 yr, in (b) 445 yr and in (c) 89 yr. A very rapid increase in the interconnectedness of the channel deposits occurs in the case of the highly asymmetrical basin because of the tendency of rivers to flow to the lowest topography, clustering close to the active tectonic margin of the basin. Other parameters used are given in Table 2 of Bridge and Leeder (1979, p. 629).

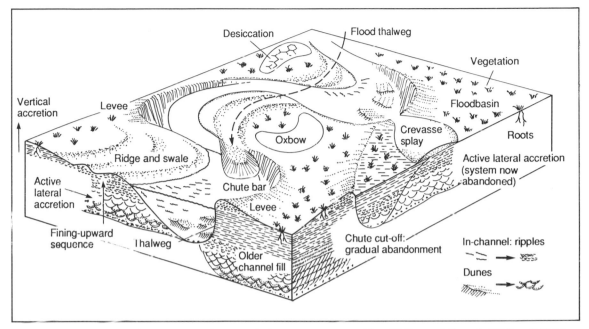

Fig. 7.8. Main morphological elements of a meandering stream system. After Walker and Cant, in Walker (1984a). The channel and bar sediments are deposited primarily by lateral accretion of the inner bank of the meanders, whereas the floodplain builds up by vertical accretion.

channel floor and in ripples higher on the point bar. The migration of the meander bend therefore commonly results in the upward passage from coarse lag to trough cross-stratification to ripple cross-lamination, all preserved by the process of lateral accretion. Highly episodic lateral accretion may result in the preservation of successive point-bar slopes dipping toward the channel axis or thalweg. These are termed *epsilon* cross-beds. Meander loops can be abandoned either (1) by chute cut-off causing gradual abandonment and the deposition of thick ripple cross-laminated sands or (2) by neck cut-off or avulsion which causes a sudden abandonment and deposition of muddy sediments by vertical accretion over the channel deposits. Beyond the main channel fine-grained sediments accumulate after floods which overtop the river banks. During lower flow stages these overbank environments are subject to pedogenesis, mediated by the action of plants and animals in humid climates and represented by calcretes in semi-arid climates.

The meandering river facies model comprises therefore, classically at least, a fining-up sequence. This serves merely as a norm by which the many possible variations can be compared. In three dimensions, the sandbodies representing point-bar sheets should be elongated in the longitudinal direction. In the transverse direction, their extent will be controlled by the rate of lateral migration of the meander loop and the duration of steady lateral migration in between abandonment phases.

Sandy braided stream model

The morphological elements of sandy braided streams are complex and dominated by bars (simple and complex) and intervening channels ornamented by sinuous crested dunes. Channels may also contain oblique or transverse bars with steep foreset slopes. Extensive sand flats appear to grow from smaller nuclei, represented by the emergent top of a cross-channel bar exposed at low flow stages (Fig. 7.9). The original nucleus accretes sand and may form sandflats several km in downstream length (e.g. Brahmaputra, Coleman 1969, and 1.2 km in the South Saskatchewan River, Cant and Walker 1978). The possible vertical sequences developed in a river dominated by channels and sandflats can be judged from Fig. 7.9.

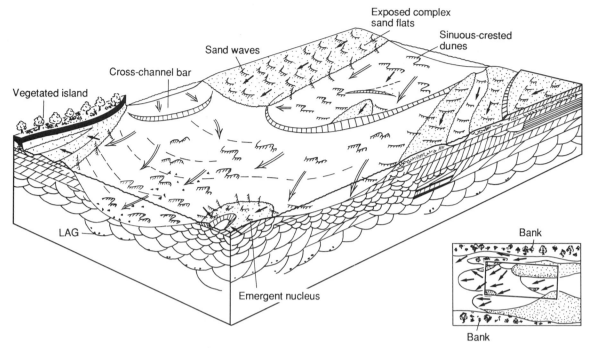

Fig. 7.9. Main elements in a sandy braided river such as the South Saskatchewan, Canada (Cant and Walker 1978). Main morphological elements include vegetated islands, exposed sandflats covered with flow-transverse dunes, and channels with large sinuous crested dunes and cross-channels bars.

Braided stream sequences lack well-developed vertical accretion deposits because:
1 The rapidly shifting channels erode any vertical accretion deposits.
2 The braided tract commonly occupies a wide zone between bounding river bluffs, so extensive, low floodplains are less common.
3 The river is generally bedload-dominated. Vegetated islands and banks may, however, accumulate peats.

Braided river deposits form extensive sheets of sandstone with poorly developed shales, in marked contrast to the three-dimensional geometry of meandering stream deposits.

Anastomosing rivers

A number of studies suggest that a river type characterized by multiple but relatively stable channels exists, although there are at present relatively few documentations of ancient examples. There is not an anastomosing river *model*, but there are some interesting and useful descriptions of individual rivers. The gravelly anastomosed rivers of western Canada, such as the North Saskatchewan, are examples (Smith and Smith 1980) (Fig. 7.10). Anastomosed rivers appear to be favoured by high rates of vertical aggradation, causing stabilization of channels between adjacent wetlands. Such conditions might be induced by the local raising of base levels. Alluvial stratigraphy produced by rivers of the western Canadian anastomosing type would be exceedingly complex in three dimensions, as indicated by Fig. 7.10.

7.1.2.2 Aeolian systems

Deserts include a variety of environments of deposition from the giant sand seas or ergs, to stony wastelands, interdune sabkhas with temporary lakes, and dried up river courses. Deserts occur in both Arctic areas, where sediment is derived from wasting glaciers, and tropical zones where ergs are concentrated. Areas of low rainfall occur as two discontinuous belts around latitudes of 20–30° and are associated with persistently high

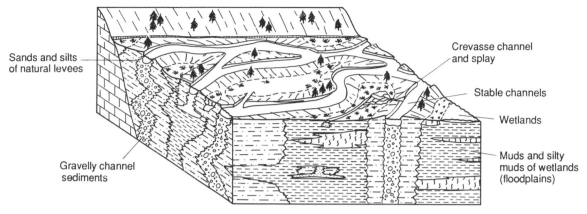

Fig. 7.10. Main elements of an anastomosed river based on rivers in the Alberta foothills of Canada (after Smith and Smith 1980). Channels bifurcate and rejoin as in a braided river, but are relatively fixed in position, leading to vertical aggradation rather than lateral accretion. This channel stabilization is promoted by the high amounts of muds and organic material in the adjacent wetlands forming the channel banks.

atmospheric pressures. Deserts also occur in the centres of large continental masses. The occurrence of aeolian deposits in the stratigraphic record is therefore to a large extent a reflection of an ancient climatic zone.

The largest present-day ergs occur in topographic basins in cratonic positions, such as the Saharan examples (Rub al Khali erg in Saudi Arabia is 560 000 km^2 in area) and those of central Australia. The average thickness of the erg may exceed 100 m with bedform (draa) heights of several tens of metres to over 200 m. Ancient examples such as the Permian Rotliegend Sandstone of northern Europe similarly occupy large continental sags. The Permian–Jurassic sequence of the Utah–Arizona–Colorado region is the best-documented ancient desert succession. It contains the Lower Jurassic Navajo Sandstone, which alone is 700 m thick.

Aeolian sequences are characterized by a hierarchy of bounding surfaces, some of which 'slice' through the entire aeolian stratigraphy and are clearly allogenic and of inter-regional importance. Some bounding surfaces are widespread and related to deflation down to a water table – these are known as 'Stokes surfaces'. Others are related to the climb of draas and smaller superimposed bedforms and are therefore autogenic.

Distinct trends have been identified in transects from the erg centre to the erg margin (Fig. 7.11). Such erg fringes may merge with ephemeral fluvial systems, playa lakes or marine environments.

These erg margin environments are characterized by a complex interaction of sedimentary processes (Porter 1987, Clemmensen and Blakey 1989, Langford 1989, Langford and Chan 1989).

Aeolian sequences are dominated by cross-stratified sandstones and facies models rely on the recognition of a hierarchy of migrating aeolian bedforms. Much recent interest, however, has focused on the finer-grained intercalations within aeolian sequences that may represent:

- interdunal environments
- playas and continental sabkhas
- products of ephemeral stream activity.

These intercalations are particularly common at the fringes of the large ancient sand seas or ergs. As a result, models of *aeolian-fluviatile interaction* (Fig. 7.12) and *aeolian-lacustrine interaction* are being developed.

Interdune areas are an integral part of the aeolian system. In sand-depleted areas, the interdunes consist of deflated surfaces lined with coarse lags or thin sand sheets and small isolated dunes. Where the water table is elevated, sabkhas with evaporites may form. A range of depositional conditions in the interdunal areas from dry to damp to wet has been recognized in the Jurassic Entrada Formation of southwestern USA (Kocurek 1981a, b). The interdune sediments are extremely important in contributing to the heterogeneity of aeolian reservoirs.

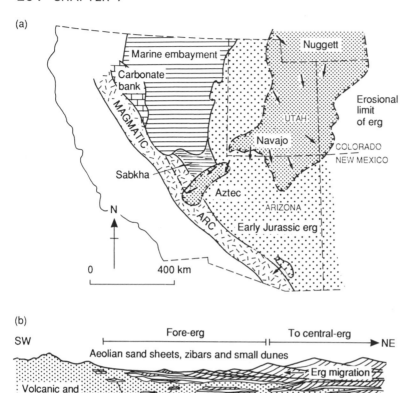

Fig. 7.11. Erg centre-margin transect from the Lower Jurassic Nugget (NE Utah, Wyoming, Colorado) – Navajo (SW Utah) – Aztec (SW Nevada) sandstones (after Porter 1987). (a) Generalized Early Jurassic palaeogeography. (b) Fore-erg and erg deposits of the Aztec erg margin. The horizontal extent of the diagram represents about 170 km. Note the decrease of aeolian cross-set thickness towards the erg margin. The fore-erg deposits consist of a 10–100 m thick succession of isolated pods, lenses and tongues of aeolian sediment encased in fluviatile and sabkha deposits. The central erg sediments consist of the deposits of extensive, large dunes and draas.

Reconstructions of dune type are based on the geometry of the preserved cross-stratification, but also on the type of lamination comprising the cross-sets (Hunter 1977):

• plane-bed lamination is produced at high wind velocities above those responsible for ripple formation

• climbing ripple lamination of two types: *translatent* where only bounding surfaces are visible, and *rippleform* where foresets are recognizable

• grainfall lamination formed from fall-out of suspended grains in the lees of bedform crestlines

• grainflow lamination caused by the avalanching of grain dispersions down steep lee faces.

The relation of these lamination styles to dune morphology has been discussed by Hunter and Rubin (1983), Rubin and Hunter (1983), Hunter (1981), Fryberger and Schenk (1981), Clemmensen and Abrahamsen (1983) and others.

7.1.2.3 Lacustrine systems

Lacustrine depositional systems are also highly sensitive to climate. There is a great diversity of lake basins and lake waters. They can be divided into lakes that have an outlet and are hydrologically open and those that lack an outlet and are hydrologically closed (Fig. 7.13) (Allen and Collinson 1986). Their hydrological status determines lakewater chemistry and therefore the terrigeneous clastic versus chemical and biochemical sedimentation in the lake. Hydrologically closed lakes commonly form in the centres of endorheic (internal drainage) regions such as the Chad Basin,

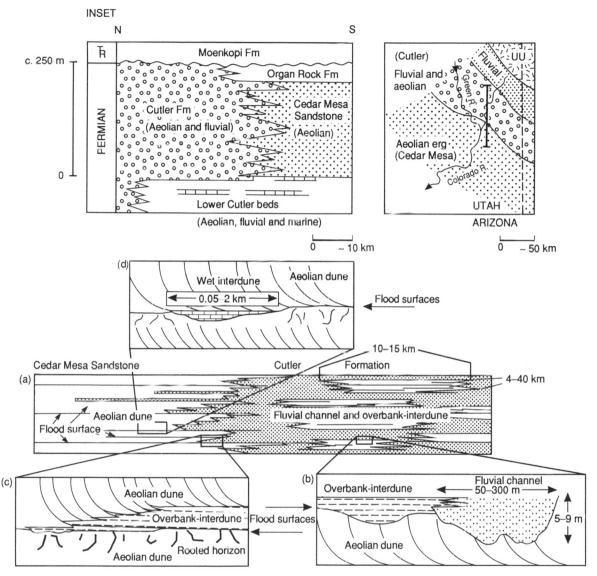

Fig. 7.12. Aeolian-fluviatile interaction in the Permian Cedar Mesa Sandstone and Cutler Formation of the Colorado Plateau, USA (after Langford and Chan 1989). Pronounced and extensive erosion surfaces are produced during river flooding (labelled flood surfaces), interdune areas being filled with overbank sediment and aeolian dunes being locally incised by fluvial channels. Insets show stratigraphic relations in the study area and the geological context, with the Palaeozoic Uncomphagre Uplift (UU) in the northeast. (a) Overall geometry of intertonguing of aeolian and fluvial deposits along flooding surfaces. (b) Geometry and facies relationships associated with the river channel deposits. (c) Relationships between overbank–interdune deposits, aeolian dune deposits and flood surfaces overlying rooted upper zone of aeolian dunes. (d) Relationships between wet interdune deposits and flooding surfaces. Carbonates may accumulate in these interdune areas starved of coarse detrital input.

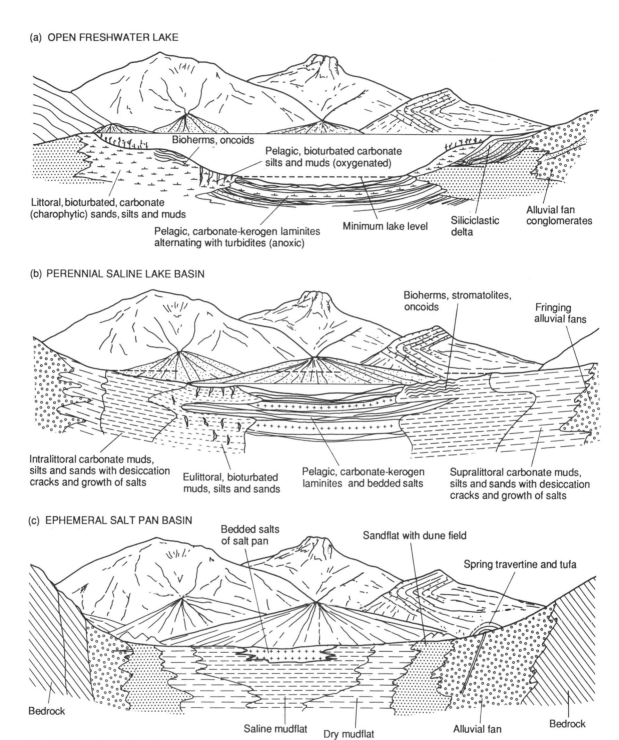

Fig. 7.13. Idealized depositional environments and facies in lacustrine systems. (a) Open freshwater lake with through drainage; (b) Hydrologically closed perennial salt lake; (c) Hydrologically closed ephemeral salt pan. After Eugster and Kelts (1983), modified from Allen and Collinson (1986).

north-central Africa (Servant and Servant 1970). They are characterized by a dominance of chemical and biochemical sediments and evaporites are common. Through-flowing lakes typify many continental rift systems such as in East Africa and the Baikal Rift of central Asia. They are dominated by terrigenous clastic input and, where river input is negligible, by alkaline earth carbonates. Apart from the hydrological status of the lake, the most important controls on lacustrine depositional systems are the slope of the nearshore zone, lake bathymetry, stratification of the water column, size and shape of the lake as well as the overriding importance of climate.

Lacustrine depositional systems may change in character rapidly as a result of climatic fluctuations. A large number of present-day lakes are shrunken, saline remnants of much larger and more dilute ancestors that existed during the pluvial period coinciding with the last glacial maximum (Flint 1971). Examples are the Dead Sea and precursor Lake Lisan, Great Salt Lake, USA and precursor Lake Bonneville, and Lake Eyre, Australia and precursor Lake Dieri. The expanded lakes, such as Lake Lisan, typically accumulated alkaline earth carbonates with an abundant flora of diatoms, whereas the present-day remnant has a lake floor covered with gypsum and halite. An analysis of Holocene/Quarternary lake levels and salinities suggests that minima and maxima recur at a frequency that is most likely a result of orbital eccentricities of the Earth around the sun. The lake level variations therefore appear to be due to Milankovitch cycles (see also Section 6.5.6).

Since lakes are extremely sensitive to allogenic changes, especially climatic variations, no facies models that serve as a synthesis, a norm and a predictor are currently available. However, there are many careful documentations of lake systems and lake deposits (review in Allen and Collinson 1986). *Hydrologically closed lake basins* are characterized by deposition in both the lake centre and on fringing flats.

Central lake facies include:
1 Saline mineral facies precipitated from highly concentrated lake brines such as the trona of the Green River Formation of the southwestern USA;
2 Organic-rich marls and oil shales representing the remains of algae and varying carbonate input

from fringing mudflats;
3 Carbonate-gypsum laminites caused by seasonal changes in water chemistry – the laminites of the Triassic Todilto Formation (Anderson and Kirkland 1960) are an example, and similar recent deposits are found in the Dead Sea (Begin, Ehrlich and Nathan 1974).

Marginal lake facies reflect periodic inundation and emergence. They include:
4 Stromatolitic limestones and oolitic-pisolitic grainstones promoted by algal activities;
5 Siliciclastic sandstones deposited in small beach zones close to river entry points;
6 Laminated marlstones with desiccation cracks;
7 Gypsiferous marlstones with nodular sabkha-like gypsum.

The vertical arrangement of these facies depends on the frequency of flood events, the hypsometry of the basin and hinterland and the time variation of climatic variables such as evaporation rate and rainfall.

Hydrologically open lake basins may also contain a wide variety of facies. *Offshore facies* include:
1 Laminites composed of clastic, organic and carbonate layers, deposited below the thermocline of stratified lakes;
2 Thinly bedded, graded silts and muds derived from rivers and transported in turbidity currents or in geostrophic flows.

Nearshore facies are variable, depending on the bathymetry of the nearshore zone, climate and terrigenous supply. They may comprise:
3 Wave rippled sandstones deposited in beach zones;
4 Cross-stratified sandstones representing distributary channel-fills or lacustrine bars;
5 Paludal sediments (lignites, silts) deposited in interdistributary bays, ponds and swampy low-gradient lake margins;
6 Bioherms, including stromatolites;
7 Coated grain facies (oncoids, pisoids, ooids);
8 Lake 'chalks' in marl benches or throughout the littoral zone.

The water level of open lakes fluctuates with a smaller amplitude than in closed basins, so facies are commonly less finely interbedded. Geological evidence suggests that the hydrological status of a lake may change completely over a short period of time (e.g. Boyer 1982).

7.1.3 Coastal and nearshore systems and facies

7.1.3.1 Siliciclastic shoreline systems

The geomorphology and oceanography of the Earth's *siliciclastic coastlines* reveals an exceedingly complex interplay between fluvial input on the one hand, and basinal parameters such as wave energy, tidal range, storm regime on the other. The two main types of coastal depositional system are deltaic and non-deltaic. Non-deltaic coastal systems may be (1) *wave-dominated*, containing beaches, microtidal barrier islands and cheniers, (2) *mixed wave-tide influenced* consisting of mesotidal barrier islands with tidal inlets and ebb- and flood-tidal deltas, and (3) *tide-dominated*, made up of tidal flats and estuaries (Hayes 1979).

Deltas develop where river systems debouch into the ocean, inland seas and lakes. Their form is controlled by a number of factors, chief of which is the relative effectiveness of river discharge compared to the tidal and wave energies of the receiving basin (Galloway 1975). Where tidal and wave energies are low, distributary channels are able to build out into the sea unhindered by coastal erosion (Fig. 7.14). This produces a typical 'birds-foot' pattern, as shown by the Mississippi delta, USA. Sand trends, representing the seaward build-out of channels and mouth bars, are orientated at a high angle to the coastline. Where wave energies are strong compared to the river inflows and tides, the sediment delivered to the sea is moulded into curved ridges at the delta's front and some is redistributed along the shore as beaches and spits (Fig. 7.14). Deltas of this type, such as the Senegal (West Africa) or the Grijalva (Gulf of Mexico) are roughly arc-shaped and prograde slowly because of the destructive nature of approaching waves. Sand trends are generally orientated parallel to shore. Deltas strongly affected by tides have tidal channels cutting deep into the coastline with associated tidal sand ridges or shoals elongated in the same direction as the tidal current pathways (Fig. 7.14). The Ganga–Brahmaputra delta in the Bay of Bengal, and the Mahakam delta, Indonesia, are of this type. Linear sand trends are generally elongated at a high angle to the shoreline.

Depositional patterns at river mouths can also be studied in relation to the dynamics of outflow dispersion (Wright and Coleman 1974, Wright 1977), an extension of the original concept proposed by Forel (1892) from his studies of the delta of the Rhône River in Lake Geneva. The dispersion of the outflow into the receiving water body falls into three categories:

1 *Inertia-dominated*, where outflows of equal density to the water of the receiving basin decelerate slowly, travelling as a jet far into the basin. Prograding clinoforms (Gilbert-type deltas) may result from inertial jet diffusion. River mouths of this type are common where high-velocity rivers enter low-energy freshwater basins (lakes), but are less common in marine environments.

2 *Friction-dominated*, where outflows enter basins with shallow nearshore waters. The jet decelerates and spreads laterally rapidly, producing bifurcating channels with a centrally located middle ground bar.

3 *Buoyancy-dominated*, where the outflow is of a different density from the water of the receiving basin. Where the outflow is less dense, the normal situation where fresh water enters a marine basin, the outflow moves as a buoyant plume over the salt wedge. Mixing between the freshwater plume and the underlying salt wedge causes deceleration of the plume and deposition of sediment. The resulting sediment body extends considerable distances from the channel mouth. River outflows may be highly charged with sediment and therefore be of greater density than the receiving waters. In such a case the outflow moves as an underflow. This is common where sediment laden streams enter lakes (Sturm and Matter 1978).

The facies models of deltas have relied on the conventional classification according to the relative importance of river, wave and tidal energy (Fisher *et al.* 1969, Galloway 1975). These end-member types of delta are as follows:

1 *River-dominated deltas*, exemplified by the present-day Mississippi, are characterized by distributaries and interdistributary bays flanked with marshes. The sequential filling by crevasse splays and abandonment of an interdistributary bay is well illustrated in West Bay on the Mississippi delta (Fig. 7.15). The vertical sequence differs considerably according to proximity to the crevasse channel. Progradation of the mouth bar produces large scale (60–150 m) coarsening-upward sequences representing the passage from prodelta muds to bar front to bar crest. Overlying sediments

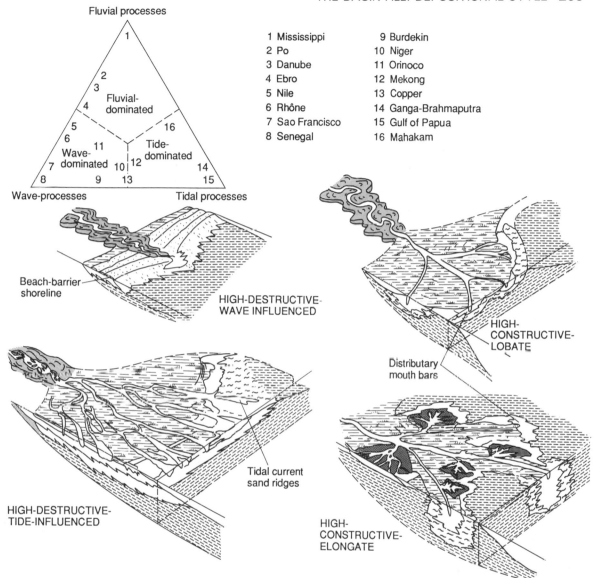

1 Mississippi	9 Burdekin
2 Po	10 Niger
3 Danube	11 Orinoco
4 Ebro	12 Mekong
5 Nile	13 Copper
6 Rhône	14 Ganga-Brahmaputra
7 Sao Francisco	15 Gulf of Papua
8 Senegal	16 Mahakam

Fig. 7.14. Delta types according to the relative importance of fluvial and basinal processes. Fisher *et al.* (1969) distinguished *high-constructive* types dominated by fluvial processes and *high-destructive* types dominated by basinal processes such as waves and tides. Inset shows the ternary plot introduced by Galloway (1975) based on the hydraulic regime of the delta front area. Sixteen modern deltas are plotted on the diagram.

may be represented by distributary channel-fills and laterally coexisting interdistributary facies.

Where the sediment supplied to the delta is predominantly fine-grained, as is the case with the modern Mississippi, the delta front may be subject to frequent failure, causing slumps and slides. These events may seriously disrupt the ideal facies sequences outlined above.

2 *Wave-dominated deltas* are characterized by beach ridge complexes with active shorelines, shallow lagoons between old beach ridges and aeolian dunes in areas of ridge reworking. Progradation should once again lead to a coarsening-upwards sequence, the uppermost part of the sequence

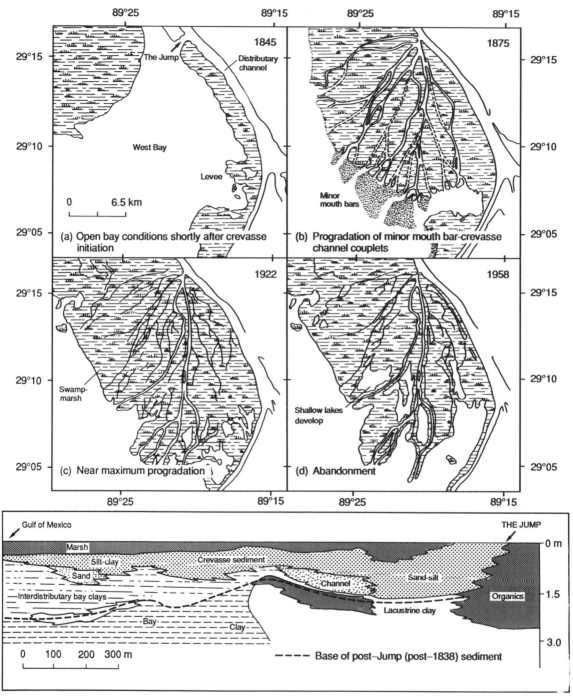

Fig. 7.15. Example of the development of a sub-delta by crevassing of a levee bordering a distributary channel. A crevasse channel called The Jump breached the levee of a distributary channel of the Mississippi delta in 1838. This caused the interdistributary West Bay to be filled with river-derived sediment. The subdelta prograded southwestwards, depositing sands and silts over the interdistributary bay clays and culminating in the development of organic-rich swamps and marshes on the subdelta top (after Gagliano and van Beek 1970).

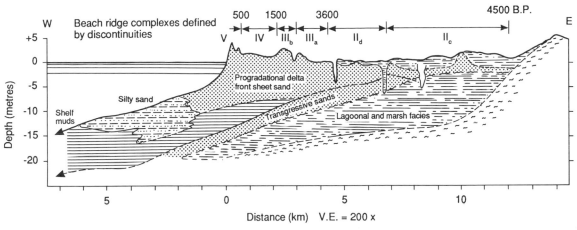

Fig. 7.16. Cross-section through the wave-dominated Costa de Nayarit delta system of Mexico. An extensive sand body has prograded following the Holocene transgression which flooded the older lagoonal and marsh facies of the delta top. The delta has prograded by the seaward accretion of beach ridge complexes, which therefore become younger as the present shoreline is approached (after Curray, Emmel and Crampton 1969).

comprising high-energy beachface deposits and rarely an aeolian capping. The Costa de Nayarit delta in Mexico is an example of a prograding wave dominated delta (Curray, Emmel and Crampton 1969) (Fig. 7.16). A variety of geometries and vertical sequences of wave-dominated deltas has been interpreted in the Cretaceous San Miguel Formation of the Gulf Coast of Texas (Weise 1980). Complete delta front sequences are truncated and modified by the effects of subsequent transgression, causing the removal of beach and aeolian dune facies and the bioturbation and diagenetic alteration of the upper shoreface sediments (Fig. 7.17). Individual wave-dominated deltas appear to change from lobate to arcuate to cuspate to elongate as the effects of wave reworking increase (or sediment input decreases).

3 *Tide-dominated deltas* are characterized by a complex mosaic of tidal current ridges, shoals and islands separated by channels carrying swift tidal flows. Progradation of the delta front gives rise to bidirectional cross-bedded sands with clay drapes at the top of coarsening upward sequences. Overlying sediments may be tidal flat deposits. Where the tidal range is lower and the sediment load is finer, as in the Mahakam delta of Indonesia, the delta plain is extensive and dominated by tidal flats crossed by tidal creeks and the delta front is a shallow-dipping platform 8 to 10 km wide of wave

reworked silts and sands (Allen, Laurier and Thouvenin 1979).

Wave-dominated shorelines may occur on delta fronts or on non-deltaic coasts. They are dominated by beaches (directly attached to the land) and barrier islands separated from the land by a shallow lagoon. Cheniers are sandy or shelly beach ridges isolated in coastal mudflats. The beachface subenvironments are controlled largely by wave approach. High-energy coastlines such as the Pacific coast of Oregon, USA (Clifton, Hunter and Phillips 1971) lack well-developed nearshore bars and are dominated by wave-generated ripples, megaripples and swash-backwash zones. Low to intermediate wave energy coastlines such as Kouchibouguac Bay, New Brunswick, Canada (Davidson-Arnott and Greenwood 1976) however, have a set of well-developed bars which shelter depressions parallel to the coast in which silts and muds accumulate. Waves begin to break on these bars, then reform before dispensing their energy on the beach.

Barrier islands on wave-dominated coasts share much in common with beaches. They are found particularly in coastal environments which have a low gradient continental shelf adjacent to a low-relief coastal plain, an abundant sediment supply and moderate to low tidal ranges (Glaeser 1978). An excellent example of a barrier is Galveston

(a)

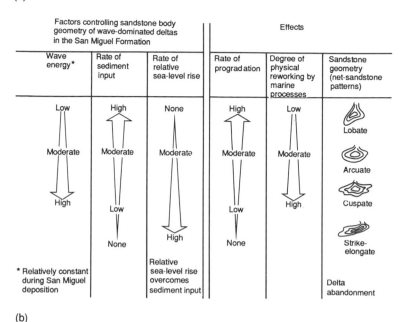

Factors controlling sandstone body geometry of wave-dominated deltas in the San Miguel Formation			Effects		
Wave energy*	Rate of sediment input	Rate of relative sea-level rise	Rate of progradation	Degree of physical reworking by marine processes	Sandstone geometry (net-sandstone patterns)
Low	High	None	High	Low	Lobate
Moderate	Moderate	Moderate	Moderate	Moderate	Arcuate
High	Low	High	Low	High	Cuspate
	None		None		Strike-elongate
* Relatively constant during San Miguel deposition		Relative sea-level rise overcomes sediment input			Delta abandonment

(b)

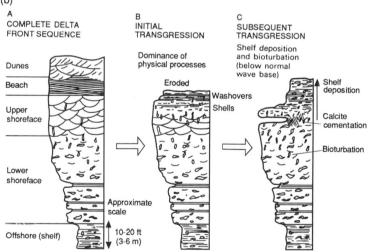

A
COMPLETE DELTA
FRONT SEQUENCE

Dunes
Beach
Upper shoreface
Lower shoreface
Offshore (shelf)

Approximate scale

10-20 ft
(3-6 m)

B
INITIAL
TRANSGRESSION

Dominance of physical processes
Eroded
Washovers
Shells

C
SUBSEQUENT
TRANSGRESSION

Shelf deposition and bioturbation (below normal wave base)

Shelf deposition
Calcite cementation
Bioturbation

Fig. 7.17. Delta geometry and vertical sequences in the wave-dominated Cretaceous San Miguel delta system of Texas (after Weise 1980). (a) Main controls and responses on the San Miguel delta system. (b) Development of delta front sequences; the complete progradational sequence (A) may be modified during subsequent transgression (B) and be capped by deeper water shelfal deposits as the relative sea level rise continues (C). The profiles show the SP response and not strictly grain size.

Island in the Gulf of Mexico (Bernard, Leblanc and Major 1962, McCubbin 1982). Tidal inlets are few and short-lived on wave-dominated barrier islands. Because of the limited connection to the open sea, the lagoons behind such barriers are prone to abnormal salinities (e.g. Padre Lagoon, Gulf of Mexico). Storms are capable of breaching the barrier, however, producing washover channels and fans.

Wave- and tide-influenced shorelines comprise barrier islands highly dissected by tidal inlets and associated tidal deltas (Fig. 7.18). Each segment of the barrier island may have a 'drumstick' shape in which the thicker portion, located on the side of the inlet upcurrent with respect to the longshore drift direction, is composed of beach spits. Tidal inlets which are dominated by tidal flows tend to be relatively stable. The channels extend seaward considerable distances producing ebb-tidal deltas. Tidal inlets with a significant wave influence,

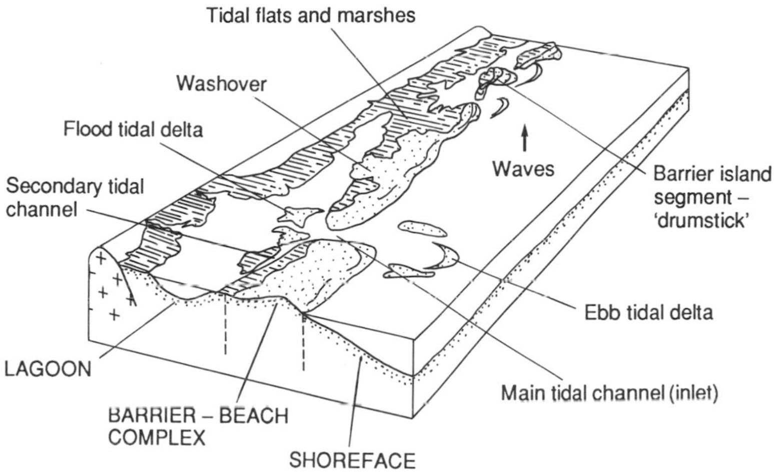

Fig. 7.18. Main geomorphological elements and subenvironments in a barrier island system. The part in the background represents a mixed wave–tide system consisting of short barrier segments with a characteristic drumstick shape (Hayes 1979) separated by numerous tidal inlets. The foreground is more wave-dominated and has a more continuous barrier-beach complex sheltering a well-developed lagoon.

however, are characterized by shallow, generally flood-dominated tidal channels with well-developed flood tidal deltas. Wave-influenced tidal inlets are highly mobile, migrating in the direction of net longshore sediment transport. They therefore progressively replace the wave-built barrier with coalesced tidal inlet sequences.

The prograding barrier model based on Galveston Island on the Texan coast (Bernard *et al.* 1969) was originally thought to represent the 'typical' barrier system model. It is now recognized to be one of a number of distinct stratigraphic possibilities – regressive model (e.g. Galveston Island), transgressive model (e.g. Delaware coast), and migrating barrier inlet model (e.g. Kiawah Island) (Reinson 1984) (Fig. 7.19).

1 Regressive model (Fig. 7.19a)

The regressive model involves a gradational coarsening-up sequence dominated by shoreface, foreshore and backshore-dune facies of the barrier-beach complex. Some ancient analogues possess the 'funnel-shaped' log character. The overriding requirement for regression is an abundant sediment supply, so regressive barriers are commonly associated with deltas.

2 Transgressive model (Fig. 7.19b)

This model is characterized by subtidal and intertidal back-barrier facies and does not show a recognizable fining-up or coarsening-up trend. Be-

cause of the landward migration of facies, the lagoonal deposits underlie the sands of the barrier and inlets.

3 Migrating barrier-inlet model (Fig. 7.19c)

This involves a fining-upward sequence with a thinning-upwards trend in cross-set thickness. The base of the sequence is strongly erosional due to the cutting of tidal channels, and the facies are dominated by sands of the tidal channels and marginal spit-beach environments. A 'bell-shaped' log character might be expected.

When tidal range is large (mesotidal to macrotidal, 2 to > 4 m) estuaries dominate the coastal geomorphology. Fine-grained sediments fringe the estuary in the form of intertidal and supratidal flats, whereas sands dominate the central zone. Here, tidal channels with bars flow between elongate tidal sand ridges, as in the lower reaches of the Ord River in Western Australia. Tidal flats can be extremely extensive on some low wave-energy mesotidal to macrotidal coasts. They are dissected by highly sinuous tidal creeks. Progradation of estuarine depositional systems should give rise to a fining upward sequence as subtidal and low intertidal sandflats are replaced by high intertidal and supratidal mudflats.

Coastal depositional systems are therefore highly sensitive to changes in wave and tidal regime. Since these regimes are themselves controlled by basin size and configuration, there is an underlying

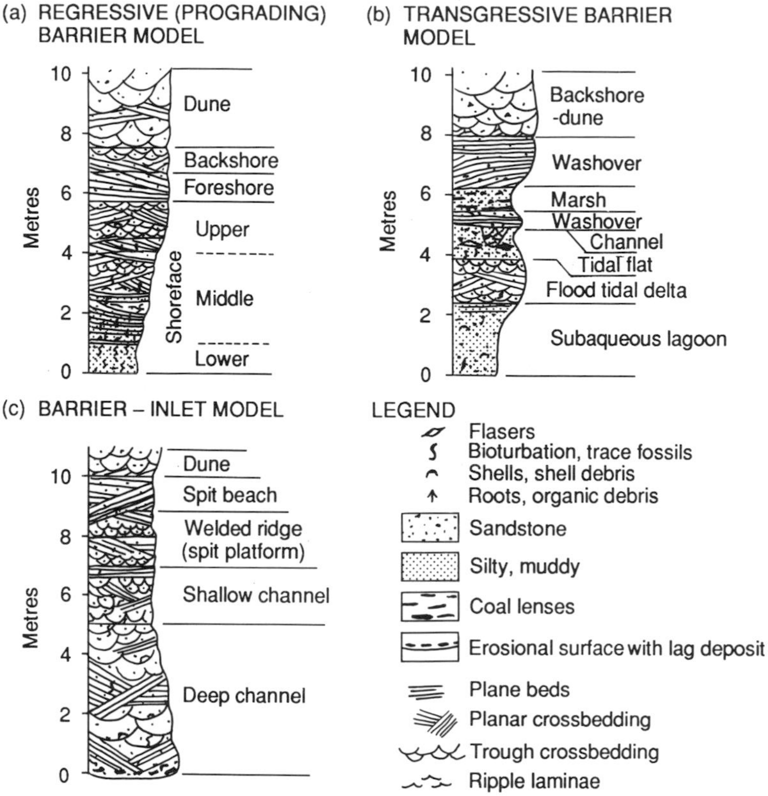

(a) REGRESSIVE (PROGRADING)
    BARRIER MODEL

(b) TRANSGRESSIVE BARRIER
    MODEL

(c) BARRIER – INLET MODEL

LEGEND

Flasers
Bioturbation, trace fossils
Shells, shell debris
Roots, organic debris
Sandstone
Silty, muddy
Coal lenses
Erosional surface with lag deposit
Plane beds
Planar crossbedding
Trough crossbedding
Ripple laminae

Fig. 7.19. Three 'end-member' facies models of barrier island stratigraphic sequences. Vertical scale of sequences is approximate only (after Reinson 1984).

tectonic influence on the development of siliciclastic coastal depositional systems.

7.1.3.2 Carbonate and evaporite shoreline systems

Arid shorelines with low terrigenous input are characterized by deposition of carbonates and evaporites. The Trucial coast, Persian Gulf is an example of a modern carbonate-rich marginal marine *sabkha* (Fig. 7.20). A wide variety of coastal geomorphological elements are present, including beaches, barrier islands, tidal channels and associated tidal deltas, intertidal and supratidal flats and aeolian dunes. The beaches, tidal deltas and bars are commonly composed of oolitic-skeletal grainstones, whereas the back-barrier lagoons accumulate pelletal muds and, where predation is restricted, stromatolites. The upper intertidal zone is dominated by algal mats, constituting the lowest part of the sabkha *sensu stricto*. The supratidal zone is the main part of the sabkha and is the site of the precipitation of evaporitic minerals in the sediment column, as surface encrustations and in small ponds. Detrital carbonate grains are transported onto the sabkha by marine flooding.

Arid shorelines may also be dominated by siliciclastic sedimentation, as in Baja California, the Gulf of Elat and some parts of the Arabian Gulf. Here the sabkhas are composed of siliceous sands and muds with a possible admixture of carbonate grains. In many ways siliciclastic sabkhas are similar to their carbonate-rich counterparts. During progradation of the sabkha, normal marine sediments are first overlain by intertidal deposits, then supratidal sabkha evaporites, giving a vertical sequence indicative of increasing salinity (see below). The thickness of the vertical sequence should approximate the tidal range of the adjacent sea, that is, 0.5 to 3 m. In some sabkhas, however, the deposits are arranged in a 'bulls-eye' pattern, suggesting the progressive filling of water bodies rather than shoreline progradation.

The preservation of very thick sabkha deposits (> a few metres) suggests that either subsidence matched the rate of deposition over a long period

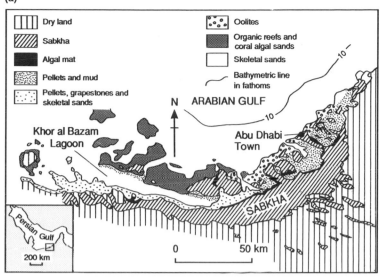

(a)

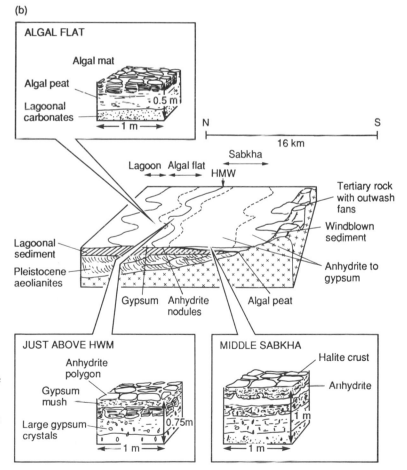

Fig. 7.20. Sedimentary facies of the arid shoreline of the Trucial Coast, Persian Gulf. (a) Map of coastal facies in the Abu Dhabi region (after Butler, Harris and Kendall 1982). (b) Intertidal and supratidal sediments. The algal flat found below high watermark is characterized by algal mats formed into polygons overlying a poorly laminated carbon-rich algal peat which in turn overlies the carbonate sands and muds of the lagoon. Shallow sections from just above high watermark possess surface anhydrite polygons with windblown carbonate and quartz grains overlying a crystal mush of gypsum and carbonate. These supratidal layers overlie the intertidal algal flat facies. The main part of the supratidal sabkha is characterized by a surface halite crust deformed into compressional polygons by evaporite growth, overlying an anhydritic layer that has replaced the gypsum crystal mush. Salt mobilization leads to the formation of upward penetrating salt pods and diapirs. Coastal progradation leads to the deposition of these sabkha facies over the high watermark and intertidal units (after Butler *et al.* 1982).

of time or that sea level was rising at the same rate as deposition.

Carbonate sediments are produced in great abundance in shallow, warm waters where the biological and physicochemical conditions are optimal for carbonate precipitation and fixation. Because the sedimentation rate commonly outstrips the rate of subsidence, the sedimentary surface shallows with time, giving a *shallowing upward sequence*. The sequence typically has four or five parts comprising a basal high-energy transgressive deposit, subtidal carbonates, intertidal stromatolitic sediments and finally supratidal to terrestrial deposits. Intertidal sediments may be of low-energy tidal-flat type or may be high-energy beaches.

1 Sequences containing low-energy tidal flats

There are a number of well-studied modern examples of carbonate tidal flats, e.g. southern coast of Persian Gulf, Bahama Banks platform and Shark Bay, Australia (see summary by Shinn 1983). The main morphological elements are:

1 A protective barrier of carbonate sand shoals, islands and reefs, dissected by tidal channels.
2 A shallow muddy lagoon.
3 Tidal flats occurring along the landward edge of the barrier, and as a wide belt attached to the mainland.

There are two contrasting sequences that may result from either progradation of the wide shore-attached tidal flats or by shoaling of the offshore barrier, termed a *muddy sequence* and a *grainy sequence* respectively (James 1984a, p. 218) (Fig. 7.21).

A variation on the muddy and grainy sequences is the development of abundant stromatolites or reefs, the former being particularly common in the Precambrian and Early Palaeozoic and locally found today in Shark Bay, Australia (Hoffman 1976). Shoaling of large bioherms in the back-reef environment of reef complexes (see Section 7.1.4.2) also results in a capping of beach carbonate sands or conglomerate and fenestral and laminated intertidal sediments.

As we have seen, in arid climates the intertidal, and especially the supratidal zone of low-energy

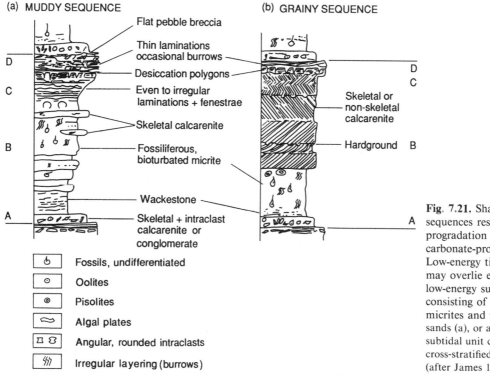

(a) MUDDY SEQUENCE

Flat pebble breccia

Thin laminations occasional burrows

Desiccation polygons

Even to irregular laminations + fenestrae

Skeletal calcarenite

Fossiliferous, bioturbated micrite

Wackestone

Skeletal + intraclast calcarenite or conglomerate

(b) GRAINY SEQUENCE

Skeletal or non-skeletal calcarenite

Hardground

⌀	Fossils, undifferentiated
◦	Oolites
◉	Pisolites
⌒	Algal plates
⊐ ⊂	Angular, rounded intraclasts
𝔰𝔥	Irregular layering (burrows)

Fig. 7.21. Shallowing-upward sequences resulting from the progradation of a carbonate-producing shoreline. Low-energy tidal flat deposits may overlie either a low-energy subtidal unit consisting of bioturbated micrites and thin skeletal lime sands (a), or a high-energy subtidal unit consisting of cross-stratified lime sands (b) (after James 1984a).

coasts is commonly the site of growth of evaporite minerals (Fig. 7.20). If the groundwaters are continuously within the field of gypsum precipitation, evaporite minerals grow as a gypsum mush in the intertidal sediments, as coalescing nodules (chicken-wire texture) or as folded layers ('enterolithic' texture). Formation of evaporite minerals is accompanied by widespread dolomitization of intertidal zone sediments.

Collapse of evaporite-bearing layers takes place if the sequence is flushed by meteoric 'fresh' waters, producing brecciated horizons. Anhydrite crystals are commonly leached, forming vugs which may subsequently be filled with quartz or chalcedony, and the dolomite may be calcitized (so-called 'dedolomitization').

2 Sequences containing high-energy intertidal units

These sequences differ from those containing extensive low-energy tidal flats in having foreshore sediments as an important component. These sediments are generally cross-stratified carbonate sands. Calcretes, karsts or soils may cap the shallowing-up sequence (Fig. 7.22).

The mechanism for producing repeated shallowing upward sequences is not well known (Wilkinson 1982, see also Chapter 6). An allocyclic eustatic mechanism involving repeated rapid sea

level rises has been suggested, as has a purely autocyclic mechanism involving changes in the rate of carbonate sedimentation controlled by the subtidal source area (Ginsburg 1971). As the tidal flat wedge progrades onto the shallow subtidal shelf, the area of source area is reduced, so that progradation is self-limiting. Continued relative sea level rise (caused by a eustatic rise or subsidence) floods the tidal flat wedge and the process of progradation begins again.

Carbonate shelf systems and reefs are discussed in Section 7.1.4.2.

7.1.4 Continental shelf systems and facies

7.1.4.1 Siliciclastic shelf systems

Continental shelf systems are extremely complex and are highly sensitive to sea level fluctuations.

Modern siliciclastic shelves contain three main sedimentary facies associations based on the physical processes operating in the nearshore-inner shelf zone, the rate of sea level fluctuation and the nature and rate of sediment supply (Curray 1964, 1965). These are:
• *shelf relict sand blanket* composed of pre-Holocene deposits out of equilibrium with present-day processes
• *nearshore modern sand prism* which thins seaward
• *modern shelf mud blanket* composed of fine sediment which has escaped from the nearshore zone into deeper water depths.

Some sediment on the shelf (perhaps 50 per cent of the Earth's shelf area), is *relict*, that is, it is remnant from an earlier environment and is now out of equilibrium with the new environment; other sediments are termed *palimpsest* which means that they are reworked and therefore possess aspects of both their present and former environments; finally, some sediment is modern and is supplied from outside the shelf area.

Modern continental shelves can therefore be classified according to the nature of the sediment (relict, palimpsest, modern) and the hydraulic regime (wave-, tide-, storm- or oceanic current-dominated) (Swift 1974, Johnson and Baldwin 1986) (Fig. 7.23). Classifying shelves according to the dominant shelf currents (Swift, McKinney and Stahl 1984), tide-dominated shelves occupy

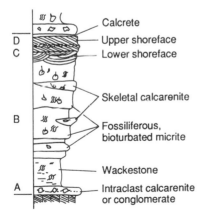

Fig. 7.22. High-energy intertidal sediments such as cross-stratified and parallel-laminated lime sands of the shoreface and beach subenvironments may cap high or low-energy subtidal units (after James 1984a).

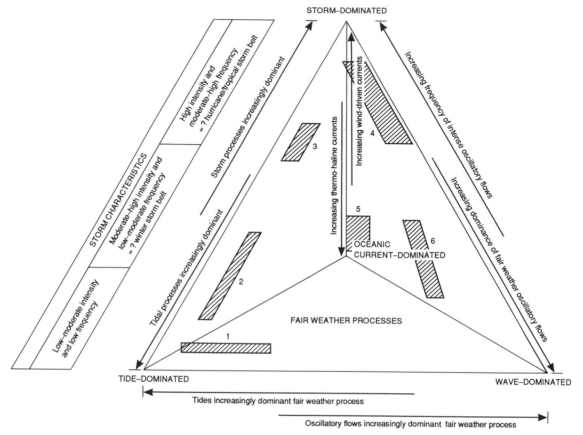

Fig. 7.23. The main types of hydraulic regime on modern shelves (modified from Johnson and Baldwin 1986). Modern shelves may be tide-dominated, ocean current-dominated or wave-dominated where fairweather processes are of greatest importance, or storm-dominated where fairweather processes are of lesser importance. Shelf currents capable of transporting large amounts of sediment occur on shelves dominated by storms, tides or intruding ocean currents.
1. Macrotidal and mesotidal embayments and estuaries (e.g. Bay of Fundy, Nova Scotia; Chesapeake Bay, US eastern seaboard; German Bight of the North Sea).
2. Tidal straits and seas (e.g. English Channel, Malacca Strait, Taiwan Strait, North Sea, Yellow Sea).
3. Winter storm and tide-influenced shelf of the NW Atlantic, off eastern North America.
4. Storm-dominated shelves (Oregon–Washington and Californian shelves; SE Bering Sea, Gulf of Mexico).
5. Ocean current-swept shelves (e.g. SE Africa, Moroccan shelf off NW Africa).
6. Low-energy embayments and shelves (Amazon–Orinoco shelf, Niger shelf, Baltic Sea, Hudson Bay).

17 per cent of the present-day shelf area, storm-dominated 80 per cent, and shelves dominated by intruding ocean currents a very small percentage.

Tide-dominated shelf sedimentation appears to take place along distinct tidal current transport paths (Belderson, Johnson and Kenyon 1982). The sedimentary facies found along these transport paths reflect both the tidal current velocity and the type of sediment available. Where sand supplies are low, proximal furrows and gravel waves pass down-tidal current paths into longitudinal sand ribbons, sandwaves and finally sand patches. Where sand supply is abundant, proximal sand ribbons pass down the tidal current path into sandbanks (or sand ridges), sandwaves and rippled sand sheets. Zones of mud accumulation are generally found at the ends of the tidal current transport paths.

The sediments deposited under the strong tidal flows of the northwestern European continental shelf (Stride 1963, 1982) show marked changes along the tidal sediment transport path. Bare rock with a gravel lag passes into longitudinal sand ribbons and then into a field of sandwaves up to 10 m high. The sandwaves have been investigated by shallow geophysical methods to reveal their internal structure (Berné, Auffret and Walker 1987).

Tidal sand ridges are up to 40 m high, 5 km wide and up to 60 km long (Houbolt 1968); they have internal reflectors dipping at up to 6°, presumably representing internal master bedding surfaces.

Although there is a handful of case studies (e.g. Jura Quartzite, Anderton 1976, Lower Greensand, Bridges 1982), there are at present no tidal shelf models that can act as a 'norm'. Facies sequences should, however, be dominated by the deposits of flow-transverse tidal sandwaves and/or flow-parallel tidal sand ridges with a stratification pattern indicative of systematic flow reversals or, more likely, systematic variations in the sediment transport rate. The deposits of the early Miocene tide-dominated seaway north of the Alps include coquina banks representing large flow-transverse bedforms constructed by highly asymmetrical tidal currents (Allen *et al.* 1985).

Storm-dominated shelves are generally dominated by the accumulation of mud derived from major river mouths, with sand being concentrated on the inner shelf, as on the southern Oregon shelf (Fig. 7.24). Repeated transgressions and regressions may clean up the muds, concentrating the sand into distinct sediment bodies. On the Atlantic shelf of North America, the inner shelf is characterized by a complex ridge topography. These ridges are oblique to the shoreline and probably originated by the detachment of formerly shoreface attached shoals during transgression.

The *storm-dominated* continental shelf of the Eastern Seaboard has been studied in considerable detail (Swift *et al.* 1972, Swift and Field 1981 and many others). The shelf is characterized by three scales of sandbody, (1) *shoal retreat massifs* which are drowned coastal and estuarine depocentres, (2) *linear sand ridges* superimposed on the shoal retreat massifs, and typically orientated at an angle of *c.* 20° to the shoreline. They are thought to be due to flooding and detachment of ridges attached to the shoreface (Swift and Field 1981), but they

are presently active. Others are approximately shore-parallel. Some workers (e.g. Stubblefield, McGrail and Kersey 1984) believe them to be transgressed and degraded former barriers which are relict (Fig. 7.25). (3) *Dunes* cover large parts of the inner shelf and migrate under storm-driven flows. They are modified during fairweather periods by benthic animals and by the formation of wave ripples.

Some of the facies sequences attributed by geologists to storm-dominated continental shelves do not match particularly well the observations of physical oceanographers (Walker 1984a for discussion). The common association is of beds of hummocky cross-stratification or graded bioclastic sandstones intercalated with bioturbated mudstones. The sharp-based beds are thought to have been formed by the offshore transport of sand in turbidity currents in the case of the Upper Jurassic Fernie–Kootenay transition in Alberta, Canada (Hamblin and Walker 1979). The beds containing hummocky cross-stratification are thought to be due to reworking by the action of storm waves, or to storm waves superimposed on an undirectional geostrophic flow. A similar mechanism is suggested for other formations in the Cretaceous of Alberta such as the Cardium Formation (Walker 1984b). The vertical sequence coarsens up from turbiditic sandstones and shales to a hummocky cross-stratified interval and then to a beach and fluviatile unit containing coals.

Other facies sequences can more easily be reconciled with the oceanographical studies. The Shannon and Sussex Sandstones of the Cretaceous Western Interior Seaway have been interpreted as due to the progradation of storm-driven linear sand ridges. The vertical sequence is strongly coarsening-up, from shelfal muds to rippled sands and finally to cross-stratified sandstones with abundant mudstone rip-up clasts thought to have been deposited on the ridge flanks. The coarsening-up sequence is capped by a pebbly sand related to wave reworking of the ridge crest (Berg 1975 and others cited in Tillman and Siemers 1984). The Duffy Mountain Sandstone of northwestern Colorado contains low-angle accretionary surfaces dipping towards the palaeoshoreline and shares many of the characters of the Sussex (Boyles and Scott 1982). The Duffy Mountain Sandstone has been interpreted as due to the migration of elongate storm-driven sandridges.

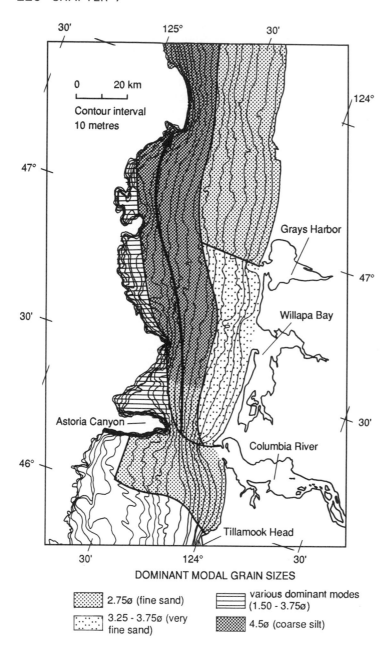

DOMINANT MODAL GRAIN SIZES

2.75ø (fine sand)

various dominant modes (1.50 - 3.75ø)

3.25 - 3.75ø (very fine sand)

4.5ø (coarse silt)

Fig. 7.24. Modal grain size of surface sediment on the Washington continental shelf based on factor analysis of box-core samples (Nittrouer and Sternberg 1981). Relatively coarse-grained samples (fine sand) are found close to river entry points, such as at the mouth of the Columbia River, and sand percentages generally decrease away from the river input. There is a distinctive mid-shelf mud deposit (consisting primarily of coarse silt) that extends northwards along the shelf at a small angle to the isobaths. This is thought to represent the pathway for transport and accumulation of the bulk of the sediment associated with the Columbia River dispersal system (after Nittrouer *et al.* 1986). The arrow is the predicted transport route of fine sediment carried in suspension for the extreme winter storm events modelled by Kachel and Smith (1986).

In all cases, however, the vertical sequence is coarsening-up with a gradational base, and there is no suggestion that the sand bodies nucleate over a transgressive erosional surface. The geophysical log pattern would be funnel shaped.

The southeastern African shelf is the best documented example of an *oceanic current dominated* shelf (Flemming 1978). The Agulhas Current flows parallel to the coast but somewhat offshore, and the irregular coastline causes the inner shelf to be broken into several sedimentary compartments. The nearshore sediment prism is therefore wave-dominated whilst the sand stream offshore is dominated by the Agulhas current. Sandwaves

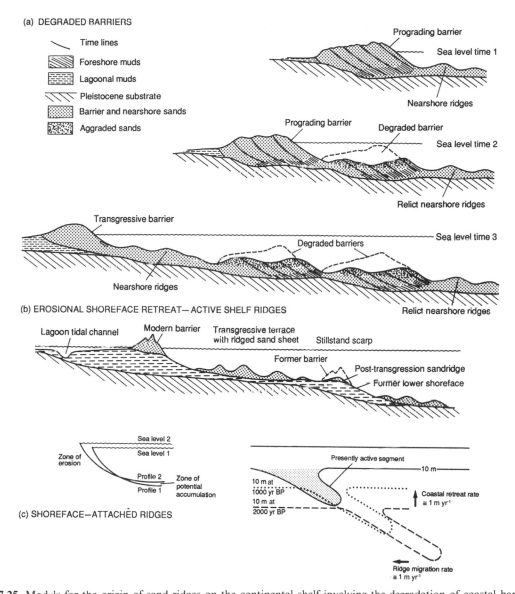

Fig. 7.25. Models for the origin of sand ridges on the continental shelf involving the degradation of coastal barriers and the active growth of sand ridges following relative sea level rise and transgression.
(a) Schematic illustration of Stubblefield *et al.*'s interpretation of the development of the sand ridges on the New Jersey Shelf. The ridges presently on the shelf are viewed as the degraded remnants of former coastal barriers drowned during transgression. After Stubblefield, McGrail and Kersey (1984). They are shore-parallel and can be termed *degraded barriers* (Stubblefield *et al.* 1984, p. 19). Nearshore shoreline-oblique ridges are viewed as active ridges in equilibrium with the present storm regime. They are termed *active shoreface-connected/nearshore ridges* by Stubblefield *et al.* (1984, p. 19). Shoreline-oblique ridges on the outer shelf are thought to be *relict* equivalents.
(b) Schematic illustration of the shelf stratigraphy resulting from erosional shoreface retreat. This involves the destruction of former coastal barriers and the growth of *active* post-sea level change ridges on the inner shelf. After Swift, McKinney and Stahl (1984). If this is correct, the lower parts of the ridges should consist of lagoonal, back-barrier or former estuary mouth bar deposits. If the ridges are degraded barriers, the lowermost deposits of the ridges should represent shoreface sedimentation during a former period of coastal progradation.
(c) As the shoreface retreats, ridges attached to the shoreface migrate along the coast, so that the ridge-generating zone is maintained at a near-constant water depth near the base of the shoreface (< 10 m). After Swift and Field (1981).

occur in extensive fields (up to 20 km long) and reach up to 17 m in height with angle of repose lee faces. Other parts of the sea bed are characterized by longitudinal ribbons and coarse lags. We are not aware of any ancient analogues, so a facies model is not available at present.

7.1.4.2 Carbonate shelf systems and reefs

Much of the continental shelf between the latitudes of 30 °S and 30 °N is an area of high organic productivity and is covered not by river-derived or relict siliciclastic sediments, but by organic carbonate material. There are two major categories of *subtropical carbonate shelf* (Ginsburg and James 1974). (1) *Rimmed shelves* sheltering protected shelf lagoons. Their margins often fall precipitously into the abyssal depths. Some rimmed shelves are attached to continental areas as in the Great Barrier Reef of Australia. Others are now *isolated platforms* as in the Bahamas. (2) *Open shelves* on the other hand, such as Yucatan, western Florida

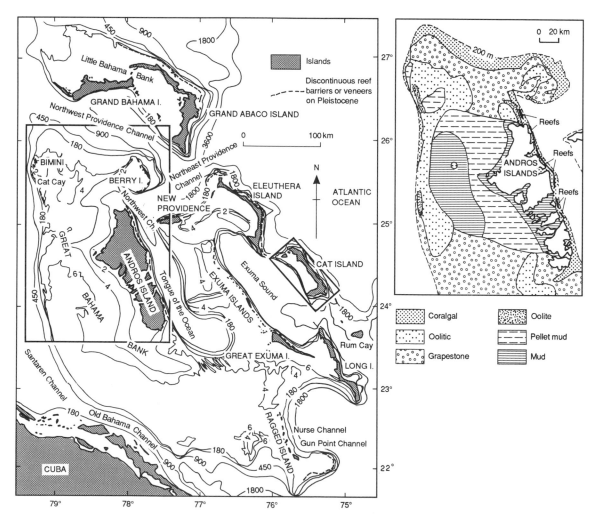

Fig. 7.26. General bathymetry of the Bahamian–Florida area, showing depths in metres (after Multer 1971, Bathurst 1971, modified from Sellwood 1986). The area is a recently flooded *rimmed* shelf. The trace of the 6 m contour gives an impression of the former extent of the subaerial portion of the shelf prior to *c.* 7000 years BP (see also Fig. 7.29). Inset shows sediment distribution on the Great Bahama Bank (after Purdy 1963a, b).

and northern Australia, slope gently towards the continental edge and are termed 'ramps'. Because of the lack of a protective rim, they are strongly affected by storm waves and tidal currents.

The Bahama Platform and South Florida shelf are the vestiges of a former extensive and continuous rimmed shelf (Fig. 7.26). The Bahama Platform is bounded on all sides by steep slopes which plunge to thousands of metres depth, but the platform itself lies under very shallow waters normally less than 6 m deep. At the edge of the platform are reefs and a zone of oolitic shoals built by waves and tides. Storms break through the oolitic shoals depositing washover lobes ('spill-overs') in the quieter water inside the highly mobile belt. These washover deposits are interbedded with the background sedimentation of carbonate mud.

The Yucatan shelf is quite different. It lacks a guarding rim of reef barriers so represents a high-energy environment with an abundance of coarse-grained shell debris. The swampy mudflats, shell beaches and dune ridges of the shoreline pass into a coarsely bioclastic inner shelf, and finer-grained carbonates are found on the deeper outer shelf where sedimentation is dominated by a rain of planktonic organisms such as foraminifera. The Persian Gulf is an example of an open shelf in an arid climate (Fig. 7.27).

(a)

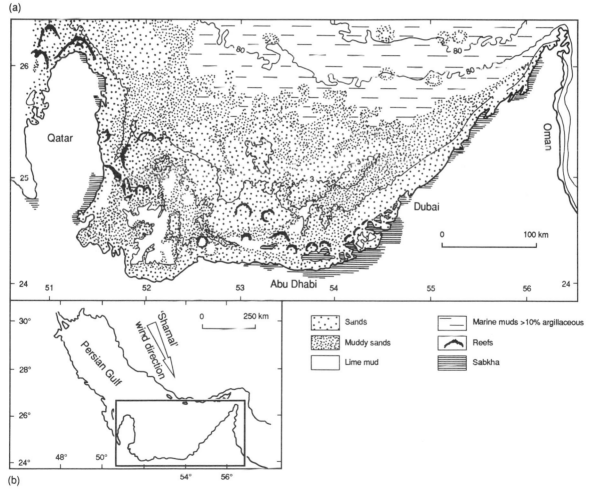

(b)

Fig. 7.27. The Persian Gulf is an example of an open shelf, unprotected by a rim of reefs. (a) Bathymetry and facies on part of the southern shelf of the Persian Gulf (after Wagner and van der Togt, 1973). Sediment transport on the shelf is influenced by the predominant 'Shamal' wind from the NNW, shown in (b).

Like siliciclastic continental shelves, facies patterns on carbonate shelves are strongly influenced by changes in relative sea level. The importance of sea level changes can be gauged from the responses to the Quaternary glaciation. The last glacial stage (~120 000–10 000 yrs BP) and its lowstand resulted in an erosional unconformity which is now overlain by unconsolidated sediment. Most recent sediment is consequently accumulating under a relative sea level rise. Shelf lagoons were karstified during the lowstand and early stages of transgression, and were abruptly inundated only during the final stages of the transgression (5000 yrs BP) (Fig. 7.28); open shelves, however, were inundated at a much earlier stage (beginning at ~20 000 yrs BP) and depth-controlled facies belts have been gradually displaced landwards during the sea level rise.

Reefs are *biogenic* constructions on the seafloor and reef facies models must successfully integrate sedimentological and palaeontological observations. Reefs can generally be divided into a (1) reef core comprising skeletons of reef-building organisms and a lime-mud matrix, (2) reef flank of bedded reef debris and (3) inter-reef of subtidal shallow marine carbonates (or siliciclastics). Where reefs form a natural breakwater on the windward sides of shelves or islands, however, they protect a back-reef environment from wave attack. Here the arrangement of reef facies is strongly asymmetrical.

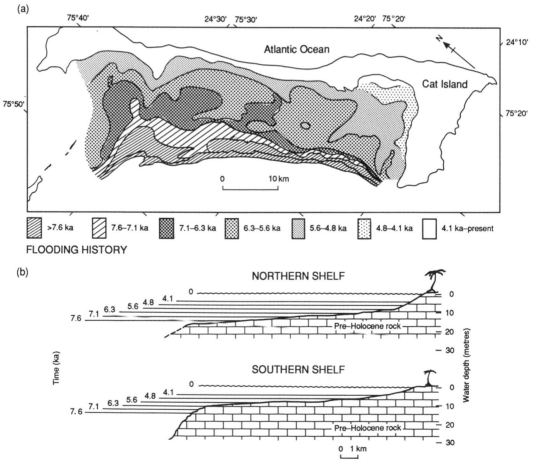

Fig. 7.28. Timing of flooding of steep shelf lagoons; an example from Cat Island, Bahamas. Cat Island platform is located on Fig. 7.26. (a) Plan view of flooding history for the Cat Island shelf. The topography of the shelf strongly controlled the timing of flooding, the northern area being flooded prior to the southern area. (b) Cross-sectional aspect of (a). After Dominguez, Mullins and Hine (1988).

• *High-energy reefs* are zoned into reef crest, reef front, reef flat, back reef and fore reef.

• *Low-energy reefs* are less well zoned and commonly occur as isolated, circular to elliptical 'patch-reefs'.

• *Reef mounds* are flat lenses to steep conical piles of poorly sorted bioclastic lime mud which accumulated in quiet water. Some reef mounds possess no large skeletons, being dominated by lime mud. These 'Waulsortian mounds', or mud mounds, occur in deeper water on carbonate slopes.

• *Stromatolite reefs* are common in the Precambrian and Palaeozoic, before the appearance of grazing metazoans. Their overall geometry and growth morphology as a function of position on the platform is similar to the distribution of metazoan reefs on Phanerozoic platforms (Hoffman 1974, Grotzinger 1989).

A facies model for reefs has been proposed (James 1984b, p. 237) which is based on four growth stages:

1 Pioneer stage when loose sediment accumulations are stabilized by organisms with roots or holdfasts.

2 Colonization stage during which reef-builders join the stabilized mound.

3 Diversification stage, representing most of the reef mass, is characterized by rapid upward growth towards sea level and a distinct zonation develops.

4 Domination stage at which point the reef has built to an elevation that produces a surf zone and reef flat.

Many reefs show stacked patterns of reef growth separated by horizons testifying to dissolution, karst and paleosol development and the formation of hardgrounds.

Reef facies models are complicated by the changing nature of the biota through geological time. James (1983) has reviewed these changes. For example, in terms of reef frame-builders, the dominant group has changed through the Phanerozoic as shown in Fig. 7.29. One should therefore expect to see important variations in the diversity of reef types and their internal constitution through geological time.

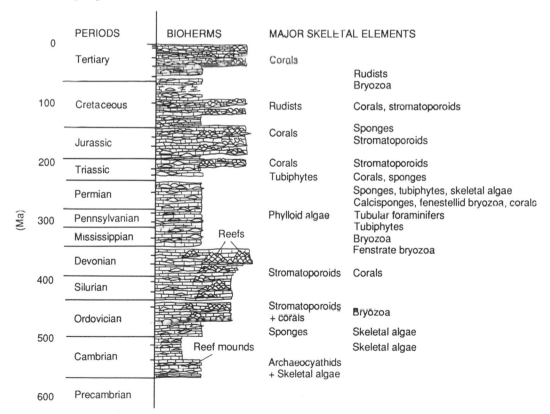

Fig. 7.29. Phanerozoic evolution of the major skeletal elements in reefs and reef mounds. Gaps in the sequence are periods when no reefs or bioherms existed (after James 1983).

7.1.5 Deep sea systems and facies

There are three fundamentally different environments of deposition of clastic sediments in the deep sea:

• *Slope-aprons,* which accumulate between the shelf and basin floor, vary in width from less than 1 km to over 200 km. Normal clastic slope aprons have a smooth convex-concave profile built by slope progradation. Faulted slope aprons typically have a highly stepped profile with perched basins alternating with steeply dipping slope segments. Carbonate slope arpons may form against reef edges or carbonate shoal margins. Where such margins are steep, sediment largely bypasses the slope apron and it is dominated by calcirudite talus wedges. Where the margins are gentle, the slope apron is more actively depositional and more like its clastic counterpart.

• *Submarine fans* are constructional bodies that build oceanward at the base of the shelf slope. They receive sediment from river mouths or from alternative feeder systems such as submarine canyons. Fans vary greatly in scale and geometry, but there appear to be two end members, radial and elongate (Stow 1985) (Fig.7.30).

Radial fans develop concentrically around a single feeder canyon or channel (examples are the La Jolla and Navy fans off California). Elongate fans often have two or more feeder channels and extend for considerable distances from the supply margin. Examples are the enormous Bengal (off the Ganga delta), Amazon, and Laurentian (eastern North America) fans, and the smaller Crati (Italy, Colella 1981, Ricci Lucchi *et al.* 1984) and Reserve (Lake Superior, Normark and Dickson 1976) fans. Radial and elongate fans correspond essentially to the low efficiency and high efficiency fans of Italian workers (e.g. Mutti 1979) (see p. 227).

• *Basin plains* are flat, relatively deep areas which act as the ultimate sediment traps for clastic sediments eroded from the continents and from submarine highs. Facies types and distributions are controlled primarily by basin geometry, tectonics and source area (Pilkey, Locker and Cleary 1980). Abyssal plains are extensive and elongated parallel to the adjacent continental margin. The Hatteras Abyssal plain on the western North Atlantic is an example. Marginal seas also have basin plains, fed from widely varied sources with a centripetal pattern (e.g. Sigsbee basin plain, Gulf of Mexico,

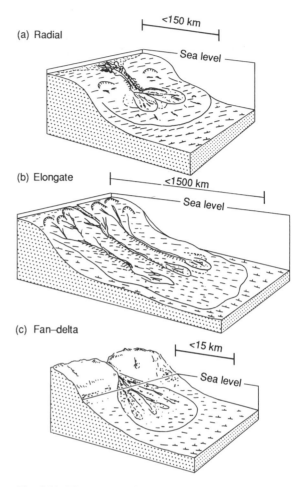

Fig. 7.30. Three types of sedimentary environment for submarine fans (after Stow 1985). Some vertical exaggeration; the steepest gradient is about 10 degrees.

Davies 1968). Deep sea trenches may consist wholly or in part of basin plains. In this case the plain parallels the highly elongate to arcuate trench orientation, and the dispersal of sediment by turbidity currents is also generally longitudinal. Strike-slip basins or borderland basins may also contain small, relatively confined basin plains.

The facies models of deep sea sedimentation of sands and gravels have been dominated by the concept of sediment gravity flows in general and turbidity currents in particular. Turbidity currents are responsible for the transport of enormous quantities of sand and mud into the deep sea. Turbidite beds can be recognized by their sharp, often erosive bases ornamented with sole marks, upward grading of grain size and a vertical sequence of sedimentary structures. These features are embodied in the *Bouma sequence* (Fig. 7.31). The vertical sequence reflects the passage of the head, then body and tail of the turbidity current and the succession of stratification types demonstrates an upward (and therefore temporal) decrease in flow energy. Early views are found in Bouma (1962) whilst Lowe (1988) provides a modern reanalysis incorporating the effects of sediment concentration.

The greatest volume of modern turbidites accumulates in submarine fans and this depositional system has received strong emphasis, perhaps overemphasis, in geological studies.

Early fan models were derived from the small submarine fans of the California Borderland (e.g. Normark 1970, 1978, Normark, Piper and Hess 1979). The essential features of these fans are a leveed valley on the upper fan, a region of aggradation of depositional lobes in the mid-fan and a flat lower fan lacking channels (Fig. 7.32). Active lobes fed by feeder channels are abandoned abruptly as a levee break in the channel initiates a new transport path to another lobe on the fan surface. The formerly active lobe is then blanketed in fine-grained sediment. This leads to a vertical alternation of lobe sands and intercalated muds. Progradation and build-up of the lobe may be responsible for *coarsening-up* and *thickening-up* sequences in the turbidite beds (Mutti and Ghibaudo 1972). Gradual channel filling and abandonment on the other hand would result in a *fining-up* and *thinning-up* sequence. However, such sequences may be produced by other autogenic processes such as lateral shifting of lobe facies (from lobe centre to lobe fringe) or of channel-levee facies associated with channel migration. Allocyclic processes are also increasingly viewed as important in determining fan sequences. As described elsewhere (Chapter 6), gradual relative sea level rise may have the effect of reducing the sediment transport to the entire fan, giving a fining upward sequence. A relative fall of sea level would promote a coarsening upward response in the depositional system.

Large submarine fans such as the Amazon Cone appear to suffer enormous slope failures producing slump and debris flow complexes (Damuth and Embley 1981). Channel-levee complexes with surprisingly sinuous channel patterns have been described from the surface of the Amazon Cone.

The concept of *transport efficiency* is useful in examining fan morphology and sequence (Mutti 1979). Low efficiency fans for example, trap coarse clastic sediment in proximal channel complexes so that lobes are starved of sand. Only in high-efficiency fans are supra-fan lobes fed sufficiently with sand-grade sediment.

Elongate but narrow troughs are commonly filled by turbidites which were deposited by longitudinal currents. In these cases traditional fan models are unlikely to closely resemble the observed facies associations.

Far-travelled, low viscosity turbidity currents characterize the basin plain. The basin-plain turbidites are interbedded with the deposits of ocean-bottom currents (Heezen and Hollister 1964). These undercurrents flow parallel to the bathymetric contours, and their deposits have been termed *contourites* (Stow and Lovell 1979).

Sedimentation in the open sea beyond the influence of the continental land masses is controlled by two major factors, the fertility of the surface water and the presence of the 'calcite compensation depth' (CCD) below which carbonate from the

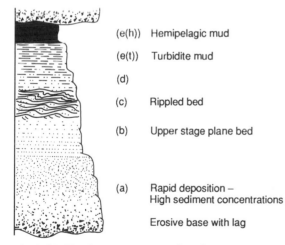

Fig. 7.31. The Bouma sequence, thought to represent the passage of first the head, then the tail of a turbidity current. The vertical sequence in a complete Bouma sequence is (a) massive or graded unit, (b) parallel laminations in sand, (c) rippled and commonly convoluted unit, (d) unit with fine parallel interlaminations of mud and silt, (e(t)) mud deposited from the tail of the turbidity current, (e(h)) mud from hemipelagic fall-out.

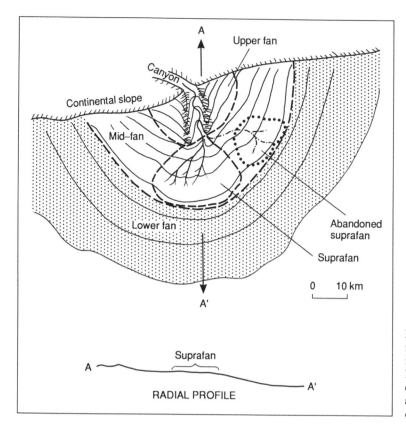

Fig. 7.32. Classical submarine fan model developed from small cones in the California Borderland (after Normark 1978). This simple model emphasizes a channelled upper fan, and active and abandoned depositional lobes in the mid-fan.

skeletons of marine organisms is dissolved. Above the CCD, calcareous oozes derived from micro-organisms such as formanifera predominate. Below this, the silica skeletons of radiolaria and diatoms produce siliceous oozes and there are also regions of sea floor dominated by red and brown clays derived from volcanoes, meteorites and dust blown from continents. The reader is referred to the review by Jenkyns (1986) and we do not go into further detail here.

7.2 RELATION OF DEPOSITIONAL STYLE TO BASIN SETTING

Sedimentary basins form by deformation of the lithosphere (stretching, cooling, bending) and their stratigraphic patterns primarily reflect the various allogenic processes, principally climate and tectonics, causing relative sea level change.

It follows therefore that basins of a similar genetic type may show a consistent pattern in their

sedimentary evolution, whilst basins of different type show correspondingly different sedimentary styles. Knowing the formative mechanism of a basin consequently has predictive power in assessing the basin-fill. In the following sections some characteristic sedimentary histories of various basin types will be sketched out. However, this should not be done in isolation from the particular 'basin-specific' tectonic and burial history. The impact of such basin-specific factors superimposed on the generalized features is therefore emphasized in the following sections.

7.2.1 Basins related to divergent motion

In Chapter 3 we considered the mechanisms of formation of continental basins associated with rifting. Some of these basins appear to be purely sags with no obvious rifted basement. Others are clearly fault-bounded rift valley basins and yet others are rift basins which have since undergone a widespread subsidence unrelated to active tectonics (failed rifts or aulacogens). Large amounts of stretching lead to the formation of passive margins. The underlying process which unites these various basin styles is lithospheric thinning and associated thermal disturbances. In this section we are not so much concerned with the relative roles of mechanical stretching (passive rifting) or thermal updoming (active rifting) as with the kinds of sedimentary infillings of these basins formed on continental lithosphere.

7.2.1.1 Intracratonic sags

Intracratonic sags are characterized by prolonged but slow subsidence and a lack of strong synsedimentary structural activity. The sedimentary systems filling intracratonic sags are most commonly continental, often in the form of rivers draining into centrally located, shallow lakes. Such endorheic systems are found in the Chad Basin of north central Africa and the Eyre basin of southern Australia. In other intracontinental sags, shallow seas are able to enter the basin, depositing marine sediments. Hudson Bay and the Palaeozoic Michigan Basin in North America are good examples.

There are at least two plausible explanations for the development of intracratonic sags. The first is that intracratonic subsidence is a response to thermal recovery of the lithosphere following a thermal disturbance. The second is that the subsidence of intracratonic sags is primarily caused by the effects of water and sediment loading.

The *Chad Basin* is an excellent example for study since it is relatively young. It is situated deep in the African interior, more than 500 km from the nearest sea. The watershed bounds a roughly square area with sides of about 200 km. The shallow (< 10m) centrally located lake occupies at present a relatively small area (30 000 km^2) compared to its precursor of 10 000 years ago, Lake Megachad (Grove and Warren 1968). Lake Megachad occupied an area of nearly 1.5 million km^2, but was unable to enlarge further because it drained to the Atlantic via the River Benue through a gap in its hinterland. The former extent of Lake Chad can be precisely determined from shoreline beach ridges (Fig. 7.33).

Most of the subsidence of the Chad Basin has taken place in the Neogene and it has been concentrated in the zone within the confines of the beach ridges represented by the maximum limit of Lake Megachad. Beyond this 'shoulder' (Fig. 7.33) there is less than 200 m of sediment. The alternating clays and sands of the Pleistocene probably represent wet-dry alternations related to northern hemisphere ice cap growth and decay. The oldest deposits are Neogene in age (Kerri Kerri Sandstones) and are located in the SW between the Jos uplift and the present Lake Chad. They are thought (Burke 1976a) to represent erosion of fringing uplifting areas and are contemporaneous with increased progradation of the Niger delta (Burke 1972) and increased rates of deposition on African coasts (Siebold and Hinz 1974). Volcanic eruptions accompanied the formation of the marginal uplifts. Continued episodic and sporadic uplift of watershed regions provided detritus for the Pleistocene Chad Formation. The influx of water and sediment into the basin would drive subsidence sufficient to accommodate moderate thicknesses of stratigraphy. The only requirement is that the peripheral uplifts continue to act as sediment sources and are not eroded to base level. In order to maintain the annulus of peripheral uplifts there must be dynamical support provided by the lithosphere, but the mechanism is unknown.

The African continent came to rest with respect to the mantle about 25 Ma (Burke and Wilson 1972). This appears to correlate with the uplift and volcanism on the periphery of the Chad Basin. The location of the Chad Basin may be ultimately related to the existence of a failed Cretaceous rift system underlying the intracratonic sag; the latter may have developed as the African plate 'settled' over an aesthenospheric hot spot.

An analogous situation of a combination of a stationary plate and a failed rift in basement may have existed in the Palaeozoic of North America to give rise to the *Michigan Basin* (Wilson and Burke 1972, Burke and Dewey 1973). The Michigan Basin has undergone subsidence at varying rates for 500 Myr, yet the greatest thickness of Phanerozoic sediments, found in the basin centre, is still only 4 km. The Cambrian to Jurassic sediments are largely unaffected by structural activity (some basement faults penetrate upwards to the Middle Devonian and very rarely into the Mississippian), but the strong positive Bouguer gravity anomaly trending NW–SE across the basin (Fig. 7.34a) suggests the existence of a Precambrian rift (~1100 Ma) underlying the basin (Hinze, Kellog and O'Hara

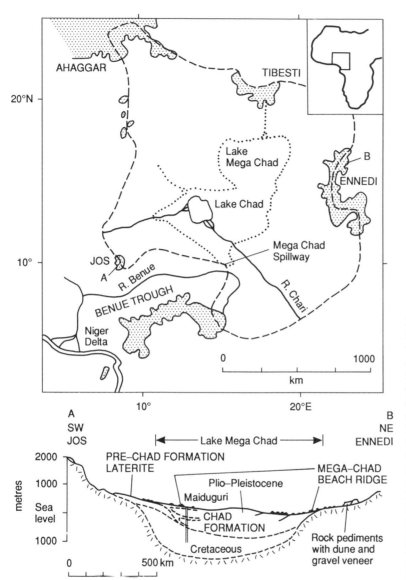

Fig. 7.33. The Chad Basin of the southern Sahara, showing the basin watershed (dashed line) enclosing an area of about 20×10^6 km, and the extent of Lake Megachad (dotted line) about 10 000 years ago. This expanded lake spilled over to the Atlantic via the River Benue. Areas higher than 1 km above sea level represent peripheral uplifts (stippled). Cross-section along A–B shows that Quaternary sediments are restricted to the limits of the ancient Lake Megachad. This depositional area is fringed by a wide annular area of pediment and high ground. A borehole at Maiduguri penetrated 600 m of Quaternary Chad Formation (after Burke 1976a).

1975). A deep borehole drilled in 1975 (Sleep and Sloss 1978) confirmed the existence of this ancestral rift valley sequence. The sedimentary fill of the rift consists of over 1500 m of turbidites (Fowler and Kuenzi 1978). They are thought to have been deposited in a submarine fan depositional system in a proto-oceanic basin (Fig. 7.34b). With time, however, the basin shoaled and the turbidites were replaced by fluvio-deltaic sedimentation.

There is some debate as to when the Michigan Basin was initiated as a distinct depocentre. The apparent lack of strong thickness variations in the Cambro–Ordovician strata led some workers to believe that the basin started in the Late Silurian (Cohee and Landes 1955) whereas others place the formation of the basin in the Cambrian (Catacosinos 1973, Haxby, Turcotte and Bird 1976).

The isopachs of the basal, high-energy Mount Simon Sandstone show two distinct depocentres to be present in Early Cambrian times. One was in the position of the present Precambrian structural low in the central Michigan area. The other

(a)

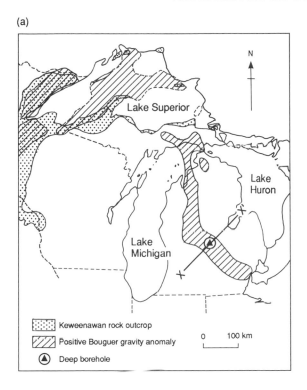

(b)

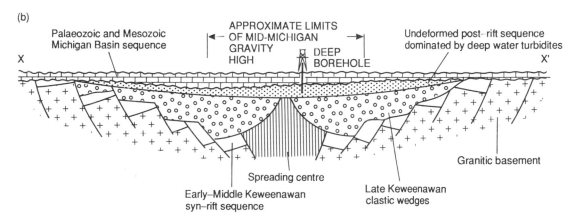

Fig. 7.34. The basement geology of the Michigan Basin, USA (after Fowler and Kuenzi 1978).
(a) Outcrop of Precambrian Keweenawan rocks and positive Bouguer gravity anomaly thought to represent the existence of a buried Precambrian rift system. Note that the N–S to NW–SE orientated anomaly passes directly under the centre of the Michigan Basin. Boundaries of positive anomaly are drawn at inflection points between gravity highs and flanking lows. X–X′ shows line of cross-section in (b).
(b) Interpretive schematic cross-section to illustrate the presence of buried Keweenawan protoceanic basin sequences beneath the Cambrian to Mesozoic Michigan basin sediments.

depocentre was in northeastern Illinois. In Michigan, no evidence of a basin configuration can be found in the younger Cambrian strata and major subsidence over a wide area did not take place until the Early Ordovician (Fischer and Barratt 1985).

Subsidence continued through the Palaeozoic with episodes of erosion producing continent-wide unconformities (Sloss 1963, Sloss and Speed 1974) (Fig. 7.35b). The bulk of the Middle Ordovician to Devonian sequence is made of carbonates, shales and evaporites. The limestones, which are also found in the Illinois Basin to the south, are well known for their build-up complexes. In the Silurian, barrier complexes grew around the perimeter of the Michigan Basin, whilst so-called 'pinnacle reefs' developed closer to the basin centre where biogenic construction kept pace with the greater rate of subsidence (Wilson 1975 for synthesis) (Fig. 7.36). Evaporites are interfingered with and overlie the carbonate build-ups. Whereas the origin of the carbonate build-ups is undisputed, the evaporites are controversial. Deep water, shallow water and supratidal environments have all been suggested. These different ideas are encapsulated in two depositional models:

1 *Barred basin model*, in which evaporites form in deep waters at the same time as coexisting reefs in shallower water.

2 *Desiccated basin model*, in which the entire basin became hypersaline during evaporitic phases, causing a cessation of reef growth.

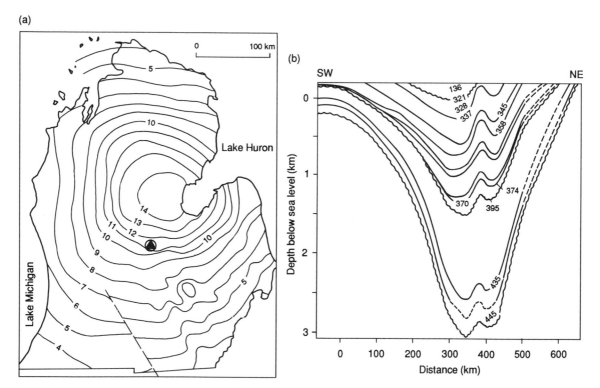

Fig. 7.35. Geometry of the Michigan Basin.
(a) Structural contours in thousands of feet on the Precambrian basement surface in the Michigan Basin (after Hinze and Merritt 1969). The basement depth increases gradually towards the centre of the basin, which is almost perfectly circular in plan view shape.
(b) Cross-section of the Michigan Basin from Middle Ordovician to Jurassic (after Sleep and Snell 1976). Major unconformities are from Sloss (1963). Youngest units are found in the centre of the basin, and some of the unconformities (e.g. at 395 Ma) are associated with a basinward shift in onlap. The anticline in the centre right (NE) is due to late Palaeozoic tectonism.

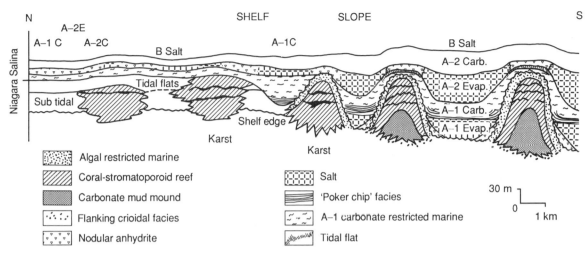

Fig. 7.36. Reconstruction of the facies and stratigraphy of the carbonate-evaporite dominated Upper Silurian of the northern part of the Michigan Basin (after Sears and Lucia 1979, 1980).

Since both the shelf carbonate reefs and the more basinal pinnacle reefs show signs of subaerial exposure (calcretes, dolomitization, travertine cements, karstic erosion and solution features), and the associated evaporites possess a range of shallow water or supratidal features, it is believed that the entire shelf region became periodically desiccated. The intensity of dolomitization of the pinnacle reefs decreases toward the basin centre, suggesting that somewhat deeper water (but never 'deep') environments existed in these regions. The choice of depositional model clearly has implications for the construction of geohistory diagrams (see Chapter 8).

A major unconformity at the top of the Mississippian is overlain by Pennsylvanian strata. A further period of major erosion occurred at the end of the Pennsylvanian. The youngest rocks to overlie the resulting unconformity are Jurassic red beds and spores in these sediments are thermally extremely immature, suggesting that the basin stabilized at this stage and has not undergone further subsidence. The presence of moderately mature (R_o 0.5 to 0.6, see Sections 9.4, 9.5, 9.7.2) coals in the Carboniferous strata led Cercone (1984) to postulate *c.* 1 km of Permian uplift. This figure satisfied both the level of organic maturity of Carboniferous coals and extrapolations of Palaeozoic subsidence rates into the Mesozoic (see also Chapter 8).

The Michigan Basin can be thought of as being more or less continuously filled, with continental and shallow marine environments dominating. The subsidence history was clearly not continuous, as shown by the important unconformities, particularly that representing the Permian–Trias. The superficial simplicity of the Michigan Basin therefore conceals a complex history. The hallmarks of intracratonic sag type basins are nevertheless present.

7.2.1.2 Rift basins

Intracratonic rifts owe their origin to either mechanical stretching of the continental lithosphere ('passive' variety) or to a thermal disturbance ('active' variety) (see Chapter 3). The location of the rift is sometimes determined by the existence of old fundamental weaknesses in the lithosphere. A number of examples of this phenomenon are:
• Opening of the modern Atlantic Ocean along the Palaeozoic Iapetus suture (Wilson 1966)
• Cretaceous separation of southeastern Africa from Antarctica along a Palaeozoic failed rift (Natal Embayment) (Tankard *et al.* 1982)
• Cenozoic East African Rift system follows Precambrian structural trends (McConnell 1977, 1980)

Secondly, the rift may be located where the slowing down or even stopping of the movement of a continental plate relative to a convecting mantle results in thermal doming over hotspots, the development of triple junctions and rifting along lines connecting the triple junctions (Burke and Dewey 1973).

Continental rifting is the initial stage of a sequence leading to complete splitting and ocean floor generation (Veevers 1981). The duration of the rift phase, accompanied by vulcanism and predominantly non-marine sedimentation, depends on the magnitude of the thermal disturbance or rate of stretching and also on the regional distribution of deviatoric stresses. For example, if the rifting continent is bordered by passive margins and the continent is kept in deviatoric compression by ridge push forces, the transition to sea floor spreading may be delayed considerably, as appears to prevail today in Africa. The rift phase in basins bordering Baffin Bay lasted from Barremian to Eocene (a period of about 60 Myr) for example (Miall, Balkwill and Hopkins 1980, McWhae 1981). In other instances, basins have experienced a whole series of phases of rifting interspersed by relatively quiescent periods, as in the case of the East Greenland rift basins which were intermittently active from Early Triassic to Palaeocene (170 Myr) (Surlyk, Clemmensen and Larsen 1981), and in the North Sea where rifting began in the Triassic and persisted episodically, with a climax in the Mid-Jurassic, until the end of Early Cretaceous, a period of over 100 Myr.

The nature of the sedimentary fill of a rift basin depends on its climatic zone, uplift pattern of the rift shoulders or arches acting as sediment sources or barriers, subsidence rate of central rift valleys determining alluvial base levels and lake water depths, and tectonic evolution of linked extensional fault systems. We can view the sedimentary fill on two scales – firstly the large scale features traceable along an entire rift, perhaps over some hundreds of kilometres, and secondly, the more detailed response of sedimentary facies to subsidence and uplift in rift compartments or in individual graben or half-graben. The broad features of the initial deposits of rifts are that they are predominantly non-marine, comprising arkosic, commonly volcaniclastic fluviatile deposits, lacustrine (freshwater or evaporitic) and aeolian deposits. These sediment types typify syn-rift sequences and can be

very extensive. For example, Triassic evaporites are found in two bands along the African and American continental edges marking the site of a former continental rift (Emery 1977, Rona 1982) (Fig. 7.37) which later served to unzip the Atlantic Ocean. Fault-controlled subsidence commonly outpaces sedimentation at later stages of rift development, encouraging marine incursions. Shallow marine sediments may be overlain by deeper marine sediments as the rift evolves towards a site of sea floor spreading.

The embryonic state of continental margin development is typified by the rifts and plateaux of the Neogene East African Rift System (Veevers 1981). We can use this example to demonstrate some typical stratigraphical and sedimentological responses to continental rifting.

The East African Rift System

The Neogene rift valleys are located within broad plateaux rising as high as 4.5 km above a background elevation of about 0.5 km. Rift valleys split the plateaux centrally, as in Ethiopia, or occur on the flanks of individual arches as in the Central Plateau of Tanzania, Kenya and Uganda (Fig. 3.1). There are also rift divergence zones where well-developed single rift valleys pass into zones of diffuse extension composed of tilted fault blocks (as in northeast Tanzania). Seismic investigations of some rift valleys such as that of Lake Tanganyika (Rosendahl *et al.* 1986) suggest that they are composed of a linked arrangement of half-graben (Fig. 7.38, 7.39). These half-graben are commonly arcuate or crescent-shaped in plan view and alternate in polarity along the strike of the rift, seldom overlapping. Individual half-graben compartments are separated by interbasinal ridges, trending oblique to the rift axis, along which rotation and shearing takes place. The interbasinal ridges serve as barriers to sediment spillover between the various rift compartments. Where overlap of half-grabens takes place so that two half-grabens effectively face each other, 'hinged highs' separate the two hangingwall depocentres. Hinged highs are therefore similar to the interbasinal highs of the non-overlapping case, but involve a slightly different linkage of compartments. Elevated rift fringes or platforms also occur, particularly at the

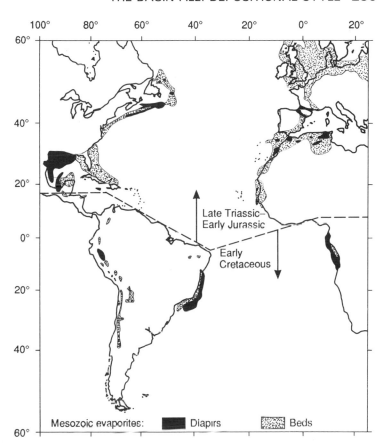

Fig. 7.37. Distribution of Mesozoic evaporites, showing the close relationship to the trailing continental edges on either side of the Atlantic Ocean. After Emery (1977).

ends of basins, and they are associated with drainage systems into the Lake Tanganyika. The existence of this structural template implies that the amounts of subsidence and extension and sediment infill are all dependent on location within the rift valley zone. Sedimentary facies within the Lake Tanganyika basin result from an interplay of
1 River input of clastics and input of fanglomerates along border fault scarps.
2 Background 'rain' of biogenic sediment from lake waters.

The entry points of clastics into the Lake Tanganyika basin are (1) the axial flowing River Ruzizi entering the lake in the north, (2) rivers flowing over platforms, e.g. Malagarasi River entering the lake from the east, (3) as conglomerates and breccias derived from slope wastage along border faults and (4) rivers entering over the shoaling sides of half-graben, i.e., over monoclines, flexures or

steps of faulted or flexed half-graben shoulders. The location of clastic facies can therefore be predicted if the structural pattern is known (Fig. 7.39). An important point is that the axial River Ruzizi has built a thick deltaic cone into the northern basin, but the platforms are regions of elevated basement and therefore act as clastic-transport pathways rather than as sites of thick deposition. Since platforms occur within facing half-graben geometries (Fig. 7.39), fluvial clastics enter facing half-graben from the non-overlapping areas, that is, near the opposing corners of the facing half graben (Fig. 7.39).

Much of the East African Rift System is in a semi-arid climatic zone. As a result, most of the deltas building into the rift lakes at river entry points are markedly ephemeral in their discharges, perhaps flowing for only a few hours each year (Frostick *et al.* 1983, Frostick and Reid 1986).

(a)

(b)

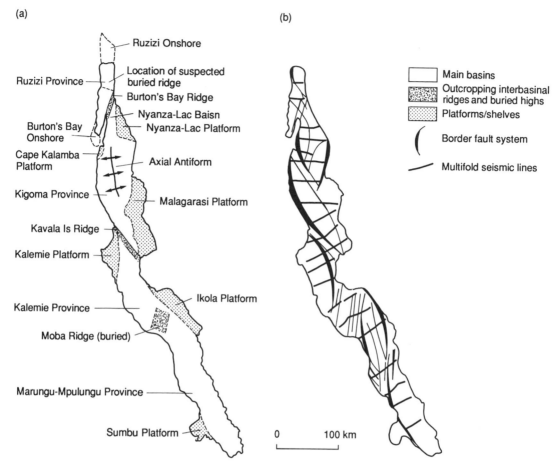

Fig. 7.38. (a) Simplified lake floor geomorphology of Lake Tanganyika. The areas termed provinces are the present-day depositional compartments of the lake basin. (b) Main tectonic elements recognized from multichannel seismic reflection lines. The structural geometry is of alternating half-graben along a sinusoidal interconnection of border faults and interbasinal ridges. The interbasinal ridges are very important elements which may extend far beyond the rift proper (after Rosendahl *et al.* 1986).

Only a few deltas are fed by large perennial streams (e.g. River Omo, Lake Turkana). Some lakes, especially in the north of the rift system have internal drainage and, under a negative water budget, precipitate evaporites.

Deltas in lakes are commonly modified by the effects of lake level changes. The largest delta in Lake Turkana for example (River Omo) owes its bird's-foot outline primarily to the drowning of interdistributary areas during lake level rise, rather than to build-out of a subaqueous distributary mouth. Local lake water circulation and wave activity are generally too weak to substantially

modify the delta front, but spits and barrier-beach complexes may be locally developed.

Lake Magadi, Kenya possesses many of the typical attributes of a rift valley lake in terms of its hydrology, hydrochemistry and sedimentation. It is a closed basin, alkaline system which receives inflow waters strongly influenced by weathering of volcanic rocks in the drainage basin. The inorganic formation of bedded chert derived from the minerals magadiite and kenyaite was first described from Magadi (Eugster 1967), and the large scale precipitation of Na–Al–Si gels, the precursors of zeolites, has also been described from the lake. The

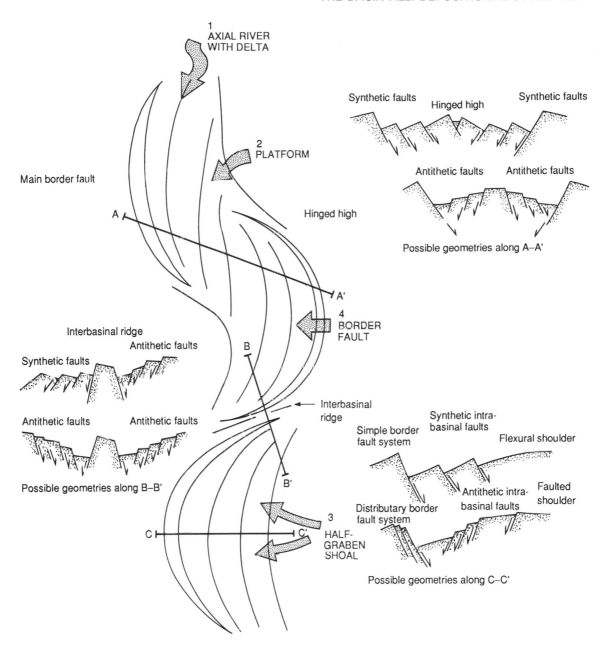

Fig. 7.39. Main entry points of clastics into a rift basin of Tanganyika type. The upper half of the diagram shows half-graben which are *overlapping*; platforms are prominent in such situations (e.g. Malagarasi platform). The lower half of the diagram shows *non-overlapping* half-graben with an interbasinal ridge. Cross-sections give an indication of the variability of structural geometries that is possible (modified from Rosendahl *et al.* 1986).

chief mineral accumulating today is trona. Layers of evaporite are interbedded with thin beds of wind-blown volcaniclastic sand and silt and black anoxic muds. Lake Magadi, and nearby Lake Natron, are both highly alkaline, owing their chemistry to the decomposition of alkaline igneous rocks and the recirculation of groundwater by hot springs. The centrally located salt pan and fringing mudflats, ephemeral streams and springs provide an arid-region closed rift-lake model which can be identified in the stratigraphic record, such as in the Eocene Green River Formation of eastern USA (Eugster 1986).

Volcanic activity has a major influence on rift sedimentation. It exerts this influence in a number of ways:

1 Diversion or damming of surface drainage by volcanic eruptions.

2 Overloading of streams with volcanic, especially pyroclastic detritus.

3 Leaching of volcanic material (e.g. peralkaline ash) leads to alkaline to hyperalkaline groundwaters which then influence lake chemistry.

4 Intercalation of subaerial and subaqueous lava flows within the terrestrial and lacustrine stratigraphy.

7.2.1.3 Failed rifts

Failed rifts are those basins in which rifting has been aborted before the onset of sea floor spreading and passive margin development. Their rift phase is identical to that outlined in the previous section. During cooling, the basin widens, post-rift sediments onlapping the previous rift shoulders, producing a steers-head geometry. A sedimentary evolution from non-marine to shallow marine in the syn-rift phase and deeper marine in the post-rift phase seems typical.

The Benue Trough of central-western Africa and the North Sea are two excellent examples of aborted rifting. The Benue trough is 1000 km long, 100 km wide and is filled with 5 km of fluvial, deltaic and marine Cretaceous sediment. At the southwestern end of the failed rift, the Tertiary Niger delta has build a wedge of fluvial, deltaic and submarine fan deposits 12 km thick into the Atlantic (Fig. 7.40).

In the northern North Sea a major period of rifting took place in the Middle Jurassic. At this time sediment was dispersed longitudinally along

graben. In the Viking Graben fluviatile deposits pass northwards into deltaic and shallow marine deposits in the Brent Group. The mid-Cretaceous saw the end of the rift phase and sediment onlapped the graben shoulders onto the East Shetland Platform. Thick deposits of Cretaceous chalks, Palaeogene submarine fan sandstones and basinal shales, and Neogene mudstones typify this post-rift phase.

In summary, intracratonic rifts and sags are characterized by continental to shallow marine depositional sequences. Sags are characterized by widespread, uniform sedimentation lacking a tectonic control. Rifts on the other hand, possess a structural configuration of half-graben which profoundly influences the location of lacustrine depocentres and entry points of clastics. Evolution of the intracratonic rift to a continental margin allows marine incursions to take place.

7.2.1.4 Passive margins

With continued extension, juvenile oceanic spreading centres develop, as in the 20 Myr-old Red Sea-Gulf of Aden, and then mature (> 40 Myr) ocean basins form. We have previously seen (Section 3.2) that fully developed passive margins, such as those bordering the Atlantic Ocean, are characterized by extensional faulting, large scale gravitational tectonics (slumps, slides, glide-sheets) and salt tectonics. Extensional faulting dies out in the post-rift phase with only minor reactivation of older normal faults. Growth faults are common in areas of high sedimentation rate (e.g. off the Niger delta, African coast). Gravitational tectonics are, however, very important during the post-rift drifting phase. The scale of the gravitational movements varies from small slumps to gigantic slides. The continental slope and rise off southwestern and southern Africa was subject to major slope instability during the Cretaceous and again in the Tertiary (Dingle 1980). The slide-units are over 250 m thick and can be traced for 700 km along strike and for nearly 50 km down palaeoslope.

Evaporites typify the closed lake basins of the rift stage and the first marine incursions during the incipient ocean phase. When buried under an overburden of passive margin sediments these evaporites become mobile. Salt diapirism frequently uses pre-existing fault surfaces, before ballooning up into the overlying cover. On the sea

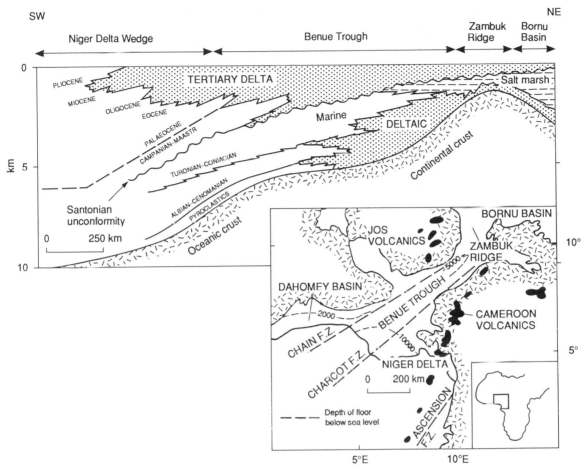

Fig. 7.40. Longitudinal cross-section through the Cretaceous Benue Trough and the Tertiary Niger delta, West Africa. The Benue is a failed rift or aulacogen (Burke, Dessauvagie and Whiteman 1972, Petters 1978). The Santonian unconformity separates the aulacogen stage from the subsidence phase associated with the opening of the Atlantic. Inset shows location of Benue Trough in relation to the Niger delta. The Benue Trough passes to the northeast into the intracratonic Chad Basin.

floor individual diapirs may form topographic highs with marginal moats, the marine sediment being ponded in these depressions and pinching out against the diapir walls. The Brazilian continental margin and the western Grand Banks, Newfoundland show examples of major diapiric activity.

At the transition from rift basin to youthful ocean basin subsidence may outpace sediment supply, leading to the deposition of a number of distinctive facies associations indicative of sediment starvation:

1 *Evaporites.* The intermittent connection of de-

veloping rifts with the sea during the incipient stage provides ideal conditions for the formation of thick evaporites. Such evaporites occur along the margins of the Atlantic Ocean (Emery 1977, Rona 1982) and, at an earlier stage of development, underlying the Red Sea (Lowell and Genik 1972).

2 *Black organic-rich shales.* High organic productivity and restricted marine circulation may allow the preservation of shales rich in organic matter. Such conditions are likely to prevail where youthful ocean basins contain submarine sills restricting the throughput of water. However, so-called oceanic anoxic events are of broader significance, being related to eustatic highstands (Arthur and Schlanger 1979, Schlanger and Jenkyns 1976). Models for the development of anoxia are described in Chapter 10 in the context of petroleum source bed deposition.

3 *Pelagic carbonates.* In new ocean basins with little clastic supply, deep water pelagic carbonate facies may directly overlie the foundered pre-rift 'basement' or newly created seafloor. The faulted basement topography controls the type of deposit, with uplifted fault block edges and seamounts accumulating shallow water carbonates, and intervening troughs being the sites of fine grained pelagic sedimentation. This pattern has been interpreted from the Tethyan realm of southern Europe. Fault block shoulders accumulated skeletal sands, pelletal muds, ooids and stromatolitic limestones during the Late Triassic whilst pelagic turbidites and resedimented reef detritus were deposited in intervening troughs. With time the fault block crests became further submerged, allowing the formation of Fe–Mn nodules and crusts and red pelagic limestones. By the Late Jurassic, most of the fault blocks had been buried beneath radiolarites, red marls and fine grained pelagic white limestones (Bernoulli and Jenkyns 1974).

The early, sediment-starved phase of passive margins is generally followed by an increase in continentally-derived sediment, building thick seaward-prograding wedges. The Atlantic margin shows great variety in the nature of this prograding wedge. The Senegal margin of western Africa contains a thick carbonate bank extending over the stretched continental crust. Further to the SE the Niger has built a thick deltaic clastic wedge, provoking growth faulting and mud diapirism. Further south off Gabon and remote from the Niger delta, oceanic muds overlie thick diapirs of evaporite. Finally, off the coast of southwestern Africa a 'normal' clastic margin exists with seaward prograding clinoforms reaching far out into the basin and overlying the oceanic crust. This latter type is also common to the highly sediment nourished North American Atlantic margin (Figs 3.33, 3.34). Examples in Fig. 7.41 show the various features of salt diapirism, carbonate banks and seaward-prograding clastic wedges.

7.2.2 Basins related to convergent motion

7.2.2.1 Morphological and tectonic elements at arc-related margins

The main components of convergent arc-related systems are, from overriden oceanic plate to overriding plate (Dickinson and Seely 1979) (Fig. 7.42),

1 An *outer rise* on the oceanic plate recognized as an arch in the abyssal plain. This is the flexural forebulge of the descending oceanic plate (Section 3.1).

2 A *trench* or deep trough, commonly more than 10 km deep situated oceanward of the arc. The sediments of trenches are dominated by fine-grained turbidites and pelagic deposits. The bathymetric expression of the trench much depends on the sediment supply into it and, associated with this, the rate of encroachment from the arc of the accretionary wedge. The trench is the bathymetric expression of the deflected (flexed) oceanic plate (Section 3.1).

3 A *subduction complex* composed of tectonic stacks of fragments of oceanic crust, its pelagic cover and arc-derived turbiditic sediments, together with *perched* or *accretionary basins* ponded on top of the accretionary wedge. The subduction complex makes up the inner slope of the trench. Where accretion rates are high, the subduction complex may rise to shelf depths or even become emergent.

4 A *forearc basin* between the ridge or terrace formed by the subduction complex and the volcanic arc.

5 The *magmatic* arc caused by partial melting of the overriding plate and possibly subducted plate when the latter reaches between 100 and 150 km depth. The volcanism is predominantly andesitic.

6 The *backarc* region floored by oceanic or continental lithosphere. Where the lithosphere is oceanic, the backarc region typically undergoes extension. Backarc basins are some of the most rapidly extending regions of the Earth's crust today, a prime example being the Aegean Sea. Where the lithosphere is continental, as in Andean-type margins the backarc (or retroarc) region is typically a zone of flexural subsidence related to major fold-thrust tectonics along the arc boundary. These

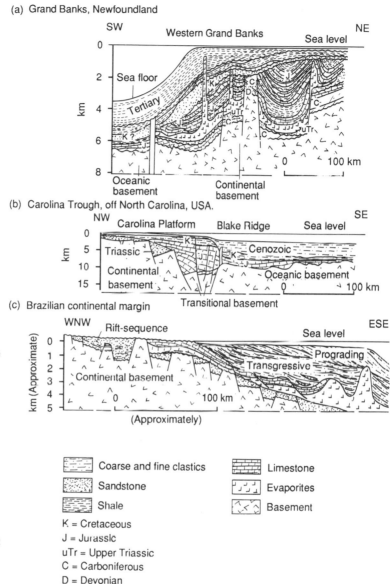

Fig. 7.41. Examples of fully developed passive margins from the western edge of the Atlantic Ocean. (a) Grand Banks, Newfoundland (McWhae 1981) shows complex unconformities and salt from diapirism from the Jurassic. (b) Carolina Trough, off Cape Fear, Carolina (Sheridan *et al.* 1981) has a thick Jurassic carbonate bank extending onto oceanic basement. (c) Brazilian continental margin (Ponte, Fonseca and Carozzi 1980) (idealized) shows salt diapirism from the Lower Cretaceous and prograding clastic wedges. Note that vertical and horizontal scales vary between (a), (b) and (c). Modified from Miall (1984).

retroarc basins are therefore discussed in Section 7.2.2.4 on foreland basins.

7.2.2.2 Brief outline of ideas on kinematics of arcs

Molnar and Atwater (1978) first attempted to answer the question why some convergent margins consisted of oceanic arcs and extensional backarcs and some consisted of magmatic arcs on continental crust with active backarc compression (sometimes termed Cordilleran-type). These two possibilities are exemplified by the western and eastern margins of the Pacific Ocean respectively.

In the western Pacific the convergent margins are generally subducting oceanic lithosphere of

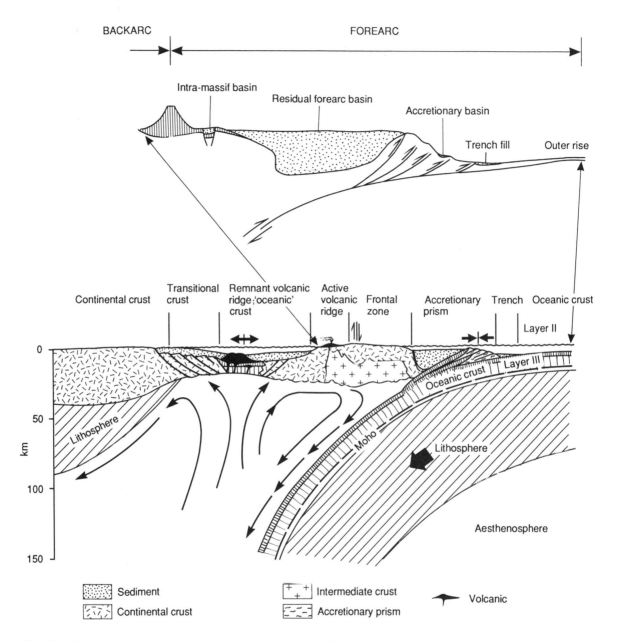

Fig. 7.42. Terminology for a convergent ocean-arc boundary with a backarc basin (modified from Toksöz and Bird, 1977, Green, 1977 and Dickinson and Seely, 1979). Upper diagram shows detail of basins in the forearc region. The intra-massif basin and the residual basin both contain forearc sediments of deep marine to non-marine facies. The accretionary basin in the subduction complex contains structural slices of abyssal plain, slope and trench deposits together with ophiolites and metamorphics.

Mesozoic age (i.e. ∼ 100 Ma) and backarc extension is widespread. In the eastern Pacific the subducting lithosphere is much younger (< 50 Ma). Here the arcs are located on the overriding continental plates, with fold-thrust belts and retroarc foreland basins in the backarc region. Since oceanic lithosphere cools and thickens with age, the older Mesozoic lithosphere of the western Pacific should be more gravitationally unstable than the < 50 Ma lithosphere of the eastern Pacific. The older lithosphere should therefore subduct more rapidly and may exceed the convergence rates of the plates. This should cause an oceanward migration of the subducting hinge together with the forearc elements. This process has been called *roll-back* (e.g. Dewey 1980). It leads to extension in the region behind the rolled-back forearc, that is, backarc spreading. Others believe that backarc extension is related to secondary mantle convection above the subducted oceanic plate (Fig. 7.42) (Toksöz and Bird 1977, McKenzie 1978b).

Dewey (1980) suggested that there are three families of arc-systems:

1 *Extensional arcs* where the velocity of roll-back exceeds the oceanward velocity of the overriding plate, producing backarc extensional basins. As the arc migrates oceanward the forearc region becomes isolated from continental sediment sources and consequently is starved of major sediment supply. *Examples*: Mariana and Tonga arcs, eastern Indonesia.

2 *Neutral arcs* where there is a balance between the rates of roll-back and oceanward movement of the overriding plate, producing well-developed subduction complexes but no backarc extension. *Examples*: Alaska–Aleutian and Sumatran (western Indonesia) arcs.

3 *Compressional arcs* where the subducting crust is young and the velocity of roll-back low. This places the forearc region in compression, causing thrusting in both overriden oceanic and overriding continental crust. *Examples*: Canadian–western USA Cordillera, Peruvian Andes.

Arc behaviour may change through time as the age of the subducted oceanic lithosphere changes. If the age of the subducted oceanic lithosphere gets progressively older for example, backarc basins may form, then be closed and margins change from western Pacific type to Cordilleran type.

7.2.2.3 Continental collision – basic scenarios

Suturing of two continental plates produces a complex amalgam of intense structural deformation, regional metamorphism, plutonism and basin formation. The ocean closing process may involve a number of variations in the manner of continent-continent locking:

• the continental plate may initially collide with an arc-backarc system before continent–continent suturing

• the continental plates may compress a number of microplates in the collision zone

• the collision may be highly irregular or oblique, triggering diachronous orogenic activity and major strike-slip displacements.

It is beyond the scope of this text to elaborate on the complex geological evolution of collisional belts, and the reader is referred to the edited series of papers by Hsü (1983) and the seminal papers by Sengör (1976) and Dewey (1977).

There are three basic scenarios for events leading to continental collision:

1 *Inversion of an extensional backarc system:* As younger oceanic crust is subducted, the backarc basin closes and the arc changes to a compressional type. Closing of the backarc basin causes major shortening of its sedimentary fill and oceanic basement. This predates the orogeny resulting from docking of the continental plates. This process has been interpreted for the southern Andes (Dalziel, De Wit and Palmer 1974) and western Cordillera of USA (Burchfiel and Davis 1972, Dickinson 1981) for example.

2 *Collision of continents with oceanic arcs:* Impingement of the continental plate on the arc-trench region can lead to involvement of continental rocks in a much thickened subduction complex. An example of this is the Timor region where a projection of the Australian plate collided with the Sunda Arc during the Neogene. Subduction is still active and Pleistocene reefs have been uplifted to 800 m above sea level as the subduction zone is choked with continental rocks of the Australian plate (Hamilton 1979).

3 *Arc–arc collision* occurs where two arcs, facing each other are subducting the same oceanic plate, as appears to occur in the Molucca Sea between Indonesia and the Philippines (Hamilton 1979).

These events lead up to the *terminal phase* of continent–continent collision. The vast amount of lithospheric shortening, thickening, metamorphism and plutonism involved in continental collision is accompanied by the formation of sedimentary basins in three tectonic settings:

1 Foreland basins (Chapter 4) both in front of and behind the overriding plate.

2 Intermontane basins caught within the megasuture.

3 Extensional and strike-slip basins located in shear zones produced by 'escape tectonics' from the collision.

In this text we have attempted to provide an explanation of sedimentary basins in terms of their mechanisms of formation and evolution. This approach works well in cases where there is one overriding lithospheric process, such as stretching or flexure. Thus backarc basins can be modelled in terms of lithospheric stretching. Other basin types such as foreland basins and oceanic trenches are united by the mechanism of lithospheric flexure. Essentially, we can make a distinction between two groups:

1 Basins lying outside of but adjacent to megasutures, belonging to the perisutural class of Bally and Snelson (1980) (Section 1.4). These basins have a dominant formative mechanism of flexure and include:

(a) *Peripheral foreland basins* developed on continental lithosphere at zones of continent–continent collision, and situated on the flexed overriden plate.

(b) *Retro-arc foreland basins* developed on continental lithosphere but associated with subduction of oceanic lithosphere (B-type subduction) and situated behind the magmatic arc relative to the subduction boundary.

(c) *Ocean trenches* developed on oceanic lithosphere at zones of ocean–continent collision or ocean–ocean collision and situated at the downbend of the overriden oceanic slab.

For background on the theory of flexure and the application to real world examples the reader is referred to Sections 2.1.4–2.1.5 and Chapter 4 respectively. The role of flexure in generating stratigraphic patterns is discussed in Section 6.2.

2 Basins lying within megasutures and belonging to the episutural class of Bally and Snelson (1980) (Section 1.4). These basins include in this context, forearc basins located between arc and subduction complex and extensional backarc basins landward of the arc.

7.2.2.4 Foreland basins (peripheral and retroarc types)

In the simplest terms foreland basins develop at the front of active thrust belts where the bulk transport direction is towards the evolving basin. Because the thrust load is inherently mobile the foreland basin itself becomes involved in the deformation. To what extent the basin becomes dissected or becomes completely detached depends on a number of variables including the propagation rate of the thrust tips, availability of subsurface easy-slip horizons underlying the basin and the angle of convergence. It is useful to differentiate between two tectonosedimentary settings (Fig. 7.43):

1 Sediment accumulates in a basin ahead of the active thrust system in a *foredeep* (or toe-trough) *sensu stricto*;

2 Sediment accumulates in a basin that rests on moving thrust sheets in a *thrust-sheet top* (Ori and Friend 1984) or *piggyback basin*.

Individual foreland basins may contain examples of both foredeep and thrust-sheet-top sedimentation. In the North Alpine Foreland Basin of Switzerland the first clastic wedges (late Eocene), composed essentially of turbidites, were shed partly into ponded basins located on top of thrust sheets and partly overspilled into foredeeps. As the foreland basin evolved through the Oligocene, thrust-sheet-top basins became far less conspicuous features of the inner margin of the basin. Postdepositional tectonics (late Miocene–Pliocene) detached the entire basin in western Switzerland as deformation progressed into the Jura province, whilst in eastern Switzerland and further east in Bavaria, there is no evidence of major detachment. The detachment of the basin in western Switzerland was facilitated by the presence of thick Triassic salt in the subsurface acting as an easy slip horizon (Homewood, Allen and Williams 1986).

A lucid picture of a linked system on inner thrust-sheet-top basins and outer foredeeps is provided by the Apenninic chain of Italy (Fig. 7.44) (Ricci-Lucchi 1986 for synthesis). As the entire system of basins migrated eastwards onto the foreland, depositional events in both thrust-sheet-top and foredeep basins were synchronized. This phase of sychronization is termed the 'Flysch stage' by Italian workers. Sedimentation was dominated by turbidites and some hemipelagics. As the deformation continued, the inner thrust-sheet-top basins

(a) Simple
e.g. Molasse Basin (NAFB)

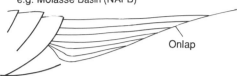

Onlap

(b) Complex – sedimentation accompanies segmentation
e.g. Padan and Adriatic Basins

(c) Complex – sedimentation follows segmentation into minor basins

(d) Associated with piggyback basin
e.g. Satellite basins on Ligurian sheet, Apennines

Piggyback basin   Foredeep

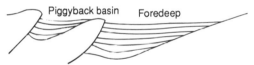

(e) Deformed after sedimentation

Fig. 7.43. Different types of foreland basin profile as suggested by seismic records of basins in the Apennine area of Italy (after Ricci-Lucchi 1986). Basins may be *simple*, asymmetrical wedges with stratigraphic onlap onto the foreland plate (a), as in the Oligo-Miocene development of the North Alpine Foreland (Molasse) Basin of Switzerland. Basins may be *complex* as a result of segmentation by thrusting, as in the Padan and Adriatic basins of northern Italy. Some complex basins were segmented contemporaneously with sedimentation (b), whereas others were deformed, and then sediment passively draped the thrusted basement topography (c). The minor basins may be subequal or of markedly different in size. Foredeeps may be associated with distinct *piggyback* or *thrust sheet top* basins (d) such as the 'satellite' basins on top of the Ligurian sheet in the northern Apennines, and the North Helvetic Flysch basins of the North Alpine Foreland Basin in Switzerland. Foreland basins may also be *deformed* after sedimentation, leading to erosion of the foreland basin sequences (e).

became decoupled from the outer foredeep at the end of the Miocene, the former being uplifted and cannibalized to provide erosional detritus for the latter. This corresponds to the so-called 'Molasse stage' of Italian workers. Sediments range from continental coarse clastics to shelfal mixed carbonate-siliciclastics and turbiditic deep water sands. The Apenninic foreland basin depocentre has now extended into the Adriatic Sea where penecontemporaneous thrust deformations have produced submarine structural culminations (Fig.

7.45). These sea floor highs have been subsequently denuded by slope failure and submarine erosion (Ori, Roveri and Valloni 1986). Three depositional seismic units have been recognized in the Plio–Pleistocene of the Adriatic foreland basin comprising initially turbiditic and hemipelagic deposits and subsequently deltaic deposits shed from the Apennines.

Tectonics also have a primary influence on sediment dispersal patterns. Uplifting thrust fronts may not act as major sediment suppliers but may

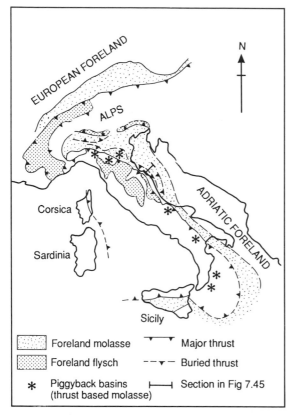

Fig. 7.44. Location map of the main foreland basin depocentres and satellite basins in Italy and contiguous regions (after Ricci-Lucchi 1986). 'Molasse' and 'flysch' are used in the sense of Italian workers (see text and Ricci-Lucchi 1986).

and if they traverse the basin itself can be responsible for large thickness variations in the sedimentary fill. In Switzerland the effect can be seen in isopachs of the stratigraphic units. For example, a transverse line closely following the present Rhône Valley south of Lake Geneva was the feeder for thick earliest Oligocene turbidites. Its effects can also be seen in the marked thickening of shallow marine and shoreline sands in this region at the end of the Rupelian (early Oligocene), and in the location of a major fluviatile fan in the Chattian (late Oligocene) and fan-delta in the Miocene.

Transverse faults had a strong influence on sedimentation in the Apennines. Some faults, such as the Sillaro Line, south of Bologna, played a predominantly vertical role and separated an eastern zone from a western zone of increased subsidence during the deposition of the Miocene *Marnoso arenacea*. However, during the late Miocene the motion on the fault was inverted. Uplift in the western zone took place in the Pliocene whilst the eastern zone subsided at a rate of about 1 mm yr^{-1}. The Forli Line is oblique to the NE-SW trending transverse structures such as the Sillaro. Although it is a neotectonic lineament with active seismicity, an earlier record of its activity lies in Messinian (late Miocene) sediments, when it separated different evaporitic basins, and at earlier times when it favoured the localization of Tortonian submarine channels. Whereas the Sillaro Line is clearly intimately related to the Apenninic thrust structures (since it represents a major discontinuity in the overthrusted ophiolite nappes), the Forli Line may be a strike-slip fault related to a change in convergence direction between Africa and Europe.

The oldest deposits of foreland basins are commonly predominantly fine-grained, often turbiditic sediments which accumulated in sub-shelf water depths. The later deposits of foreland basins are, in contrast, predominantly shallow-water or continental and typify the term 'Molasse'. This kind of vertical megasequence is found in the North Alpine Foreland Basin (NAFB) of Switzerland. The sequence of stratigraphic units can be summarized as follows (Fig. 7.47):

7 Upper Freshwater Molasse, the final choking of the NAFB by coarse continental clastics.
6 Upper Marine Molasse, representing shallow marine and estuarine depositional systems.
5 Lower Freshwater Molasse, the first fluviatile and lacustrine deposits of NAFB.

instead form barriers to basinward sediment transport. This is well illustrated in the Miocene of the Southern Pyrenees. Alluvial dispersal patterns in the foreland basin south of the Pyrenees are controlled by the position of the frontal and lateral ramps of the thrust front (Fig. 7.46):
• the apices of major fluviatile systems are located at structural lows or re-entrants in the thrust front
• small, locally developed fans with highly restricted drainage basins typify the structural salients in the thrust front.

Structural re-entrants are commonly provided by lateral or oblique ramps or through-going strike-slip faults unrelated to the boundaries of thrust sheets. They act as conduits for the removal of erosional detritus from the orogenic belt to the foreland basin,

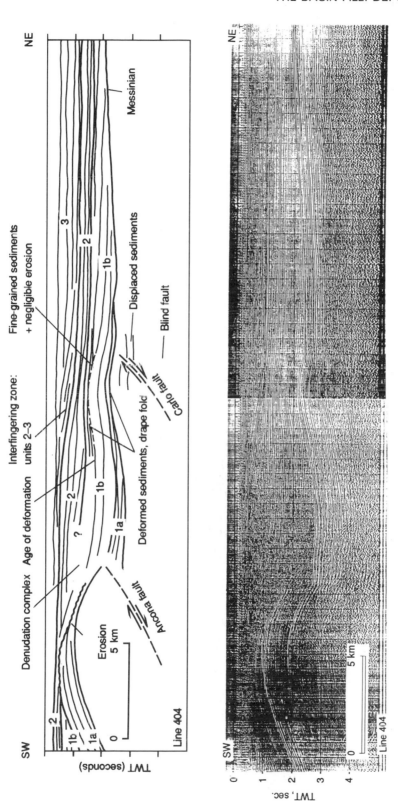

Fig. 7.45. Seismic line and line drawing interpretation of the section shown in Fig. 7.44. The Adriatic Plio-Pleistocene foredeep is well developed, but faulting (Carlo thrust and Ancona thrust) have caused structural highs to develop in the basin, focusing denudation complexes over the strongly erosional culminations, and fine-grained sediment cappings over the more subdued, less erosional culminations (after Ori, Roveri and Valloni 1986).

(a)

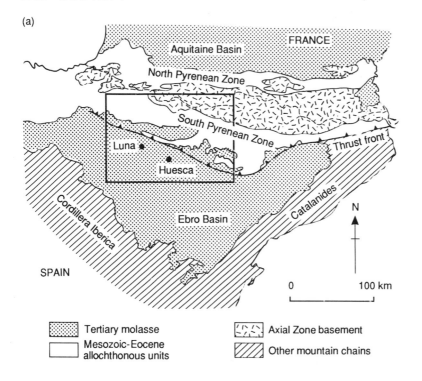

(b)

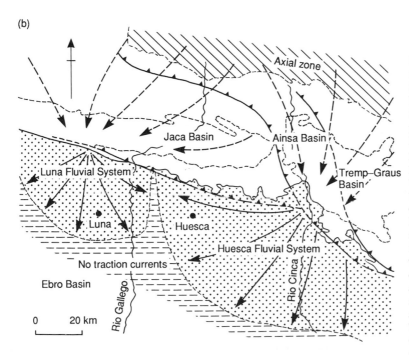

Fig. 7.46. Palaeo-drainage patterns related to thrust activity in the Southern Pyrenees (after Hirst and Nichols 1986). (a) Location map of Southern Pyrenees and the Tertiary peripheral foreland basin of the Ebro. (b) The Miocene drainage pattern. Rivers from the western part of the Axial Zone and the thrust-sheet-top Jaca Basin entered the Ebro Basin at the western end of the frontal ridge of the Pyrenees known as the External Sierras, spreading out to form the Luna fluvial system. The Huesca system was sourced in a more easterly part of the Axial Zone, the Ainsa Basin and the Tremp–Graus Basin. This system had a less well-developed apical system in the structural low between the oblique ramps of the Gavarnie–Boltana and Cotiella–Montsec thrust units.

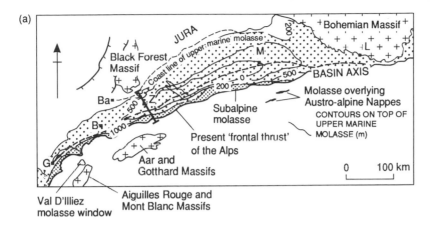

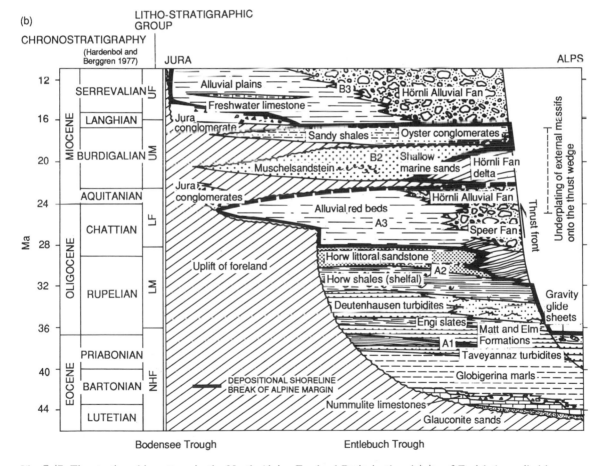

Fig. 7.47. The stratigraphic pattern in the North Alpine Foreland Basin in the vicinity of Zurich (compiled by Sinclair *et al.*) The stratigraphy is made of two megasequences, each one shallowing and coarsening upwards. Coarse clastic wedges fringe the Alpine thrust front, whereas the feather edge along the Jura margin is relatively passive with the exception of some localized pockets of conglomerates. (a) Location of NAFB; Z, Zürich; M, Munich; L, Linz; B, Berne; G, Geneva; Ba, Basel. (b) Chronostratigraphy with lithological ornament added: A1, A2 and A3 comprise the lower megasequence, B2 and B3 the upper.

4 Lower Marine Molasse, the transition from shelf to shoreline sedimentation.
3 North Helvetic units representing turbiditic depositional systems shed from the active orogenic wedge.
2 Nummulitic limestones and Globigerina shales representing foreland drowning.
1 Basal unconformity due to regional uplift of foreland lithosphere.

This foreland basin-fill is composed of a basal foreland collapse sequence (1), followed by two shallowing-upwards megasequences (3–5 and 6–7). The shoreline unit at the top of the Lower Marine Molasse can be thought of as the pivot point between an early underfilled stage and a later steady-state stage. During the underfilled stage the topography of the orogenic wedge is subdued, sediment delivery rates are low and the basin retains deep water conditions. After the mountain belt has grown to a steady state, rapid erosion counterbalances the uplift and the basin is filled to the spill point with detritus. During this phase any excess sediment may be removed from the foreland basin by fluvial and/or shallow marine processes, so that a constant basin geometry is established. Covey (1986) has documented such a process from the western Taiwan foreland basin (Fig. 7.48). We have previously (Section 6.2.2) commented on some of the geodynamic reasons why the early history of foreland basins is characterized by deep water sedimentation, one being the loading of an initially stretched lithosphere (Stockmal, Beaumont and Boutilier 1986) and another the rapid advance of a sub-critically tapered orogenic wedge (Sinclair *et al.* 1990) (Section 6.2.2).

Retroarc foreland basins such as the Cretaceous Rocky Mountains foreland basin in western USA and the series of basins east of the Andes do not fundamentally differ from the peripheral foreland basins. Their main distinguishing characteristic is that they commonly evolve from regions of backarc extension, and the composition of the sedimentary fill reflects the large amounts of plutonic and volcanic rocks in the orogenic belt. An excellent example is the Magallanes Basin of Argentina documented by Biddle *et al.* (1986) (Fig. 6.6).

7.2.2.5 Ocean trenches and accretionary basins

Ocean trenches are one of a series of sedimentary basin types found at convergent arc-related or ocean-continent boundaries. Here we draw the distinction between (1) *trenches* situated on the down-bent oceanic lithosphere (2) *accretionary basins* perched on the accretionary subduction complex, (3) *forearc basins* located between the arc and the subduction complex, and (4) *backarc* basins found on the landward side of the arc (Fig. 7.42).

The association of *trench* and *accretionary basins* perched on the subduction complex of accreted slices of oceanic basement and cover is mechanically analogous to the foredeeps and thrust-sheet-top basins of continental collision zones. Although the mechanics may be similar between the two cases, the sedimentary fills are markedly different.

The tectonic deformation of subduction complexes is generally intense including (Dickinson and Seely 1979):
- isoclinally folded and thrusted bedded sequences
- metamorphic tectonites

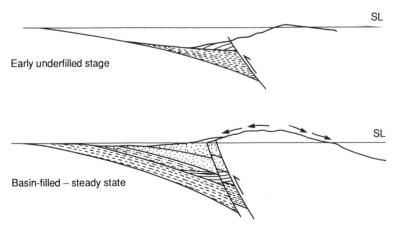

Fig. 7.48. The concept of underfilling and steady state as proposed by Covey (1986). Early in foreland basin evolution, sediments are deep marine with localized submarine fans. As the orogenic belt becomes subaerially exposed, erosion reaches a maximum, causing basin filling and eventually a steady state. Excess sediment is then bypassed out of the basin.

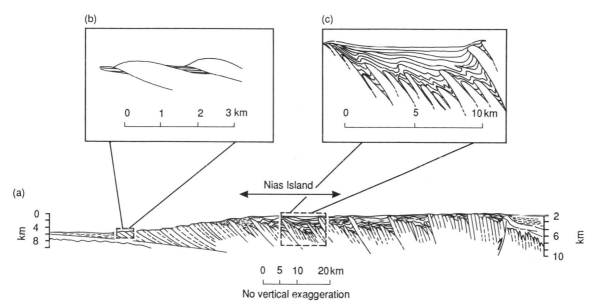

Fig. 7.49. Structural geometry based on seismic profiles across the subduction complex of the Sunda arc in SE Asia (after Moore and Karig 1976, Karig *et al.* 1978). The section is entirely below sea level, but the equivalent (along strike) position of a subaerial portion at Nias Island is indicated on (A). (B) and (C) show details from the platform of the complex and from the leading edge.

• mclanges consisting of tectonic inclusions of original rock in a sheared matrix.

Much of the deformation probably takes place at very high water saturations, and the existence of mud volcanoes on top of several Indonesian subduction complexes and at the toe of the Barbados complex in the eastern Caribbean testifies to large scale dewatering taking place.

The structural style of an uplifted subduction complex has been investigated by seismic reflection profiling across Nias Island off southwestern Sumatra (Moore and Karig 1976, 1980, Karig *et al.* (1978) (Fig. 7.49). The subduction complex is made of a series of imbricate thrust sheets composed of a melange of sheared sedimentary rock debris. The composition of the debris reflects the varied provenances of the clasts:
• basalt fragments, rare periodotite, dunite and serpentinite derived from the oceanic crust
• chert, red shale and pelagic limestone derived from the sedimentary cover of the oceanic crust
• turbiditic conglomerates, sandstones and mudstones derived from the adjacent arc and especially the Bengal Fan.

The longitudinal supply of sediment from the Bengal Fan in the north has produced marked differences in the nature of the trench and subduc-

tion complex around the Sunda arc. In Java, remote from the sediment source, the trench is 7 km deep and the accretionary subduction complex is under 1–3 km of water. Off northern Sumatra the trench is at 4 km depth and the subduction complex is partly emergent (Hamilton 1979). High sedimentation rates therefore appear to favour the growth of submarine fans or accretionary wedges completely across the trench, whereas sediment starved margins have deep trenches with well-developed bathymetric profiles. Some active accretionary complexes can accumulate sediment in shelf and coastal environments, such as off the Alaskan arc, in the Makran, Pakistan, and Hawke Bay, New Zealand.

Large slope failures produce olistostromes on the forearc slope.

7.2.2.6 Forearc and backarc basins

Three types of sedimentary basin occur in the forearc region:

1 Accretionary basins mentioned in the previous section.

2 Intra-arc or intra-massif extensional basins, common where a broad forearc region has developed over continental crust, as in the Andes. These elongate basins typically follow volcanic lines or fundamental tectonic lineaments and are filled mainly with non-marine fluviatile and lacustrine sediments dominated by volcanic constituents.

3 Large basins located between the subduction complex and the magmatic arc, of two sub types:

(a) *'residual'* basins with a basement of stretched continental crust or obducted oceanic crust, and

(b) *'constructed'* basins underlain by the landward portion of the subduction complex.

The oldest sediments of 'residual' forearc basins are generally deep water pelagic deposits, whereas those of 'constructed' types may be shallower. Submarine fans build into the basins transversely from the magmatic arc. Intra-oceanic forearcs tend to be sediment starved and remain deep marine, whereas sediment-nourished examples near major continental catchments may rapidly become shallow marine.

Depositional environments in *backarc basins* on oceanic crust are also deep marine, except along their margins where fluviatile, coastal and shallow marine depositional environments may exist. Karig and Moore (1975) presented a model for the evolution of backarc basins based on the western Pacific examples. Initially, volcaniclastic wedges shed from the arc interfinger with a background of pelagic clays. As subsidence continues and outstrips sediment supply, the sea floor commonly descends below the *carbonate compensation depth* (CCD), so that the more evolved basin accumulates siliceous rather than calcareous clays. The adjacent continent may contribute clastic wedges into the landward edge of the basin.

Backarc extension is less common on continental crust because the arc commonly becomes compressional. However, some continental areas behind convergent ocean–continent boundaries are undergoing or have undergone widespread extension. The Basin and Range province of western USA is an example. The area has been substantially uplifted to a regional elevation of 2 to 3 km and half-graben and graben contain up to 3 km of non-marine sediment. The Pannonian Basin, located to the south of the Carpathians (Burchfiel and Royden 1982), may also be due to backarc spreading.

7.2.3 Strike-slip basins

The tectonic style and evolutionary sequence of strike-slip (especially pull-apart) basins were discussed in Chapter 5 and this information is not repeated here. The sedimentary fills of strike-slip basins have a number of features in common (Miall 1984, p. 411):

1 The geometry of the basin is deep but relatively narrow, and there is evidence of syndepositional relief, such as the occurrence of conglomerates and breccias banked up against faulted basin margins. Sedimentation rates are rapid.

2 Lateral facies changes are rapid, so that marginal breccias may pass laterally directly into lacustrine mudstones.

3 Fault movements cause syndepositional unconformities to form in individual basins and different stratigraphies to develop in closely adjacent basins, making correlation difficult.

4 The basin sediments are commonly offset from their source. This may be proved by a mismatching between size of depositional system and drainage area, or between sediment petrography and hinterland geology.

5 In modern basins there may be offsets of geomorphological features such as rivers, alluvial fans or submarine canyons.

The best known intracontinental transform is the San Andreas system, and one of the best documented pull-apart basins in this system is *Ridge Basin*, California. It shows many of the elements indicated above. The basin was initiated in the Miocene and continued to accumulate sediment during the Pliocene, after which it was uplifted. It contains over 13.5 km of sediment deposited at an estimated rate of 3 mm yr^{-1}, and is located between the San Andreas and San Gabriel faults (Fig. 7.50). During the late Miocene the San Gabriel fault was a major active strand of the San Andreas system. The Ridge Basin formed to the east of a releasing bend in the fault. During the late

---

Fig. 7.50. (*Opposite*) The Ridge Basin, California. (a) Outcrop map of northern end of the Ridge Basin (modified from Link and Osborne 1978 after Crowell 1954, 1975) and generalized sediment transport directions (after Link 1982). Note that younger stratigraphic units occur progressively to the north of the basin. (b) Generalized cross-section showing the sedimentary facies and principal faults. The fine-grained lacustrine depocentres occur close to the main border fault (San Gabriel) (after Crowell and Link 1982).

(a)

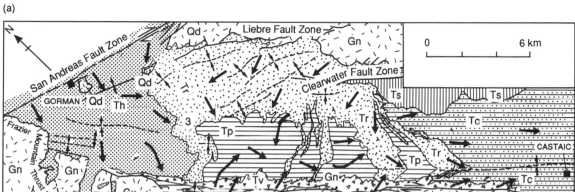

(b)

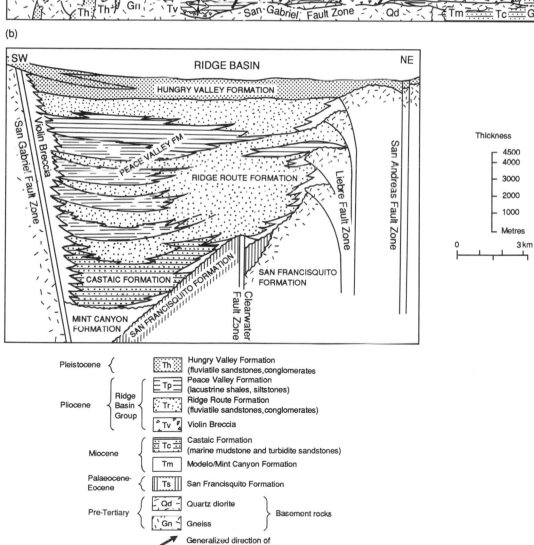

Pleistocene		Th	Hungry Valley Formation (fluviatile sandstones, conglomerates
Pliocene	Ridge Basin Group	Tp	Peace Valley Formation (lacustrine shales, siltstones)
		Tr	Ridge Route Formation (fluviatile sandstones, conglomerates)
		Tv	Violin Breccia
Miocene		Tc	Castaic Formation (marine mudstone and turbidite sandstones)
		Tm	Modelo/Mint Canyon Formation
Palaeocene-Eocene		Ts	San Francisquito Formation
Pre-Tertiary		Qd	Quartz diorite — Basement rocks
		Gn	Gneiss
		↗	Generalized direction of sediment transport

Miocene–Pliocene over 60 km dextral strike-slip took place along the San Gabriel fault, but in the Pleistocene slip was transferred to the San Andreas fault along the northeastern flank of the basin.

The sedimentary fill of the basin is made up of the following units (Crowell and Link 1982, Link 1982):

1 A basal non-marine unit (Mint Canyon) overlain by the 2.2 km thick, upper Miocene Castaic Formation, consisting of marine mudstones and turbidites.

2 9–11 km thick, mostly non-marine Ridge Basin Group, with marine deposits in the lowermost 600 m. The Ridge Basin Group comprises four formations:

(a) marginal breccias along the active western fault scarp (Violin Breccia)

(b) central lacustrine deposits, chiefly mudstones (8 km), of the Peace Valley Formation

(c) fluviatile clastic wedges of the Ridge Route Formation (9 km) along the eastern margin of the basin, and

(d) a basin-wide, final basin-filling of alluvial sands and gravels (1.1 km) of the Hungry Valley Formation.

The marginal alluvial cones and talus of the Violin Breccia were derived from the SW and pass very rapidly (within 1.5 km) into lacustrine shales and siltstones of the Peace Valley Formation, or sandstones of the Ridge Route Formation. The thick clastic wedges of the Ridge Route Formation were shed from source areas to the NE of the basin, but the younger Hungry Valley Formation was derived from the N, NW and W. This demonstrates the complexity of sourceland switching in strike-slip basins. Dextral strike-slip along the San Gabriel fault displaced the source region for the Violin Breccia northwestwards with time. As a result, the successive alluvial fans that form the Violin Breccia become younger northwestward, and form an overlapping or shingled pattern (Fig. 7.51). Within the axial part of the basin, sediments were transported southeastward down the axis of the basin concurrently with northwestward migration of the depocentre.

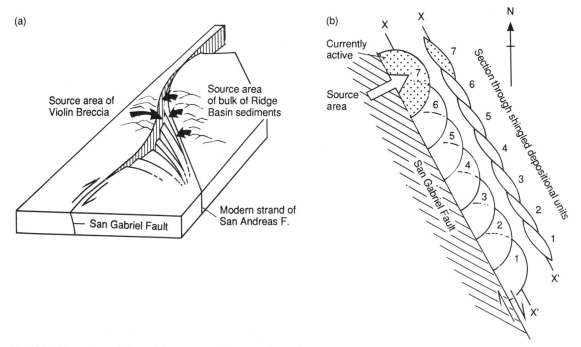

Fig. 7.51. Tectonic and depositional model for the origin of the Ridge Basin as a pull apart on the releasing bend of the San Gabriel Fault (Crowell 1974b), and its filling during active dextral strike slip, resulting in a highly shingled pattern to the depositional packages (After Crowell 1982b).

The Ridge Basin has been compared, in terms of its structural and sedimentological development with the larger Hornelen Basin, Norway, and the smaller Little Sulphur Creek group of basins, southern California (Nilsen and McLaughlin 1985) (Fig. 7.52). Each basin is characterized by marginal fans located tight up against the active strike-slip fault, axial lacustrine facies and streamflow-

dominated fans along the opposite margin. These streamflow-dominated fans contribute most of the sediment to the basin, sometimes filling the basin completely and spreading alluvial deposits across to the active fault scarp to interfinger with the talus fans.

The present-day submarine equivalents of the Ridge Basin are found in the California Borderland

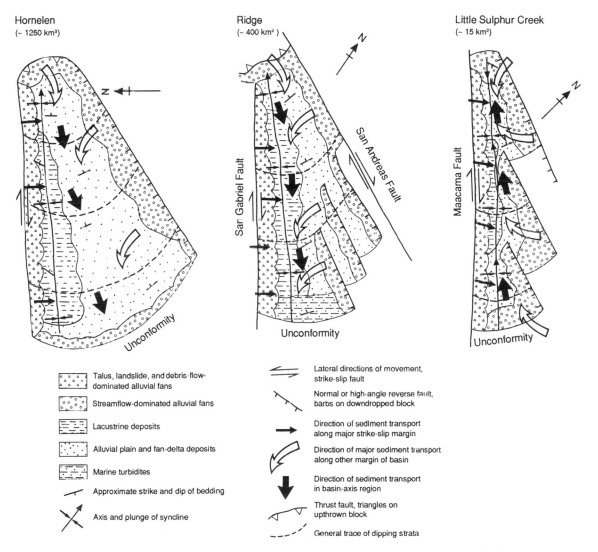

Fig. 7.52. Comparison of the Hornelen Basin (Devonian) of Norway, the Tertiary Ridge Basin, California, and the Plio-Pleistocene Little Sulphur Creek Basin, California (after Nilsen and McLaughlin 1985). Although the scales are variable in the three basins, there are many sedimentological and tectonic similarities. The main border faults shed coarse-grained conglomerates and breccias, and fine-grained lacustrine facies are found close to these marginal screes and debris flow-dominated fans. The other margins provide the bulk of the sediment into the basins in the form of streamflow-dominated alluvial fans, fluviatile and deltaic deposits.

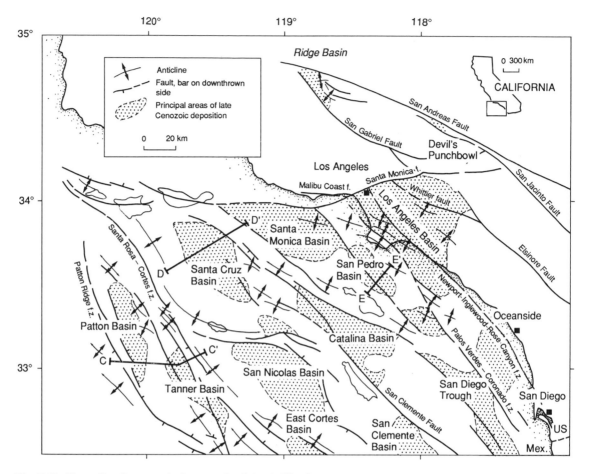

Fig. 7.53. The strike-slip tectonic framework of the California Borderland (Moore 1969, Junger 1976). C–C′, D–D′ and E–E′ are the lines of profiles shown in Fig. 7.54.

basins. This area, to the west of the San Andreas fault, is underlain by an arc complex formed during subduction of the Pacific plate in the Mesozoic to Early Cenozoic (Howell and Vedder 1981). A large number of small basins filled with submarine fans formed during the Palaeogene, and Oligocene dextral strike-slip faulting fragmented the region into en echelon ridges and rhomboidal basins (Fig. 7.53). Sedimentation is dominated by turbidites fed from the nearby American coast into a background environment of fall-out of fine-grained terrigenous and pelagic material. The region has several well-studied deep sea fans (La Jolla, Navy, etc.), recent investigations of which emphasize the importance of submarine slides in fan development (e.g. Sur Submarine slide on

Monterey fan, described by Normark and Gutmacher 1988). Sedimentation rates increase towards the coast, so that the Los Angeles and Ventura Basins located close to the continental source contain as much as 8 km of Neogene sediment, whereas the more offshore basins such as the Tanner and Patton Basins (Fig. 7.54) have much reduced sedimentary thicknesses (Howell *et al.* 1980, Fig. 11, p. 56).

The sedimentation in a classic pull-apart in an arid climate is well illustrated by the Dead Sea. Movement along the Dead Sea fault commenced in the Miocene in response to the opening of the Red Sea. It has continued to move up to the present day. The basin contains over 10 km of fluvial clastics, lacustrine limestones and evaporites. Like

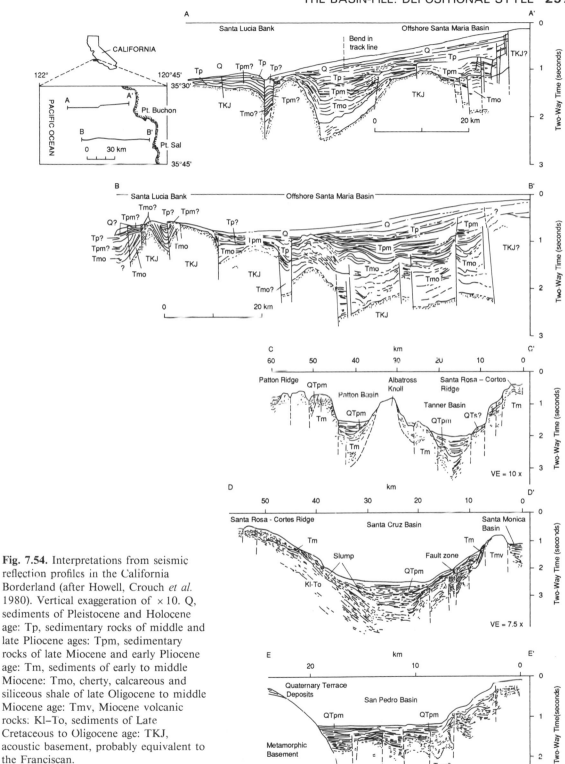

Fig. 7.54. Interpretations from seismic reflection profiles in the California Borderland (after Howell, Crouch *et al.* 1980). Vertical exaggeration of × 10. Q, sediments of Pleistocene and Holocene age: Tp, sedimentary rocks of middle and late Pliocene ages: Tpm, sedimentary rocks of late Miocene and early Pliocene age: Tm, sediments of early to middle Miocene: Tmo, cherty, calcareous and siliceous shale of late Oligocene to middle Miocene age: Tmv, Miocene volcanic rocks: Kl–To, sediments of Late Cretaceous to Oligocene age: TKJ, acoustic basement, probably equivalent to the Franciscan.

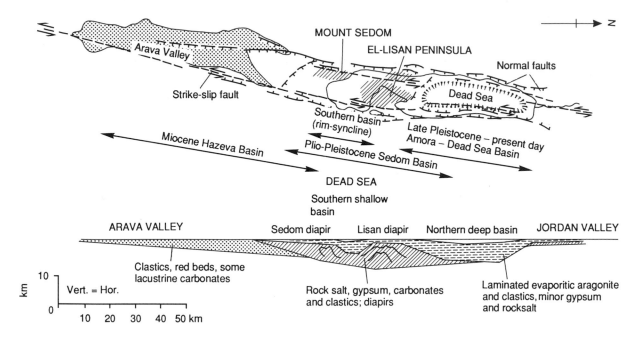

Fig. 7.55. Map and section of the Dead Sea–Arava depression, illustrating the northerly migration of the depocentre. Early Miocene (25–14 Ma) strike-slip of about 60–65 km opened the Arava Basin, which was filled with about 2 km of red beds during a pause in the strike-slip displacement. Later movement in the last 4.5 Myr has allowed the deposition of over 4 km of marine to lacustrine rock salt of the Sedom Formation, followed by 3.5 km of lacustrine evaporitic carbonates and clastics (after Zak and Freund 1981).

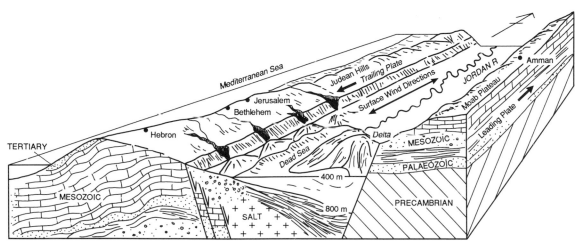

Fig. 7.56. Block diagram of the northern end of the Dead Sea. Sediment enters the lake via high discharge ephemeral streams feeding fan deltas along the western edge of the basin (Judean Hills), and the perennial, axial-flowing River Jordan, whose drainage area lies in a humid region, depositing a muddy, fine-grained delta at its entry point. Autochthonous supply to the lake is of aragonite and gypsum precipitates (after Manspeizer 1985).

the Ridge Basin, depocentres have moved considerably, producing a highly diachronous fill. The Miocene Hazeva Formation consists of continental clastics and some lacustrine carbonates and is found in the Arava Valley in the south. The Pliocene–early Pleistocene Sedom Formation consists mainly of lacustrine salts, gypsum, carbonates and some clastics and occurs in the central section. The Pleistocene to Recent Amora and younger formations consist of laminated evaporitic (gypsum) and aragonite sediments which are accumulating today in the modern Dead Sea in the northern sector of the basin (Fig. 7.55, 7.56) (Zak and Freund 1981).

SECTION 4
EVOLUTION OF THE BASIN-FILL

8 Subsidence history

In Nature's infinite book of secrecy
A little I can read

(Shakespeare, *Antony and Cleopatra*)

SUMMARY

Present-day stratigraphic thicknesses are a product of cumulative compaction through time. A quantitative analysis of subsidence rates through time, sometimes called *geohistory analysis,* relies primarily on the decompaction of stratigraphic units to their correct thickness at the time of interest. Two other corrections must also be made in order to plot subsidence relative to a fixed datum such as present sea level. These are (1) corrections for the variations in water depth through time, and (2) corrections for absolute fluctuations of sea level ('eustasy') relative to the present sea level datum.

The decompaction of stratigraphic units requires the variation of porosity with depth to be known. Estimates of porosity from borehole logs (such as the sonic log) suggest that normally pressured sediments exhibit an exponential relationship of the form

$$\phi = \phi_0 e^{-cy}$$

where ϕ is the porosity at any depth y, ϕ_0 is the surface porosity and c is a coefficient that is dependent on lithology and describes the *rate* at which the exponential decrease in porosity takes place with depth. Overpressuring of porefluids trapped in a formation has the effect of inhibiting compaction. This causes strong deviations from the expected porosity–depth curve in the overpressured units.

Information on changing palaeobathymetry through time may come from sedimentary facies and distinctive geochemical signatures, but principally it comes from micropalaeontological studies. Benthic microfossils are especially useful. Eustatic corrections are hazardous to apply and a simple transferral from the Vail/Haq curve is not recommended. Correction for the widely accepted long-term eustatic variation of sea level caused by changes in mid-ocean ridge volume is justified. However, the amplitude of the variation is in question, the Vail/Haq curve possibly overestimating the long-term sea level variation.

The sediment deposited in a marine basin replaces water, and so drives further subsidence of the basement. The exercise of partitioning the subsidence due to tectonics and that due to sediment loading is termed *backstripping*. If the lithosphere is in local Airy isostasy, the decompacted subsidence, corrected for palaeobathymetric and eustatic variations, can be simply used to calculate the tectonic component. This requires the average bulk density of the sediment column as a function of time to be known. However, if the lithosphere supports the sediment load by a regional flexure, the separation of the tectonic and sediment contributions is complex. The flexural loading of the sedimentary basin can only be accounted for if both the flexural rigidity and wavelength of the sediment load are known. This is generally not the case in geological applications.

A worked example is provided which goes through the method of backstripping for a North Sea borehole.

8.1 INTRODUCTION TO GEOHISTORY ANALYSIS

Improvements in the dating of stratigraphic units and in estimates of palaeobathymetry, largely brought about by advances in micropalaeontology, have allowed the development of quantitative techniques in geological analysis of sedimentary basins. Van Hinte (1978) termed this quantitative approach 'geohistory analysis'. The first qualitative attempts at plotting subsidence/uplift and palaeowater depth as a function of time date back at least to Lemoine's *Géologie du Bassin de Paris* (1911). Quantitative geohistory analysis was developed in the 1970s, principally in response to a vastly improved palaeontological commercial data base.

Geohistory analysis aims at producing a curve for the subsidence and sediment accumulation rates through time. In order to do this, three corrections to the present stratigraphic thicknesses need to be carried out:

1 *Decompaction*: present-day compacted thicknesses must be corrected to account for the progressive loss of porosity with depth of burial.

2 *Palaeobathymetry*: the water depth at the time of deposition determines its position relative to a datum (such as present-day sea level).

3 *Absolute sea level fluctuations*: changes in the palaeosea level relative to today's also needs to be considered (see Section 6.3).

Having made these corrections, comparisons between boreholes or other sections are readily made possible. In addition, the subsidence curves give an immediate visual impression of the nature of the driving force responsible for basin formation and development (see Chapters 3 to 5).

The time–depth history of any sediment layer can be evaluated therefore if the three corrections above can be applied. Such a time–depth history can also be tested from independent methods. These fall into two main classes (Chapter 9):

- organic thermal indicators
- mineralogical thermal indicators.

The first class includes measurements of, for example, vitrinite reflectance or coal rank, spore coloration or fluorescence, atomic ratios of kerogens and so on (Héroux, Chagnon and Bertrand 1979). The second class evaluates the abundance of certain index minerals, such as illite compared to expandable clay minerals, the crystallinity of illite and the position of the (001) reflection of smectite. These are all thermal maturation indices which allow geohistory curves to be calibrated.

The addition of a sediment load to a sedimentary basin causes additional subsidence of the basement. This is a simple consequence of the replacement of water (ρ = 1000 kg m^{-3}), or less commonly air, by sediment (ρ = 2500 kg m^{-3}). The total subsidence is therefore partitioned as follows:

- tectonic driving force
- sediment load.

The way in which this partitioning operates depends on the isostatic response of the lithosphere (Chapter 2). The simplest assumption is that any vertical column of load is compensated locally (Airy isostasy). This implies that the lithosphere has no strength to support the load. Alternatively, the lithosphere may transmit stresses and deformations laterally by a regional flexure (Chapters 2 and 4). The same load will therefore cause a smaller subsidence in the case of a lithosphere with a strength sufficient to cause flexure. The technique whereby the effects of the sediment load are removed from the total subsidence to obtain the tectonic contribution is called *backstripping*. Backstripped subsidence curves are especially useful in investigating the basin-forming mechanisms.

Burial history and thermal history can be used to determine the oil and gas potential of a basin and to estimate reservoir porosities. Burial history curves from a number of locations can also be used to construct palaeostructure maps at specific time slices. Combined with information on thermal maturity, this can be a powerful tool in evaluating the timing of oil migration and likely migration pathways in relation to the development of suitable traps (see Section 5 for fuller discussion).

8.2 DECOMPACTION

The decompaction technique seeks to remove the progressive effects of rock volume changes with

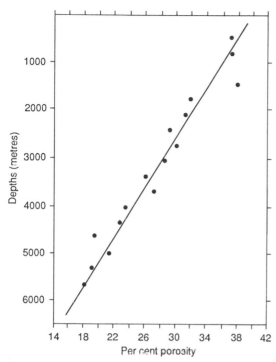

Fig. 8.1. The effect of the presence of ductile rock fragments on compaction of uncemented Tertiary sands in the subsurface of southern Louisiana (after Blatt in Schluger 1979). Compaction is reflected in the decrease in average porosity with depth. Based on over 17 000 cores, averaged at every 1000 ft (*c*. 300 m).

time and depth. Any compaction history is likely to be complex, being affected by lithology, overpressuring, diagenesis and other factors. Consequently, what are needed are some general relationships which hold good over large depth ranges.

To give an indication of the influence of detrital mineralogy on compaction history, the porosity of a number of sandstone samples from Gulf Coast wells has been measured. Whereas quartz arenites suffered very limited mechanical compaction over burial depths of kilometres, sandstones containing even small quantities of ductile clasts such as mudstones or micas, suffered considerable reductions in porosity over the same depth range (Fig. 8.1).

8.2.1 Porosity–depth relations in normally pressured and overpressured sediments

General relationships can be obtained using basic principles of soil mechanics. For a water-saturated

clay for example, the overlying weight of a layer of sediment is supported jointly by the fluid pressure in the pores and the grain to grain mechanical strength of the clay aggregates:

$$\text{Effective stress } \sigma = \text{vertical compressive stress } s \\ - \text{ fluid pressure } p$$

$$(8.1)$$

As the amount of compaction increases, the effective stress also increases. Since the vertical compressive stress is determined by the weight of the overlying water-saturated sediment column,

$$s = \bar{\rho}_b \, gy \tag{8.2}$$

($\bar{\rho}_b$ is average water-saturated bulk density, g is acceleration due to gravity and y is depth), this increasing vertical load must be divided between p and σ.

If the ratio of fluid pressure to overburden pressure $p/s = \lambda$, it must vary between zero where the fluid pressure is nonexistent, to 1 where the sediment layer is effectively 'floating' on the highly pressured pore-filling fluid. We can therefore write

$$p = \lambda s = \lambda \, \bar{\rho}_b \, gy \tag{8.3}$$

and

$$\sigma = (1 - \lambda) \, \bar{\rho}_b \, gy \tag{8.4}$$

As the sediment load is increased, the extra vertical stress is taken up initially by an increase in pore fluid pressure so that both p and λ increase. With time, however, water is expelled from the pores, reducing fluid pressures but increasing effective stresses. The increase in σ results in compaction of the grains supporting the stresses, reducing porosity. The lowest pressure that the pore fluid can attain is that due to the hydrostatic column, that is

$$p = \rho_w \, gy, \tag{8.5}$$

in which case

$$\lambda = \frac{\rho_w}{\bar{\rho}_b} \tag{8.6}$$

This is called the normal pressure since the pore fluids are at a pressure equivalent to the head of a static body of water and the grain–grain contacts are supporting the formation. If $\lambda > \rho_w/\bar{\rho}_b$, the pore fluids are at a higher pressure than hydrostatic.

Assuming that water is free to be expelled from the sediment pore space and is not trapped, increasing burial should lead to the equilibrium state where p is hydrostatic. If this is the case, from (8.4) and (8.6)

$$\sigma = (\bar{\rho}_b - \rho_w)\,gy \qquad (8.7)$$

which states that the magnitude of the stress causing compaction is a function of depth and the difference between the water-saturated sediment and water densities.

Porosity can be estimated from downhole electrical logs. For example, sonic, neutron and density logs are sensitive to formation lithology and porosity.

The *sonic log* is a recording of the time taken (interval transit time, Δ_t) for a compressional sound wave emitted from a sonic sonde to travel across 1 foot of formation to a receiver. The interval transit time is the inverse of velocity, which is a function of lithology and porosity. When lithology is known from other data such as a drill cuttings log, Δ_t may be used to calculate porosity from the Wyllie time-average equation:

$$\Delta_t = \underbrace{\Delta_{t\,ma}(1 - \phi)}_{\text{matrix}} + \underbrace{\phi\,(\Delta_{t\,f})}_{\text{fluid}} \qquad (8.8)$$

where $\Delta_{t\,ma}$ = transit time through the solid rock matrix (51.3–55.5 µs/ft for sandstones, 43.5-47.6 µs/ft for limestones)

$\Delta_{t\,f}$ = transit time through the pore space fluid (dependent on fluid salinity, 189 µs/ft for fresh water)

Rearranging the equation, a solution is obtained for porosity:

$$\phi = \frac{\Delta_t - \Delta_{t\,ma}}{\Delta_t - \Delta_{t\,f}} \qquad (8.9)$$

Instead of using this equation for each porosity determination, it may be more convenient to use a chart such as those provided by Schlumberger (1974). It has been found that in uncompacted, geologically young sands, the Wyllie time-average equation needs to be corrected by a compaction factor.

The principle of the *density log* is that gamma rays are emitted by a radioactive source in the logging tool, and are scattered and lose energy as a result of collisions with electrons in the formation. The number of scattered gamma rays recorded at a detector, also on the logging tool, depends on the density of electrons in the formation. This electron density is virtually the same as the formation bulk density for most common minerals, although for some evaporite minerals such as rock salt and sylvite (KCl), and coal, there is a significant difference.

The Formation Density (FDC) Sonde is normally calibrated in fresh-water filled limestone formations, so that the log reads the actual bulk density for limestone and for fresh water.

The bulk density is a function of the average density of the substances making up the formation, that is, both the rock matrix and fluid-filled pores, and the relative volumes occupied, as shown by the equation

$$\rho_b = \phi\rho_f + (1 - \phi)\,\rho_{ma} \qquad (8.10)$$

where ρ_f = average density of the fluid occupying the pore space, which is a function of temperature, pressure and salinity, and ρ_{ma} = average density of the rock matrix. By rearranging equation (8.10) we may very easily arrive at a means to determine porosity:

$$\phi = \frac{\rho_{ma} - \rho_b}{\rho_{ma} - \rho_f} \qquad (8.11)$$

ρ_{ma} may itself be decomposed into its constituent parts. When clay minerals are present, a correction which accounts for the density and amount of clay may be needed in order to avoid inaccurate porosity determination.

$$\rho_b = \phi\rho_f + V_{clay}\cdot\rho_{clay} + (1 - \phi - V_{clay})\,\rho_{ma} \qquad (8.12)$$

where ρ_{clay} is the average density of the clay and V_{clay} is the fraction of the total rock occupied by clay. The clay correction is substantial when ρ_{ma} and ρ_{clay} are very different. This is usually at shallow depths where the clay is particularly uncompacted.

Hydrocarbons may also lower the density log recording. Generally, the presence of oil in the invaded zone pore space has a negligible effect on the density log, but residual gas has a considerable effect which must be corrected for.

The principle of the *neutron log* is that neutrons emitted from a radioactive source collide with formation nuclei and are captured. A detector, or detectors, counts the returning neutrons. Because neutrons lose most energy when they collide with a *hydrogen* nucleus in the formation, the neutron log gives a measure of the hydrogen content, that is, the liquid-filled content, of the formation. It may be used, therefore, to determine porosity.

The most commonly used neutron tool is the Compensated Neutron Log (CNL). The CNL is calibrated to read true porosity in clean limestone formations. Its unit of measurement, as presented on the log, is therefore the 'limestone porosity unit'. When measuring in lithologies other than limestone, for example in a quartzose sandstone, a correction needs to be made.

The neutron log responds to *all* the hydrogen present in the formation, including the water bound up in clay minerals. As a result, it is very sensitive to clay content. In shales, the neutron log normally shows a very high reading, and needs correction:

$$\phi_N = \phi + V_{clay}\phi_{N.clay} \tag{8.13}$$

Although oil has a hydrogen content close to water and has only a small effect on the CNL, the hydrogen content of gas is considerably lower. As a result, the neutron response is low in formations which contain gas within the depth of investigation of the tool (generally < 30 cm).

Of all the porosity tools, the sonic log is most widely used (e.g. Magara 1976). This is largely because the neutron and density logs are normally only run in the deeper zones of hydrocarbon interest in a borehole, and the sonic log may therefore be the only porosity log available in the shallow sections of the borehole.

Porosity from the sonic log can be determined if the lithology is known. The most common method of obtaining porosity from the density and neutron logs is to cross-plot them. The techniques and corrections necessary are given in manuals such as Schlumberger (1974).

For normally pressured sediments, the variation of porosity with depth has long been thought to follow an exponential path (Athy 1930, Hedberg 1936, Rubey and Hubbert 1960) (Fig. 8.2). If this is the case, the porosity at any depth is

$$\phi = \phi_0 e^{-cy} \tag{8.14}$$

where c is a coefficient determining the slope of the ϕ-depth curve, y is the depth and ϕ_0 is the porosity at the surface. In other words, the surface porosity declines to $1/e$ of its original surface value at a depth of $1/c$ km (Fig. 8.3). On a depth versus log porosity graph, the value of c is the inverse of the rate of change of porosity with depth. The coefficient c can therefore be estimated if a number of porosity measurements can be made, for example from a sonic log from a representative borehole in the basin. Each lithology has its own characteristic value of c. There is a considerable literature on the compaction of sandstone. Some workers have preferred a linear relationship between porosity and depth. However, the North Sea data of Sclater and Christie (1980) when constrained by field studies of the surface porosities of Holocene sand-

Fig. 8.2. Plots of log porosity against depth for shales and sandstones (after Sclater and Christie 1980). The North Sea shale data are from sonic log values in normally pressured sections, porosities being calculated from the sonic velocity/porosity relation proposed by Magara (1976). The North Sea sandstone data are from the data of Seeley (1978) supplemented by data from sonic logs. The best-fit lines for the North Sea data and for the south Louisiana data of Atwater and Miller (1965) are constrained to pass through the surface porosity values of Pryor (1973).

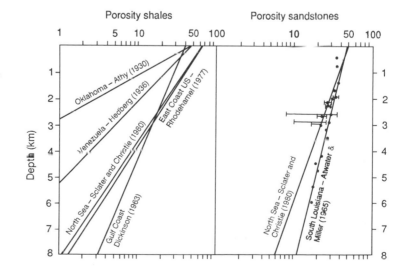

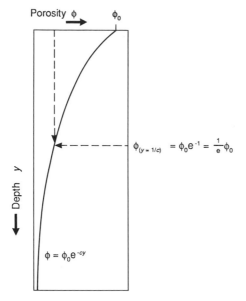

Fig. 8.3. Schematic diagram illustrating the use of the porosity–depth coefficient c. If a porosity–depth curve is known and it is exponential, c can be found by determining the depth at which the porosity has decreased to $1/e$ of its surface value. This should be repeated for all lithologies.

stones (Pryor 1973) can equally well be matched by an exponential curve. An exponential relationship is also found in chalks (Scholle 1977) from the North Sea and from Deep Sea Drilling Project sites. Sclater and Christie (1980) for example found that in the North Sea, the coefficient values given in Table 8.1 prevailed in the normally pressured sections.

Substituting (8.7) into (8.14) to eliminate depth y,

$$\phi = \phi_0 \exp\left[-\left\{ \frac{c}{(\bar{\rho}_b - \rho_w)g} \right\} \sigma \right] \tag{8.15}$$

Table 8.1 Porosity-depth parameters for lithologies in the North Sea basin

Lithology	Surface porosity ϕ_0	c (km^{-1})	Sediment grain density ρ_{sg} (kg m^{-3})
Shale	0.63	0.51	2720
Sandstone	0.49	0.27	2650
Chalk	0.70	0.71	2710
Shaley sandstone	0.56	0.39	2680

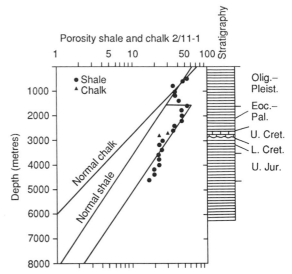

Fig. 8.4 Estimated porosity versus depth plot for an overpressured borehole in the Central Graben of the North Sea (Amoco 2/11-1). The increase in estimated porosity between about 1500 and 2000 m is thought to be due to overpressuring. It causes a marked offset of the exponential segments of the porosity–depth curves (after Sclater and Christie 1980). The porosity–depth relations were calculated assuming total sealing occurred at the end of the Eocene, with overpressured units below and normally pressured above. The wavy line in the stratigraphic column on the right represents the Albian–Aptian unconformity.

This equation expresses a fundamental relation between the effective stress on clay grains and the resultant porosity. In overpressured sections this equation can be modified by use of the ratio between fluid pressure and overburden pressure λ (eq. 8.4). This gives

$$\phi = \phi_0 \exp\left[-c\left\{ \frac{\bar{\rho}_b(1 - \lambda)y}{(\bar{\rho}_b - \rho_w)} \right\} \right] \tag{8.16}$$

The example from a North Sea borehole shown in Fig. 8.4 illustrates the effect of overpressuring on the porosity–depth curve. It has been argued that the exponential porosity–depth relationship does not fit shallower depth data particularly well (Falvey and Middleton 1981). An alternative general porosity–depth function is based on the assumption that an incremental change in porosity is proportional to the change in the load and the ratio of void space to skeletal (grain) volume. This relationship is given by

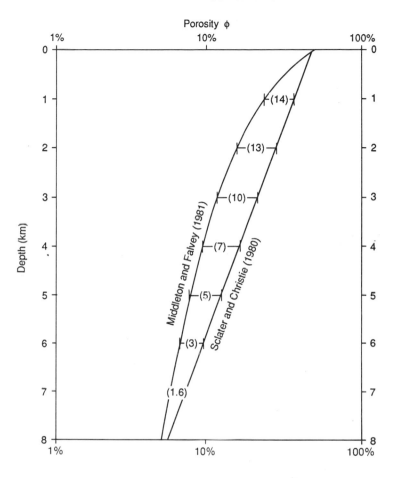

Fig. 8.5. Comparison of the porosity–depth relations of a sandstone with a surface porosity of 49 per cent according to the techniques advocated by Sclater and Christie (1980), with c = 0.27, and Falvey and Middleton (1981) with k = 2.17. Numbers in brackets are the percentage differences in porosity between the two methods, showing that differences are most marked at shallow depths, and the differences are negligible at depths of greater than 6 km.

$$1/\phi = 1/\phi_0 + ky \tag{8.17}$$

where k is a coefficient related to lithology. A comparison between the two porosity–depth relations for a sandstone with ϕ_0 = 0.49 (c = 0.27 Sclater and Christie 1980, p. 3732, k = 2.18, Falvey and Middleton 1981) is shown in Fig. 8.5.

8.2.2 Decompacted thicknesses

To calculate the thickness of a sediment layer at any time in the past, it is necessary to move the layer up the appropriate porosity–depth curve: this is equivalent to sequentially removing overlying sediment layers and allowing the layer of interest to decompact. In so doing, we keep mass constant and consider the changes in volumes and therefore thicknesses.

Consider a sediment layer at present depths of y_1 and y_2 which is to be moved vertically to new

shallower depths y'_1 and y'_2 (Fig. 8.6). From equation (8.14) the amount of water-filled pore space between depths y_1 and y_2 is simply the porosity integrated over the depth interval,

$$V_w = \int_{y_1}^{y_2} \phi_0 \, e^{-cy} \, dy \tag{8.18}$$

which on integration gives

$$V_w = \frac{\phi_0}{c} \{\exp(-cy_1) - \exp(-cy_2)\} \tag{8.19}$$

Since the total volume of the sediment layer (V_t) is the volume due to pore-filling water (V_w) and the volume of the sediment grains (V_s),

$$V_s = V_t - V_w$$

and from (8.19), considering a unit cross-sectional area,

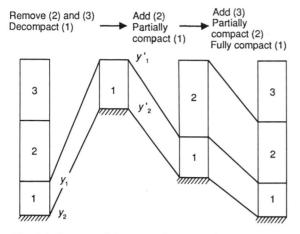

Remove (2) and (3)
Decompact (1) → Add (2)
Partially
compact (1) → Add (3)
Partially
compact (2)
Fully compact (1)

Fig. 8.6. Concept of the successive stages in a decompaction exercise.

$$y_s = y_2 - y_1 - \frac{\phi_0}{c}\{\exp(-cy_1) - \exp(-cy_2)\}$$
$$(8.20)$$

On decompaction the sediment volume remains the same, only the volume of water expanding. The weight of the water in a unit area sedimentary column lying between depths y'_1 and y'_2 is from (8.19)

$$y'_w = \frac{\phi_0}{c}\{\exp(-cy'_1) - \exp(-cy'_2)\} \qquad (8.21)$$

The new decompacted thickness of the sediment layer is the sum of the thickness due to the sediment grains (8.20) and that due to the water (8.21). That is,

$$y'_2 - y_1 = y_s + y'_w \qquad (8.22)$$

which becomes

$$y'_2 - y'_1 = y_2 - y_1 - \frac{\phi_0}{c}$$
$$\times\{\exp(-cy_1) - \exp(-cy_2)\}$$
$$+ \frac{\phi_0}{c}\{\exp(-cy'_1) - \exp(-cy'_2)\}$$
$$(8.23)$$

This is the general decompaction equation which represents mathematically the exercise of sliding the sediment layer up the exponential porosity–depth curve. Its solution is by numerical iteration, which makes it ideal for solving by computer. A program for use with a NIMBUS microcomputer is found in Appendix A. It can be relatively easily modified for use with other PCs.

Using the alternative porosity–depth relation in equation (8.17), the skeletal or solid grain volume in a unit cross-sectional area of a sediment layer between depths y_1 and y_2 is

$$y_s = \int_{y_1}^{y_2}(1 - \phi(y))\,dy \qquad (8.24)$$

which with reference to (8.17) and evaluating between the limits y_1 and y_2 is

$$y_s = (y_2 - y_1) - \int_{y_1}^{y_2}\left(\frac{1}{\phi_0} + ky\right)^{-1}dy$$
$$= (y_2 - y_2) - \frac{1}{k}\ln\left(\frac{1}{\phi_0} + ky_2\right)$$
$$- \frac{1}{k}\ln\left(\frac{1}{\phi_0} + ky_1\right)$$
$$= (y_2 - y_1) - \frac{1}{k}\ln\frac{\dfrac{1}{\phi_0} + ky_2}{\dfrac{1}{\phi_0} + ky_1}$$
$$(8.25)$$

Sliding this rock volume up the porosity–depth curve to some new depth, the correct thickness is now $y'_2 - y'_1$, which is

$$y'_2 - y'_1 = y_s + \frac{1}{k}\ln\left(\frac{1}{\phi_0} + ky'_2\right)$$
$$- \frac{1}{k}\ln\left(\frac{1}{\phi_0} + ky'_1\right) \qquad (8.26)$$

From (8.25) this becomes

$$y'_2 - y'_1 = (y_2 - y_1) - \frac{1}{k}\ln\frac{\dfrac{1}{\phi_0} + ky_2}{\dfrac{1}{\phi_0} + ky_1}$$
$$+ \frac{1}{k}\ln\frac{\dfrac{1}{\phi_0} + ky'_2}{\dfrac{1}{\phi_0} + ky'_1}$$
$$(8.27)$$

It is immediately apparent that this solution has a similar form to that of (8.23) which uses an exponential porosity–depth relation.

Either (8.23) or (8.27) calculates the thicknesses of a sediment layer at any time from the time of deposition to the present day. A decompacted subsidence curve can therefore be plotted. The sources of the data points of the subsidence curve are the stratigraphical boundaries of presumed known absolute age defining stratigraphical units of known present-day thickness. All depths are, however, in relation to a present-day datum, normally taken as mean sea level. Consequently it is necessary to correct the decompacted subsidence curve for firstly the difference in height between the depositional surface and the regional datum (*palaeobathymetric correction*), and secondly, for past variations in the ambient sea level compared to today's (*eustatic correction*). Finally, the sediment weight drives basement subsidence. In order to calculate the true tectonic subsidence, it is necessary to remove the effects of the excess weight of the sediment compared to water.

A worked example which calculates decompacted thicknesses as a function of time is provided on p. 275.

8.3 TECTONIC SUBSIDENCE

8.3.1 Palaeobathymetric corrections

The estimation of water depth for a given stratigraphic horizon is generally far from easy, yet it is essential in order to accurately study burial history. As an example of the potential problems, consider two sedimentary basins A and B. In basin A the water depth is 5 km. If the basement subsides tectonically by 1 km over a certain time period and during this time sediment is supplied to the basin such that it becomes filled to the brim, about 15 km of sediment will have accumulated (since the sediment load drives further subsidence). In basin A, therefore, a stratigraphic thickness of 15 km reflects a very small (1 km) driving subsidence. In basin B, however, a tectonic subsidence of 1 km takes place in a basin which has its depositional surface already at sea level. The resultant sediment thickness is barely 3 km at the end of the same time period. Clearly, for the same rate of tectonic subsidence, enormously different stratigraphic thicknesses can result depending on the initial and ensuing palaeobathymetry.

Information on palaeobathymetry comes from a number of sources, chief of which are:
- benthic microfossils, and less importantly,
- sedimentary facies and
- distinctive geochemical signatures.

Although some organisms inhabit a particular depth range as an adaptation to hydrostatic pressure, most palaeodepth estimates are indirectly obtained. For example, qualitative estimates can be obtained by a comparison with the modern occurrences of certain species or assemblages and by recognition of the ecological trends of benthonic and planktonic organisms through time. Estimates can also be obtained by quantitative methods, using ratios of, for example, plankton/benthos, arenaceous/calcareous foraminifers, per cent radiolarians or ostracods, or alternatively using species dominance and diversity, morphological characters and so on. These many techniques allow the palaeontologist to make meaningful interpretations of environmental factors such as the chemical environment (salinity, pH, oxygen and CO_2 contents, nutrient availability), the physical environment (temperature, light, energy level, type of substrate, turbidity) and the biological environment. As much information as possible needs to be synthesized to produce a reliable depth estimate. An example given in Fig. 8.7 shows the type of depth information that can be obtained palaeontologically, together with the ranges over which the estimate may span.

Sedimentary and geochemical data are, by comparison, far less useful. Sedimentary facies reflect supply and process and are not therefore particularly diagnostic of depth. Although some structures, such as wave ripple marks, are restricted to particular depth ranges (<200 m), the sedimentary facies more likely will provide a back-up to the palaeontological observations in marine sediments. In continental environments the potential for using sedimentary facies is greater. If shoreline facies can be identified for example, the calculation of floodplain slopes from fluviatile sediments can give valuable information on heights above sea level (Homewood, Allen and Williams 1986, p. 207; Fasel 1986). The most obvious geochemical data relate to the carbonate dissolution depth (CCD) below which calcareous material is dissolved. Because most of this material is in the form of calcareous microfossils, it falls within the realm of the palaeontologist. Some mineral species such as glaucony and phosphates may provide useful information on palaeowater depth, but estimates are likely to be far from precise.

8.3.2 Eustatic corrections

We have previously investigated in Chapter 6 the stratigraphical evidence for global sea level

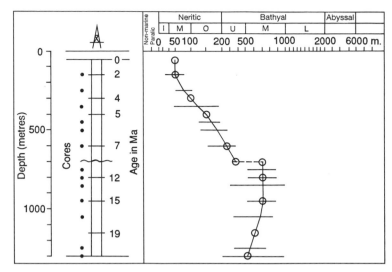

Fig. 8.7. Stratigraphy penetrated in a borehole showing position of core data (dots) and palaeobathymetric interpretations for core data given as horizontal lines. As much palaeontological information as possible is used to guide the choice of the position of the palaeobathymetric curve through the ranges indicated by the horizontal lines. The mid-points of the ranges need not be used. After van Hinte (1978).

fluctuations and the controversy surrounding the precise significance of the first-, second- and third-order cycles of Vail, Mitchum and Thompson (1977) and the short- and long-term curves of Haq, Hardenbol and Vail (1987). Bearing in mind these uncertainties, it is advisable firstly to decompact ignoring any possible global sea level fluctuations. Following this, the sea level changes associated with first-order cycles (most likely due to volume changes in the ocean ridge system) and those confidently related to glaciations/deglaciations can be included. Even in these cases, the amplitude of the sea level fluctuation is a key question (see Section 6.3).

8.3.3 Sediment load

The true tectonic subsidence is obtained after the removal of the subsidence due to the sediment load and after corrections for variations in water depth and eustatic sea level fluctuations. The influence of the sediment load can be evaluated as follows. The porosity of the sediment layer at its new depth is

$$\phi = \frac{\phi_0}{c} \frac{\exp{(-cy'_1)} - \exp{(-cy'_2)}}{y'_2 - y'_1} \qquad (8.28)$$

Since the bulk density of the new sediment layer (ρ_s) depends on the porosity and the density of the sediment grains (ρ_{sg})

$$\rho_s = \phi\rho_w + (1 - \phi)\rho_{sg} \qquad (8.29)$$

the bulk density of the entire sedimentary column ($\bar{\rho}_s$) made up of i layers is

$$\bar{\rho}_s = \sum_i \left\{ \frac{\bar{\phi}_i\rho_w + (1 - \bar{\phi}_i)\rho_{sgi}}{S} \right\} y'_i \qquad (8.30)$$

where $\bar{\phi}_i$ is the mean porosity of the ith layer, ρ_{sgi} is the sediment grain density of the same layer, y_i is the thickness of the ith sediment layer, and S is the total thickness of the column corrected for compaction.

The loading effect of the sediment can then be treated as a problem of a local (Airy) isostatic balance. Where sediment is replacing a column of water,

$$Y = S \left(\frac{\rho_m - \bar{\rho}_s}{\rho_m - \rho_w} \right) \qquad (8.31)$$

where Y is the depth of the basement corrected for sediment load and $\rho_m, \bar{\rho}_s, \rho_w$ are mantle, mean sediment column and water densities.

8.3.4 Flexural support for sediment loads

We can now incorporate all of the corrections to obtain the true tectonic subsidence (Bond and Kominz 1984),

$$Y = \Phi \left\{ S \left(\frac{\rho_m - \bar{\rho}_s}{\rho_m - \rho_w} \right) - \Delta_{SL} \left(\frac{\rho_w}{\rho_m - \rho_w} \right) \right\} + (W_d - \Delta_{SL}) \qquad (8.32)$$

where Δ_{SL} is the palaeosea level relative to the present

W_d is the palaeowater depth

Φ is a basement function equal to unity for Airy isostasy.

If the lithosphere supports the sedimentary basin by a long-wavelength bending, the basement function Φ varies from $(\rho_m - \rho_w)/(\rho_m - \bar{\rho}_s)$ to 1 depending on the degree of compensation of the load (see below).

The significance of the basement response function can be appreciated by considering the way in which the lithosphere responds to a simple periodic load, as represented for example by topography. Assuming the periodic load to be sinusoidal (Fig. 8.8) and therefore given by:

$$h = h_o \sin(2\pi x/\lambda) \tag{8.33}$$

the load actually exerted on the lithosphere has the form of pressure or stress, $\rho g h$,

$$q_a(x) = \rho_s g h_o \sin(2\pi x/\lambda) \tag{8.34}$$

where ρ_s is the density of sediments constituting the load. Assuming no horizontal applied forces (i.e. $P = 0$), and returning to the general flexural equation, we have

$$D\frac{d^4w}{dx^4} + (\rho_m - \rho_s)gw = \rho_s g h_o \sin(2\pi x/\lambda) \tag{8.35}$$

Since the load is sinusoidal and periodic, the deflection of the lithosphere will also be sinusoidal and periodic. The solution for the deflection can therefore be assumed to be of the type

$$w = w_o \sin(2\pi x/\lambda) \tag{8.36}$$

Substituting (8.36) into (8.35), the amplitude of the deflection of the lithosphere is

$$w_o = \frac{h_o}{\left\{ D\dfrac{d^4w}{dx^4} \cdot \dfrac{1}{\rho_s g w} + \dfrac{\rho_m}{\rho_s} - 1 \right\}} \tag{8.37}$$

The fourth differential of w from (8.36) is $(2\pi/\lambda)^4 w$, so (8.37) simplifies to

$$w_o = \frac{h_o}{\dfrac{\rho_m}{\rho_s} - 1 + \dfrac{D}{\rho_s g}\left(\dfrac{2\pi}{\lambda}\right)^4} \tag{8.38}$$

If the wavelength of the load is short, the deflection is very small compared to the maximum

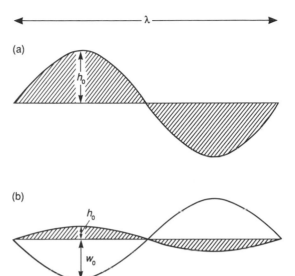

Fig. 8.8. Deflection of the lithosphere under a periodic (sinusoidal) load. In (a) the wavelength of the load is short and there is no deflection of the lithosphere. In (b) the wavelength of the load is long, leading to isostatic compensation of the load by a deflection of the lithosphere. w_0 is the maximum deflection, h_0 is the maximum elevation of the load (after Turcotte and Schubert 1982).

height of the load ($w_o \ll h_o$). The lithosphere therefore appears to behave very rigidly to loads of this scale. If, however, the wavelength of the load is long, the deflection can be written

$$w = w_{o\infty} = \frac{\rho_s h_o}{(\rho_m - \rho_s)} \tag{8.39}$$

which is the result obtained for a purely vertical isostatic balance (see Section 2.1). This means that for sufficiently long wavelengths, the lithosphere appears to have no rigidity and any sediment load should be in hydrostatic equilibrium. It is clearly desirable to be able to predict where we are in this range of complete to no compensation. The degree of compensation C of the load is the ratio of the actual deflection compared to the maximum (Airy) or hydrostatic deflection

$$C = \frac{w_o}{w_{o\infty}} \tag{8.40}$$

Substituting (8.38) and (8.39) into (8.40)

$$C = \frac{(\rho_m - \rho_s)}{\rho_m - \rho_s + \dfrac{D}{g}\left(\dfrac{2\pi}{\lambda}\right)^4} \tag{8.41}$$

$(2\pi/\lambda)$ is sometimes termed the wave number (e.g. Watts 1988).

The relationship is sketched in Fig. 8.9. It is possible therefore to estimate the flexural loading of the overlying sediment if and only if the following pieces of information are available (and we make the simplifying assumption of a sinusoidal load in this case):

• the equivalent elastic thickness or flexural rigidity of the lithosphere
• the wavelength of the sediment load.

Taking the example of a sedimentary basin 200 km wide ($\lambda/2 = 200$ km) with a sinusoidal sediment load and an underlying lithosphere of flexural rigidity 10^{24} Nm, $(\rho_m - \rho_s) = 800$ kg m^{-3}, the degree of compensation C is about 0.12. This suggests that the lithosphere behaves very rigidly to this kind of wavelength of load. Changing the wavelength of the load such that $\lambda/2$, the width of the basin, is now 400 km, $C = 0.68$, indicating that the sediment load is only weakly supported. In this case of large compensation, Airy-type isostasy is approached. In order to perform flexural rather than Airy backstripping, we must know the value of the basement response function, or degree of compensation. This will depend on the exact form of the load, for example, whether it can be approximated by a rectangular block, a right angled triangle or a harmonic expression such as the one previously discussed (equation 8.41). Where the load is sinusoidal therefore, we can write for the tectonic subsidence

$$Y = S\left\{1 - C\left(1 - \frac{\rho_m - \rho_{\bar s}}{\rho_m - \rho_w}\right)\right\} \tag{8.42}$$

where C is given by equation (8.41) and sea level and water depth changes are ignored (cf equation (8.32)).

It must be stressed, however, that rarely can C be estimated accurately in geological situations. An error of one order of magnitude in the estimate of the flexural rigidity can cause very considerable differences in the calculated value of C. For example, in the above example, if $D = 10^{23}$ Nm, $C = 0.95$, indicating almost full compensation (Airy isostasy), whereas if $D = 10^{25}$ Nm, $C = 0.17$ indicating a strongly supported load. Flexural backstripping may, however, be attempted where other data-sets such as gravity measurements are available to help constrain flexural rigidity. An example is the backstripping of the Baltimore Canyon Trough region of the American Atlantic margin by Watts (1988).

1 Worked example on decompaction

Conoco 15/30–1, situated in the north of the Central Graben of the North Sea, penetrated the stratigraphy shown in Fig. 8.10.

Construct a decompacted subsidence plot from the above stratigraphic column. Assume the values of c and surface porosity (from Sclater and Christie 1980) appearing in Table 8.1. Assume that the relation between porosity and depth is an exponential of the form

$$\phi = \phi_0 e^{-cy}$$

where ϕ is the porosity at any depth y, ϕ_0 is the surface porosity and c is a coefficient describing the slope of the curve.

We assume that the water column has a negligible effect on the compaction of underlying sedimentary layers. Therefore, subtract the present-day water depth of 0.172 km from the present strati-

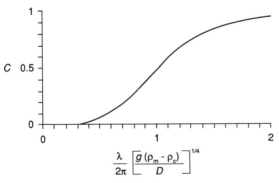

Fig. 8.9. Dependence of the degree of compensation, C, on the non-dimensional wavelength of periodic (sinusoidal) topography (after Turcotte and Schubert 1982). D is the flexural rigidity, λ the wavelength of the load, ρ_m and ρ_c are the mantle and crustal densities and g is the gravitational acceleration.

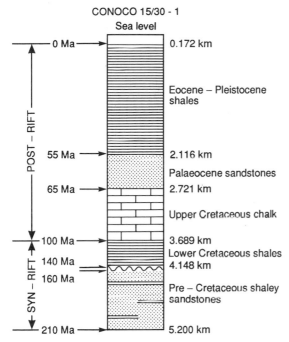

Fig. 8.10. Summary stratigraphic column from Conoco 15/30–1 in the Central Graben of the North Sea.

graphic thicknesses before carrying out the decompaction exercise.

The syn-rift sequence contains unconformities. Although the amount of erosion represented by the unconformities is unknown, their duration is rather better constrained. For the purposes of this exercise, assume an unconformity at the base of the Cretaceous representing the time gap 140–160 Ma (cutting down into Middle Jurassic rocks).

Solution to decompaction problem

The data set for Conoco 15/30–1 can be summarized as in Table 8.2, where stratigraphic depths are compensated for present-day water depth.

The decompacted depths to the stratigraphic boundaries are as shown in Table 8.3.

The decompacted subsidence history of Conoco 15/30–1 can be plotted on a diagram of depth versus age (Fig. 8.11). The bottom line represents the subsidence of the top of the basement. Note that this plot ignores any palaeobathymetric or eustatic corrections. These will be investigated in the following worked example.

2 Worked example on the sediment load

The total subsidence is made up of two components, the tectonic driving force and the sediment load. Sediment filling a basin and therefore replacing water exerts an additional force on the basement, amplifying the tectonic driving force for subsidence. The effects of the sediment load can be removed using an isostatic model.

The porosity of each sedimentary unit during its compaction history is given by equation (8.28).

Table 8.2 Input parameters for decompaction problem

Unit	Top (km)	Base (km)	Thickness (km)	ϕ-depth coefficient (km^{-1})	Surface porosity	Age (Myr)
						210
1 Pre-Cretaceous	3.976	5.028	1.052	0.39	0.56	
						160
2 Unconformity	3.976	3.976	0	—	—	
						140
3 Lower Cretaceous	3.517	3.976	0.459	0.51	0.63	
						100
4 Upper Cretaceous	2.549	3.517	0.968	0.71	0.70	
						65
5 Palaeocene	1.944	2.549	0.605	0.27	0.49	
						55
6 Eocene–Pleistocene	0	1.944	1.944	0.51	0.63	
						0

Table 8.3 Decompacted depths, 15/30–1

	Age (Ma)					
Unit	160	140	100	65	55	0
1	1.623	1.623	2.229	3.282	3.666	5.028
2	0	0	0.851	2.098	2.524	3.976
3	0	0	0.851	2.098	2.524	3.976
4	0	0	0	1.544	2.005	3.517
5	0	0	0	0	0.793	2.549
6	0	0	0	0	0	1.944

The bulk density of the sedimentary column is then given by equation (8.30), where ρ_{sgi} is $2720\,kg\,m^{-3}$ for shales, $2650\,kg\,m^{-3}$ for sandstone, $2710\,kg\,m^{-3}$ for chalk and $2680\,kg\,m^{-3}$ for shaley sand.

The loading effect of the sediment, assuming Airy isostasy, can be removed to obtain the 'tectonic' subsidence from equation (8.31) where ρ_m can be taken as $3330\,kg\,m^{-3}$.

Palaeobathymetry and absolute sea level fluctuations Marine microfossils, particularly benthonic foraminifera, provide information on palaeowater depths. These assemblages indicate a change from bathyal (150–400 m deep) in the Palaeocene to middle neritic (100–200 m) from late Eocene to the present day. Assume that the palaeowater depth was zero at the time of the Albian/Aptian boundary.

The subject of eustatic sea level fluctuations versus tectonic effects is currently being debated. For the purposes of this exercise, assume that sea levels were elevated at 150 m above their present level at 100 Ma, and have fallen steadily since that time.

Remove the effects of changes in bathymetry and absolute sea level from the post-rift phase of subsidence using the data in Table 8.4.

Find the tectonic subsidence taking into account sediment and water loading, palaeobathymetry and eustatics using equation (8.32).

Table 8.4 Paleobathymetric and absolute sea level data for sediment load problem

Age (Ma)	Water depth (km)	Absolute sea level relative to present (km)
0	0.172	0
55	0.300	+ 0.080
65	0.250	+ 0.100
100	0	+ 0.150

Solution to sediment load problem

In this problem we are concerned with the post-rift phase of so-called 'thermal subsidence'. Bulk densities of the sediment columns are therefore averaged over only three units – Upper Cretaceous, Palaeocene and Eocene–Pleistocene. The Lower Cretaceous–Upper Cretaceous boundary is therefore regarded as the top of 'basement'.

The average porosities of the individual sediment layers as they undergo compaction/

Table 8.5 Backstripped porosity and bulk density, 15/30–1

	Age (Ma)			
Unit	100	65	55	0
Upper Cretaceous	–	0.425	0.267	0.083
Palaeocene	–	–	0.441	0.267
Eocene–Pleistocene	–	–	–	0.400
Bulk density	–	1996	2106	2230

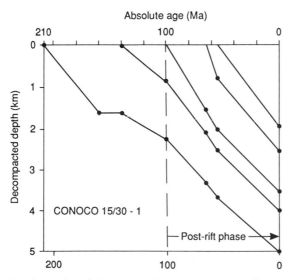

Fig. 8.11. Plot of decompacted depth versus time for the stratigraphy penetrated in Conoco 15/30-1.

Table 8.6 Backstripped tectonic subsidence using Airy isostasy

	Age (Ma)			
	100	65	55	0
Y(km)	0	0.896	1.067	1.681

Table 8.7 Tectonic subsidence corrected for paleobathymetry and eustasy

	Age			
	100	65	55	0
$Y_{corrected}$(km)	− 0.217	1.031	1.251	1.854

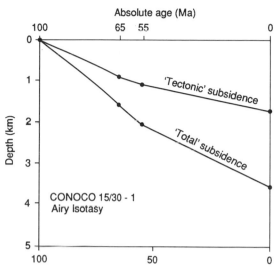

Fig. 8.12. Backstripped subsidence plot for 15/30-1 using an Airy isostatic model.

decompaction and the bulk density averaged over the entire sediment column are as given in Table 8.5.

Assuming Airy isostasy, the loading effect of the sediment can be removed to obtain the tectonic driving force for subsidence shown in Table 8.6 (Fig. 8.12).

The tectonic subsidence corrected for palaeobathymetry and eustasy is as in Table 8.7.

Note that these corrections have little impact on the shape of the tectonic subsidence curve in this particular case. We shall use the results above in the worked examples on stretch factors and heat flows found in Chapter 4.

APPENDIX : BACKSTRIPPING PROGRAM LISTING

```
10  REM backstrip version 2
11  CLS : HOME
12  PRINT TAB (25); "***************************"
13  PRINT TAB (25); "*                         *"
15  PRINT TAB (25); "*"; TAB (30); "BACKSTRIP VERSION 2"; TAB (53); "*"
16  PRINT TAB (25); "*                         *"
17  PRINT TAB (25); "***************************"
18  PRINT ! ! !
20  GLOBAL N, I, J, Ratio, X
30  GLOBAL Zt(), Zb(), T(), C(), Surpor(), Ma(), Tect()
35  GLOBAL Nzt(), Nzb(), Por(), Dt()
36  GLOBAL H(), V()
37  GLOBAL Rhoa(), Rhos()
40  INPUT "number of stratigraphic units "; N
50  DIM L$(N)
60  DIM Zt(N), Zb(N), T(N), C(N), Surpor(N), Ma(N + 1)
65  DIM Rhoa(N), Rhos(N), Tect(N)
70  DIM Nzt(N, N), Nzb(N, N), Por(N, N), Dt(N, N)
75  DIM H(N + 1, N + 1), V(N + 1, N + 1)
90  PRINT ! !
100 PRINT "Give bases of stratigraphic units "
101 PRINT "from bottom to top in kilometres"
110 PRINT : PRINT
120 FOR I := 1 TO N
130    INPUT "name of stratigraphic unit "; L$(I)
140    INPUT "depth of base of stratigraphic unit in km "; Zb(I)
150    INPUT "depth to top of stratigraphic unit in km "; Zt(I)
160    T(I) := Zb(I) - Zt(I)
170    INPUT "porosity-depth coefficient "; C(I)
180    INPUT "estimated surface porosity "; Surpor(I)
182    INPUT "sediment grain density in kg m-3 "; Rhos(I)
185    PRINT ! !
190 NEXT I

200 PRINT : PRINT TAB (25); "Stratigraphic Data Set"
210 PRINT ! "unit"; TAB (20); "top"; TAB (30); "base"; TAB (40);
"thickness"; TAB (50); "coef"; TAB (60); "sur por"
220 FOR I := 1 TO N
230    PRINT ! L$(I); TAB (20); Zt(I); TAB (30); Zb(I); TAB (40); T(I); TAB
(50); C(I); TAB (60); Surpor(I)
240 NEXT I
300 FOR J :- 1 TO N
310    FOR I := J TO N
315       X := I - J + 1
320       Ratio := Surpor(X) / C(X)
330       IF J = 1 THEN Nzt(I, J) := 0
340       Nzt(I, J) := Nzb(I, J - 1)
350       Solution
370    NEXT I
375    PRINT ! !
380 NEXT J
400 PRINT TAB (30); "Decompacted Depths"
410 PRINT ! !
420 FOR J := 1 TO N
430    FOR I := 1 TO N
440       PRINT Nzb(I, J); TAB (I * 15);
460    NEXT I
465    PRINT !
```

```
 470 NEXT J
 500 INPUT "graphical output (Y or N)? "; P$
 510 PRINT ! !
 520 IF P$ = "Y" OR P$ = "y" THEN GOTO 540
 530 IF P$ = "N" OR P$ = "n" THEN GOTO 650
 540 PRINT "Give absolute ages in Ma "
 550 FOR I := 1 TO N + 1
 551    IF I > 1 THEN GOTO 560
 552    PRINT "age of start of subsidence (Ma) "
 553    INPUT Ma(I) : GOTO 580
 560    PRINT "top of unit "; L$(I - 1)
 570    INPUT Ma(I)
 580    PRINT ! !
 590 NEXT I
 600 Graphics
 610 FOR I := 1 TO 5000
 620    REM pause
 630 NEXT I
 640 CLS
 650 PRINT !
 660 PRINT TAB (30); "Decompacted Thicknesses"
 665 PRINT ! !
 670 FOR I := 1 TO N
 680    FOR J := 1 TO I
 690       Dt(I, J) := Nzb(I, J) - Nzb(I, J - 1)
 700    NEXT J
 720 NEXT I
 730 FOR J := 1 TO N
 740    FOR I := 1 TO N
 750       PRINT Dt(I, J); TAB (15 * I);
 760    NEXT I
 765    PRINT !
 770 NEXT J
 780 Porosity
 790 Density
 800 PRINT !
 805 INPUT "Screen Dump? (Y or N) "; P$
 810 IF P$ = "N" OR P$ = "n" THEN GOTO 840
 820 PRINT "Type <CTRL> and <PRT SC> together to send"
 825 PRINT "screen contents to printer"
 830 PRINT !
 832 FOR I := 1 TO 1000
 834    REM pause
 836 NEXT I
 840 Isostasy
 850 Geohistory
 900 END
1000 PROCEDURE Solution
1010    GLOBAL I, J, Ratio, X
1015    GLOBAL C(), Surpor(), Zt(), Zb(), Nzt(), Nzb(), T()
1020    A := EXP(- C(X) * Zt(X)) - EXP(- C(X) * Zb(X))
1030    B := EXP(- C(X) * Nzt(I, J))
1040    Lhs := Nzt(I, J) + T(X) - (Ratio * A) + (Ratio * B)
1050    Nzb(I, J) := 1
1060    R := Nzb(I, J) + Ratio * EXP(- C(X) * Nzb(I, J))
1070    IF R >  1.0001 * Lhs THEN GOTO 1110
1080    IF R <  0.9999 * Lhs THEN GOTO 1130
1090    IF R >  0.9999 * Lhs AND R <  1.0001 * Lhs THEN GOTO 1180
1110    Nzb(I, J) := Nzb(I, J) - ABS(Lhs - R)
```

```
1120    GOTO 1060
1130    Nzb(I, J) := Nzb(I, J) + ABS(Lhs - R)
1140    GOTO 1060
1180 ENDPROC
1200 PROCEDURE Graphics
1210    GLOBAL I, J, N
1220    GLOBAL Nzb(), Ma()
1222    GLOBAL H(), V()
1230    SET ORIGIN 100, 20
1232    SET POINTS STYLE 3
1240    CLS
1250    LINE 0, 0; 0, 200; 450, 200; 450, 0; 0, 0
1260    PLOT "Absolute Age (Ma)", 150, 210
1270    PLOT "Decompacted", - 90, 110
1280    PLOT "Depth (Km)", - 80, 100
1285    POINTS 0, 200
1290    FOR J := 1 TO N
1300      FOR I := J TO N
1310        H(I, J) := (Ma(1) - Ma(I + 1)) * 450 / Ma(1)
1320        V(I, J) := 200 - (Nzb(I, J) * 40)
1330        POINTS H(I, J), V(I, J)
1340      NEXT I
1350    NEXT J
1360    REM draw lines between data points
1380    FOR J := 1 TO N
1385      FOR I := J TO N
1390        LINE (Ma(1) - Ma(I)) * 450 / Ma(1), 200 - Nzb(I - 1, J - 1) * 40;
(Ma(1) - Ma(I + 1)) * 450 / Ma(1), 200 - Nzb(I, J) * 40
1395      NEXT I
1400    NEXT J
1450 ENDPROC
1500 PROCEDURE Porosity
1510    GLOBAL I, J, N
1520    GLOBAL Nzb(), Surpor(), C(), Por(), Dt()
1530    FOR I := 1 TO N
1540      FOR J := 1 TO I
1550        X := I - J + 1
1560        Ratio := Surpor(X) / C(X)
1570        D := EXP(- C(X) * Nzb(I, J - 1)) - EXP(- C(X) * Nzb(I, J))
1580        Por(I, J) := Ratio * D / Dt(I, J)
1590      NEXT J
1600    NEXT I
1610 ENDPROC
1700 PROCEDURE Density
1710    Rhow := 1030
1715    GLOBAL N, I, J
1720    GLOBAL Rhos(), Por(), Dt(), Rhoa(), Nzb()
1740    FOR I := 1 TO N
1745      Rhoa(I) := 0
1750      FOR J := 1 TO I
1760        Rhoa(I) := Rhoa(I) + ((Por(I, J) * Rhow) + ((1 - Por(I, J)) *
Rhos(I))) * Dt(I, J) / Nzb(I, I)
1790      NEXT J
1800    NEXT I
1810 ENDPROC
1900 PROCEDURE Isostasy
1910    GLOBAL I, J, N
1920    GLOBAL Nzb(), Rhoa(), Tect()
1930    Rhom := 3300
```

```
1940    Rhow := 1030
1950    FOR I := 1 TO N
1960      Tect(I) := Nzb(I, I) * (Rhom - Rhoa(I)) / (Rhom - Rhow)
1970    NEXT I
1980  ENDPROC
2000  PROCEDURE Geohistory
2010    GLOBAL I, J, N
2020    GLOBAL Nzb(), Ma(), Tect()
2030    GLOBAL H(), V()
2040    SET ORIGIN 100, 20
2050    SET POINTS STYLE 3
2060    CLS
2070    LINE 0, 0; 0, 200; 450, 200; 450, 0; 0, 0
2080    PLOT "Absolute Age (Ma) ", 150, 210
2100    POINTS 0, 200
2110    FOR J := 1 TO N
2120      FOR I := J TO N
2130        H(I, J) := (Ma(1) - Ma(I + 1)) * 450 / Ma(1)
2140        V(I, J) := 200 - (Nzb(I, J) * 40)
2150        POINTS H(I, I), V(I, I)
2160      NEXT I
2170    NEXT J
2180    FOR J := 1 TO N
2190      FOR I := J TO N
2200        V(I, J) := 200 - (Tect(I) * 40)
2210        POINTS H(I, I), V(I, I)
2220      NEXT I
2230    NEXT J
2240    REM draw lines for tectonic and decompacted subsidence of basement
2250    FOR J := 1 TO N
2260      FOR I := J TO N
2270        LINE (Ma(1) - Ma(I)) * 450 / Ma(1), 200 - Nzb(I - 1, I - 1) * 40;
(Ma(1) - Ma(I + 1)) * 450 / Ma(1), 200 - Nzb(I, I) * 40
2280        LINE (Ma(1) - Ma(I)) * 450 / Ma(1), 200 - Tect(I - 1) * 40;
(Ma(1) - Ma(I + 1)) * 450 / Ma(1), 200 - Tect(I) * 40
2290      NEXT I
2300    NEXT J
2320  ENDPROC
```

9 Thermal history

Accuse not Nature, she hath done her part;
Do thou but thine

(Milton, *Paradise Lost*)

SUMMARY

Subsidence in sedimentary basins causes thermal
maturation in the progressively buried sedimentary
layers. Indicators of the thermal history include
organic parameters and mineralogical parameters.
The most important factors in the maturation of
organic matter are temperature and time, pressure
being relatively unimportant. This temperature
and time dependency is described by the *Arrhenius
equation* which states that the reaction rate increases exponentially with temperature; the rate of
the increase, however, slows with increasing temperature. The cumulative effect of increasing temperture over time can be evaluated by integrating
the reaction rate over time. This is called the
maturation integral. It can be related directly to
measurable indices of burial. Other parameters
making use of the first-order kinetics of the Arrhenius equation are the Level of Organic Metamorphism (LOM) of Hood, Gutjahr and Heacock
(1975) and the Time–Temperature Index (TTI) of
Lopatin (1971) and Waples (1980). These last two
parameters make use of the erroneous assumption
that the reaction rate continues to increase at the
same rate over the entire temperature range to
250 °C, causing overestimates of maturity at high
formation temperatures.

Palaeotemperatures are controlled by the basal
heat flow history of the basin (which in turn reflects
the lithospheric mechanics), but also by 'internal'
factors such as variations in thermal conductivities, heat generation from radioactive sources
within the sediments, and regional water flow
through aquifers. The latter has profound consequences for heat flow in basins with important
recharge areas of water in uplift areas, such as in
foreland basins and intracratonic sags.

Formation temperatures can be estimated from
borehole measurements, with a correction applied
to account for the cooling that takes place during
the circulation of drilling fluids. These corrected
formation temperatures allow geothermal gradients
to be calculated.

Vitrinite reflectance is the most widely used indicator of thermal maturity. Other organic indicators are also used. Mineralogical parameters include the quantity and nature of smectite, percentage of illite in mixed layer clays, illite crystallinity and other clay mineral transformations such as smectite → corrensite → chlorite in sediments containing volcanic detritus or evaporites.

Vitrinite reflectance measurements plotted against depth – termed R_o profiles – provide a great deal of information on the thermal history of the basin. The 'normal' pattern is a sublinear relationship between log R_o and depth, indicating a continuous, time-invariant geothermal gradient. R_o profiles with distinct kinks between two linear segments (doglegs) indicate two periods of different goethermal gradient separated by a thermal event. R_o profiles with a sharp break or jump (offsets) indicate the existence of an unconformity with a large stratigraphic gap. R_o profiles in sedimentary basins can therefore be used to test between time-constant and time-variant geotherms if the subsidence history is known.

Studies of present-day heat flows and ancient geothermal gradients estimated from thermal indicators such as vitrinite reflectance suggest that thermal regime closely reflects tectonic history. In particular, *hypothermal* (cooler than average) basins include ocean trenches and outer forearcs and foreland basins. *Hyperthermal* (hotter than average) basins include oceanic and continental rifts, some strike-slip basins with mantle involvement, and magmatic arcs in collisional settings. Mature passive margins which are old compared to the thermal time constant of the lithosphere (Section 3.1.5.1 and accompanying worked example on uniform extension) tend to have near-average heat flows and geothermal gradients.

Case studies of the Anadarko Basin, Oklahoma and the Michigan Basin, northern USA, demonstrate the problems and uses of organic maturity measurements in constraining mechanical models of basin development.

9.1 INTRODUCTION

Subsidence in sedimentary basins causes material initially deposited at low temperatures and pressures to be subjected to higher temperatures and pressures. Sediments may pass through diagenetic, then metamorphic regimes and may contain indices of their new pressure–temperature conditions. Thermal indices are generally obtained from either dispersed organic matter or from mineralogical trends. A great deal of effort has been spent in attempting to find an analytical technique capable of unambiguously describing thermal maturity, and an equal amount of effort attempting to correlate the resulting proliferation of indicators.

Numerical values of the organic geochemical parameters are dependent on time, thermal energy and type of organic matter (e.g. Weber and Maximov 1976 and many others). The evolution of clays and other minerals is controlled by temperature and by chemical and petrological properties. The scale of maturation to which a given organic or mineralogical phase can be calibrated is that of *coal rank*. Any analytical technique must be able to make use of very small amounts of dispersed organic matter in order to be valuable in basin analysis. Vitrinite reflectance and elemental analyses enable coal rank to be related to hydrocarbon generation stages. The stages are described in Chapter 10. The objective of this section is to describe the use of a number of thermal indicators in constraining and calibrating geohistory plots.

9.2 THEORY – THE ARRHENIUS EQUATION

It is now believed that the effects of depth *per se* on the maturation of organic matter are of minor importance, the most important factors being *temperature* and *time*. Pressure is relatively unimportant. Phillipi (1965) assessed the effect of pressure by studying hydrocarbons in two Californian basins. In the Los Angeles basin, hydrocarbons were generated at about 8000 ft (~ 2.4 km) whereas in the Ventura basin, generation did not take place until about 12 500 ft (3.8 km) of burial. Since pressure is directly related to depth of burial ($\sigma = \rho g h$) this suggests that pressure does not play a major role in hydrocarbon generation. However, the generation of hydrocarbons in the two basins took place at the same temperature, strongly suggesting that subsurface temperature was the overriding control.

The relationship between temperature and the rate of chemical reactions is given by the *Arrhenius equation*:

$$K = A \exp(-E_a/RT) \qquad (9.1)$$

where K is the reaction rate, A is a constant sometimes termed the frequency factor (it is the maximum value that can be reached by K when given an infinite temperature), E_a is the activation energy, R is the Universal Gas Constant and T is the absolute temperature (°K). The constants in the Arrhenius equation can be estimated from compilations of organic metamorphism (e.g. Hood *et al.* 1975, Shibaoka and Bennett 1977). The activation energies of each individual reaction involved in organic maturation are not known, but, for each organic matter type a distribution of activation energies may be established from laboratory and field studies.

The Arrhenius equation suggests that reaction rates should increase exponentially with temperature, and a 10 °K rise in temperature (from 50 °C to 60 °C) causes the reaction rate to double. This result is widely known, but it is less widely realized that the rate of increase in reaction rate slows down with increasing temperature, so at 200 °C the reaction rate increases by a factor of 1.4 for a 10 °C rise in temperature (Robert 1988). Clearly, both time and temperature influence organic mat-

uration, a view supported by the occurrence of shallower oil generation thresholds as the sediments containing the organic matter become older (Dow 1977) (Fig. 9.1). Connan (1974) believed that the threshold of the principal zone of oil generation was related to the logarithm of the age of the formation, further supporting a time–temperature dependence obeying the laws of chemical kinetics (Arrhenius equation).

The cumulative effect of increasing temperature can be evaluated from the *maturation integral*, the reaction rate integrated over time,

$$C = A \int_0^t \exp(-E_a/RT) + C_o \qquad (9.2)$$

where C_o is the original level of maturation of the organic material at the time of deposition ($t = 0$). The maturation integral for any nominated horizon can be calculated if the decompacted burial history, heat flow through time and thermal conductivities of the sediments and basement are known or can be assumed. For the less mathematically inclined, equation (9.2) shows that when palaeotemperatures are plotted on an exponential scale, the area under the curve from deposition to

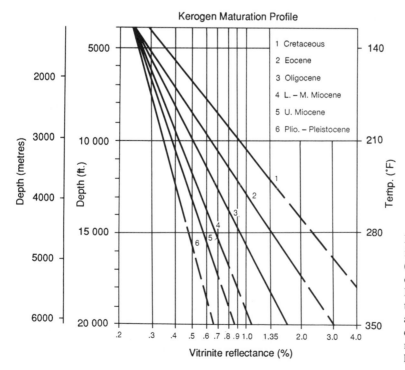

Kerogen Maturation Profile

1 Cretaceous
2 Eocene
3 Oligocene
4 L. – M. Miocene
5 U. Miocene
6 Plio. – Pleistocene

Fig. 9.1. Plot of depth versus an index of organic maturation (vitrinite reflectance) for kerogens of various ages (after Dow 1977). For the same depth of burial (and therefore for the same temperature at a constant geothermal gradient), older kerogens are significantly more mature than younger kerogens.

a given time is proportional to the maturation integral at that time at the nominated horizon (plus the value of C_0). The reader is referred to worked example 2 in Section 3.1.5.1 and to Section 9.3 below. Some authors believe that the maturation integral is related to measurable values of vitrinite reflectance (see Sections 9.4.2, 9.5) (Royden, Sclater and Von Herzen 1980, Falvey and Middleton 1981).

Hood *et al.* (1975) devised an artificial maturation parameter, the *Level of Organic Metamorphism* (LOM), based on a rank progression of coals, from lignites to meta-anthracites. Hood's diagram shows the relationship between the 'effective heating time' and the maximum temperature attained. The effective heating time is defined as the length of time the temperature remains within a 15 °C range of the maximum temperature. This method, although not stated explicitly by Hood *et al.* (1975), is based on the first-order chemical kinetics outlined above (Robert 1988, p. 27).

Another application of the Arrhenius relationship is the *Time–Temperature Index* (TTI) (Lopatin 1971, Waples 1980). This index is based on the view that the reaction rate doubles for every 10 °C rise in temperature over the entire range from 50 °C to 250 °C (we have seen above that this is not the case). The temperature can therefore be expressed as a power of two, where the power $n = (T\,°K - 373)/10$. The temperature factors for the following temperatures are 80 °C = 2^{-2}, 90 °C = 2^{-1}, 100 °C = 2^0 = 1, 110 °C = 2^1. The time spent by a geological horizon within a particular 10 °C temperature range, multiplied by the temperature factor accounts for both the effects of temperature and time. Summing all of these time–temperature values gives the TTI. Because the method assumes that the reaction rate continues to double over 10 °C intervals over the entire temperature range to 250 °C, it tends to overestimate maturity. First-order kinetics suggest instead that the reaction cannot continue indefinitely because the materials undergoing thermal maturation are used up.

Other techniques, such as those of Tissot (1969), Tissot and Espitalié (1975) and Mackenzie and Quigley (1988), have been developed that enable masses of petroleum generated during thermal maturation of organic matter to be calculated. The Mackenzie and Quigley model is described in relation to petroleum source rocks in Section 10.3.2.2.

9.3 PALAEOTEMPERATURES

Chapter 2 contains some basic concepts about heat flow. The specific problem of one-dimensional (vertical) heat flow in basins due to stretching is addressed in Chapter 3. In essence, the basal heat flow into a sedimentary basin of this type is determined by the amount of lithospheric stretching β, since β controls the amount of aesthenospheric upwelling. Sections 3.1.7 and 3.2.3 indicate qualitatively the effects on heat flow of secondary convection in the upwelled region and of melt segregation. Here, we are concerned with the various 'internal' factors that influence the temperatures within sedimentary basins – (1) variations in thermal conductivity, (2) internal heat generation and (3) convective/advective heat transfer within sediments.

9.3.1 Effects of thermal conductivity

The distribution of temperature with depth (geotherm) in the continents is determined by conductive heat transport. We know the relation between heat flux and temperature gradient as given by Fourier's law (eq. 2.30). This law states that conductive heat flux is related to the temperature gradient by a coefficient, K, known as the coefficient of thermal conductivity. If two measurements of temperature are known, one T_y at depth y and another T_0 at the surface ($y = 0$), Fourier's law can be restated as

$$q = -K(T_y - T_0)y \qquad (9.3)$$

which by rearrangement becomes

$$T_y = T_0 + \left(\frac{-qy}{K}\right) \qquad (9.4)$$

where q is the heat flux (negative for y increasing downwards). We are here initially ignoring internal heat production with the sedimentary pile (see following section).

Ignoring for the moment lithological variations, thermal conductivities of sediments vary as a function of depth because of their porosity loss with burial (see Section 8.2.1). Equation (9.4) can be modified to account for the different thermal conductivities of the sedimentary layers,

$$T_y = T_0 + (-Q)\left\{\frac{l_1}{K_1} + \frac{l_2}{K_2} + \frac{l_3}{K_3} + \ldots\right\} \quad (9.5)$$

where l_1 to l_n are the thicknesses of the layers with thermal conductivities K_1 to K_n, and $l_1 + l_2 + l_3 \ldots$ must of course be equal to y. Falvey and Middleton (1981) recommended the use of a function that assumed an exponential relation between porosity and depth

$$K = K_d - \{(K_d - K_0) \exp(-\gamma y)\} \quad (9.6)$$

where K_d is the thermal conductivity deep in the sedimentary section, K_0 that at the sediment surface and γ is a constant for a given section. Since K varies with depth, temperature gradients must also vary with depth in order to maintain a constant heat flow. If present-day heat flow can be calculated from a borehole by measurement of conductivities and surface and bottom hole temperatures (Section 9.4.1), equations (9.4) and (9.6) can be used to find the temperature at any depth. If palaeoheat flow is then assumed to be constant with depth, the temperature history of any chosen stratigraphic level can be estimated. The assumption of a constant heat flow with depth is a condition of any one-dimensional steady-state heat conduction model. Measurements in some sedimentary basins such as the North Sea aulacogen (Andrews-Speed, Oxburgh and Cooper 1984) however, suggest that this is not a good assumption, deep circulation of water most likely being responsible for the departure from the steady-state assumption.

The thermal conductivities can be estimated if the lithology and pore-filling fluid is known. Thermal conductivity depends on the framework mineralogy (quartz, felspar, calcium carbonate etc.), the type and amount of fines in the matrix (usually clay minerals), and the type and volume of pore-filling fluid (usually water). The individual conductivities of framework, matrix and pore-fluid are also dependent on temperature. The variation of *effective thermal conductivity* of a quartzose sandstone with pore-filling water is shown in Fig. 9.2a for a temperature gradient of $30\,^\circ\text{C km}^{-1}$ and a surface temperature of $20\,^\circ\text{C}$. The effective thermal conductivity is almost invariant with depth. This is because the decrease of the conductivity of the

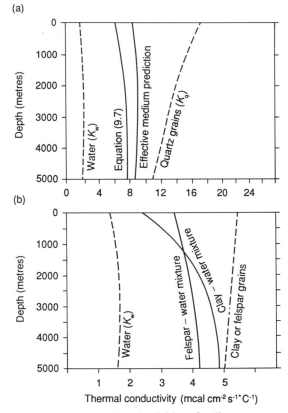

Fig. 9.2. The thermal conductivities of sediments as a function of depth. (a) Quartz–water mixtures. The effective medium prediction (eq. 9.8) and the empirical relation in equation (9.7) closely agree, demonstrating a negligible increase in thermal conductivity of the mixture with depth, despite the fact that the quartz grains decrease in thermal conductivity considerably with depth. (b) Felspar–water and clay–water mixtures, showing that the thermal conductivities increase markedly with depth, especially for clay-rich sediments. This is principally due to the effects of compaction. After Palciauskas (1986).

quartz grains due to increasing temperature compensates for the effects of compaction which increase conductivity (Palciauskas 1986).

Felspar and certain clays do not show such a marked effect of temperature on thermal conductivity, so the effects of compaction may dominate. A clay–water mixture (shales) increases in conductivity rapidly with depth because of compaction (see Section 8.2), whereas a feldspar–water mixture, because it compacts similarly to a sand, increases in conductivity much more slowly with depth (Fig. 9.2b).

The bulk conductivity of a sediment layer can therefore be thought of as being made up of the contributions of the pore-fluid and the grain conductivities. An empirical relation for the bulk conductivity of the form given below has been used

$$K = K_s (K_w/K_s)\phi \qquad (9.7)$$

where K_s and K_w are the thermal conductivities of sediment grains and water respectively and ϕ is the porosity, assumed to be filled with water. This is useful where K_w/K_s is neither very small nor very large. An alternative method, termed the *effective medium* theory calculates an effective bulk thermal conductivity for a randomly inhomogeneous medium made of constitutents with volume fractions V_i and thermal conductivities K_i. The basic result of the theory is

$$K^{-1} = \sum_{i=1}^{n} 3V_i (2K + K_i)^{-1} \qquad (9.8)$$

This expression is particularly useful where mixed components are present in the sediment layer. For example, for the water–quartz mixture mentioned above, if the quartz framework (K_q = 13 mcal cm^{-2} s^{-1} °C^{-1} at T = 100 °C) occupies 0.7 of the rock volume and water (K_w = 1.7 mcal cm^{-2} s^{-1} °C^{-1} at T = 100 °C) occupies 0.3 of the rock volume, the bulk conductivity from effective medium theory (9.8) is approximately 8 mcal cm^{-2} s^{-1} °C^{-1}. From the empirical result in (9.7), the bulk conductivity is approximately 7 mcal cm^{-2} s^{-1} °C^{-1}.

9.3.2 Effects of internal heat generation in sediments

Heat generation by radioactive decay in sediments may significantly affect the heat flow in sedimentary basins (Rybach 1986). Although all naturally occurring radioactive isotopes generate heat, the only significant contributions come from the decay series of uranium and thorium and from ^{40}K. As a result, heat production varies with lithology, generally being lowest in evaporites and carbonates, low to medium in sandstones, higher in shales and siltstones and very high in black shales (Rybach 1976, Haack 1982, Rybach and Cermak 1982).

In the continents, crustal radioactivity may account for a large proportion (20–60 per cent) of the surface heat flow. For a purely conductive, one-dimensional (vertical) heat flow, the temperature at any depth y is

$$T(y) = T_0 + \frac{q_b + HL}{K} \cdot y - \frac{H}{2K}y^2 \qquad (9.9)$$

where T_0 is the surface temperature, q_b is the basal heat flow at $y = L$, K is the average thermal conductivity of the sediments and H is the internal heat production (estimated from natural gamma-ray logs). The effect of the internal heat generation is greatest at large depths, as can be seen from the third term in equation (9.9) (Fig. 9.3). The temperature increase after a time t as a result of the internal heat generation depends on the value of H, but the net temperature change also depends on the rate of conductive heat loss. Over geological time scales of >10 Myr the temperature rise may be considerable (Ryback 1986, p. 317). Internal heat generation in sediments may therefore strongly affect the temperature field in the basin if it is deep (>5 km) or long-lived (> 10 Myr).

9.3.3 Effects of water flow

The temperatures in sedimentary basins may also be affected by the advective flow of heat through regional aquifers. Such processes may cause anomalously low surface heat flows at regions of recharge, and anomalously high surface heat flows in regions of discharge. The heat flow distributions of the Great Plains, USA (Gosnold and Fischer 1986) and the Alberta Basin (Majorowicz and Jessop 1981, Majorowicz *et al.* 1984) have been explained in this way. Smith and Chapman (1983) provide a review of the effects of fluid flow on heat flow in regional-scale systems. Luheshi and Jackson (1986) have applied the theory of Smith and Chapman (1983) to the Alberta Basin. Using a permeability and thermal conductivity structure for the basin, they were able to explain the raised temperatures at discharge points of fluid flow and lowered temperatures at the recharge areas in the fringing hills (Fig. 9.4). The model results suggest that the temperature distribution is dominated by convection above the Palaeozoic, while the heat flows in the Precambrian can be explained simply by conduction. Andrews-Speed *et al.* (1984) similarly found that heat flow measurements strongly suggested a deep

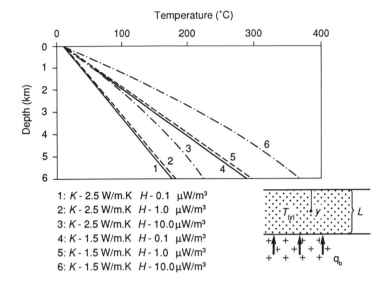

Fig. 9.3. The influence of internal heat generation in the sedimentary column H and the thermal conductivity K on the distribution of temperature with depth ($T(y)$). The different curves were calculated by Rybach (1986) for a thickness of the heat-producing zone of 6 km, a basal heat flux q_b of 70 mW m^{-2} and a surface temperature T_0 of 10 °C.

1: K - 2.5 W/m.K H - 0.1 μW/m³
2: K - 2.5 W/m.K H - 1.0 μW/m³
3: K - 2.5 W/m.K H - 10.0 μW/m³
4: K - 1.5 W/m.K H - 0.1 μW/m³
5: K - 1.5 W/m.K H - 1.0 μW/m³
6: K - 1.5 W/m.K H - 10.0 μW/m³

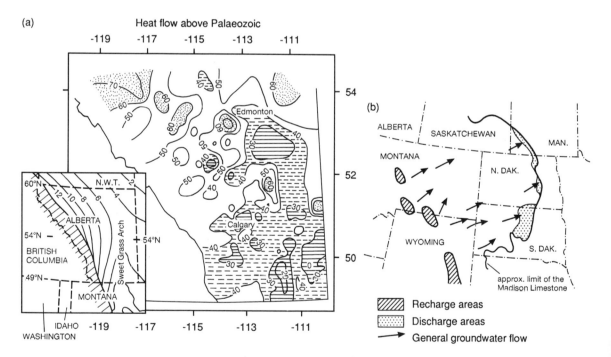

Fig. 9.4. Effects of groundwater flow on surface heat flows in sedimentary basins. (a) Heat flow map of southern and central Alberta, Canada based on estimated heat flow values (in mW m^{-2}) above the top of the Palaeozoic. These values are based on 33 653 bottom hole temperature data from 18 711 wells analysed by Majorowicz *et al.* (1984). The heat flows are strongly influenced by groundwater flow from recharge areas in structurally high regions, such as the Sweet Grass Arch (inset), to discharge areas. (b) Pattern of recharge and discharge in the Great Plains, USA in the Mississippian (Lower Carboniferous) Madison Limestone aquifer (after Downey 1984).

water circulation, possibly controlled by the configuration of faults, in the North Sea aulacogen. The implications of detailed studies such as this are that simple one-dimensional conductive heat flow models may be very poor predictors of actual heat flows in some sedimentary basins. The most strongly affected basins are likely to be continental basins with marginal uplifts, such as foreland basins and some intracratonic rifts and sags.

9.4 INDICATORS OF FORMATION TEMPERATURE AND THERMAL MATURITY

9.4.1 Estimation of formation temperature from borehole measurements

Formation temperatures from boreholes are used in thermal modelling studies to calculate the geothermal gradient and basal heat flow to the sedimentary section. The temperature in the borehole is recorded on each logging run, using a suite of maximum recording thermometers. Because the circulation of drilling fluid tends to cool the formation, it is necessary to analyse the rate at which temperature restores itself to its original true formation value using temperatures recorded on each successive logging run within a suite of logs. These temperatures may be plotted on a 'Horner'-type plot, as described by Dowdle and Cobb (1975).

The form of the temperature build-up plot is shown in an example from the Gulf Coast in Fig. 9.5. Temperature measured on each logging run is plotted against a dimensionless time factor, $(t_c + \Delta t)/\Delta t$, where

t_c = Cooling time. This is the duration of mud circulation from the time the formation opposite the thermometer was drilled to the time circulation of the drilling mud stopped.

Δt = Thermal recovery time. This is the time since mud circulation stopped to the time the logging sonde is in position at the bottom of the borehole.

A fully recovered or stabilized formation temperature T_f is obtained by extrapolation to the ordinate, where $(t_c + \Delta t)/\Delta t = 1$.

9.4.2 Vitrinite reflectance

Vitrinite reflectance is the most widely used indicator of maturity of organic materials. It is an optical parameter and is denoted by R_o – reflectance in oil. Section 10.3.1.3 provides more details.

Drawbacks in the use of vitrinite reflectance measurements are outlined by Héroux, Chagnon and Bertrand (1979, p. 2134), Kübler *et al.* (1979) and Durand *et al.* (1986). These arise from reflectance measurements being taken from maceral types other than vitrinite (especially in lacustrine and marine sediments), the possibility of reworking of organic material (especially in sandstones), and the lack of higher plants yielding vitrinite in pre-Devonian strata. Vitrinite reflectance tends to be unreliable at low levels of thermal maturity (R_o less than 0.7 or 0.8 per cent).

9.4.3 Other burial indices

Although vitrinite reflectance has become pre-eminent in its use in basin studies, it is not the only index of thermal maturity (see Section 10.3.1.3). Other optical parameters on organic material include sporinite microspectrofluorescence and spore, pollen and conodont colouration scales. Fluorescence and reflectance studies are complementary, fluorescence intensity and reflectance being inversely proportional.

A range of specialized geochemical measurements have also been developed in the oil industry to estimate the thermal maturity of source rocks and oils (Héroux *et al.* 1979 for summary). Organic parameters are discussed at greater length in Sections 9.1 (theory) and 9.5 to 9.7 (applications). The reader is also referred to Section 10.3.1 on source rocks.

Mineralogical parameters are controlled by the temperature and chemical properties of the diagenetic environment of the sediment. A number of diagenetic models now exist (excellent review in Burley, Kantorowicz and Waugh, 1985) which allow an interpretation of the sequence of authigenic minerals in terms of their relationship to:
• their depositional environment or surface chemistry (*eogenesis*)
• burial or subsurface conditions (*mesogenesis*)
• weathering or re-exposure to surface conditions (*telogenesis*).

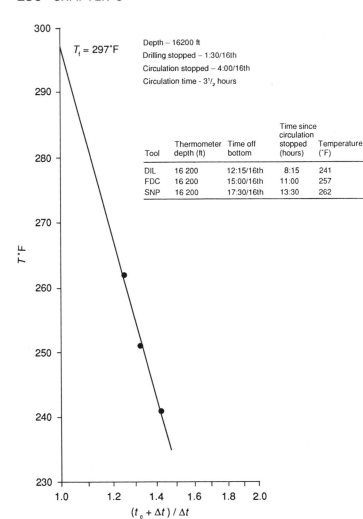

Depth – 16200 ft
Drilling stopped – 1:30/16th
Circulation stopped – 4:00/16th
Circulation time – 3$\frac{1}{2}$ hours

$T_f = 297°F$

Tool	Thermometer depth (ft)	Time off bottom	Time since circulation stopped (hours)	Temperature (°F)
DIL	16 200	12:15/16th	8:15	241
FDC	16 200	15:00/16th	11:00	257
SNP	16 200	17:30/16th	13:30	262

$(t_c + \Delta t) / \Delta t$

Fig. 9.5. The determination of true formation temperature from a 'Horner'-plot (after Dowdle and Cobb 1975). This example is from a high-temperature well in the Gulf Coast. The depth at which measurements were taken was 16 200 feet. Temperatures increased from 241 °F measured 8 h 15 min after circulation of mud stopped, to 262 °F taken 13 h 30 min after circulation stopped. The estimated formation temperature T_f is 297 °F.

Since eogenetic changes are related to depositional environment and associated pore water chemistry, they are of limited use in thermal modelling. However, mesogenesis marks the removal of the sediment from the predominant influence of surface agents in the interstitial pore water. One of the most important mineral transformations taking place in the new geochemical environment during burial is the conversion of smectites to illites (e.g. Boles 1981, Hower 1981). This conversion in mudrocks releases ionic species which are transported in pore waters to neighbouring porous sediments such as sandstones where cementation may take place in the form of quartz

overgrowths and chloritization of kaolinite, albitization of plagioclase and the growth of ferroan dolomite or ankerite from pre-existing calcites. The burial of any contained organic matter leads to thermal decarboxylation and the formation of acidic pore waters. These acidic solutions are capable of generating widespread *secondary porosity* by the removal of carbonate cements. Physical processes accompany these chemical changes during burial diagenesis. The most important result is *compaction* due to the weight of the overlying sediments (Chapter 8). In sandstones compaction brings about a number of porosity-reducing adjustments including initial mechanical compaction which simply compresses grains together, rotation, grain slippage, brittle grain deformation and fracturing and plastic deformation of ductile grains.

The best-documented mineral transformations of use in evaluating thermal maturity are from shaly mudstones, where the clay mineral assemblages, the position of the (001) reflection of smectite, the percentage of illite layers in the mixed layer illite – 2:1 expandable, and the illite crystallinity index (width at half height of illite peak on diffractogram) are used.

Fine-grained volcanic detritus contains a number of index minerals that are useful in suggesting thermal development. For example, smectite evolves to corrensite and then toward chlorite. Very similar mineral assemblages are, however, found in evaporitic sediments and those derived from hydrothermal alteration.

Clay mineral transformations during burial have been used to investigate the thermal history of the North Alpine Foreland Basin (Molasse Basin) in Switzerland (Monnier 1982). The average composition of the Oligo-Miocene Molasse sediments conforms to a feldspathic sandstone with a calcareous–argillaceous cement. The fine clay fraction (<2 μm) contains smectite, mixed layers, illite-mica, chlorite and, in the western part of the basin where Triassic evaporites are present, corrensite. Two mineralogical indices were studied from borehole samples:

1 Disappearance of smectite with burial.
2 Illite crystallinity in relation to smectite transformation.

Smectite abundance was studied in three boreholes. In two of them smectite disappeared by 2300 m to 2700 m via transition zones (2000–2300 and 2400–2700 m) characterized by the presence of interstratified clays. In the other well a different set of mineral changes took place, following partly an evolution towards chlorite via regular interstratification to give corrensite, and in part an evolution towards illite.

Illite crystallinity indices are highly variable in the Swiss Molasse, and cannot in all cases be used to characterize the level of diagenesis. In the upper zone of diagenesis where smectite is abundant, the index reflects the detrital origin of the clay fraction and the inheritance of mica crystallinity. Newly formed illite in the zone where mixed layer silicates

predominate gives more reliable information on diagenetic stage. This suggests that there is a critical illite/mica threshold above which illite crystallinity values are of some significance.

The beginning of oil generation (R_o = 0.5 per cent) has been correlated with the disappearance of smectite (e.g. Kübler *et al.* 1979, Powell *et al.* 1978).

9.5 APPLICATION OF VITRINITE REFLECTANCE MEASUREMENTS

9.5.1 R_o profiles

Vitrinite reflectance measurements can be plotted as a function of depth to give R_o profiles. The slope of the R_o curves gives an indication of the geothermal gradients in the history of the basin. Although many profile shapes are possible (Fig. 9.6), they generally indicate an exponential evolution of the organic matter with time (Dow 1977), as expected from the kinetics described in Section 9.2. In basins largely unaffected by major unconformities, young dip-slip faulting and localized

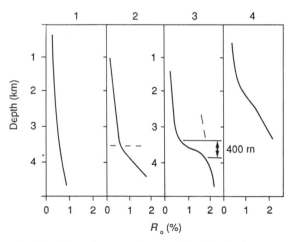

Fig. 9.6. The main types of R_o profile (after Robert 1988). 1, normal sublinear: 2, two periods with different geothermal gradients: 3, strong thermal perturbation, then returning to normal: 4: intermediate between 2 and 3.

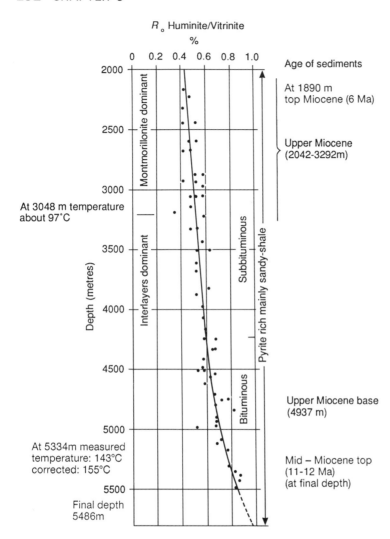

At 3048 m temperature about 97°C

At 5334m measured temperature: 143°C corrected: 155°C

Final depth 5486m

Age of sediments

At 1890 m top Miocene (6 Ma)

Upper Miocene (2042-3292m)

Upper Miocene base (4937 m)

Mid – Miocene top (11-12 Ma) (at final depth)

Fig. 9.7. Vitrinite reflectance profile for Terrebonne Parish, Point au Fer well in Louisiana. The profile is sublinear and continuous, suggesting a near constant geothermal gradient through time (after Heling and Teichmüller 1974).

igneous activity, there should therefore be a linear relationship between depth and log R_o.

An example of this simple, sublinear profile is the Terrebonne Parish well in Louisiana (Heling and Teichmüller 1974) (Fig. 9.7). R_o is 0.5 per cent at 3 km and 1 per cent at 5 km. It indicates a normal and constant geothermal gradient through time.

Other R_o profiles are more complex. A dogleg pattern of two linear segments of different slope indicates that two periods of different geothermal gradient have occurred. This may result from a thermal 'event' occurring at the time presented by the break in slope. Such an interpretation is plausible for the R_o profiles from boreholes in the Rhine Graben (Teichmüller 1970, 1982, Robert 1988) (Fig. 9.8).

R_o profiles may consist of two sublinear segments offset by a sharp break or jump in R_o values. The jump may correspond to an unconformity with a large stratigraphic gap. This is well illustrated in the Mazères 2 borehole in the Lacq area of the Aquitaine Basin, France (Fig. 9.9), where R_o values jump from c. 0.8 per cent to c. 2.4 per cent at the level of an unconformity separating Aptian–Albian rocks from underlying Kimmeridgian.

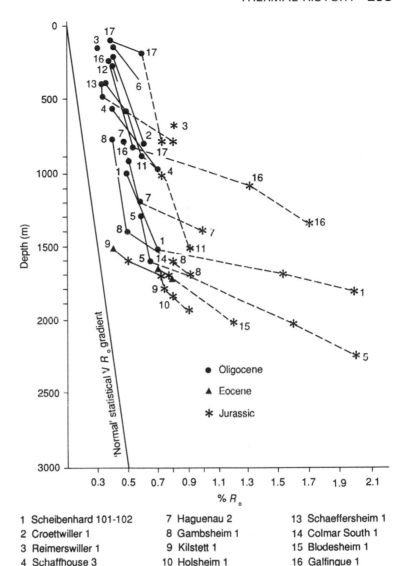

Fig. 9.8 Reflectance profiles from a number of wells in the Alsace region of the Rhine Graben. In general, there are pronounced dog-legs in the R_o profiles at about the age of the Eocene–Oligocene boundary. The post-Oligocene history shows a near 'normal' gradient, whereas the pre-Oligocene sediments have high reflectance values in relation to their depth of burial (after Teichmüller 1970).

1 Scheibenhard 101-102	7 Haguenau 2	13 Schaeffersheim 1
2 Croettwiller 1	8 Gambsheim 1	14 Colmar South 1
3 Reimerswiller 1	9 Kilstett 1	15 Blodesheim 1
4 Schaffhouse 3	10 Holsheim 1	16 Galfingue 1
5 Roeschwoog 1	11 Eschau 1-11	17 Knoeringue 1
6 Donau 2	12 Meistratzheim 1	

If there is a known (logarithmic) relationship of R_o with depth and the subsidence history of a sedimentary basin is known, the R_o values give an indication of the variation of the geothermal gradient through time. This then allows different tectonic histories to be tested. The following histories were tested by Middleton (1982):

A Constant geothermal gradient through time;
   1 Subsidence at a constant rate, then quiescence, then erosion of thickness h

2 subsidence proportional to $t^{1/2}$, then quiescence, then erosion of thickness h.

B Decreasing geothermal gradient with time, subsidence proportional to $t^{1/2}$. This corresponds to a basin formed by stretching, with a well-developed thermal subsidence phase.

For models of constant geothermal gradient G, the plot of theoretical log R_o values versus depth is linear. In case B, however, the theoretical R_o as a function of depth does not follow the simple

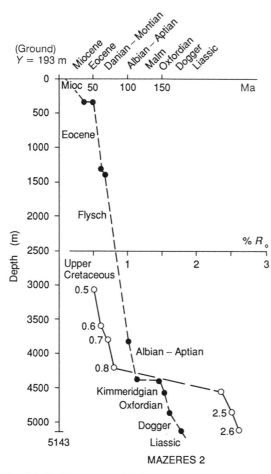

Fig. 9.9. Reflectance profile for the Mazères borehole in the Lacq region of southern France. The sharp increase in R_o marks an unconformity between the Lower Cretaceous and the Late Jurassic (after Robert 1988).

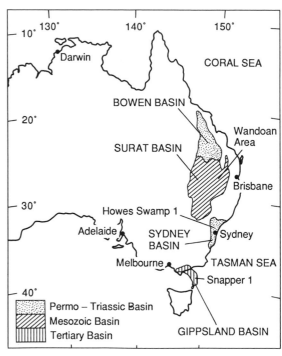

Fig. 9.10. Location of the Surat-Bowen, Sydney and Gippsland basins in eastern Australia (after Middleton 1982).

expression of Dow (1977). Middleton (1982) considered three Australian basins (Fig. 9.10).

The *Sydney Basin* of Australia was initiated in the Permian and underwent uplift and erosion 60 to 100 Myr ago. The Howes Swamp 1 well shows the subsidence to be constant and to average 0.073 mm yr^{-1} (Fig. 9.11). The tectonic model of case A(1) fits the vitrinite reflectance data from this well with a constant geothermal gradient of 50 °C km^{-1}. Although this seems high, the intermittent igneous activity during deposition and until the Miocene may have elevated the average geothermal gradient.

The *Bowen Basin* in Queensland is also Permo-

Triassic in age, and is superimposed by the Jurassic-Cretaceous *Surat Basin*. Subsidence continued from the Permian to ~100 Ma, and the subsidence approximates $t^{1/2}$ proportionality. Uplift terminated sedimentation at about 70 Ma. The tectonic model is case A(2) and the vitrinite reflectance data can be fitted by choosing suitable values of geothermal gradient, duration of tectonic subsidence, duration of quiescence and amounts of erosion (Fig. 9.12).

Finally, the *Gippsland Basin* on the southeastern Australian continental margin was formed in response to sea floor spreading in the Tasman Sea 70 Myr ago. Its subsequent development as a passive margin basin follows the case B for cooling lithosphere. The cooling model is consistent with the observed reflectance data using a present-day geothermal gradient of 34 °C km^{-1} (obtained from an offshore borehole temperature measurement – 116 °C at 3.15 km – and an assumed sea bed temperature of 10 °C) after 70 Myr of cooling (Fig. 9.13).

Fig. 9.11. Subsidence history (a) and reflectance profile (b) of the Howes Swamp 1 borehole in the Sydney Basin (Mayne *et al.* 1974). The subsidence curve gives a constant rate of 0.073 mm yr^{-1}. In (b) black dots are reflectance measurements, and solid line is the curve resulting from the tectonic model A1 with a constant geothermal gradient of 50 °C km^{-1} (after Middleton 1982).

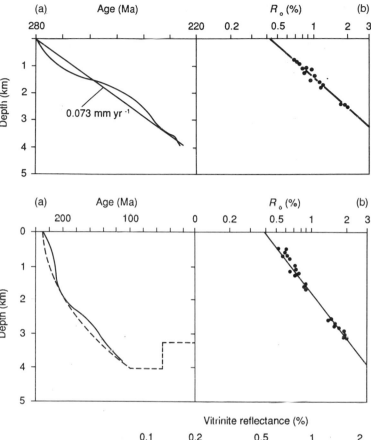

Fig. 9.12. Subsidence history (a) and reflectance profile (b) for the Windoan region of the Bowen Basin. The solid line in (a) is the observed basement subsidence, showing a rough $t^{1/2}$ relationship. The dashed line is the curve for the tectonic model A2 with a constant geothermal gradient of 44 °C km^{-1}. (Shibaoka, Bennett and Gould 1973). In (b) the black dots are vitrinite reflectance measurements and the solid line is the curve for the A2 tectonic model (full details of parameter values in Middleton 1982).

These three examples show that there is considerable scope for the use of vitrinite reflectance data in interpreting basin tectonic history, palaeotemperatures and geothermal gradients. This subject is revisited in the two case studies in Section 9.7.

9.6 GEOTHERMAL AND PALAEOGEOTHERMAL SIGNATURES OF BASIN TYPES

We have previously seen that vitrinite reflectance measurements can be used to constrain palaeotemperatures and palaeogeothermal gradients. This then helps to determine the formative mechanism of the basin. Robert (1988) suggested three main types of palaeogeothermal history:

1 Basins with normal or near-normal palaeogeothermal history.

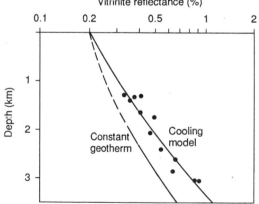

Fig. 9.13. Plot of vitrinite reflectance with depth for the Snapper 1 well, Gippsland Basin (Kantsler, Smith and Cook 1978). The R_o measurements fall on a line corresponding to the tectonic model B for a basin with a cooling geotherm with time (34 °C km^{-1} after 70 Myr (present day)). The other curve is for a constant geotherm constrained to pass through the surface R_o value. The cooling geotherm is by far the best fit (after Middleton 1982).

2 Cooler than normal (*hypothermal*) basins.
3 Hotter than normal (*hyperthermal*) basins.
• Old passive margins have present-day geothermal gradients of *c.* 25–30 °C km^{-1} (Congo 27 °C km^{-1}, Gabon 25 °C km^{-1}, Gulf Coast USA 25 °C km^{-1} at the Terrebonne Parish well (Fig. 9.7)). The vitrinite reflectance profiles show that R_o is about 0.5 per cent at a depth of 3 km (Fig. 9.7) and the shape of the curve is sublinear (Section 9.5). These mature margins therefore have near-normal geothermal gradients.
• *Hypothermal basins* include oceanic trenches,

outer forearc and foreland basins. Ocean trenches are cold, with surface heat flows often less than 1 HFU. In the Japanese archipelago Eocene–Miocene coals occur in two regions, one in Hokkaido in the north along a branch of the present-day Japan trench, and the other in Kyushu in the south is situated in a volcanic arc position relative to the Ryu-Kyu trench (Fig. 9.14). Figure 9.15 shows the R_o profiles for the two different regions. The Hokkaido region is cold with poorly evolved coals (sub-bituminous coals with R_o = 0.5 per cent still occurring at a depth of 5 km) whereas the

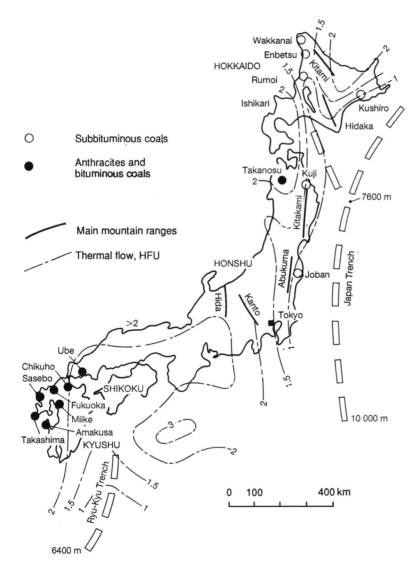

Fig. 9.14. The deposits of Tertiary coals in Japan and the surface heat flows in HFU (Aihara 1980, in Robert 1988). Note that anthracites and bituminous coals are found in Kyushu and the extreme SW of Honshu whereas sub-bituminous coals are found in the north of Honshu and on Hokkaido.

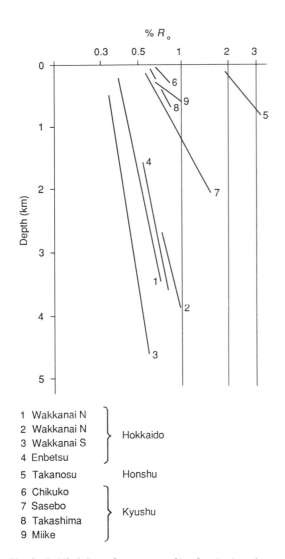

% R_o

Depth (km)

1 Wakkanai N
2 Wakkanai N } Hokkaido
3 Wakkanai S
4 Enbetsu

5 Takanosu Honshu

6 Chikuko
7 Sasebo } Kyushu
8 Takashima
9 Miike

Fig. 9.15. Vitrinite reflectance profiles for the locations shown in Fig. 9.14. The first set of locations 1–4 belong to a branch of the oceanic Japan trench system running along the east of Japan; they are characterized by low reflectances and are associated with low present-day heat flows. The second group of locations 6–9 corresponds to the internal arc relative to the Ryu-Kyu trench. These locations are highly evolved at shallow depths of burial and are associated with high present-day heat flows. The single location, 5, from the NW of Honshu occurs in the present volcanic arc area where the heat flows exceed 2 HFU, and the vitrinite reflectance is very high. There is thus a very clear relationship between tectonic environment and geothermy. After Robert (1988).

volcanic arc in Kyushu is hot, containing anthracites (>2 per cent R_o). The Mariana trench, which is a southward continuation of the Japan trench, and its forearc region are also cold, with surface heat flows of less than 1 HFU.

Foreland basins are also characterized by low present-day geothermal gradients, 22 °C km^{-1} to 24 °C km^{-1} being typical of the North Alpine Foreland Basin in southern Germany (Teichmüller and Teichmüller 1975, Jacob and Kuckelhorn 1977). The Anzing 3 well near Munich penetrates the autochthonous molasse, undisturbed by Alpine tectonic events. At the base of the Tertiary at 2630 m depth the R_o is still only 0.51 per cent. The Miesbach 1 well cuts through about 2 km of thrust sheets of the frontal thrust zone of the Alps (the subalpine zone), before penetrating the autochthonous sediments to a depth of 5738 m (Fig. 9.16). Even at this great depth, the R_o is still only 0.6 per cent, indicating an abnormally low geothermal gradient during the Tertiary. The greater subsidence rate at Miesbach 1 (nearly 0.3 mm yr^{-1}) compared to Anzing 1 (0.1 mm yr^{-1}) may have been responsible for the very low geothermal gradient in the former. In summary, the low present-day geothermal gradients (Anzing 3, 22.8 °C km^{-1}, Miesbach 1, 23.5 °C km^{-1}) may have been even lower in the past during the phase of rapid subsidence related to continental collision and flexure (Chapter 4).

• *Hyperthermal basins* are those found in regions of lithospheric extension such as backarc basins, oceanic and continental rift systems, some strike-slip basins and the internal arcs of zones of B-type subduction. This follows from the mechanics of basin formation in stretched regions (Chapter 3), involving the raising towards the surface of isotherms.

Oceanic rifts are zones of very high heat flows, 3 to 4 HFU being typical and values occasionally reaching 5 to 6 HFU. Some *Californian strike-slip basins* (Chapter 5) have very high geothermal gradients (*c.* 200 °C km^{-1} in Imperial Valley), so that very young sediments can be highly mature. Continental rifts (Chapter 3) have high present-day heat flows (>50 °C km^{-1} in the Red Sea, up to 100 °C km^{-1} in the Upper Rhine Valley) and ancient continental rifts have intense organic maturations in their contained sediments.

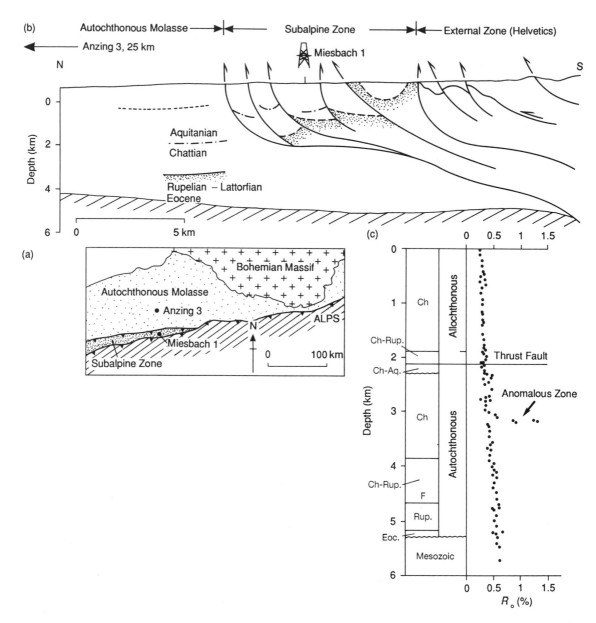

Fig. 9.16. (a) Location of the Bavarian part of the North Alpine Foreland Basin in southern Germany. Anzing 3, near Munich and Miesbach 1 are boreholes discussed in the text. (b) Cross-section of the southernmost part of the Bavarian section of the North Alpine Foreland Basin, showing the location of Miesbach 1 in the tectonically imbricated subalpine zone (after Teichmüller and Teichmüller 1975). (c) The R_o profile at Miesbach 1 (Jacob and Kuckelhorn 1977) shows that the autochthonous molasse under the basal subalpine thrust is poorly evolved, not exceeding 0.6 per cent even at 5738 m depth. This is indicative of a very low geothermal gradient during the period of rapid sedimentation in the Oligocene.

Oceanic measurements and deep boreholes in the *Red Sea* (Girdler 1970) suggest that high surface heat flows (generally >3 HFU) occur in a broad band at least 300 km wide centred on the axis of the rift. The organic maturation shown by R_o profiles and the occurrence of oil, gas and condensate fields (see Chapter 10) suggests that the highest maturity is found in the south of the Red Sea, intermediate values are found in the north of the Red Sea, and the lowest occur in the Gulf of Suez (Fig. 9.17). This can be correlated with different amounts of extension, the largest amount being in the south of the Suez–Red Sea system. The former elevated heat flows in the Oligo-

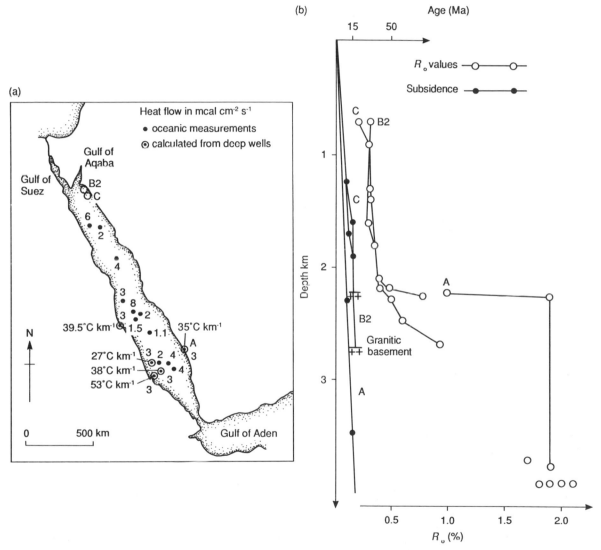

Fig. 9.17. (a) The present heat flow of the Red Sea rift, with location of boreholes illustrated in the reflectance profiles in (b). The black dots in (b) refer to subsidence history of prominent evaporitic layers (late Miocene) in the stratigraphy, open dots to R_o values. Note the elevated reflectance values in borehole A located in the south of the Red Sea, where surface heat flow values at the present day are also high. The Red Sea rift therefore becomes more hyperthermic from the north (Gulf of Suez) to the south (after Girdler 1970, Robert 1988).

Miocene of the Gulf of Suez have now diminished to near-normal values, while the southern Red Sea, which is still actively rifting, still has very high heat flows.

There are many other examples of high organic maturation in ancient continental rift basins: 2–3 per cent R_o in the Lower Cretaceous of the Congo; 3.3 per cent R_o in the Upper Cretaceous of Cameroon; 3.5 per cent in the Coniacian of the Benue Trough, Nigeria; 5 per cent R_o in the Permian of the Cooper Basin, Australia.

Internal arc heat flows are elevated because of magmatic activity. The Tertiary anthracites of Honshu, Japan (see above) (2–3 per cent R_o) are an example. Similar patterns are found in ocean–continental collision zones such as the Andean Cordillera, and hyperthermal events may also affect parts of continent–continent collision zones such as the Alps – the 'Black Earths' of southeastern France have R_o values of over 4 per cent – but the precise origin of the thermal event is unknown (Robert 1988, p. 261).

The heat flows of the main genetic classes of sedimentary basin are summarized in Fig. 9.18.

9.7 CASE STUDIES OF USE OF VITRINITE REFLECTANCE IN THERMAL HISTORY INVESTIGATIONS

9.7.1 Anadarko Basin, Oklahoma, USA

The Anadarko Basin in western Oklahoma has some of the deepest exploratory wells in the world, penetrating to more than 7900 m (~ 26 000 ft), and is therefore an excellent case study for thermal maturation. One of the most important source rocks is the Upper Devonian–Lower Mississippian *Woodford Shale*; it is the oldest rock in Oklahoma that contains vitrinite.

The Anadarko Basin is thought to be a failed rift (aulacogen) which has undergone late compression (Ham, Denison and Merritt 1964). Crustal extension commenced in the Cambrian (or possibly late Precambrian) and was associated with igneous activity. Subsidence, which occurred in the basin in several phases from Late Cambrian to Early Mississippian (Brewer *et al.* 1983), is thought to be due to thermal contraction. By Early Pennsylvanian, the tectonics had changed to intense crustal shortening (Wichita orogeny) with reverse faults with

throws of more than 9 km (30 000 ft) defining the frontal Wichita fault zone to the south of the Anadarko Basin (Fig. 9.19).

The Woodford Shale was sampled at depths ranging from 1542 m (5060 ft) in the northwestern part of the basin, to 7655 m (25 115 ft) in the deepest part of the basin in the southwest (Cardott and Lambert 1985). Vitrinite reflectance measurements were obtained (standard procedures given by Bostick and Alpern 1977, Bostick 1979, Hunt 1979, Dow and O'Connor 1982, Stach *et al.* 1982, van Gijzel 1982, Tissot and Welte 1984) from 28 boreholes, which permits the construction of an isoreflectance map (Fig. 9.20). Mean R_o values range from <0.6 per cent in the northern areas and uplifted fault blocks of the frontal Wichita fault zone, to >3.0 per cent in the deeper parts of the Anadarko Basin (peak value of 4.89 per cent R_o). These high reflectances (>2.5 per cent R_o) are characteristic of anthracite coal rank, indicating high palaeotemperatures of >200 °C, even with a long duration of heating of 300 Myr (see Section 9.2). The very high values of >4 per cent R_o are not usually found in regions of normal geothermal gradient (15–50 °C km^{-1}) (Teichmüller and Teichmüller *in* Stach *et al.* 1982, p. 55).

The present geothermal gradient in the Anadarko Basin is low (20–24 °C km^{-1}) but nonlinear with depth, a change in gradient taking place in the Lower Pennsylvanian from *c.* 22 °C km^{-1} in older rocks to *c.* 18 °C km^{-1} in younger rocks. If burial alone produced the observed rank of organic metamorphism, the present-day temperatures would also represent the maximum temperatures reached. However, higher palaeotemperatures may have existed, perhaps reaching a maximum during the Pennyslvanian and decreasing since the Permian to the present day. This goes some way to explaining the very high reflectance values.

A thermal anomaly is interpreted to explain the isoreflectance contours which cut sharply across depth contours in the western part of the basin (Figs. 9.19, 9.20), and cause the offset of the highest R_o values from the deepest part of the basin. The most likely source of the thermal anomaly is from crustal and lithospheric thinning (Garner and Turcotte 1984). This relatively localized anomaly also drove increased subsidence during the Pennsylvanian.

The maturity of the Woodford Shale can be predicted by studying the variation of R_o with

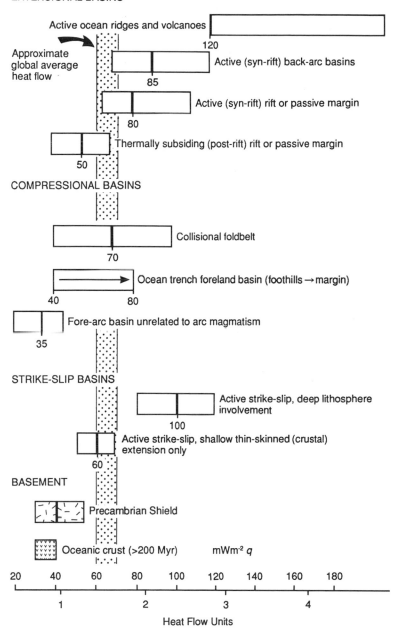

Fig. 9.18. Summary of the typical heat flows associated with sedimentary basins of various types.

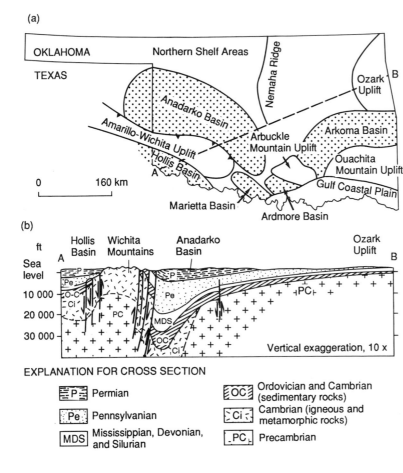

(a)

OKLAHOMA   Northern Shelf Areas

TEXAS

Nemaha Ridge

Anadarko Basin

Amarillo-Wichita Uplift

Hollis Basin

Arbuckle Mountain Uplift

Ozark Uplift   B

Arkoma Basin

Ouachita Mountain Uplift

Gulf Coastal Plain

0   160 km

A

Marietta Basin   Ardmore Basin

(b)

ft   Hollis   Wichita   Anadarko   Ozark
    Basin   Mountains   Basin   Uplift
Sea   A   B
level

10 000

20 000

30 000

Vertical exaggeration, 10 x

EXPLANATION FOR CROSS SECTION

P Permian

Pe Pennsylvanian

MDS Mississippian, Devonian, and Silurian

OC Ordovician and Cambrian (sedimentary rocks)

Ci Cambrian (igneous and metamorphic rocks)

PC Precambrian

Fig. 9.19. The Anadarko Basin, USA. (a) Location of the Anadarko Basin north of the Wichita Mountains in Texas–Oklahoma (after Johnson *et al.* 1972). The Anadarko Basin forms one of a series of basins (Marietta, Ardmore) comprising the aulacogen of southern Oklahoma (Ham, Denison and Merritt 1964).
(b) Cross-section showing thick Palaeozoic sedimentary fill of the Anadarko Basin, and the overthrust margin of the basin against the Wichita Mountains. Wichita faulting did not affect Permian sedimentation, since Permian deposits drape the basement faults.

depth for one stratigraphic level (Fig. 9.21). The increase in R_o with depth is exponential, so when plotted in log R_o versus depth space, the relationship becomes linear. The intersection of the per cent R_o line at zero depth at $R_o = 0.2$ per cent indicates the amount of maturation that the vitrinite had undergone prior to deposition – it is therefore an index of the amount of erosion of the sourceland of the vitrinite since the end of the Wichita orogeny. This erosion varies from 460 m to 1200 m based on extrapolating the reflectance values.

Maximum temperatures in the Woodford Shale probably occurred soon after the compressional deformation in the Pennsylvanian (Cardott and Lambert 1985) and the maximum depth of burial is assumed to be that observed today. However, uplift in the Wichita mountains and Wichita fault zone removed the Woodford Shale from the zone of post-orogenic heating in these areas.

The construction of isoreflectance maps and reflectance gradient plots (R_o profiles) allows both the timing and nature of maturation of the Woodford Shale at any depth in the Anadarko Basin to be assessed.

9.7.2 Michigan Basin, USA

The Cambrian to Jurassic Michigan Basin (Figs 7.34, 7.35, 7.41) is discussed in terms of sedimentary evolution in Section 7.4.1.1. It is thought to be an intracratonic sag caused principally by flexure of the lithosphere during thermal contraction. The thermal contraction may have followed the emplacement of dense rocks into the lower crust, rather than a phase of uniform stretching (Haxby, Turcotte and Bird 1976, Nunn and Sleep 1984).

Nunn, Sleep and Moore (1984) estimated palaeotemperatures in the Michigan basin by assuming an 'equilibrium temperature' based on a present-

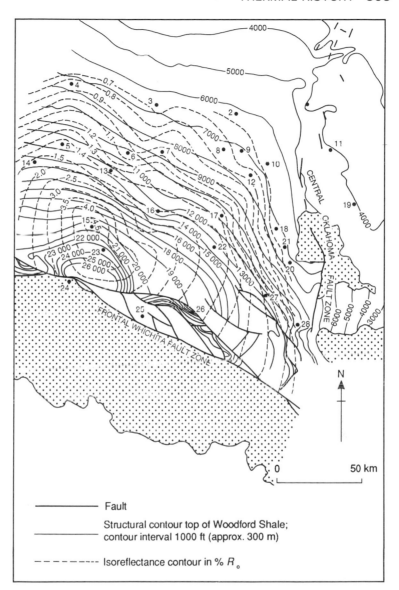

Fig. 9.20. Combined isoreflectance and structure map of the Woodford Shale in the Anadarko Basin of Oklahoma (after Cardott and Lambert 1985). Vitrinite reflectance values in general increase with depth of burial, but strong cross-cutting relationships of the isoreflectance and structure contours suggest that there may have been local thermal disturbances superimposed on the burial-related maturation.

day geothermal gradient of 22 °C km^{-1} (Pollock and Watts 1976), and an 'excess temperature' caused by the postulated thermal disturbance. The excess temperature depends on depth and time (cf. Chapter 3). For example, for the Middle Ordovician level, the excess temperature first of all rises because subsidence brings the horizon closer to the heat source, then after about 30 Myr, it falls because the effects of the decay of the thermal anomaly become predominant (Fig. 9.22). The sum of the excess temperature and the equilibrium temperature for the given geothermal gradient gives the actual palaeotemperature. This technique predicted that only Middle Ordovician and older units in the centre of the basin should be thermally mature enough to liberate Type II kerogen (see Section 10.3), yet oil is found in Middle Devonian sediments, implying extensive lateral and upward migration of petroleum. This presents difficulties because the Silurian is composed of nearly 2 km of impermeable salts. In addition, coals in Carboniferous strata are mature, with R_o values of 0.5 to

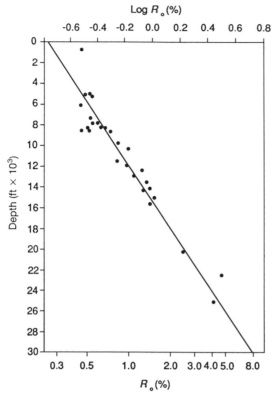

Fig. 9.21. Log of vitrinite reflectance versus depth for the Woodford Shale. Data from Cardott and Lambert (1985). The lower two R_o values of 4.29 per cent and 4.89 per cent come from the region of the postulated thermal anomaly.

0.6 per cent. A post-Pennsylvanian (post Upper Carboniferous) overburden of 300 m (Sleep, Nunn and Chou 1980) to 1 km (Cercone 1984) has therefore been suggested to increase the palaeotemperatures in the Palaeozoic fill of the basin. However, the thermo-mechanical model of Nunn, *et al.* (1984), which uses a very conservative (uneventful) thermal history, still cannot explain generation of hydrocarbons in rocks younger than Ordovician.

Cercone (1984) explained this difficulty by suggesting that there was a high, constant geothermal gradient during the main basin subsidence between Cambrian and Carboniferous (35 to 45 °C km^{-1}) followed by a linear decrease from the Permian to the present value of 20 to 25 °C km^{-1} (Fig. 9.23). This temperature history would strongly violate a basin model involving exponential cooling following thermal disturbance (Haxby *et al.* 1976, Nunn

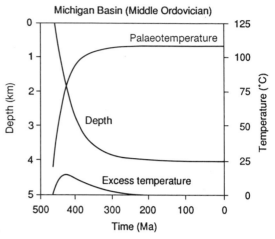

Fig. 9.22. Model values for depth, excess temperature and palaeotemperature for the Middle Ordovician time-stratigraphic horizon (*c.* 462 Ma) using the tectonic model of Nunn, Sleep and Moore (1984) for an elastic lithosphere with an effective flexural rigidity of 2×10^{28} dyn cm. The curves rcfer to a position in the centre of the basin experiencing the maximum subsidence.

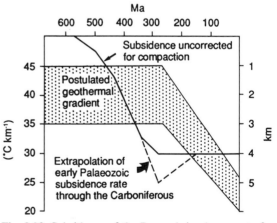

Fig. 9.23. Subsidence of the Precambrian basement of the Michigan Basin (after Sleep and Snell 1976) with an extrapolated portion through the Carboniferous following the suggestion of Cercone (1984). Continued subsidence through the Carboniferous is followed by about 1 km of uplift in order to explain the high organic maturity of Carboniferous coals. Cercone (1984) postulated a constant geothermal gradient in the Palaeozoic followed by a linear fall to present-day levels. This is in strong disagreement with the thermal-mechanical model of Nunn, Sleep and Moore (1984). See text for further details.

et al. 1984). Similar discrepancies have, however, been noted from the Sydney Basin (Middleton and Schmidt 1982, also Section 9.5.1) and the aulacogen of southern Oklahoma (Feinstein 1981).

Such a large increase in geothermal gradient in the Palaeozoic (from 20 °C km^{-1} to 35–45 °C km^{-1}) should have produced an additional heat flow of about 1×10^{-6} cal cm^{-2} s^{-1} or 1 HFU. To sustain this additional heat flow throughout the Palaeozoic development of the basin, an extremely large surface area of thermal anomaly is required – Nunn *et al.* (1984) liken it to the equivalent of a 50 km thick batholith beneath the basin. There is no igneous expression of this kind. Secondly, using the uniform extension model (McKenzie 1978a), a stretch factor β of 2.6 would be required to sustain the increased heat flow. Such large amounts of stretching should

produce a 10 km deep basin, yet the Michigan Basin contains only 4km of sediment. Thirdly, with such a high geothermal gradient, deeper units in the centre of the basin should have been overheated or 'burnt-out', yet they produce 60° API condensate from Cambro-Ordovician sediments at 3.4 km (Nunn *et al.* 1984).

The Michigan Basin therefore provides an interesting example of conflict between organic maturity and oil measurements on the one hand (see Cercone 1984; Daly and Lilly 1985; Illich and Grizzle 1985) and plausible thermal models on the other (Nunn *et al.* 1984). The incorporation of depth- and time-dependent thermal conductivites, particularly important in the evaporitic basin-fill, goes some way to explain the discrepancy, but the subject must still be regarded as controversial at present.

SECTION 5
APPLICATION TO PETROLEUM
PLAY ASSESSMENT

10 The petroleum play

'Ah! *Vanitas vanitatum*! Which of us is happy in this world? Which of us has his desire? or, having it, is satisfied? – Come, children, let us shut up the box and the puppets, for our play is played out.'

(William Makepeace Thackeray (1811–1863))

SUMMARY

A play is a perception or model of how a producible reservoir, petroleum charge system, regional topseal and traps may combine to produce petroleum accumulations at a specific stratigraphic level. Prediction of source rocks, reservoirs, topseals and traps requires an understanding of the structural and stratigraphic evolution of the depositional sequences within a basin. This understanding may be achieved through basin analysis, which serves as a springboard for the assessment of petroleum plays.

Correct identification and interpretation of the fundamental tectonic and thermal processes controlling basin formation, and the geometry and sedimentary facies contained in the basin's depositional sequences, is the first and most important step towards building the geological models that underpin play assessment.

The first requirement for a play is that there is a *petroleum charge*. The petroleum charge system comprises *source rocks*, which must be capable of *generating* and *expelling* petroleum, and a *migration* pathway into the reservoir unit. Source rocks are sediments rich in organic matter derived from photosynthesizing marine or lacustrine algae and land plants which contain chemical compounds known as lipids. Lipids are preserved when sediments are deposited under anoxic conditions. Lakes, deltas and marine basins are the main depositional settings of source beds.

Organic matter buried in sediments is in an insoluble form known as kerogen. Petroleum is *generated* when kerogen is chemically broken down as a result of rising temperature. For typical rates of heating, a stage of oil generation at approximately 100 to 150 °C is followed by a stage of oil cracking to gas (150 to 180 °C) and finally by dry gas generation (150 to 220 °C). Petroleum *expulsion* probably occurs as a result of the build-up of overpressure in the source rock as a consequence of hydrocarbon generation. For lean (i.e. organic-poor) source rocks, petroleum expulsion is probably very inefficient. *Secondary migration* carries expelled petroleum towards sites of accumulation, and is driven by the buoyancy of petroleum fluids with respect to formation pore waters. Migration stops when the capillary pressure of small pore systems exceeds the upward-directed buoyancy force.

A further requirement for a play is a porous and permeable *reservoir rock*. The porosity and mineralogy of potential reservoir rock intervals may be determined from wireline logs, notably the sonic, formation density and neutron logs. Reservoir rocks may result from deposition in almost any of a very wide range of depositional environments. Reservoir prediction requires careful interpretation of sedimentary facies within each depositional sequence. A number of scales of heterogeneity exist from kilometre scale to microscopic that affect the distribution of porosity and permeability in the gross reservoir unit. Particular basin tectonic settings have associated with them particular types of reservoir geometry and composition.

A *regional topseal* or caprock is needed to seal petroleum in the gross reservoir unit. The mechanics of sealing are the same as those that control secondary migration. The ideal caprock is of a fine-grained lithology, and is ductile and laterally persistent. Thickness and depth of burial do not appear to be critical. Two of the most successful reservoir–caprock associations are where marine shales transgress over gently sloping clastic shelves, and where sabkha evaporites regress over shallow-marine carbonate shelves.

The final requirement for the operation of a petroleum play is the presence of *traps*. Traps are local subsurface concentrations of petroleum and may be classified into structural, stratigraphic and hydrodynamic traps. Structural traps represent the habitat of most of the world's already discovered petroleum, and are formed by tectonic, diapiric and gravitational processes. Stratigraphic traps are those inherited from the original depositional morphology of, or discontinuities in, the basin fill, or from subsequent diagenetic effects. Much of the world's future undiscovered petroleum resource is likely to be found in stratigraphic traps.

10.1 INTRODUCTION: THE PLAY CONCEPT

Basin analysis is a stepping-stone towards assessment of the undiscovered petroleum potential of an area. Assessments of this kind guide the exploration programmes of the petroleum industry. An understanding of the distribution and evolution of depositional sequences and facies allows rational and realistic predictions to be made of petroleum source rocks, reservoir rocks and caprocks – the building-blocks of a petroleum play. The associated structural development of the basin is primarily responsible for the formation of petroleum traps.

A play may initially be defined as a perception or model in the mind of the geologist of how a number of geological factors might combine to produce petroleum accumulations at a specific stratigraphic level in a basin. These geological factors must be capable of providing the essential ingredients of the petroleum play, namely:
• a *reservoir unit*, capable of storing the petroleum fluids and yielding them to the well bore at commercial rates;
• a *petroleum charge system*, comprising thermally mature petroleum source rocks capable of expelling petroleum fluids into porous and permeable carrier

beds, which transport them towards sites of accumulation (traps) in the gross reservoir unit;

• a *regional topseal* or caprock to the reservoir unit, which contains the petroleum fluids at the stratigraphic level of the reservoir;

• petroleum *traps*, which concentrate the petroleum in specific locations, allowing commercial exploitation;

• *the timely relationship* of the above four ingredients so that, for example, traps are available at the time of petroleum charge.

Thus, a play may further be defined as a family of undrilled prospects and discovered pools of petroleum that are believed to share a common gross reservoir, regional topseal, and petroleum charge system.

A brief description of a play might be: 'Mid-Jurassic submarine-fan sandstone reservoirs in Late Jurassic fault blocks, sealed by Lower Cretaceous marine mudstones, and charged during the Early Tertiary from Upper Jurassic marine source rocks (Fig. 10.1)'

The geographical area over which the play is believed to extend is the *play fairway*. The extent of the fairway is determined initially by the depositional or erosional limits of the gross reservoir unit, but may also be limited by the known absence of any of the other factors. The mapping-out of the play fairway is discussed later in Section 11.2.

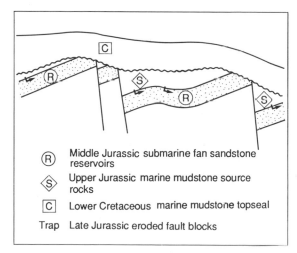

Fig. 10.1. Schematic illustration of a petroleum play. Plays must be carefully defined in terms of their reservoir, charge system and regional topseal.

A play may be considered *proven* if petroleum accumulations (pools or fields) are known to have resulted from the operation of the geological factors that define the play. These geological requirements are thus known to be present in the area under investigation, and the play may be said to be 'working'.

In *unproven* plays, there is some doubt as to whether the geological factors actually do combine to produce a petroleum accumulation. One of the objectives of play assessment is to estimate the probability of the play working; this is known as *play chance*. Play fairway risk estimation is discussed in Section 11.4.

10.2 FROM BASIN ANALYSIS TO PLAY CONCEPT

Play concepts are founded on an understanding of the stratigraphic and structural evolution of the basin. The geological models upon which predictions of source, reservoir and caprocks and their evolution through time are based, are outcomes of this understanding.

The validity of these models, and therefore of the plays that are generated from them, is dependent on a correct interpretation of the boundaries and overall *geometry* of the depositional sequence or sequences involved in the play, and on a correct interpretation of the *sedimentary facies* within the sequence(s). Basin analysis provides the means of making these interpretations.

The previous chapters of this book have shown that the location and overall form of major depositional sequences may be understood in terms of the mechanical processes of basin formation. These are in turn governed by the behaviour of the underlying lithosphere. Thus basins due to lithospheric stretching, basins due to flexure, and basins in strike-slip zones, each exhibit characteristic locations, geometries and evolutions which may be understood in terms of the controlling broad plate tectonic processes. Sets of depositional sequences may be identified that are related to particular phases of basin formation. These packages, termed *megasequences* by Hubbard (1985) and Hubbard *et al.* (1988), and broadly equivalent to the Vailian *supersequences* (Chapter 6), are bounded by major regional unconformities that mark the onset and end of a major basin-forming process. Thus, a rift megasequence formed by lithospheric stretching, may be overlain by a passive-margin post-rift megasequence formed during the subsequent thermal contraction phase, and perhaps by a foreland basin megasequence caused by flexure adjacent to a thrust belt.

Knowledge of the underlying basin-forming process will also tend to imply a particular tectonic and thermal development for the basin, which is an important input to the thermal modelling of potential source rock intervals.

Correct identification and interpretation of the megasequences present in a province is the first step towards building the geological models for play assessment.

The stratigraphic fill of each megasequence is controlled by the interplay of tectonic subsidence, sedimentation rate and sea level changes. Each megasequence may be broken down into a series of depositional sequences and systems tracts representing discrete phases of the basin infill. Chapters 6 and 7 describe the processes that control sediment facies and distribution within depositional sequences. This forms the basis for the prediction of source, reservoir and caprocks.

The type, amount and quality of data available will limit the confidence held in any stratigraphic interpretation. The goal is to achieve a reliable *chronostratigraphic* interpretation of a sequence, so that the distribution and nature of sedimentary facies may be understood in terms of geological processes operating at a specific time. The chronostratigraphic interpretation must, however, be built up from interpretations of lithostratigraphy, biostratigraphy and seismic-stratigraphy. All of these, on their own, are potentially unreliable.

Lithostratigraphy is dependent on outcrop and well information. Lithostratigraphic boundaries tend to be diachronous, and may cause serious chronostratigraphic miscorrelations. Reliability is limited by low outcrop exposure, sparse well control, and our general inability to reliably determine lithology from seismic data. *Biostratigraphy* is again restricted to outcrop and well data. Environmental conditions at the time of deposition may strongly influence the existence and likelihood of preservation of sensitive fossil groups, and diagenesis/metamorphism may destroy important biostratigraphic information. Both lithostratigra-

phy and biostratigraphy provide important stratigraphic 'fence-posts'. It is only *seismic stratigraphy*, however, that provides the inter-well and inter-outcrop information that forms the 'fences' of the stratigraphic interpretation. As a result, seismic stratigraphy has had an enormous impact on the interpretation of depositional sequences.

Miscorrelation of information, and misinterpretation of sedimentary facies, is a serious danger unless lithostratigraphic and seismostratigraphic interpretations are integrated. The confidence held in a depositional sequence interpretation may be assessed by considering the adequacy of each of the required data types – outcrop, well and seismic data – in terms of both quantity and quality. Deficiencies in the data base must be recognized. Normally, more than one interpretation fits the observable data. Figure 11.12 shows three possible interpretations (models) of some seismic data. Each has different implications for petroleum plays. Each geological model carries an associated risk of not being valid. This is called *model-risk*. The way in which model-risk is incorporated in the play assessment is explained in Chapter 11.

A *chronostratigraphic diagram* is a useful way of illustrating the relationships between sedimentary facies in a sequence (Fig. 10.2). Combined with a sequence isopach map, the chronostratigraphic diagram may be used to make sedimentary facies predictions for the entire sequence. Given these sedimentary facies, the next step is to make predictions of potential source, reservoir and caprocks. The thermal maturity of the source rock (cf. Chapter 9), and the presence and timing of traps must also be addressed. Sections 10.3 to 10.6 discuss the factors that control their development. There is a risk that these elements of the petroleum play do not exist, even though the geological model is valid. This additional element of risk may be termed *conditional play risk*. It is discussed further in Section 11.4.

A suite of maps showing the distribution of potential reservoir, source and caprock facies may be drawn. It is important to note that caprocks and sources may be external to the sequence containing the reservoir. A single source rock horizon may charge a number of separate reservoir-defined plays, and a single reservoir-defined play may be charged from a variety of separate source rock

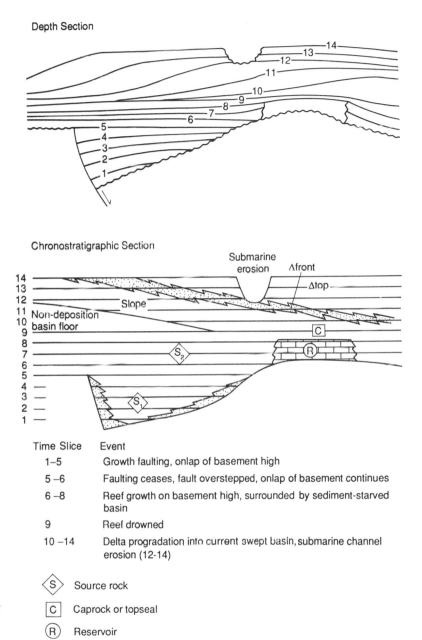

Depth Section

Chronostratigraphic Section

Time Slice	Event
1–5	Growth faulting, onlap of basement high
5–6	Faulting ceases, fault overstepped, onlap of basement continues
6–8	Reef growth on basement high, surrounded by sediment-starved basin
9	Reef drowned
10–14	Delta progradation into current swept basin, submarine channel erosion (12-14)

Ⓢ Source rock

Ⓒ Caprock or topseal

Ⓡ Reservoir

Fig. 10.2. Schematic chronostratigraphic diagram. The chronostratigraphic diagram is a very useful way of showing the relationships between sedimentary facies in a sequence, and the overall development of the basin.

horizons. The objective of play assessment is to anticipate all of the possible combinations of potential reservoirs, sources and caprocks that may produce petroleum plays in the basin. For each reservoir-defined play, a single map may be pro-duced that shows the distribution of the potential reservoir facies, the source 'kitchen(s)' needed to charge the reservoir, and the potential caprock facies. The use of play fairway maps is discussed in more detail in Section 11.2.

The final step in play assessment is the evaluation of individual traps in the fairway.

Chapter 11 shows how quantitative estimates of the undiscovered potential of the play can be made.

10.3 THE PETROLEUM CHARGE SYSTEM

10.3.1 Source rocks

Summary

There is now a wealth of geochemical evidence that petroleum is sourced from biologically-derived organic matter buried in sedimentary rocks. Organic-rich rocks capable of expelling petroleum compounds are known as *source rocks.*

In order to understand and predict the distribution and type of petroleum source rocks in space and time, it is necessary to consider the biological origin of petroleum. Source beds form when a very small proportion of the organic carbon circulating in the Earth's carbon cycle is buried in sedimentary environments where oxidation is inhibited.

In the world oceans, simple photosynthesizing algae (phytoplankton) are the main primary organic carbon producers. Their productivity is controlled primarily by sunlight and nutrient supply. The zones of highest productivity are in the surface waters (euphotic zone) of continental shelves (rather than open ocean) in equatorial and mid latitudes, and in areas of oceanic upwelling or large river input.

The productivity of land plants is controlled primarily by climate, particularly rainfall. Coals have formed in the geological past predominantly in the equatorial zone and in the cool wet temperate zone centred at about 55° (N and S).

All living organic matter is made up of varying proportions of four main groups of chemical compounds. These are *carbohydrates, proteins, lipids* and *lignin.* Only lipids and lignin are normally resistant enough to be successfully incorporated into sediment and buried. Lipids are present in both marine organisms and certain parts of land plants, and are chemically and volumetrically capable of sourcing the bulk of the world's oil. Lignin is found only in land plants and cannot source significant amounts of oil, but is an important

source of gas. Geochemical studies of coal macerals have shown a very significant oil potential amongst the exinite group, comprising material derived from algae, pollen and spores, resins and epidermal tissue.

The organic compounds provided to sea bottom sediments by primitive aquatic organisms have probably not changed dramatically over geological time. In contrast, important evolutionary changes have taken place in land plant floras. As a result, a distinction can be made between the generally gas-prone Palaeozoic coals, and the coals of the Jurassic, Cretaceous and Tertiary, which may have an important oil-prone component.

Anoxic conditions (oxygen-depleted) are required for the preservation of organic matter in depositional environments, because they limit the activities of aerobic bacteria and scavenging and bioturbating organisms which otherwise result in the destruction of organic matter. Anoxic conditions develop where oxygen demand exceeds oxygen supply. Oxygen is consumed primarily by the degradation of dead organic matter; hence, oxygen demand is high in areas of high organic productivity. In aquatic environments, oxygen supply is controlled mainly by the circulation of oxygenated water, and is diminished where stagnant bottom waters exist. The transit time of organic matter in the water column from euphotic zone to sea floor, sediment grain size, and sedimentation rate also affect source bed deposition.

The three main *depositional settings* of source beds are lakes, deltas and marine basins.

Lakes are the most important setting for source bed deposition in continental sequences (see also Section 7.1.2.3). Favourable conditions may exist in deep lakes, where bottom waters are not disturbed by surface wind stress, and at low latitudes, where there is little seasonal overturn of the water column and a temperature-density stratification may develop. In arid climates, a salinity stratification may develop as a result of high surface evaporation losses. Source bed thickness and quality is improved in geologically long-lasting lakes with minimal clastic input.

Organic matter on lake floors may be autochthonous, derived from fresh water algae and bacteria, which tends to be oil-prone and waxy, or allochthonous, derived from land plants swept in from the lake drainage area, which may be either gas-prone or oil-prone and waxy.

The Eocene Green River Formation of the western USA, and the Palaeogene Pematang rift sequences of Central Sumatra, Indonesia are examples of rich, lacustrine source rock sequences.

Deltas may be important settings for source bed deposition (see also Section 7.1.3.1). Organic matter may be derived from freshwater algae and bacteria in swamps and lakes on the delta-top, marine phytoplankton and bacteria in the delta-front and marine pro-delta shales, and, probably most importantly, from terrigenous land plants growing on the delta plain. On post-Jurassic deltas in tropical latitudes, the land plant material may include a high proportion of oil-prone, waxy epidermal tissue. Mangrove material may be an important constituent. Examples of deltaic source rocks include the Upper Cretaceous to Eocene Latrobe Group coals of the Gippsland Basin, Australia.

Much of the world's oil has been sourced from *marine* source rocks. Source beds may develop in *enclosed basins* with restricted water circulation (reducing oxygen supply), or on open shelves and slopes as a result of *upwelling* or *impingement of the oceanic midwater oxygen-minimum layer*.

Examples of modern enclosed marine basins include the Black Sea and Lake Maracaibo. Source bed deposition is favoured by a positive water balance, where the main water movement is a strong outflow of relatively fresh surface water, leaving denser bottom-waters undisturbed. The Upper Jurassic Kimmeridge Clay Formation of the North Sea, and the Jurassic Kingak and Aptian–Albian HRZ Formations of the North Slope, Alaska, are examples of source rocks deposited in restricted basins on marine shelves.

The upwelling of nutrient-rich oceanic waters may give rise to exceptionally high organic productivity. Oxygen-depletion may occur in the underlying bottom-waters as oxygen supply is overwhelmed by the demand created by degradation of dead organic matter. Upwelling coastlines tend to be arid, and the organic matter in upwelling deposits is almost entirely of marine origin and strongly oil-prone.

Upwelling may have played a part in the formation of source rocks such as the Permian Phosphoria Formation of the western USA, the Triassic Shublick Formation of the North Slope, the Cretaceous La Luna Formation of Venezuela and Colombia, and the Miocene Monterey Formation of California.

In open oceans whose floors are swept by cold, dense currents originating in the polar regions, an oxygen-deficient layer develops at depths of 100 to 1000 metres. At times in the geological past, during periods of warmer climate and higher sea level, this layer may have intensified and impinged on large areas of the continental shelves and slope. The 'global anoxic events' of the Mid-Cretaceous may have resulted from this process. The Toarcian source rocks of western Europe may also be an example.

The organic matter buried in sediments is in a form known as kerogen. *Geochemical measurements* may be used to determine the presence, richness and stage of thermal maturity of a petroleum source rock, as well as the range of compounds likely to be generated and expelled. The richness or petroleum-generating potential of a source rock can be determined by measurements of Total Organic Carbon (TOC) and the pyrolysis yield. Rocks with pyrolysis yields of greater than approximately 5 kg tonne^{-1} have the potential to be effective source rocks. More sophisticated geochemical techniques, such as gas-chromatography and isotope studies can be used to determine likely petroleum products, and in a range of other applications, including the correlation of source rocks with oils. Visual (optical) descriptions of kerogen may also give a useful guide to petroleum potential and petroleum type. From microscopic examination in reflected light, kerogen may be classified into the *exinite, vitrinite,* and *inertinite* groups. The exinite group comprises macerals with significant oil potential, while the vitrinite group are gas-prone. Inertinites have no petroleum-generating potential.

Measurements of the *reflectance of vitrinite* are used as an index of thermal maturity (see Sections 9.4 to 9.7).

10.3.1.1 The biological origin of petroleum

Since the discovery of Treibs in 1934 of a porphyrin 'biological marker' compound in rock material, a wealth of geochemical evidence has accumulated to show that petroleum is sourced from biologically-derived organic material buried in sediments. In order to understand the distribution in space and time of source rocks for oil and gas, it is necessary to first consider the characteristics of the biomass from which the organic material is originally derived. This section briefly discusses:

a) source bed deposition in the context of the overall carbon cycle

b) the main components of the biomass

c) geographical variations in organic productivity in the world's environments at the present day, and the main factors controlling these variations

d) changes in the composition of the biomass through geological time

e) the chemical composition of living organic matter and its likely hydrocarbon products.

The carbon cycle

Figure 10.3 illustrates the main elements of the carbon cycle. The cycle is initiated by photosynthesizing land plants and marine algae which convert carbon dioxide present in the atmosphere and seawater into carbon and oxygen using energy from sunlight. Carbon dioxide is recycled back in

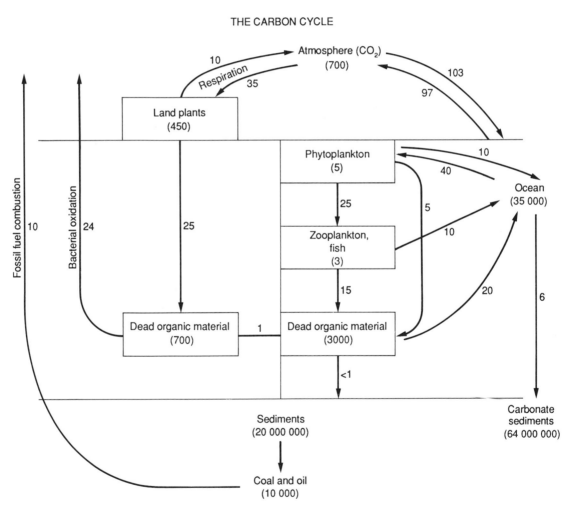

Fig. 10.3. The main elements of the carbon cycle. Numbers represent quantities in billions of metric tons. Numbers in parentheses represent stored quantities, numbers without parentheses are yearly fluxes. A relatively small amount of organic carbon escapes the carbon cycle to form organic-rich sediments, the source rocks for oil and gas. (Data provided by Bolin, *The Carbon Cycle*, Scientific American Inc., from Waples 1981).

many ways, the most important of which are

a) animal and plant respiration – back to the atmosphere

b) bacterial decay and natural oxidation of dead organic matter

c) combustion of fossil fuels – both natural and by man.

However, the importance of the carbon cycle to the petroleum geologist is that a small proportion of carbon escapes from the cycle as a result of deposition in environments where oxidation to carbon dioxide cannot occur. These environments are generally depleted in oxygen (for example, some restricted marine basins and deep lakes), or toxic for bacteria (swamps).

The proportion of organic material buried in sediments in this way, relative to that originally produced, is very small (less than 1 per cent), but over geological time is significant. Because it is preferentially concentrated in specific environments, it results in commercially significant petroleum source bed development. Petroleum is sourced therefore from organic carbon that has dropped out of the carbon cycle, at least temporarily. It rejoins the cycle when extracted by man and combusted.

Organic production

The nature of organic production is quite different in continental and marine ecosystems. Continental ecosystems are dominated by land plants in low-lying coastal plain environments and by fresh water algae in lakes. Marine ecosystems are overwhelmingly dominated by phytoplankton.

MARINE ECOSYSTEMS

Simple photosynthesizing algae are the primary organic carbon producers in the world's oceans, and are the start of a complex food chain. Phytoplankton are responsible for over 90 per cent of the supply of organic matter in the world's oceans. The phytoplankton group includes the diatoms, dinoflagellates, blue green algae and nannoplankton.

Figure 10.4 illustrates the fate of the 26.6×10^9 tons supply of organic carbon per year. Only a

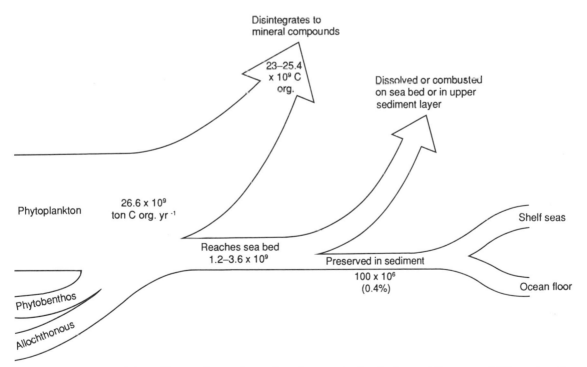

Fig. 10.4. Fate of the 16.6×10^9 tons of annual organic carbon production in the world's oceans. Of the order of 0.4 per cent is preserved in the bottom sediments of shelf seas and ocean floors.

small percentage (0.4 per cent according to Romankevich (1984)) of the net carbon production in the world's seas and oceans is transferred to and preserved in sea bottom sediment.

Apart from phytoplankton, other organisms such as zooplankton, benthos, bacteria and fish may also be important elements of the biomass. In the Black Sea (Romankevich, 1984), annual bacteria production far exceeds even the phytoplankton. The main function of bacteria is to break down dead organic matter, but the bacteria may themselves also contribute to the organic content of the sediment.

Geographical variations in phytoplankton production in the world's marine environments are shown in Table 10.1 and Fig. 10.5. Although the open ocean accounts for a large percentage of the organic carbon produced, the concentration of organic carbon per square metre in open ocean water is relatively low (the red pelagic oozes of the deep ocean basins are typically very lean in organic content). In contrast, the continental shelves are very rich, particularly in some specific environments of enhanced organic activity such as the algally-dominated intertidal zone and in reefs and estuaries. Upwelling zones, such as those off the Peruvian coast at the present day, are also areas of relatively high organic productivity.

At a global scale, several trends in organic productivity may be recognized:

1 Primary productivity decreases from coastal/marine shelf into open ocean.
2 Mid-latitude humid and equatorial latitudes are more productive than tropical latitudes.
3 Lowest productivity is in polar and arid tropical areas.

The factors controlling organic productivity include:

- *Sunlight.* The zone of highest productivity is the top 200 m of the world seas, especially the upper 60–80 m. This is the photic zone.
- *Nutrient supply.* Nutrients, particularly nitrates and phosphates, are required to sustain high organic productivity. These are supplied by water circulation. Stagnant seas are not very productive. Ocean bottom currents set up by the sinking of very cold water in the polar regions may cause upwelling along the western coasts of continents in tropical latitudes. The best known examples are offshore Peru and West Africa. A rich nutrient supply is provided, and organic productivity is very high. Nutrient supply is also locally increased in areas of large river input and coastal abrasion.
- *Turbidity.* Productivity is limited in areas with turbid coastal waters.
- *Salinity.* Extremes of salinity (high or low) reduce the diversity of species present, though productivity of certain groups may still be very high.
- *Temperature.* Temperature also influences the composition of the phytoplankton population, rather than net productivity. Dinoflagellates for example, require high water temperatures of >25 °C. Diatoms and radiolarians prefer 5–15 °C.

Figure 10.6 illustrates the concentration of organic carbon in sea-bottom sediment, and shows that a large proportion is in continental shelf environments. Organic productivity in surface waters is an important, but not sole factor controlling the concentration of organic matter in bottom sediments. Indeed, Demaison and Moore (1980) were unable to find a convincing systematic correlation between these two factors at the present day.

The critical factors for source bed development are the *deposition* and *preservation* of organic matter in significant quantities in sediments, rather

Table 10.1 Global net primary production (Woodwell *et al.* 1978, Nienhuis 1981)

Ecosystem type	Area $10^6 km^2$	Total net primary production		Total plant mass of carbon, $10^9 tC_{org}$
		$10^9 tC_{org} yr^{-1}$	$gC_{org} m^{-2} yr^{-1}$	
Marine ecosystems including:	361.0	24.7	68.7	1.74
Algal bed and reef	0.6	0.7	1166.7	0.54
Estuaries	1.4	1.0	714.3	0.63
Upwelling zones	0.4	0.1	250.0	0.004
Continental shelf	26.6	4.3	161.6	0.12
Open ocean	332.0	18.7	56.3	0.45

Note: The productivity per square metre of the open oceans is notably low compared to shelf environments.

THE PETROLEUM PLAY 319

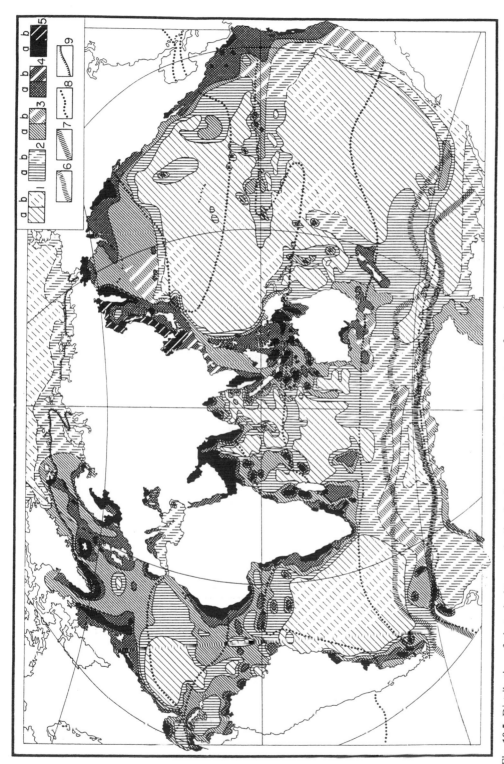

Fig. 10.5. Distribution of phytoplankton production in the ocean in units of mg C m^{-2} day^{-1}: 1 = 100; 2 = 100–150; 3 = 150–250; 4 = 250–500; 5 = >500; a, direct measurements; b, indirect data; 6 boundaries of climatic zones; line with teeth, 7 Antarctic divergence; 8 boundaries of climatic zones; line with teeth, the ice boundary in the Arctic (from Romankevich 1984). The highest phytoplankton productivities are in shelf areas rather than in oceans, particularly in algally-dominated intertidal zones, reefs, estuaries and upwelling zones. (Reproduced with permission from Romankevich, 1984.)

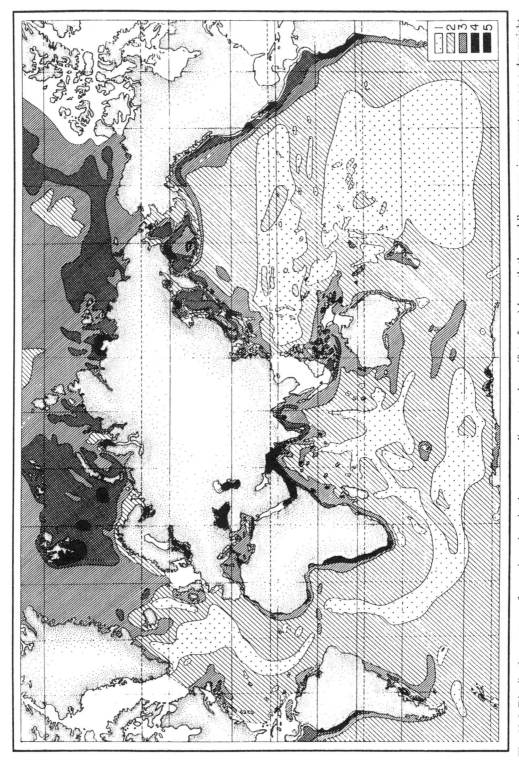

Fig. 10.6. Distribution pattern of organic carbon in the upper sedimentary layer (0 to 5 cm) beneath the world's oceans, in percentages on a dry weight basis: 1 = <0.25; 2 = 0.25–0.50; 3 = 0.51–1.00; 4 = 1.01–2.00; 5 = >2.00 (from Romankevich 1984). The highest organic carbon concentrations in bottom sediments are also in shelf seas. Organic productivity is a major controlling factor but there are a number of additional factors that control organic matter preservation. (Reproduced with permission from Romankevich, 1984.)

than organic productivity *per se*. Factors affecting the deposition and preservation of organic matter in sediment are discussed in Section 10.3.1.2.

CONTINENTAL ECOSYSTEMS

Organic productivity in continental ecosystems is dominated by land plants and freshwater algae.

The productivity of land plant material is controlled primarily by climate. Land plant material may be swept by rivers into lakes and adjoining marine areas, constituting an allochthonous organic supply. The most important allochthonous land plant deposit is peat, which forms below the water table in swamps or stagnant lakes where a wet climate allows luxuriant plant growth and topography causes poor drainage. A balance is necessary between the rate of accumulation of dead plant matter and the rate of subsidence. Accumulated peat may be preserved where bacterial decay of the dead organic matter is inhibited by anoxic or toxic conditions and where net subsidence takes place.

Peat swamps form in the lower delta plain environment, typically in the lagoonal areas behind coastal spits and barriers, and in bays between vertically-accreting distributary channels.

Ancient coal occurrences can be predicted using palaeoclimatic maps such as those of Parrish, Ziegler and Scotese (1982). Palaeolatitude studies show that coals ranging in age from Early Triassic to mid-Miocene are concentrated in the equatorial zone and in the cool wet temperate zone centred on about 55° N and S.

Freshwater algae make an important contribution to the organic matter supply in lakes (see also Section 7.1.2.3). An example is the present day alga, *Bottryococcus*. Ancestors of *Bottryococcus* have been identified in ancient lake sediments, and derived geochemical compounds have been recognized in many lake-sourced oils.

Chemical composition of living organic matter

The chemical compounds that make up all living organic matter fall into four groups. These are:
1 Carbohydrates.
2 Proteins.
3 Lipids.
4 Lignin.
   Carbohydrates are compounds that function as

sources of energy and as supporting tissue in plants and some animals. Examples are sugar, such as glucose and fructose, starch, cellulose and chitin. Cellulose is an important supporting tissue in land plants, while chitin is the material manufactured by crustaceans to form a hard protective exoskeleton.

Proteins are organic compounds made up of amino acids, and perform a variety of biochemical functions vital to life processes. Examples are enzymes, haemoglobins and antibodies. Proteins also make up most of the organic matter in shells and substances such as hair and nails.

Lipids are a range of organic substances that are insoluble in water, and include animal fats, vegetable oils and waxes. They are similar in chemical composition to petroleum. Lipids are abundant in marine plankton, and are present in the seeds, fruit, spores, leaf coatings and barks of land plants. A range of lipid-like substances, for example sterols, are important biological markers in crude oils.

Lignin and *tannins* are compounds common in higher plants. Lignin is the substance that gives strength to plant tissue, for example in trees, providing a much firmer support than cellulose. Tannins are found in some tree barks, seed coats, nut shells, algae and fungi.

Other important organic compounds are *resins* and *essential oils*. Resins are found in the wood and leaf coatings of trees, and are particularly resistant to chemical and biological attack.

The relative amounts of these groups of organic compounds in living organisms varies enormously, as shown in Table 10.2. Factors such as food supply and overcrowding are also known to affect lipid content. Of the four groups of compounds, proteins and carbohydrates are very susceptible to degradation, and tend to be dissolved, oxidized or bacterially degraded, without being incorporated in sediment beyond its surface layers. In contrast, lipids and lignin are much more resistant to mechanical, chemical and biochemical breakdown, and under the conditions to be discussed later, will be buried successfully in sediment.

Lipids are closest in chemical composition to petroleum. A relatively small number of chemical changes are involved in transforming lipids into petroleum, and more petroleum can be produced from lipids than from any of the other substances. The lipid content of organic matter buried in

Table 10.2 Composition of living matter (Hunt 1980)

	% Weight (ash-free basis)			
	Proteins	Carbohydrates	Lignin	Lipids
Plants				
Land plants				
Spruce wood	1	66	29	4
Oak leaves	5	44	32	4
Scots-pine needles	7	41	15	24
Diatoms	29	63	0	8
Lycopodium spores	8	42	0	50
Animals				
Zooplankton	53	5	0	15
Copepods	65	22	0	8
Higher invertebrates	70	20	0	10

Note: The relative amounts of proteins, carbohydrates, lignin and lipids vary widely amongst the plant and animal groups listed here. It is the lipid component that provides oil potential. Lignin may source gas.

sediments is probably sufficient to source all of the world's known oil.

If the biological precursors of petroleum can be microscopically identified, we can go a long way towards predicting the kind of hydrocarbons, if any, that are likely to be generated. Marine organic matter in sediments tends to be amorphous, but much of our understanding of the generating potential of land plant source rocks is derived from coal petrography. A large number of coal 'macerals' have been identified. The three main groups are listed below. Kelley *et al.* (1985) have geochemically analysed the generating potential of some macerals.

1 *Vitrinite* is derived from the lignin and cellulose component of plant tissues. It is normally the largest constituent of the so-called humic coals, and generates predominantly gas.

2 *Inertinite* is also derived from the lignin and cellulose of plants, but has been oxidized, charred, or biologically attacked. The consensus is that it has negligible hydrocarbon generating potential. What potential it has is for gas. Dispersed inertinite has, however, been proposed as one of the sources of the liquid hydrocarbons in the Permian Gidgealpa Group of the Australian Cooper Basin.

3 *Exinite* is a diverse group of macerals including:
(a) *Alginite* is derived from algae, and when abundant forms a boghead or cannel coal. It is the main constituent of the torbanites of Scotland. This type of coal is quite rare, being most common in the Permo-Carboniferous. The algae responsible for these alginites are similar to the modern fresh water alga *Botryococcus*. Alginite is strongly oil prone.
(b) *Sporinite* is derived from the spores and pollen of plants, and may be abundant in coals from the Devonian through to the present day. The sporinite in Palaeozoic coals is derived mainly from spores, whereas in Mesozoic and Tertiary sporinite, pollen predominates. Spores and pollen are extremely lipid-rich (50 per cent) and may give rise to excellent oil prone source rocks. Since spores and pollen are often transported large distances by wind or water, oxidation before burial in sediment often occurs. The best source rocks are therefore those in which the spores and pollen are autochthonous to the depositional environment, as for example in a coal swamp.
(c) *Resinite* is derived from the resins and essential oils of land plants. It is a prolific source of napthenic and aromatic hydrocarbons.
(d) *Cutinite* is derived from the protective surface coating or cuticle of higher plants. The cuticle occurs on the outside of the epidermal tissue. It is rich in hydrocarbon waxes, and is thus an important oil source. Poor preservation may result in a gas-prone cutinite, with very little remaining oil potential.

Thus, all of the exinite macerals have oil generating potential. Preservation of the maceral is, however, critical. This is a function of the transport route and distance, and the depositional environment. The oil generating potential of sporinite and cutinite is particularly strongly affected by poor conditions for preservation.

Of the four groups of compounds found in living organic matter, therefore, only lipids and lignins are likely to be substantially incorporated into sediment beyond the surface layer. Lipids are present in both marine organisms and parts of land plants, are chemically suited to sourcing petroleum, and could account for all of the world's known oil. Lignin is found only in land plants. It is unlikely to source oil, but is an important source of gas.

Changes in the composition of the biomass through geological time

The ancestors of the primitive aquatic organisms that comprise the phytoplankton, zooplankton and bacteria may be traced back into the Precambrian with little apparent evolutionary change. The kind of aquatic organic matter buried in marine sediment has therefore probably changed very little over geological time.

For land plants, however, important floral changes have taken place since their appearance in the Late Silurian–Devonian that have had a major impact on the hydrocarbon generating characteristics of the source rocks in which they are buried (Fig. 10.7).

The Carboniferous coals of the northern hemisphere and the Permian coals of the southern hemisphere contain floras dominated by early plant groups (mostly lycopods) without extensive foliage. The resulting coal macerals are gas-prone vitrinite and its oxidation product inertinite, with minor amounts or local concentrations of sporinite, resinite, cutinite and alginite. Palaeozoic coals, therefore, tend on the whole to be gas-prone.

Important evolutionary changes took place in floras in the Jurassic and Cretaceous. Conifers became dominant in the Jurassic, and angiosperms (flowering plants) appeared in the Cretaceous. Both these plant groups are rich in waxy epidermal tissue and resin, and have significant oil generating potential. Large volumes of gas are also generated from coals of this type, since vitrinite is normally still abundant. Most of Australia's waxy oils have been sourced from coals of this type, most notably

the Gippsland basin oils from the Early Tertiary Latrobe Group, and the Eromanga and Surat basin oils from the Jurassic (Thomas, 1982).

Thus, evolutionary changes in land plant floras through geological time are responsible for the oil-prone component of Mesozoic and Tertiary coals, while Palaeozoic sediments are more typically sources solely for gas.

10.3.1.2 Source rock prediction

Introduction: anoxia

Anoxic conditions are critical to the preservation of organic matter in sediments. Source rock prediction is therefore concerned primarily with predicting where and when in the geological past 'anoxic' conditions are likely to have existed. Questions addressed in this section are, 'what causes anoxic conditions', and 'in what geological environments are anoxic conditions likely to develop?'

'Anoxic' means 'devoid of oxygen', but the term is frequently used in the sense of 'depleted in oxygen' – dysaerobic. 'Anaerobic' means that insufficient oxygen is available for aerobic biological processes. This critical oxygen concentration is different for different organisms. Below 1.0 millilitres of oxygen per litre of water there is a serious reduction in biomass, but deposit-feeding organisms (those responsible for bioturbation of the sediment) can persist down to concentrations of 0.3 ml/l. As a general guide, 0.5 ml/l may be taken as the oxic/anoxic threshold. Anoxic conditions are critical for source bed deposition because they prevent the scavenging of dead organic matter and bioturbation of the surface sediment by benthic fauna and degradation of organic matter by bacteria, which would otherwise destroy the organic matter prior to burial.

Anoxic conditions develop where *oxygen demand* exceeds *oxygen supply*.

Oxygen demand is caused primarily by the degradation of dead organic matter. Large amounts of organic matter are supplied to the sea floor in areas of high surface organic productivity (photosynthesizing algae in the euphotic zones of seas and lakes) and/or where there is a large terrigenous supply of organic matter. Oxygen is consumed as the dead organic matter is degraded.

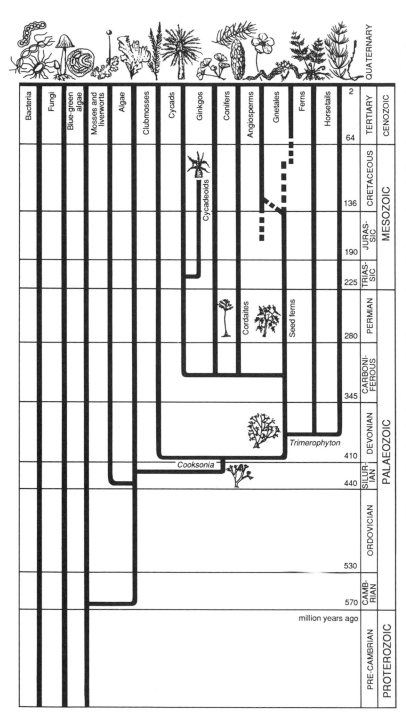

Fig. 10.7 Plant evolution over the last 600 Myr. The first land plants (primitive vascular varieties) such as Cooksonia appeared in the Silurian. A large number of new groups evolved in the Devonian. Important changes in land plant floras have taken place since the Silurian which have affected their hydrocarbon-generating characteristics. Only Mesozoic and Tertiary land plants have significant oil-generating potential. Today, two main groups dominate the land; the *conifers*, which cover over more than $10 \times 10^6 km^2$ of the Earth's surface, mainly in the cooler, drier areas, and *flowering plants* (angiosperms) which occur everywhere. The conifers may date back as far as the Carboniferous. The oldest definite flowering plants in the geological record are Barremian (Early Cretaceous), but some authors place their origin in the Late Jurassic. In the sea, organisms adapted to life in water, such as the algae, have undergone their own evolution.

Oxygen supply is controlled by the circulation of oxygenated water. This may be a downward movement of oxygen-saturated surface waters as a result of mixing by waves, or a movement of cold, oxygen-bearing ocean bottom currents. Cold water can dissolve more oxygen than warm water. A feature of the world's oceans at the present day is that cold dense water carrying large amounts of oxygen descends in the polar regions and moves over the ocean floor towards the equator, bringing oxygenated conditions to almost all parts of the oceans. It is reasonable to expect, however, that there were times in the past when such a circulation was less well developed. This has important implications for source bed deposition.

A sea or lake floor, therefore, is prone to anoxic conditions primarily under the following two sets of circumstances: (1) when organic productivity in the overlying water column is very high and the system becomes overloaded with organic matter, and (2) when stagnant bottom water conditions exist, causing a restriction in the supply of oxygen. These factors largely determine the geological settings in which source beds are deposited.

Source beds may, under exceptional circumstances, be deposited under *oxic* conditions. This sometimes occurs when sedimentation rate is very high. A special case is when mass gravity flows deposit an anoxic sediment almost instantaneously into oxic waters. As a rule, however, a source rock is not expected to be developed under oxic conditions.

The factors which affect the development of anoxia are discussed in more detail in the next section.

Factors affecting source bed deposition

An understanding of anoxic environments and their importance in petroleum exploration has rapidly developed over the past decade. Demaison and Moore (1980) is one of the most important contributions to the subject.

We have seen that anoxic conditions are a prerequisite for source bed deposition primarily because they prevent the bacterial degradation of dead organic matter and the scavenging and bioturbation of the surface sediment by benthic fauna. These and other factors will now be discussed in greater detail. Figure 10.8 illustrates Demaison and Moore's models for the degradation of organic matter under anoxic and oxic water columns.

BACTERIAL DEGRADATION

Degradation of organic matter by bacteria takes place in both the water column and in sediment pore waters, under both aerobic and anaerobic conditions. Organic matter is oxidized by aerobic bacteria using the available oxygen in the environment until there is no more organic matter to oxidize or there is no more oxygen. If the latter, the environment becomes anoxic. Anaerobic bacteria derive oxygen first of all from nitrates, and then from sulphates. It is thought that they can degrade organic matter just as fast as aerobic bacteria. An important difference, however, is that anaerobic degradation appears to result in a greater preservation of lipid-rich, oil-prone material. Furthermore, under anoxic conditions, the bacteria population itself may contribute significantly to the preserved organic matter.

SCAVENGING AND REWORKING BY BENTHIC FAUNA

The role of benthic metazoans such as worms, bivalves and holothurians is critical to the preservation of organic matter. Their activity is important in two respects:

1 They consume particulate organic matter in the water just above the sea or lake floor and in the surface sediment itself.
2 Burrowing metazoans churn up the sediment to a depth of 5 to 30 cm, allowing the penetration of oxygen and sulphates into the sediment column, thus promoting bacterial degradation.

Bioturbation seems to take place at all water depths under oxic water columns. Below oxygen concentrations of 0.3 ml/l this activity is virtually eliminated. Sediments remain laminated and organic-rich. Even the activity of anaerobic sulphate-reducing bacteria is limited because oxidants cannot easily penetrate the surface. Figure 10.9 shows that the occurrence of unbioturbated laminated muds in the Gulf of California is closely correlated with low oxygen concentrations in bottom waters. For these reasons, organic matter stands a much better chance of being preserved in the absence of benthic fauna, that is, in anoxic environments.

ANOXIC ENVIRONMENT

Better o.m. preservation
(1–25% TOC)
Higher quality o.m.

Biological reworking
is slowed by:

• The absence of animal
  scavengers

• Restricted diffusion of
  oxidants (SO_4) into
  undisturbed sediment

• Lesser utilization
  of lipids by
  anaerobic bacteria ?

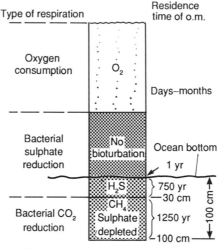

Sedimentation rate $U = 0.5$ mm yr^{-1}

OXIC ENVIRONMENT

Poorer o.m. preservation (0.2–4% TOC)
Lower quality o.m.

Biological reworking
is enhanced by:

• Presence of animal
  scavengers at interface

• Bioturbation
  facilitates diffusion
  of oxidants (O_2, SO_4)
  in sediments

• Lesser organic
  complexation with
  toxic metals.

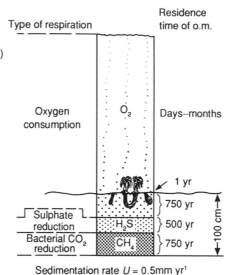

Sedimentation rate $U = 0.5$ mm yr^{-1}

Fig. 10.8. Degradation of organic matter under anoxic and oxic conditions (after Demaison and Moore 1980). Anoxic environments are the primary sites of organic matter preservation because scavenging and bioturbation by benthic fauna and aerobic bacterial degradation are inhibited.

TRANSIT TIME OF ORGANIC MATTER IN THE WATER COLUMN

Almost all marine organic matter is formed by photosynthesis in the euphotic zone. Before it can accumulate on the sea bed it has to fall through a water column of up to 6 km (in deep ocean areas). The smallest particles take the longest to fall, faecal pellets falling the fastest.

Organic matter is scavenged by fauna during its transit through the water column. Preservation of organic matter is therefore favoured by shallow water depths and large organic particle size. Scavenging in the water column is probably one of the factors that contributes to the general lack of source bed deposition in deep ocean areas. Another factor is low organic productivity due to remoteness from nutrient supply.

SEDIMENT GRAIN SIZE

The low permeability of fine-grained sediments inhibits the diffusion of oxidants from the water

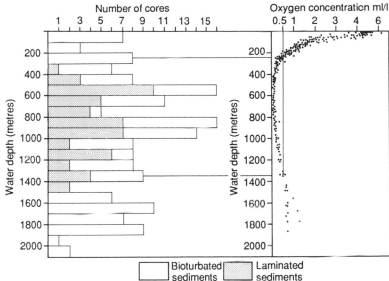

Fig. 10.9. The correlation of laminated sediments with low oxygen concentrations in the bottom waters of the Gulf of California demonstrates that bioturbation is limited by oxygen concentrations of less than 0.5 ml/l.

column into the sediment, and, as a result, bacterial activity is lower than in coarse-grained sediment. Coarse-grained sediment is usually associated with high energy environments which are in any case likely to be well oxygenated.

SEDIMENTATION RATE

Under oxic conditions, high sedimentation rates favour source bed deposition because they reduce the period during which organic matter is subject to metazoan grazing, bioturbation and aerobic bacterial attack. Sufficient organic matter may be preserved even under oxic conditions to form a source bed. The organic matter becomes diluted, however, by the large amount of mineral matter in the sediment and the resulting source rock usually has a low organic matter concentration. It may not be capable of generating sufficient liquid hydrocarbons to saturate the source rock and expel oil. Hydrocarbons may instead be expelled at high maturity as gas. Rapidly deposited sequences, therefore, rarely contain good oil sources. Examples are thick, muddy pro-delta sequences.

Under anoxic conditions, high sedimentation rates are likely to have only an adverse diluting effect on organic content. Deposition of rich source beds is therefore favoured by low sedimentation rates.

The ideal conditions for oil source bed deposition are therefore:

- anoxic conditions with high organic productivity and restricted oxygen supply (poor water circulation)
- shallow water depths
- fine-grained sediment.

Under oxic conditions, moderate sedimentation rates favour source bed deposition.

Depositional settings of source beds

The main depositional settings of source beds are:

- Lakes
- Deltas
- Marine basins.

There are a number of other less important settings, including freshwater swamps, non-deltaic shorelines, and continental slopes and rises. These settings appear to have sourced a relatively small proportion of the world's oil and, in terms of source prediction, provide only a relatively low probability of source bed presence. There are, of course, individual exceptions to this, notably paralic source rocks developed along non-deltaic but sheltered tropical shorelines. These may have the same characteristics as the lower delta plain. These less important settings, however, are not discussed further.

LAKES

Lakes are the most important depositional setting for source beds in continental sequences (Section

7.1.2.3). In order to form volumetrically significant source beds, lakes must be geologically long-lasting. Anoxic conditions develop in 'permanent' lakes when the water column becomes stratified. This is most likely to occur in the following circumstances (Allen and Collinson 1986 for summary):

1 *In deep lakes.* Wind stress causes the mixing of the whole water column in shallow lakes causing oxygenation of bottom waters and sediments. Deep lakes are usually tectonically controlled. They frequently develop in rapidly subsiding extensional continental rift systems but may also occur in areas of compressional tectonics. The 1500 m deep Lake Tanganyika in the East African rift system is anoxic below 150 m. TOCs (total organic carbon) of 7 to 11 per cent have been recorded from bottom sediment in the anoxic part of the lake. The shallower Lake Mobutu (Albert) and Lake Victoria are oxic. Beadle (1974) is an excellent source of information on African lakes.

2 *At low latitudes.* Wide seasonal variations in weather cause overturn of the water column. Cold, dense river waters carrying large amounts of dissolved oxygen sink to the bottom of temperate lakes, causing oxygenation. All temperate lakes at the present day, even the 1620 m deep Lake Baikal, are oxygenated for at least part of the year. In warm, tropical, equable climates river water is less dense, does not have a tendency to form high-density flows, and carries less oxygen. These conditions favour the development of anoxic conditions. In addition, the temperature–density behaviour of water (Ragotzkie 1978) means that more work is required to mix two layered water masses at elevated temperatures (e.g. 29 °C and 30 °C) than at low temperatures (e.g. 4 °C and 5 °C), so tropical lakes tend to stratify easily. On the other hand, the slightest cooling in a tropical lake may initiate convection currents which may eventually affect the entire water body, causing mixing and oxygenation of bottom waters.

Abundant water supply in a wet climate also ensures that the lake is kept filled. In arid climates, lakes may intermittently dry up, resulting in oxidation of its surface sediment. Provided this does not happen, however, high evaporation losses may encourage anoxia, by producing a *salinity stratification*. Salinity-stratified lakes may form important source environments in low latitudes. Ancient examples are the Devonian Orcadian basin of the UK (review in Allen and Collinson 1986) and the Eocene Green River Formation of western USA (Eugster and Hardie 1975).

Hydrothermal solutions in areas of volcanic activity, and run-off over peralkaline volcanic products, may produce *alkaline lakes.* Strongly reducing conditions may develop at the lake floor. Distinctive mineral assemblages are characteristic of alkaline lakes. The Lakes Magadi and Natron of East Africa (Section 7.2.1.2) are today precipitating trona, and the Wilkins Peak Member of the Eocene Green River Formation of Colorado and Utah has a similar evaporite mineralogy.

Clastic input to lakes

Clastic input is a function of the *relief* of the drainage (catchment) area of the lake, the *lithologies outcropping* in the drainage area, and the *climate.* Hilly or mountainous relief usually causes rapid erosion and the rapid infilling of a lake basin by coarse detritus. If the hinterland is an area of carbonate outcrop, however, much of the weathering is chemical rather than mechanical. As a result, the suspended particulate clastic input to the lake is small. Chemical weathering is dominant in humid climates, rocks quickly breaking down under the combination of high temperatures and abundant water. The transport of such weathering products depends on the rainfall intensity but also on the vegetation type (see Fig. 2–4 in Blatt, Middleton and Murray 1972). Lush vegetation, particularly grasses, in the drainage area will tend to reduce the amount of surface erosion, and hence the clastic input to the lake.

Source beds will be richer and thicker and more likely to develop if a deep lake can be maintained for a long time with the minimum of clastic input. Low relief vegetated areas in humid climates with carbonate bedrock geology provide these conditions. Small lakes may be completely swamped by clastic input, whereas centres of large lakes may see very little terrigenous influence. Deltas may build out where rivers enter the margins of lakes with relatively steep nearshore bathymetries (see Section 7.1.3.1).

Organic matter input to lake sediments

Organic matter input is of two types:

1 *Produced within the lake itself ('autochthonous').* This comprises algae and bacteria. It produces strongly oil-prone kerogen. Ancestors of the present day alga *Botryococcus braunii* have been identified in ancient lake sediments, and its biomarker compound Botryococcane has been recognized in many lake-sourced oils. An example is the oil of the Minas field in central Sumatra (Williams *et al.* 1985).

2 *Swept into the lake from the drainage area ('allochthonous').*
This comprises organic matter from higher plants. Much of it will be lignin, which is gas prone. There may also be a contribution of oil-prone waxy epidermal tissues, spores or pollen. As discussed previously, these oil-prone components are only likely to be important in tropical areas since the Jurassic. Terrigenous organic matter is likely to dominate a small lake. In large lakes, it may be concentrated around the margins (particularly around river mouths) leaving the centre the site of algal and bacterial organic matter deposition.

Petroleum composition

Lake sediments may source oil, gas-condensate or gas, depending on the factors discussed above. Lacustrine oils tend to be very variable in density, low in sulphur, and have a very variable wax content, ranging up to 40 per cent. Wax is derived from land plant cuticles and from wax-secreting freshwater algae.

Some ancient examples of lacustrine source rocks

The lacustrine oil-shales of the *Eocene Green River Formation* of Utah, Wyoming and Colorado accumulated in palaeo-lakes Uinta and Gosiute, now preserved in the Uinta and Piceance Creek sub-basins (Lake Uinta) and the Green River and Washakie sub-basins (Lake Gosiute). A range of lacustrine environments are represented by the members of the Green River Formation. Some of the most important oil source rock units (e.g. Laney Shale, Mahogany Oil Shale) were deposited in deep, anoxic, density-stratified, generally *saline* lakes, while *hypersaline* shallow lakes (Wilkins Peak and Parachute Creek members) and their fringes were also sites of significant organic matter accumulation. Hypersalinity and lack of sulphate inhibited bacterial oxidation of the organic matter. Arid conditions precluded significant land plant input to the lake systems; organic matter is, as a

result, autochthonous and oil-prone. The formation has been intensively studied; a large number of studies are available in the literature, including early contributions by Bradley (1964) and Bradley and Eugster (1969), and more recent papers by Surdam and Wolfbauer (1975), Cole and Picard (1981) and Smoot (1983).

In contrast to the saline and hypersaline units of the Green River Formation, the Pematang lake sediments of the *Palaeogene Rifted Basins of Central Sumatra, Indonesia*, are deposits of low salinity lakes. Williams *et al.* (1985) have recently provided a good description of these sequences, including their depositional facies, setting and source characteristics. The Pematang reaches up to 1800 m in thickness and was deposited in structurally controlled half-grabens, under humid tropical conditions. The most important oil source rock unit is the Brown Shale Formation, which represents the deposits of deep lakes formed by rapid syn-rift subsidence. These are well-laminated, reddish-brown to black non-calcareous mudstones. Geochemical correlations demonstrate a convincing match between Brown Shale algal source rocks and crude oils, including those of the giant Minas and Duri fields. In contrast to the deep-lake oil-prone Brown Shale Formation, gas-condensate prone land plant source rocks predominate in the shallow-lake sequences of the Coal Zone Formation.

Albian–Turonian lacustrine source rocks have been described from the *Songliao Basin of eastern China*. These represent the deposits of deep thermally-stratified freshwater lakes.

Small-scale examples of rich oil-prone lacustrine source rocks have been described from the Tertiary Mae Sot and Mae Tip Basins of northwest Thailand (Gibling 1985a, b). They were deposited in shallow fresh to brackish lakes.

DELTAS

Deltas may be an important setting for source rock deposition (see Section 7.1.3.1). In SE Asia and Australasia, deltas appear to have sourced a large proportion of the discovered oil.

The different delta types are described in Section 7.1.3.1. Constructive deltas (fluvial or tide dominated) are characterized by persistent low-energy environments on the delta top which favour source bed deposition. Destructive or static deltas (wave dominated) generally provide less favourable environments for source bed deposition. Migration of shoreline bars tends to rework the sediments in the lower delta plain, and organic matter is usually degraded to an inert state.

Sources of organic matter in deltas

Organic matter in delta sequences may be of three types:

1 *Freshwater algae (phytoplankton) and bacteria* present in lakes, swamps, and abandoned channels on the delta top. This material is oil-prone. It will only be preserved if anoxic conditions exist at the sediment surface. Fluvial channel migration results in the infilling of lakes in deltaic environments. Lakes are most likely to persist, therefore, in the upper part of the delta plain. High subsidence rates on the lower delta plain may allow lakes to persist despite rapid sedimentation rates, as occurs on the Niger and Nile deltas at the present day.

2 *Marine phytoplankton and bacteria* in the delta front and pro-delta areas. Abundant nutrients provided by river input frequently stimulate high organic productivity in the marine basin into which the delta debouches, but conditions for preservation are generally poor. Preservation requires anoxicity in the marine basin. The delta front is normally a high-energy, oxic environment. High accumulation rates allow the preservation of some organic matter, but it is strongly diluted with mineral matter, and source rocks are usually lean with potential to expel only gas.

3 *Terrigenous land plant material.* Vegetation growing on the delta plain contributes a large amount of organic matter to depositional environments. On tropical deltas since the Jurassic, the land plant material is likely to comprise both oil-prone (waxy epidermal tissues, resins, spores) and gas-prone (lignin) material. Pre-Jurassic and temperate land plants are predominantly gas prone. The most prolific sites of accumulation are the interdistributary peat swamps, where *in-situ* coals may form. Terrigenous organic matter may, however, be dispersed across the entire delta top, and into the pro-delta environment where preservation will largely depend on high sedimentation rate. In peat swamps, the high accumulation rates, highly acidic conditions, and presence of bactericidal phenol compounds released from lignin, enhance the preservation of organic material. In the upper part of the delta plain, freshwater swamps dominate. On the lower delta plain, where waters are brackish to saline as a result of some marine influence, mangrove swamps dominate. Mangroves trap plant material drifted in from the freshwater upper delta plain. Terrigenous organic matter may also be reworked onto low energy tidal flats.

Mangrove-dominated shorelines may be important source depositional environments. A modern example has been described by Risk and Rhodes (1985) in Missionary Bay, north Queensland, Australia. Mangrove litter originating on intertidal mudflats is swept into the adjoining anoxic bay bottom sediments, providing organic matter with a very high lipid content. The mangrove swamps are sites of prolific organic productivity. A thin intertidal strip of mangrove swamp may produce a vast quantity of lipid-rich organic matter and spread it over a large offshore area. The high oxygen demand caused by the abundant influx of mangrove detritus may cause anoxia in the surrounding depositional environments. Mangrove material is also relatively resistant to degradation, both physical and chemical. Under fungal and bacterial attack, the waxy lipid-rich cuticle which coats mangrove epidermal tissue appears to be preferentially preserved. Mangrove material is likely to source oils that have a high wax content.

Reworking of upper delta plain plant material in brackish conditions appears to result in a selective bacterial (or fungal?) degradation of cellulose and lignin (to humic acids), leaving a relative enrichment in oil prone waxy and resin components (Thomas, 1982).

Shanmugam (1985) has described the coals within the Upper Cretaceous–Eocene *Latrobe Group* of the Gippsland Basin of SE Australia. These units have sourced about 3000 million barrels of oil. The coals are dominantly vitrinitic but contain up to 15 per cent exinite macerals, comprising cutinite, sporinite and resin. These components are derived from the cones, bark, seeds, leaves and resin bodies originating in the adjacent coniferous forests. The climate was temperate and wet. Oils have been geochemically matched with these coals, and have typical coal-source characteristics, including high wax content (up to 27 per cent).

MARINE BASINS

Marine source rocks may form in enclosed, silled basins such as the Black Sea, Baltic Sea and Lake Maracaibo, or on open marine shelves and continental slopes. The mechanisms for source bed development in each of these settings are quite different:

Enclosed basins	*Water stratification* reduces oxygen supply
Open shelves/slopes	1 *Oceanic upwelling* causes high organic productivity and hence high oxygen demand.
	2 Impingement of *oceanic midwater oxygen minimum layer.*

(a) Enclosed basins

These marine basins are physically restricted to some extent, by land or by chains of islands, but retain some connection with the open sea. Water exchange is limited, however, and the basin is prone to water stratification and hence anoxic conditions. The nature of the water exchange with the open sea is important, since not all enclosed basins become anoxic.

A *positive water balance* is where the outflow of freshwater (as a surface layer) exceeds the relatively small inflow of deeper saline water. Most of the water movement takes place in the surface layers, allowing stratification of deeper waters. This process characterizes the Black Sea, Baltic Sea and Lake Maracaibo. The Black Sea is one of the best documented anoxic silled marine basins. Total organic carbon is up to 15 per cent in sediment 7000 to 3000 years old. At the present day, it is anoxic below 150 to 250 m water depths. Figure 10.10 illustrates the geometry and TOC distribution in modern sediments of the Black Sea.

In contrast, a *negative water balance* is where the inflow of oceanic water dominates over a relatively meagre freshwater input. This often develops in

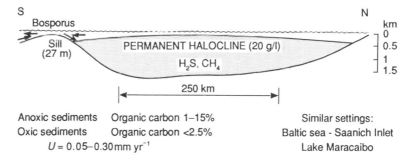

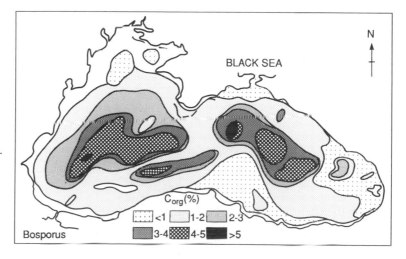

Fig. 10.10. Geometry and total organic carbon concentrations in modern sediments of the Black Sea. This is an example of a silled marine basin with a positive water balance. Organic carbon concentrations are locally up to 15 per cent in the deeper parts of the basin (after Demaison and Moore 1980).

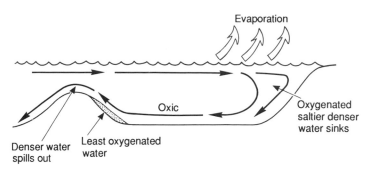

Fig. 10.11. Model for a silled marine basin with a negative water balance (after Demaison and Moore 1980). Oceanic inflow dominates over freshwater fluvial input, a situation that commonly develops in arid climates. Dense, salty, oxygenated waters resulting from surface evaporation may sink and sweep the basin floor, preventing anoxia.

arid climates where high evaporation losses at the surface cause the sinking of oxygenated waters (Fig. 10.11). Examples of oxic enclosed basins include the Red Sea, Mediterranean Sea and Persian Gulf. The Mediterranean is the world's largest silled marine basin, but it is well oxygenated and organic contents in bottom sediment are very low.

Size and depth of enclosed basins do not appear to be critical. Lake Maracaibo, for example, is only 30 metres deep. The danger of water mixing by wind-generated waves, however, renders shallow basins less favourable source bed environments. Enclosed basins may range in size up to the South Atlantic during the Aptian. Small basins tend to be short-lived, particularly when there is high clastic input.

Organic matter type in enclosed marine basins depends on the amount of terrigenous land plant material brought into the basin by rivers. Wet climates imply a positive water balance and hence a tendency to water stratification. However, high terrigenous organic input may produce gas-prone source rocks. In arid climates, the organic matter is made up largely of marine phytoplankton and, when source beds are developed (often in association with carbonates and evaporites), they are predominantly oil-prone. The Devonian of the western Canada basin is believed to be an example.

Source bed deposition is sensitive to changes in the water balance of the basin, and tends to be periodic and of varying lateral extent. Deep, big, enclosed marine basins in areas of wet palaeoclimate offer the highest probability of source bed occurrence.

Examples of source rocks deposited in restricted marine basins probably include the Upper Jurassic Kimmeridge Clay Formation of the North Sea, and the Jurassic Kingak and Aptian–Albian HRZ Formation of the North Slope, Alaska.

(b) Open marine shelves

Ocean upwelling Upwelling occurs along coastlines where wind-driven currents flowing parallel to the coast are deflected offshore by the Earth's rotational (Coriolis) force. It is most common on the east side of oceans. Upwelling occurs today along the coasts of Peru–Chile, California, Namibia and Morocco.

The deep ocean water drawn into the upwelling cell to replace the offshore moving surface water is rich in nutrients such as phosphates and nitrates, and can give rise to exceptionally high organic productivity in the near-surface photic zone. Degradation of dead organic matter creates a high demand for oxygen, and anoxicity may develop in the underlying waters. Underneath the Benguela current offshore Namibia, for example, there is a 340 km by 50 km oxygen depleted zone. Under this zone, the sediment contains 5 to 24 per cent total organic carbon (Fig. 10.12) (Demaison and Moore, 1980). The organic matter is almost entirely made up of marine plankton. There is very little terrestrial input from the arid hinterland; this is a feature of many upwelling coastlines at the present day.

Not all upwelling zones develop anoxic conditions. Oxic examples include off SE Brazil, the northern Pacific and bordering Antarctica. The most likely reason for oxicity is that the upwelling is only seasonal.

A diagnostic feature of sediments deposited under upwelling currents is a distinctive mineral assemblage including phosphorites and uranium minerals, as in the Neogene deposits of the Californian basins.

Prediction of ancient upwelling zones depends on the accurate reconstruction of atmospheric circulation and palaeogeography. Occurrence of

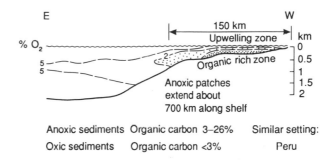

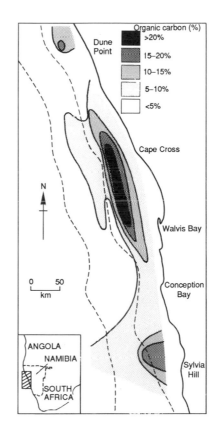

Fig. 10.12. Upwelling zone, offshore Namibia, showing oxygen-depleted zone, and total organic carbon concentrations of up to 26 per cent (after Demaison and Moore 1980). Upwelling causes high phytoplankton productivity in surface waters. The sea bottom sediments are anoxic under the highly productive waters.

upwelling means significantly improved chances of source bed presence. Examples of sourced beds thought to have been deposited as a result of upwelling include the Permian Phosphoria Formation of western USA, the Triassic Shublick Formation of the North Slope, Alaska, the Cretaceous La Luna Formation of Venezuela/Colombia, the Upper Cretaceous Brown Limestone of Egypt and the Miocene Monterey Formation of California.

The oceanic midwater oxygen minimum layer A layer of oxygen-deficient water occurs in the world's oceans at depths of about 100 to 1000 metres. It is caused by the degradation of organic matter that has fallen from the overlying highly productive photic zone. Below this midwater zone, oxygen contents rise again because of the influence of cold, dense currents which originate in the polar regions and sweep along the ocean floors towards tropical latitudes. These currents supply oxygen which prevents the midwater oxygen minimum layer becoming anoxic, except in rare examples. The present-day Atlantic Ocean is well oxygenated because there is virtually no obstruction to the passage of cold polar waters from both its northern and southern ends. In contrast, in the eastern Pacific and northern Indian Oceans, oxygen levels drop to less than 0.5 ml/l. Ocean floor currents reaching these areas have lost much of their oxygen, and the midwater oxygen minimum layer is free to intensify. TOC's in these areas range up to 11 per cent. Figure 10.13 shows midwater oxygen concentrations and Total Organic Carbon concentrations in bottom sediment of the Indian Ocean.

The presence of strong ocean currents derived from the poles is a relatively recent feature of the world's oceans. At the present day, the Earth is in an interglacial phase. At times in the past, for example during most of the Mesozoic (particularly during the Late Jurassic and Mid-Cretaceous), the global climate was warmer, and the shape of the world's oceans was quite different from that today. At this time, oceanic circulation may have been much more sluggish, and an intense midwater oxygen minimum zone may have developed. The Mid-Cretaceous 'global anoxic events' probably occurred under such circumstances.

During times of high sea level, the midwater oxygen minimum zone may impinge on the continental shelf over wide areas. As a result, source beds are deposited on the continental shelf in association with reservoir and carrier beds, and a hydrocarbon play may be produced. Oxygen deficiency could be reinforced in bathymetric depressions in the broad epicontinental seas produced at this time. Sediments deposited in midwater anoxic zones may be finely laminated, unbioturbated, organic-rich, diatomaceous mudstones (Gulf of California) or olive grey muds (northern Indian Ocean).

Examples of source rocks thought to have been deposited as a result of impingement of the midwater oxygen minimum zone on the continental shelf include the Toarcian source rocks of Western Europe.

The optimum conditions for marine source bed deposition occur when one of the mechanisms is reinforced by another. For example, a silled geometry may combine with the midwater oxygen minimum zone, or the midwater oxygen minimum may be intensified by upwelling.

Petroleum type

The gas or oil-proneness of a marine source rock depends primarily on the presence or absence of gas-prone terrigenous plant material. Enclosed marine basins close to a major clastic source may be gas-prone. Oil-prone organic matter of truly marine origin occurs in upwelling zones offshore from arid land areas.

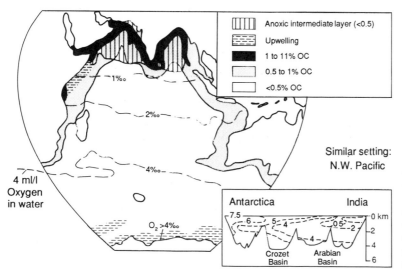

Fig. 10.13. Midwater oxygen concentrations and total organic carbon (TOC) concentrations in the Indian Ocean. TOCs are the highest around the Indian coastline where oxygen levels in the midwater layer fall to less than 0.5 ml/l and impinge on the continental slope and shelf.

10.3.1.3 Detection and measurement of source rocks

A range of geochemical techniques have been developed to identify and measure source rocks and petroleum fluids. These techniques may be used initially to establish simply whether a source rock exists in the sediments being sampled, but a series of other important questions about the petroleum charge system operating in an area may also be addressed. The richness of a source rock and the petroleum type or composition likely to be expelled may be determined. The thermal maturity may also be measured. Source rocks and fluids may be correlated geochemically, so that migration routes can be interpreted.

An understanding of these parameters is necessary for efficient exploitation of most known plays, and for the recognition of new conceptual plays.

Attempts have been made to identify mature source rock horizons on petrophysical wireline logs, such as the gamma ray, sonic and resistivity logs. Detection generally depends on the occurrence of non-conductive petroleum in the pore space of the mature source rock, which makes it abnormally resistive, or on the overpressure that tends to be created by actively generating source rocks, which causes abnormally long sonic transit times. Source rocks have also been known to be abnormally radioactive compared to surrounding non-source shales, and may therefore be detected on gamma ray logs. The Kimmeridgian 'hot' shale of the North Sea is an example. There are numerous pitfalls, however, in the identification of source rocks on wireline logs, and potential source horizons should always be confirmed where possible by correlation with geochemical indicators.

The following is a brief description of some routine geochemical and visual microscopic measurements on source rocks. First, however, it is necessary to understand the nature of the petroleum-bearing matter contained in source rocks, and the nature of petroleum itself.

We have seen that petroleum is derived mainly from lipid-rich organic material buried in sediments. Most of this organic matter is in a form known as *kerogen. Kerogen* is that part of the organic matter in a rock that is insoluble in common organic solvents. It owes its insolubility to its large molecular size. Different types of kerogen can be identified, each with different concentrations of the five primary elements, carbon, hydrogen, oxygen, nitrogen and sulphur, and each with a different potential for generating petroleum.

The organic content of a rock that is extractable with organic solvents is known as *bitumen.* It normally forms a small proportion of the total organic carbon in a rock. Bitumen forms largely as a result of the breaking of chemical bonds in kerogen as temperature rises.

Petroleum is the organic substance recovered from wells and found in natural seepages. Bitumen becomes petroleum at some point during migration. Important chemical differences often exist between source rock extracts (bitumen) and crude oils (petroleum).

Crude oil is naturally occurring petroleum in a liquid form. The term *black oil* is sometimes used to indicate petroleum that is liquid at both reservoir and surface temperatures and pressures.

Natural gas is petroleum occurring in the gaseous phase. *Wet gas* is differentiated from *dry gas* in that it yields significant volumes of liquid (*condensate*) on changing from reservoir to surface conditions. When the condensate yield is potentially high, the fluid is called a *gas-condensate.*

Under conditions of very low temperature and high pressure, *gas hydrates* may form. These are solid crystalline structures, usually containing methane. Methane hydrates may be found in arctic permafrost regions but also under the deep sea floor even in tropical latitudes.

Natural gas resulting from the thermal breakdown of kerogen is known as *thermogenic. Biogenic gas,* however, is a natural gas formed solely as a result of bacterial activity in the early stages of diagenesis (up to 60 or 70 °C). It normally occurs at shallow depths, and is always very dry.

TOTAL ORGANIC CARBON (TOC)

TOC is a measure of the carbon present in a rock in the form of both kerogen and bitumen. Typical values of TOC for different lithologies are shown in Table 10.3. Shales tend to be more rich in organic matter than carbonates. TOC values in source rocks may be quite low, and are frequently less than 2 per cent. Coals, however, may have TOCs of over 50 per cent. 0.5 per cent TOC is frequently taken as the minimum organic content for a shale source rock; a slightly lower value applies for carbonates. Below 0.5 per cent TOC, not enough

Table 10.3 Average values of hydrocarbons, non-hydrocarbon bitumen, and *organic carbon* in ancient nonreservoir sediments (from Tissot and Welte, 1978)

Origin	Type of rock	Number of samples	Extractable bitumen (ppm)		Organic carbon (%)	Average HC Average org. C (mg g)	Author
			HC	non HC			
200 formations from 60 sedimentary basins	shales	791	300	600	1.65	18	Hunt (1961)
	carbonates	281	340	400	0.18	151	Hunt (1961)
Rocks of various origins	shales	very large	180		0.9	20	Vassoevich *et al.* (1967)
	silts		90		0.45	20	Vassoevich *et al.* (1967)
	carbonates		100		0.2	50	Vassoevich *et al.* (1967)
Source rocks from 18 sedimentary basins	all types	668	860	780	1.82	47	Institut Français du Pérole (unpublished)
	shales and silts	418	930	810	2.16	43	
	calcareous shales	97	1260	1220	1.90	66	
	carbonates	118	335	440	0.67	50	
9 source rocks formation	not specified	not specified	267 to 2360		0.53 to 3.67	34 to 101	Philippi (1965)

Note: Relatively low (0.5%) concentrations of organic carbon may be sufficient to make a source rock, if it is of the right type. In this sample, shale source rocks are more organic-rich, on average, than carbonate source rocks, and the average organic content of all types is 1.82 per cent.

petroleum can possibly be generated to saturate the source rock; saturation must take place before expulsion can occur. Rocks with greater than 0.5 per cent TOC are not, however, guaranteed as source rocks. If the organic carbon is inert, no amount of it will form a source rock.

AMOUNT OF SOLUBLE EXTRACT (BITUMEN)

The soluble extract of source rocks reflects oil content, and hence varies strongly with maturity. Figure 10.14 shows its variation with maturity for source rocks in four different basins; in each case it is strongly correlated with the onset of oil generation.

ROCK PYROLYSIS

Espitalié developed a standard procedure for the pyrolysis of rock samples, known as Rock-Eval pyrolysis. About 100 mg of finely ground rock sample is placed into a furnace at 250 °C in an inert atmosphere and then raised to a temperature of 550 °C. The amount of hydrocarbon products

evolved is recorded by a flame ionization detector (FID) as a function of time.

Three peaks are typically recorded, known as the S_1, S_2 and S_3 peaks (Fig. 10.15). The S_1 peak represents hydrocarbons evolved at low temperatures – these represent free or adsorbed hydrocarbons (bitumen) that were already present in the rock before pyrolysis. The S_2 peak is produced at higher temperatures by the thermal breakdown of kerogen. Oxygen-bearing volatile compounds (carbon dioxide and water) are passed to a separate (thermal conductivity) detector, which produces an S_3 response. S_1, S_2 and S_3 are expressed as milligrams per gram of original rock, (mg g^{-1}) or kilograms per tonne (kg t^{-1}).

The temperature at which the S_2 generation peak occurs is also recorded, and is an indicator of source maturity.

Rocks with ($S_1 + S_2$) values of less than 2 kg t^{-1} are considered as insignificant source rocks. Between 2 kg t^{-1} and at 5 kg t^{-1} a significant amount of petroleum may be generated in the source

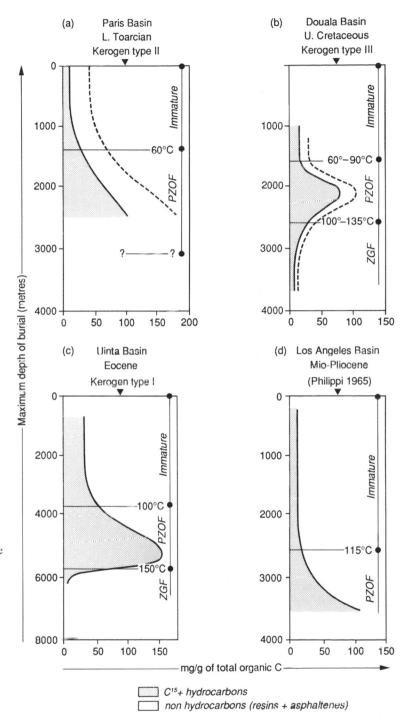

Fig. 10.14. Formation of hydrocarbons and non-hydrocarbons (resins and asphaltenes containing N, S, O) as a function of burial depth in different basins. For each basin, the temperatures that mark the upper and lower limits of the main oil generation zone (PZOF) are marked. Bitumen content clearly rises as the oil generation zone is entered, and diminishes as the underlying gas zone (ZGF) is approached. Bitumen content represents the oil content that can be dissolved from a rock, and this shows a strong correlation with its thermal maturity.

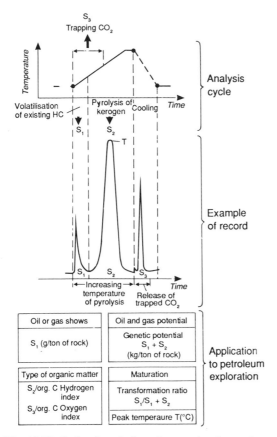

Fig. 10.15. Cycle of analysis and example of record obtained by the pyrolysis method of Espitalié *et al.* (1977) (after Tissot and Welte 1978). Three peaks are normally observed. The S_1 peak represents already-existing bitumen. S_2 represents hydrocarbons generated from the thermal breakdown of kerogen. The S_3 response is produced by oxygen-bearing compounds released at high temperature. S_1 and S_2 can be used to assess oil-generating potential, whereas S_2 and S_3 can be used to calculate *hydrogen index* and *oxygen index*, which indicate kerogen type. The temperature corresponding to the S_2 peak, and the $S_1/(S_1 + S_2)$ ratio indicate the level of thermal maturation.

rock but it may be too small to result in expulsion. If the source rock is raised to higher maturity, generated oil may be cracked to gas and expelled in the gas phase.

Source rocks with potentials of 5 to $10 \, \text{kg} \, \text{t}^{-1}$ have the potential to expel a proportion of their generated oil. Source rocks with greater than $10 \, \text{kg} \, \text{t}^{-1}$ are considered rich; oil generated will almost certainly be in sufficient quantities to ensure expulsion. Exceptionally, yields of several hundred $\text{kg} \, \text{t}^{-1}$ are measured; these are usually from coals or oil shales.

The *type* of kerogen in the source rock is indicated by the *hydrogen index*, which combines the S_2 pyrolysis peak with TOC:

$$\text{hydrogen index} = \frac{S_2 \times 100}{\text{TOC}} \, \text{mg} \, \text{g}^{-1} \, {}^\circ\text{C}^{-1} \quad (10.1)$$

It expresses the 'usable' or pyrolysable fraction of the organic content. What is left is inert carbon, which is incapable of sourcing petroleum.

Hydrogen indices of less than 50 imply that the kerogen is made up predominantly of inert kerogen. Values of greater than 200 suggest the presence of significant amounts of hydrogen-rich (oil prone) kerogen. The hydrogen index may be as high as 900 in strongly oil prone oil shales. The ratio of S_3 to TOC is known as the oxygen index. Tissot and Welte (1978) cross-plot hydrogen index and oxygen index in order to classify source rocks into three types: I, II and III (Fig. 10.16). Each has different petroleum generating characteristics.

GAS-CHROMATOGRAPHY

Gas-chromatography is a technique which separates the individual petroleum compounds in a petroleum mixture, according to increasing carbon number. An example of a gas-chromatogram is shown in Fig. 10.17. Gas chromatography can be performed on the products from pyrolysis (pyrolysates), on soluble extracts or on crude oil samples. From pyrolysis gas-chromatography (PGC), the broadest application is in estimating the oil-versus gas proneness of the kerogen. This can be determined by dividing the pyrolysate into gas ($C_1–C_5$) and liquid (C_{6+}) fractions. At a greater level of refinement, the gas chromatogram determines the detailed composition of the fluid for use in rock-rock, rock-oil, and oil-oil correlations.

VISUAL KEROGEN DESCRIPTIONS

The size, shape, structure and colour of kerogen fragments, once isolated from the rock, can be microscopically examined in *transmitted light*. Kerogen that contains weakly translucent to opaque,

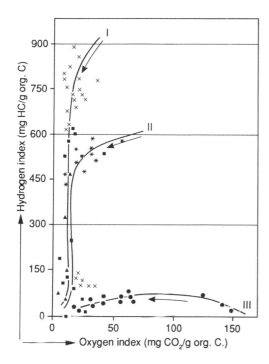

Fig. 10.16. Classification of kerogen types using hydrogen and oxygen indices. The diagram is readily comparable with the van Krevelen diagram plotted from elemental analyses of kerogen (Espitalié *et al.* 1977). Each kerogen type has different hydrocarbon-generating characteristics. Type I is the most oil prone whereas Type III produces mostly gas (after Tissot and Welte 1978).

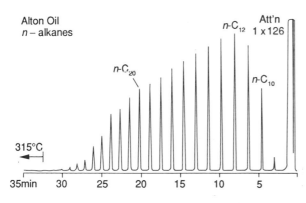

Fig. 10.17. Gas chromatogram of *n*-alkanes of an Australian crude (from Matthews *et al.* 1971). Each *n*-alkane compound in the oil is identified by a peak in the gas chromatogram.

structured fragments recognizable as higher plants is sometimes referred to as *humic*, while translucent amorphous kerogen is called *sapropelic*.

Translucent algae, spores, cuticles, pollen and resin bodies may also be recognized on the basis of internal structure, shape and colour. The colour of spores, pollen and other microfossils has been found to be broadly related to thermal maturity: spore colour changes from yellow, through orange to brown and eventually black, as thermal maturity increases.

Humic kerogen was formerly equated with gas prone source rocks, and sapropelic kerogen with oil prone source rocks, but this correlation is now known to be exceedingly unreliable. Certainly, not all sapropelic material is oil prone.

Examination of kerogen particles in *reflected light* allows the definition of three major groups. In order of increasing reflectance, these are:
• the *exinite* group
• the *vitrinite* group
• the *inertinite* group.

These groups were described in Section 10.3.1.1
Fig. 10.18 shows reflectance measurements on the organic particles in a mid-Liassic shale, illustrating the differing reflectances of the three main maceral groups.

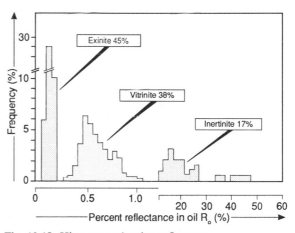

Fig. 10.18. Histograms showing reflectance measurements on different groups of kerogen particles. The exinite, vitrinite and inertinite groups each have different ranges of reflectance. The reflectance of vitrinite is used as an index of thermal maturity (Ch. 9). Middle Liassic shale, Luxembourg, modified by Tissot and Welte (1978) from Hagemann (1974).

The *fluorescence* of organic particles under ultra-violet light allows the identification of the liptinite group, which fluoresce strongly. The vitrinite and inertinite groups do not usually fluoresce.

VITRINITE REFLECTANCE MEASUREMENTS

Vitrinite reflectance is the most widely used indicator of source rock maturity. It is denoted by R_o(reflectance in oil). R_o values measured on the different vitrinite particles present in a sample tend to vary widely (Fig. 10.18). To ensure reliable results, a reasonably large number of determinations (e.g. 20) need to be carried out on the same sample, and the mean calculated ($\overline{R_o}$). Where a distribution of vitrinite reflectances is strongly bimodal, reworking of the higher reflectance group has probably taken place.

The vitrinite reflectance scale has been calibrated by other maturity parameters and by field studies in oil and gas provinces, so that R_o may be correlated with the main zones and thresholds of petroleum generation as follows:

$R_o < 0.55$	Immature
$0.55 < R_o < 0.80$	Oil and gas generation
$0.80 < R_o < 1.0$	Cracking of oil to gas
(gas condensate zone)	
$1.0 < R_o < 2.5$	Dry gas generation

Vitrinite reflectance is a very good maturity indicator above about 0.7 or 0.8 per cent R_o.

Processes operating during the thermal maturation of kerogen will be further discussed in Section 10.3.2.2.

An important use of vitrinite reflectance measurements in basin analysis is in calibrating thermal and burial history models with present-day maturity data. This was described in Chapter 9.

10.3.2 The petroleum charge

Summary

A petroleum charge occurs when petroleum is generated in a source rock, is expelled, and migrates through a carrier bed to a trap.

Petroleum is chemically a mixture of saturated and aromatic hydrocarbons and NSO (nitrogen–sulphur–oxygen) compounds. A wide range of geochemical analyses can be carried out on petroleum and source rock extracts that aim to relate the petroleum back to its original source rock and

depositional environment. Important physical properties of petroleum are its density, formation volume factors and boiling points; these influence secondary migration processes, subsurface volume changes and phase behaviour.

Petroleum generation takes place as a result of the chemical breakdown of kerogen with rising temperature. As hydrocarbons are released, the remaining kerogen evolves towards a carbon residue. Temperature and time are the most important factors affecting the breakdown of kerogen. The rate of breakdown can be calculated from the Arrhenius equation given in Chapter 9. The reactive fraction of kerogen can be subdivided into a labile portion, which yields chiefly oil, and a refractory portion, which yields mainly gas. Labile kerogen breaks down over approximately the 100 to 150 °C range, followed by refractory kerogen from 150 to 220 °C. Over the 150 to 180 °C range, oil is rapidly cracked to gas. Thus a stage of oil generation is succeeded by a stage of wet gas/gas-condensate generation, and finally by a stage of dry gas generation. *Petroleum expulsion* is probably caused by microfracturing of the source rock after overpressure has built up as a result of hydrocarbon generation. Lean source rocks may not generate sufficient oil to cause expulsion. If raised to higher maturity, generated oil may be cracked to gas which will be efficiently expelled. For rich source rocks (>5 kg t^{-1}) efficiency of oil expulsion may be quite high (60 to 90 per cent). Source rocks can be classified into three types on the basis of their richness and expulsion products.

Secondary migration carries petroleum from the site of expulsion through porous and permeable carrier beds to sites of accumulation (traps) or seepage. The main driving force behind secondary migration is buoyancy, caused by the density difference between oil (or gas) and formation pore waters. The main restricting force is capillary pressure, which increases as pore sizes become smaller. During secondary migration, petroleum flows as slugs through the interconnected network of largest pores in the carrier bed, rather than sweeping its whole volume. Movement is stopped when a smaller pore system is encountered whose capillary pressure exceeds the upward-directed buoyancy of the petroleum column. This pore system constitutes a seal. The maximum petroleum column height that can be supported by a seal can be calculated.

Petroleum will tend to move in the true dip direction of the top of the carrier bed. Thus structural contour maps can be used to model migration pathways (orthocontours). During long-distance migration, for example in some foreland basins, petroleum flow may be strongly focused along regional highs. Losses of petroleum during secondary migration are difficult to quantify.

Finally, petroleum may be physically and chemically altered while it is in the trap by the processes of biodegradation, water-washing, de-asphalting and thermal alteration.

10.3.2.1 Some chemical and physical properties of petroleum

In order to understand petroleum generation, expulsion and migration, and the chemical changes that may take place in the trap, we need to know a little more about the chemical and physical properties of petroleum. The following discussion is very basic and very brief. Kinghorn (1983), Hunt (1979) and Tissot and Welte (1978) have provided excellent texts on the geochemistry of petroleum.

Hydrocarbons are compounds made up solely of hydrogen and carbon. Petroleum is usually a mixture of hydrocarbon compounds and other compounds containing additional substantial amounts of nitrogen, sulphur and oxygen, and other minor elements.

There are three main groups of compounds found in petroleum:
1 *Saturated hydrocarbons*
2 *Aromatic hydrocarbons*
3 *NSO compounds (resins and asphaltenes).*

Saturated hydrocarbons are compounds in which each carbon atom is completely saturated with respect to hydrogen. Structures include simple straight chains of carbon atoms (the normal paraffins or normal alkanes), branched chains (the iso-alkanes), and rings (cyclic hydrocarbons). Methane and ethane are examples of simple normal alkanes, and are commonly referred to as C_1 and C_2. For all normal alkanes, the subscript after the C refers to the number of carbon atoms in the molecule.

The *aromatic* hydrocarbons are a group of unsaturated hydrocarbons with cyclic structures, and include several important biomarker compounds that allow oils and source rocks to be correlated.

NSO compounds contain atoms other than carbon and hydrogen, predominantly nitrogen (N), sulphur (S) and oxygen (O). They are known as heterocompounds and are subdivided into the *resins* and the *asphaltenes*.

Figure 10.19 shows the gross composition of 636 crude oils in terms of their saturates, aromatics and NSO contents. Average compositions are shown in Table 10.4.

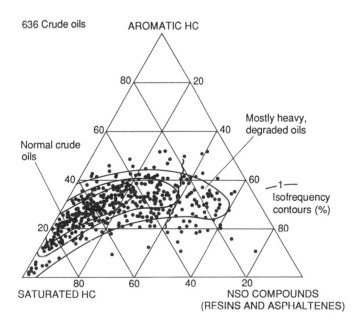

Fig. 10.19. Gross composition of 636 crude oils (from Tissot and Welte 1978) in terms of the three main groups of compounds found in petroleum – saturates, aromatics and NSO compounds. Normal (non-degraded) crudes typically contain 60–80 % saturates, and less than 20 per cent NSO compounds.

Table 10.4 Average composition of hydrocarbons (wt. %) for a large number of crude oils (number of samples in brackets) (Tissot and Welte 1978, p. 342)

	Normal producible crude oils (517)	All crude oils including tars (size 141)	Disseminated bitumen (668)
n + iso-Alkanes	33.3	31.7	27.7
Cyclo-alkanes	31.9	32.1	29.3
Aromatics	34.5	36.2	43.0
Saturated: aromatics*	2.8	2.7	1.8
Alkanes: saturated*	0.49	0.48	0.47

*Average value of the ratio

The composition of source rock bitumens tends to be different from crude oils – they contain fewer aromatic and saturated hydrocarbons and more resins and asphaltenes. These differences are probably due to important chemical changes that appear to take place during migration.

Biomarkers are compounds found in crude oils and source rock extracts that can be unmistakably traced back to living organisms. Mackenzie (1984) gives a recent account of this topic. The nature of the biological input to a sediment, and the chemistry of the depositional environment, give source rock extracts and expelled oils a characteristic 'finger-print', which is superimposed by the effects of diagenesis and maturation. These fingerprints may be geochemically recognized. Biomarker compounds may be used to assess thermal maturity (at low levels), and to correlate oils with source rock extracts.

Carbon isotope (δC_{13}) values of crude oils and source rock extracts may be used to distinguish marine from freshwater/terrestrial sources, and biogenic from thermogenic gases. Figure 10.20 shows δC_{13} values for various sources of organic (and inorganic) carbon.

Crude oils may be *classified* to enable specific oil types to be directly related back to their source

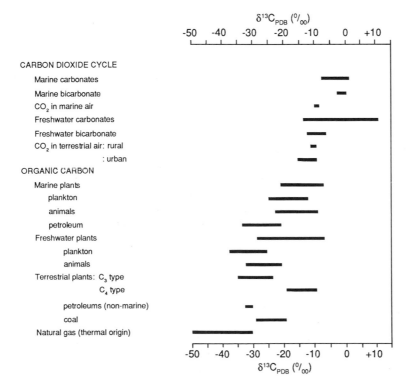

Fig. 10.20. Ranges of $\delta^{13}C$ values for various sources of organic and inorganic carbon versus PDB standard (after Waples 1981). Carbon isotope values can be used to distinguish marine from freshwater/terrestrial sources, and biogenic from thermogenic gas.

rocks. Such classification schemes use parameters such as oil density, sulphur content, metals content, wax content, carbon isotope value and pristane/phytane ratio. High sulphur contents (>1 per cent) indicate marine sources, high wax content indicates land plant or freshwater algae sources, and high pristane/phytane ratios (>3) indicate land plant material as the source.

Oil density is normally quoted as an API gravity

$$\text{where } °API = \frac{141.5}{\text{Specific gravity at 60 }°F} - 131.5 \tag{10.2}$$

A fluid with a specific gravity of $1.0\,\text{g cm}^{-3}$ has an API gravity of 10 degrees. *Heavy oils* are those with API gravities of less than 20 (sp. gr. >0.93). These oils have frequently suffered chemical alteration as a result of microbial attack (biodegradation) and other effects. Not only are heavy oils less valuable commercially, but they are considerably more difficult to extract. API gravities of 20 to 40 degrees (sp. gr. 0.83 to 0.93) indicate *normal oils*. Oils of API gravity greater than 40 (sp. gr. <0.83) are *light*.

At surface conditions, normal oils are clearly less dense than water. However, under subsurface conditions, this density difference is much greater. Oil has a great capacity to contain dissolved gas at elevated temperatures – as a result, subsurface oil densities are typically in the 0.5 to $0.9\,\text{g cm}^{-3}$ range. Subsurface pore water densities, in contrast, are typically 1.0 to $1.2\,\text{g cm}^{-3}$, and largely dependent on salinity. This density difference is the main driving force behind the secondary migration of petroleum.

Gas densities vary markedly between surface and subsurface conditions. At atmospheric pressure, the density of methane (C_1) is only $0.0003\,\text{g cm}^{-3}$, but at subsurface pressures of 5000 psi (equivalent to depths of 3 to 4 km), typical natural gas mixtures have densities of approximately 0.2 to $0.4\,\text{g cm}^{-3}$. In the subsurface, therefore, oils and gases take on more similar physical properties. Methane is the lightest of the hydrocarbon gases, and is normally the most abundant. Dry gases typically have methane concentrations of over 95 per cent.

Formation volume factors relate the subsurface volumes of oil or gas to the volumes occupied at surface under standard conditions of temperature and pressure (60 °F and 14.7 psia). These factors must be estimated in order to calculate the potential recoverable reserves of an exploration prospect.

Gas expands enormously on release of subsurface pressure, and may occupy several hundred times its subsurface volume at surface. The relationship can be calculated by:

$$\text{Gas expansion factor} = \frac{(P_r)\,(T_s)\,Z}{(T_r)\,(P_s)} \tag{10.3}$$

where

P_r = pressure at reservoir (subsurface) conditions (psia)

P_s = pressure at standard (surface) conditions (psia)

T_r = temperature at reservoir conditions (°Rankine = °F + 460)

T_s = temperature at standard conditions (°Rankine)

Z = gas deviation factor (frequently close to 1.0)

Oil shrinks on movement to the surface, owing to the release of dissolved gas. Oil shrinkage factors vary from nearly 1.0 for shallow oils with almost no dissolved gas, to 2.0 or more for extremely gassy oil in deep reservoirs.

The boiling point of a petroleum compound is the temperature above which it is in a vapour state. At temperatures below its boiling point, the compound is in a liquid state. C_1 (methane) to C_4 (butane) are the only hydrocarbon compounds that are gases (vapours) at surface temperature and pressure. The other compounds are liquids.

Phase changes may take place in the subsurface during the processes of hydrocarbon generation, migration and entrapment. These may be important in prospect and play assessment. Liquids may condense out of a petroleum vapour as it migrates through a carrier bed into areas of lower pressure. These valuable liquids may be lost as a residual oil saturation in the pores of the carrier bed. If an oil accumulation is uplifted as a result of, say, inversion tectonics or thrust fold belt tectonics, large quantities of gas may be exsolved from the oil. This may cause displacement of oil from the trap.

10.3.2.2 Petroleum generation

Chemical changes to kerogen during source rock maturation

At shallow depths of burial of only a few hundred metres kerogen remains relatively stable. At greater depths of burial, however, under conditions of higher temperature and pressure, it becomes unstable and rearrangements take place in its structure in order to maintain thermodynamic equilibrium. Structures which prevent the parallel arrangement of cyclic nuclei are progressively eliminated. By this process, a wide range of compounds are generated (heteroatomic compounds, hydrocarbons, carbon dioxide, water, hydrogen sulphide, etc.), as the kerogen evolves towards a highly ordered graphite structure. Petroleum generation is, therefore, a natural consequence of the adjustment of kerogen to conditions of increased temperature and pressure.

Kerogen is a complex macromolecule composed of nuclei linked by heteroatomic bonds or carbon chains that are successively broken as temperature increases. As breakdown occurs, the first products released are heavy heteroatomic compounds, carbon dioxide and water. These are followed by progressively smaller molecules, including hydrocarbons. The kerogen left behind becomes progressively more aromatic and evolves towards a carbon residue.

Mackenzie and Quigley (1988) classify kerogen into *reactive kerogen* and *inert kerogen* (Fig. 10.21). Reactive kerogen is transformed into petroleum at elevated temperatures, whereas inert kerogen rearranges towards graphite-like structures without the generation of petroleum. Reactive kerogen is subdivided into a *labile* portion which is transformed into petroleum that is chiefly oil at surface, and a refractory portion that generates chiefly gas.

Kinetic models of kerogen breakdown

We have seen in Section 9.2 that temperature and time are the most important factors in controlling the maturation of organic matter. The complex series of consecutive reactions that cause kerogen breakdown proceed at varying rates, governed primarily by temperature and the activation energy of the particular reaction, and expressed in the *Arrhenius equation* (equation (9.1), Section 9.2).

If the constants in the Arrhenius equation are known for a particular petroleum-forming reaction, the rate at which it will proceed may be determined as a function of temperature, and the temperature range over which the bulk of the reaction will take place (before the raw material is used up) can be calculated. Masses of petroleum generated as a result of kerogen breakdown can also be calculated as a function of temperature, and hence, if the

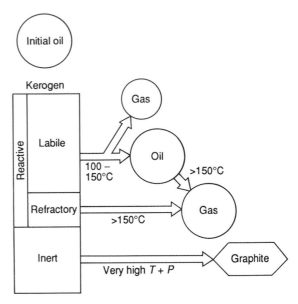

Fig. 10.21. Classification and fate of organic matter in source rocks. Kerogen is divided into reactive and inert portions. Inert kerogen rearranges towards graphite-like structures at very high temperatures (*T*) and pressures (*P*) without generating petroleum. Reactive kerogen is subdivided into a refractory part that yields mainly gas, and a labile portion that is transformed into petroleum, which is chiefly oil at the surface. Initial oil corresponds to bitumen normally present in immature source rocks. Relative amounts of initial oil, labile, refractory and inert kerogen are determined by the nature of the precursory organisms and the depositional setting of the host source rock (after Mackenzie and Quigley 1988).

subsidence and thermal history of an area are known, as a function of geological time.

The activation energies of each individual reaction are not known, but for each kerogen type a distribution of activation energies can be established from laboratory and field studies. These distributions, together with the other parameters in the Arrhenius equation, can be used to model kerogen breakdown and the associated generation of petroleum products for each kerogen type.

Figure 10.22 shows the diminishing concentrations of kerogen with increasing temperature for a range of heating rates, calculated from the kinetic model of Mackenzie and Quigley (1988). Separate sets of curves are shown for labile and refractory kerogen. The heating rate parameter incorporates the time factor in source rock maturation, which is known to be important from the occurrence of

(a)

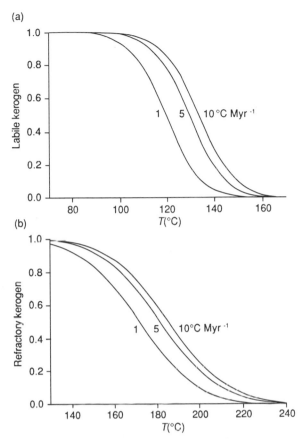

(b)

Fig. 10.22. Calculated concentrations of reactive labile (a) and refractory (b) kerogens, relative to initial amount of reactive kerogen, as a function of maximum temperature for the range of heating rates shown. Mean heating rates of about 0.5 °C Myr^{-1} occur in old stretched basins. Mean heating rates of about 10 °C to 50 °C Myr^{-1} occur in young (<25 Myr) stretched basins. Labile kerogen breaks down generally over the 100 °C to 150 °C range. Refractory kerogen, however, breaks down at much higher temperatures, from about 150 °C to 220 °C. After Mackenzie and Quigley (1988).

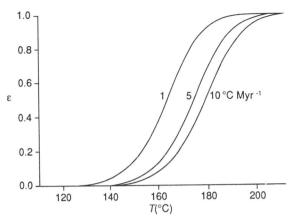

Fig. 10.23. Predicted fraction of generated oil that has been cracked to gas (ε = gas/[oil + gas]) as a function of maximum temperature for a range of geological heating rates. At 180 °C, for heating rates of less than 10 °C Myr^{-1}, over 60 per cent of generated oil would have been cracked to gas. Cracking takes place mainly over the 150 °C to 180 °C range (after Mackenzie and Quigley 1988).

generally shallower oil generation thresholds in older basins (Dow, 1977). Depending on heating rate, kerogen breakdown into petroleum takes place largely over the 100 to 150 °C range for labile kerogen. For refractory kerogen, the range is approximately 150 to 220 °C.

Any oil left in a source rock from the breakdown of labile kerogen will be cracked to gas if temperatures continue to rise above 150 °C. According to

Mackenzie and Quigley (1988), most cracking reactions take place over the 150 to 180 °C range. Figure 10.23 shows the fraction of generated oil that is predicted to be cracked to gas as a function of temperature, for a range of heating rates. An alternate presentation is shown in Fig. 10.24, which shows the time required to crack half the mass of oil to gas, if held at a constant temperature. At 180 °C, an oil has a half-life of less than a million years. At temperatures above 160 °C, oil would not be expected to exist for 'geological' periods of time. The cracking process applies, of course, to oil accumulations as well as to oil remaining in source rocks. As a result, oil fields will not exist at depths greater than those corresponding approximately to the 160 °C isotherm.

The kinetic model predicts that the following stages of petroleum formation succeed each other without significant overlap:

1 Immature stage preceding petroleum generation (the *diagenesis stage* of Tissot and Welte, 1978).

2 Stage of oil and gas generation from labile kerogen containing lipid material (exinitic macerals).

3 Stage of wet gas/gas-condensate generation as a result of the cracking of previously generated oil. Stages 2 and 3 make up the *catagenesis* stage of Tissot and Welte.

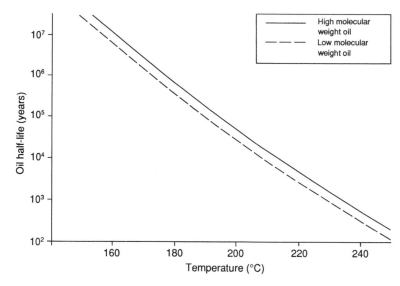

Fig. 10.24. Oil half-life as a function of temperature. At 180 °C, oil has a half-life of less than 1 million years. At temperatures above 160 °C, oil would not be expected to exist for 'geological' periods of time.

4 Stage of dry gas generation from refractory (vitrinitic) kerogen. This stage is called the *metagenesis stage* by Tissot and Welte (1978). Methane is the main petroleum product.

A general scheme of petroleum generation is shown in Fig. 10.25. Figure 10.26 shows the chemical evolution of each of Tissot and Welte's kerogen types during petroleum generation.

Some heavy heteroatomic (NSO) compounds, together with carbon dioxide and water, are generated in the immature (diagenesis) stage of kerogen evolution, but there is effectively no hydrocarbon generation. Hydrocarbons present in rocks of this maturity are inherited from their precursor organisms (biomarkers, or geochemical fossils) and have not been generated from kerogen.

During the main zone of oil formation, hydrocarbon compounds (normal and iso-alkanes, cyclo-alkanes and aromatics) are generated; the proportions of each depends on kerogen type. As maturity increases, low molecular weight hydrocarbons become most abundant, until only methane is present.

10.3.2.3 Expulsion from the source rock

Mechanics

Expulsion is also known as primary migration.

As a result of the compaction of source rocks during burial, pore sizes may become smaller than the size of some petroleum molecules (Fig. 10.27). This presents a difficulty in explaining how petroleum migrates out of the source rock.

Of all the mechanisms of primary migration debated in the geological literature, the most likely appears to be as a discrete phase through microfractures caused by the release of overpressure. The cause of the overpressure in the source rock may be a combination of oil or gas generation, fluid expansion on temperature increase, compaction of sealed source rock units, or release of water on clay mineral dehydration.

The conversion of kerogen to petroleum results in a significant volume increase. This causes a pore pressure build-up in the source rock. The pressure build-up is sometimes large enough to result in microfracturing. This releases pressure, and allows the migration of petroleum out of the source rock and into adjoining carrier beds, from which point secondary migration processes take over. Cycles of petroleum generation, pressure build-up, microfracturing, petroleum migration and pressure release continue until the source rock is exhausted.

The implication of this theory is that mature source rocks will always expel petroleum as long as they are rich enough. In this sense, primary migration is not a major concern for the practising petroleum geologist. Primary migration may clearly take place both upwards and/or downwards out of the source beds, as governed by local pressure gradients.

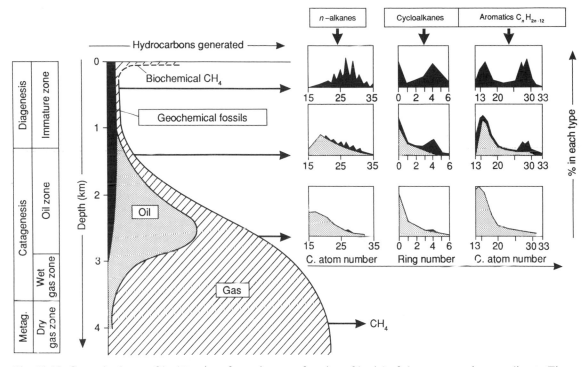

Fig. 10.25. General scheme of hydrocarbon formation as a function of burial of the source rock, according to Tissot and Welte (1978). As formation temperature rises on progressive burial, an immature stage is succeeded by stages of oil generation, oil cracking (wet gas stage) and finally dry gas generation. Typical distributions of *n*-alkanes, cycloalkanes and aromatics at three points in this general evolution are shown.

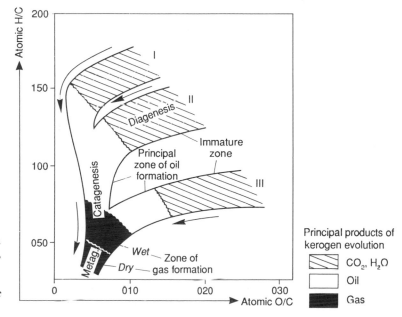

Fig. 10.26. General scheme of kerogen evolution presented on van Krevelen's diagram. The diagenesis, catagenesis and metagenesis stages are indicated and the principal products generated during that time are shown (after Tissot 1973).

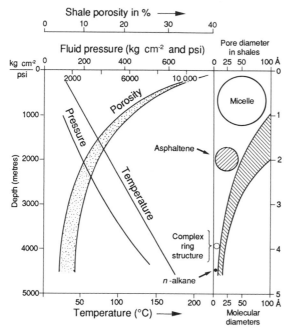

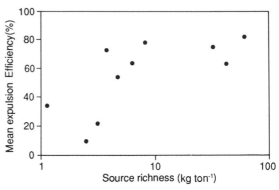

Fig. 10.28. Mean petroleum expulsion efficiency as a function of initial petroleum potential for 10 source rocks (from Cooles, Mackenzie and Quigley 1986). These data indicate that for rich source rocks (>5 kg ton^{-1}), very efficient expulsion of oil may take place from the source rock. Lean source rocks (<5 kg ton^{-1}) are characterized by inefficient expulsion. Oil remaining in the source rock may, however, subsequently be expelled as gas, provided sufficiently high temperatures for oil cracking are achieved.

Fig. 10.27. Interrelationship of various physical parameters with increasing depth of burial for shale-type sediments, showing shale pore diameters in relation to the molecular diameters of the petroleum. At moderate depths of burial, shale pore diameters typically become very small in relation to the larger petroleum molecules.

A large volume expansion takes place when petroleum liquids are cracked to gas within the source rock. A lean oil prone source rock may not generate sufficient hydrocarbons to cause micro-fracturing. As a result, no expulsion will occur. If raised to higher maturity, however, the oil that has remained in the source rock will be cracked to gas. The resulting volume increase and overpressure may allow expulsion to occur. Thus, lean oil prone source rocks tend to expel gas-condensate once they are raised to sufficiently high maturity.

Efficiency of expulsion

How much of the generated (plus initial) petroleum is likely to be expelled from the source rock? Cooles, Mackenzie and Quigley (1986) have shown that, between 120 and 150 °C, petroleum expulsion efficiency is strongly dependent on the original richness of the source rock. Figure 10.28 shows that for some rich source rocks (potential greater than 5 kg ton^{-1}, TOC > 1.5 per cent) oil expulsion may be very efficient with 60 to 90 per cent of the total petroleum generated being expelled. There is a lag, however, between petroleum generation and petroleum expulsion. It appears that a certain minimum petroleum saturation (probably about 40 per cent) in the source rock is required before efficient expulsion takes place. In leaner source rocks (<5 kg ton^{-1}, <1.5 per cent TOC) expulsion efficiency is much lower, and most of the oil generated remains in the source rock. As we have seen, if raised to higher maturity, it may be cracked to gas and expelled. Expulsion appears to be very efficient for gas or gas-condensate, irrespective of original source richness.

Mackenzie and Quigley (1988) have classified source rocks into three end-member classes on the basis of initial kerogen concentration and kerogen type. These parameters determine the timing and composition of the petroleum expelled. These are illustrated in Fig. 10.29 for the three source rock classes, for a mean heating rate of 5 °C per million years. On Fig. 10.29, Petroleum Generation Index (PGI) is the fraction of petroleum prone organic matter that has been transformed into petroleum, and is thus a measure of source maturity. Petroleum Expulsion Efficiency (PEE) is the fraction of

Fig. 10.29. Petroleum Generation Index (PGI) and Petroleum Expulsion Efficiency (PEE) as a function of maximum temperature for three classes of source rock, according to Mackenzie and Quigley (1988). Principal petroleum phases expelled over relevant temperature ranges are shown. Curves were constructed assuming a mean heating rate of 5 °C Myr^{-1}. PGI is the fraction of petroleum-prone organic matter that has been transformed into petroleum. PEE is the fraction of petroleum fluids generated in the source rock that have been expelled. Class I are rich source rocks containing mainly labile kerogen. Class II are lean source rocks comprising labile kerogen. Class III source rocks contain mostly refractory kerogen.

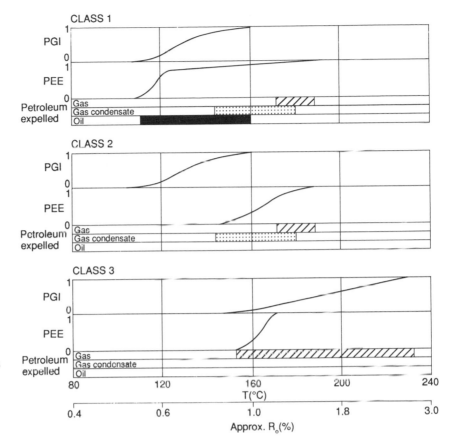

petroleum fluids formed in the source rock that have been expelled.

A *Class 1* source rock has predominantly labile kerogen at concentrations of greater than 10 kg ton^{-1}. Generation starts at about 100 °C as the labile kerogen generates an oil-rich fluid. This rapidly saturates the source rock, and between 120 and 150 °C, 60 to 90 per cent of the petroleum is expelled as oil with dissolved gas. The remaining fluid is cracked to gas at higher temperatures and expelled as a gas phase initially rich in dissolved condensate. Examples of Class 1 source rocks are the North Sea Kimmeridge Clay, and the Bakken Shale of the Williston Basin.

Class 2 source rocks are a leaner version of Class 1, with initial kerogen concentrations of less than 5 kg ton^{-1}. Expulsion is very inefficient up to 150 °C because insufficient oil-rich petroleum is generated. Petroleum is expelled mainly as gas-

condensate formed by cracking above 150 °C, followed by some dry gas.

Class 3 source rocks contain mostly refractory kerogen. Generation and expulsion takes place only above 150 °C, and the petroleum fluid is a relatively dry gas.

Some formations contain mixtures of different source rock classes.

10.3.2.4 Secondary migration: through carrier bed to trap

Introduction

Secondary migration concentrates subsurface petroleum into specific sites (traps) where it may be commercially extracted. The main difference between primary migration (out of the source rock) and secondary migration (through the carrier bed) is the porosity, permeability and pore size distribution of the rock through which migration takes place. These parameters are all much higher for carrier beds. As a result, the mechanics of migration may be quite different.

The end points of secondary migration are the trap or seepage at surface. If a trap is disrupted at some time in its history, its accumulated petroleum may remigrate either into other traps, or leak to the surface. The same processes of secondary migration apply to the remigration as to the original migration into the trap.

A knowledge of the mechanics of secondary migration is important in the general understanding of active charge systems, but specifically in the following ways:
• in tracing and predicting migration pathways, hence in defining areas receiving a petroleum charge
• in interpreting the significance of subsurface petroleum shows and surface seepages
• in estimating seal capacity in both structural and stratigraphic traps.

The mechanics of secondary migration are now well studied and well understood (Hubbert 1953, Gussow 1954, Berg 1975, and Schowalter 1976). The following section describes:
• the *main driving forces* behind secondary migration. These are *buoyancy*, caused by the density difference between oil (or gas) and the pore waters of carrier beds, and *pore pressure gradients* which attempt to move all pore fluids (both water and petroleum) to areas of lower pressure. The latter is known as a *hydrodynamic* condition.
• the main *restricting forces* to secondary migration. This is *capillary pressure*, which increases as pore sizes become smaller. When capillary pressure exceeds the driving forces, entrapment occurs.

Driving forces in secondary migration

Buoyancy is a vertically directed force caused by the difference in pressure between some point in a continuous petroleum column and the adjacent pore water. It is a function of the density difference between the petroleum and the pore water, and the height of the petroleum column (Fig. 10.30):

Buoyant Force $\Delta P = Y_p g (\rho_w - \rho_p)$ (10.4)

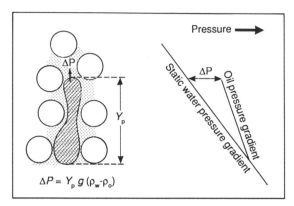

Fig. 10.30. Buoyancy as a driving force in secondary migration. Buoyancy is the pressure difference between a point in the petroleum column and the surrounding pore water. It is a function of the petroleum/water density difference and the height of the petroleum column. A large buoyancy pressure may develop at the tops of large, low density (gas) petroleum columns. Pressure measurements at points throughout the petroleum column define a petroleum pressure gradient; this intersects the hydrostatic gradient at the petroleum–water contact.

where Y_p = height of petroleum column
  g = acceleration due to gravity
  ρ_w = subsurface density of water
  ρ_p = subsurface density of petroleum

Under hydrostatic conditions, buoyancy is the only driving force in secondary migration. Under *hydrodynamic conditions*, however, (that is, when water flows through a carrier bed) the driving force is modified. Hydrodynamics may either assist or inhibit secondary migration, depending on whether it acts with or against the buoyancy force. Hydrodynamics may be important in the following respects:
1 by affecting the directions and rates of secondary migration
2 by increasing or decreasing the driving pressures against vertical or lateral seals, thus reducing or increasing the heights of the petroleum columns that the seals can withstand
3 by tilting petroleum water contacts and displacing petroleum accumulations (for example, off the crests of structural closures).

In connection with (1), secondary migration, hydrodynamics may be safely ignored except in basins where there is good evidence of hydrodynamics operating at the present day. Without this evidence, it is difficult to support a strong argument for hydrodynamics having operated during secondary migration. Geologically long-lasting hydrodynamic conditions are most likely to have existed in foreland basins (see also Section 9.3.3). In these basins, the modification of secondary migration directions and rates by hydrodynamics should be taken into account.

Restricting forces in secondary migration

When a petroleum globule or slug moves through the pores of a rock, work has to be done to distort the globule and squeeze it through the pore throats. The force required is called capillary pressure (or displacement, or injection pressure), and it is a function of the size (radius) of the pore throat, the interfacial surface tension between the water and the petroleum, and the wettability of the petroleum–water–rock system:

$$\text{Displacement pressure} = \frac{2\gamma \cos \Theta}{R} \qquad (10.5)$$

where γ = interfacial tension between petroleum and water (dyne cm^{-1})

Θ = wettability, expressed as the contact angle of the petroleum–water interface against the rock surface (degrees)

R = radius of the pore (cm)

These parameters are illustrated in Fig. 10.31, for a simple water-filled cylindrical pore. Higher pressures are needed to force petroleum globules through smaller pores.

Interfacial tension depends on the properties of the petroleum and water, and is independent of the rock characteristics. It is a function primarily of the composition of the petroleum (it is smaller for light, low viscosity oils), and temperature (interfacial tension generally decreases with increasing temperature). The effects of pressure and water chemistry are less important. For a given petroleum composition, therefore, interfacial tension may be considered effectively constant over large parts of the migration pathway, unless considerable vertical migration takes place.

Gas–water interfacial tensions are in fact generally higher than those for oil–water. This means that, for the same rock, displacement pressures are higher for gas than for oil. The buoyancy pressures, however, are normally greater for gas.

Wettability is a function of the petroleum, water and rock. Most rock surfaces are 'water-wet' and Θ may be taken to be zero. In carrier beds along secondary migration routes, and in lateral and vertical seals to petroleum accumulations, the cos Θ term in equation (10.5) can be ignored so that:

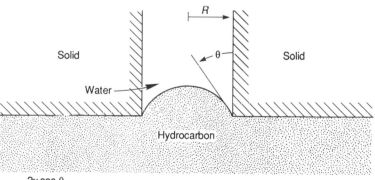

$$p_d = \frac{2\gamma \cos \theta}{R}$$

where p_d = displacement pressure

γ = oil – water interfacial tension

θ = contact angle of oil and water against the solid

R = radius of the pore throat

As γ increases p_d increases

As θ decreases p_d increases

As R decreases p_d increases

Fig. 10.31. Resistant forces in secondary hydrocarbon migration. Higher pressures are needed to force petroleum globules through smaller pores (after Purcell 1949 in Schowalter 1976).

$$\text{Displacement pressure} = \frac{2\gamma}{R} \qquad (10.6)$$

Some of the grains of oil-filled reservoir rocks may be oil-wet, and organic-rich source rocks may also be partly oil-wet. Displacement pressures in these cases could, as a result, be considerably smaller than in water-wet rocks. This would assist oil migration.

Pore sizes are the most important control on secondary migration and entrapment. Pore sizes can be estimated visually (in thin section, or by scanning electron microscope, for example) in reservoir/carrier bed lithologies. Ideally, displacement pressure can be measured directly by mercury injection techniques for both reservoir and potential sealing lithologies. The principle of this technique is that a non-wetting fluid (mercury) is injected into a core plug and its saturation as a percentage of pore volume (cumulative volume of mercury injected) is recorded as a function of steadily increasing injection pressure. Figure 10.32 is an example of a mercury capillary pressure test. The pressure at which mercury first begins to saturate the pores of the rock is the displacement pressure. This takes place when the *largest* pores are invaded. Thereafter, at higher pressures, the smaller pores are successively invaded. Mercury injection pressure can be easily converted to petroleum–water displacement pressure.

Once the displacement pressure has been overcome, and a connected petroleum slug is established in the largest pores of the rock, secondary migration may take place. One of the interpretations made from mercury capillary test data is that the petroleum saturation required to produce this connected petroleum slug is surprisingly small. In a series of experiments reported by Schowalter (1976), this critical saturation for a range of rock types varied from 4.5 to 17 per cent, and averaged 10 per cent. Active secondary migration pathways may therefore be characterized by petroleum saturations of only about 10 per cent. Such low saturations provide weak shows that frequently go undetected or are considered of no significance.

Since it is the network of largest pores that controls displacement pressure (and hence secondary migration and seal potential), care should be taken that the rock material analysed is representative of the carrier bed or potential seal as a whole. If larger pore systems exist in the carrier bed

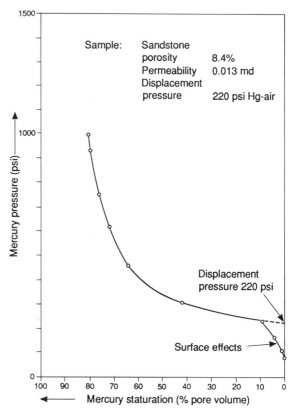

Fig. 10.32. Typical mercury capillary pressure curve. Mercury first begins to saturate the pores of the rock when the largest pores are invaded (the displacement pressure). A relatively small further increase in pressure commonly results in rapid saturation of the pore space and the establishment of a connected petroleum slug which may migrate through the rock. As little as 10 per cent oil saturation may be required before secondary migration takes place (after Schowalter 1976).

or potential seal than were analysed in the injection tests, displacement pressures may be seriously overestimated.

So far we have considered only the entry of petroleum into a pore network from an infinite body, as illustrated in Fig. 10.31. Within the pore network of a rock, the pore throat radii at both upper and lower ends of the oil globule need to be considered, and the capillary pressure equation is modified to:

$$\text{Capillary pressure} = 2\gamma\left(\frac{1}{r_\text{t}} - \frac{1}{r_\text{b}}\right) \qquad (10.7)$$

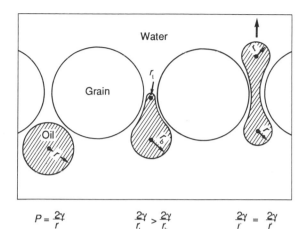

$$P = \frac{2\gamma}{r} \qquad \frac{2\gamma}{r_t} > \frac{2\gamma}{r_b} \qquad \frac{2\gamma}{r} = \frac{2\gamma}{r}$$

Fig. 10.33. Transport of an oil globule through pore throats in a water-wet subsurface environment. Capillary pressure opposes the buoyant force until the radius of curvature inside the distorted oil globule is equal at its lower and upper ends. The oil globule may then pass through the pore throat. (After Berg 1975, in Tissot and Welte 1978).

where r_t = pore radius of upper pore throat
r_b = pore radius of lower pore throat

Figure 10.33 illustrates the conditions required for the transport of an oil globule through the pore throats of a porous rock.

Petroleum column heights and seal potential

Once a petroleum slug has entered a pore system of a constant size it will continue to move. Its rate of movement is governed by the driving forces and the permeability of the rock. When a smaller pore system is encountered, the driving forces may not be sufficient to overcome the increased capillary pressure. In this case, movement into the smaller pore network will not take place. The slug will either migrate away laterally (in a dipping carrier bed), using the larger pore system, or remain trapped. If joined by a large number of other petroleum slugs, a sufficient vertical column of petroleum may build up to provide a large buoyant force which is enough to cause invasion of the finer pore network. Thus, a seal may be effective only up to a critical petroleum column height, at which point it leaks.

The critical petroleum column height (Y_{pc}) can be calculated by:

$$Y_{pc} = 2\gamma / \left(\frac{1}{r_t} - \frac{1}{r_b} \right) g\, (\rho_w - \rho_p) \qquad (10.8)$$

When the pore size of a sandstone reservoir is very large in relation to the very small pore sizes of a shale caprock, the $1/r_b$ term tends towards zero. The equation may then be simplified to:

$$Y_{pc} = \frac{2\gamma}{rg\,(\rho_w - \rho_p)} \qquad (10.9)$$

Since the subsurface density of gas is less than that of oil, it is clear that seals can support much larger oil columns than gas columns. This has important implications for migration and entrapment. For example, it may prevent the formation of gas caps overlying oil columns. It also suggests that gas should migrate vertically more successfully than oil.

In order to calculate seal potential, we need to know the pore radius that is relevant to leakage. This should be the smallest pore throat in a network of large pores that, if penetrated, will allow an interconnected petroleum slug to be established. This is clearly not the largest pore throat, nor the smallest, but somewhere in between! It may be approximated by a mean hydraulic radius, r_h, where

$$r_h = 2.8\,(K/\phi) \qquad (10.10)$$

where K is permeability and ϕ is porosity.

The difficulty of estimating the critical pore radius of the caprock renders seal capacity calculations subject to wide ranges of error. They may frequently give only order of magnitude estimates.

Faults and fractures

Fault zones can act as both conduits and barriers to secondary migration. The material crushed by the frictional movement of the fault, the fault gouge, is frequently impermeable and does not allow the passage of petroleum. Clay smeared along fault planes, as in the growth faults of the Niger delta (Weber *et al.* 1978) and in the Louisiana Gulf Coast region (Smith 1980), also blocks petroleum migration. Fractures formed in either the footwall or hangingwall, if they remain open, may form effective vertical migration pathways. This is unlikely except at shallow depths, but may occur in the uplifted hangingwalls of contractional (thrust) faults on release of compressive stresses. Tensional fractures in the crestal zones of anticlinal structures may also allow migration of petroleum.

Lateral migration will tend to be inhibited by the presence of faults, since they interrupt the lateral continuity of the carrier bed.

The sealing qualities of fault zones are discussed further in Section 10.6 on petroleum traps.

Migration pathways

Since the driving force behind secondary migration (in the absence of hydrodynamics) is buoyancy, it is clear that petroleum will tend to move in a homogeneous carrier bed in the direction which has the steepest slope. This is perpendicular to its structural contours, that is, in the true dip direction. Lines drawn at right angles to the structural contours of the top carrier bed/base seal horizon are known as *orthocontours*. Orthocontour maps illustrate the focusing and de-focusing effects of structural features in prospect drainage areas (Fig. 10.34). When lateral migration is long-distance, as for example in foreland basins, where prospects may be remote from areas of mature source rock (source 'kitchens'), these effects may strongly influence the pattern of hydrocarbon charge. It is important in play assessment to recognize those parts of the fairway that are located on petroleum migration routes. A petroleum flow may be split when encountering a low, and concentrated along regional highs. The geometry of the kitchen also affects petroleum charge volumes; prospects located close to the ends of strongly elongate source kitchens will receive relatively little charge.

It is important that orthocontour maps are constructed for the actual time of secondary migration. Present-day structure maps may be used to model present-day migration. Isopaching (or 3-D decompaction, Section 8.2) allows the production of palaeostructure maps for use in modelling palaeomigration routes. For example, an isopach of the base Upper Cretaceous to base Miocene interval will allow the modelling of migration routes in a carrier bed situated at the base of the Upper Cretaceous sequence at the beginning of Miocene time.

Other factors should also be considered, such as:
• sealing faults, which may deflect petroleum flow laterally

• non-sealing faults which allow petroleum to flow across the fault plane into juxtaposed permeable units at a different stratigraphic level. From this point a different structure map needs to be used for migration modelling
• communication between carrier beds caused by lateral stratigraphic changes (e.g. by the sanding-out of a shale seal). The orthocontour map should be constructed only as far as a seal persists.

These factors affect the likelihood of petroleum charge into specific segments of the play fairway, and into specific prospects within them.

Secondary migration losses

Volume losses occur along secondary migration pathways. These losses are in two distinct habitats:
• In miniature traps – dead-ends along the migration route – produced by faulted and dip-closed geometries, and stratigraphic changes. The traps may be observable but of no commercial interest, or they may not be observable at all, for example, if they are below the resolution of the seismic tool.
• As a residual petroleum saturation in the pores of the carrier rock, trapped by capillary forces in dead-end pores and absorbed onto rock surfaces. This may represent up to 30 per cent of the pore volume through which the petroleum migrates (and a greater percentage of the hydrocarbon saturation achieved during active migration), so major losses may occur in this way. Losses are minimized when petroleum flows through a relatively small volume of carrier rock. This is achieved in high permeability strata where migration is rapid and takes place without large petroleum columns building up.

The petroleum volumes expelled, lost and trapped can be related by the following simple equation:

$$V_{\text{expelled}} = V_{\text{lost}} + V_{\text{trapped}} \qquad (10.11)$$

The aim of play assessment and prospect evaluation is to estimate V_{trapped}. Volumes expelled can be calculated after geochemical source rock evaluation, taking into account source rock richnesses, thicknesses, source rock kitchen sizes and maturities, and expulsion efficiencies. However, at the basin or play scale, volumes lost are almost impossible to quantify. Moreover, they are likely to be

(a)

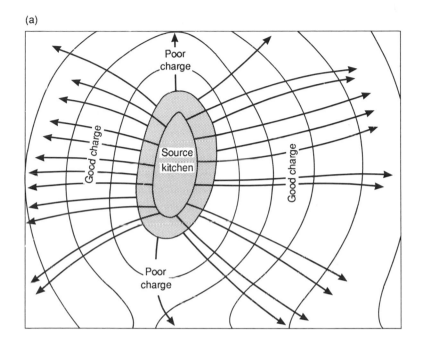

(b)

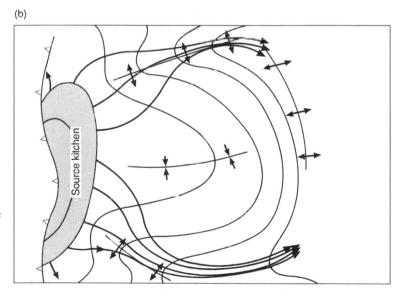

Fig. 10.34. Orthocontours illustrating secondary migration routes.
(a) Effect of an elongate kitchen on migration directions. Areas along the long axis of the kitchen may receive a relatively poor charge.
(b) Migration is focused along regional highs in the basin drainage area, and may penetrate large distances away from the source kitchens. Some foreland basins show examples of this type of migration.

very large in relation to volumes trapped. As a result, it is very difficult to estimate the volumes trapped in a play fairway through geochemical source rock volumetrics.

It is conceivable that a prospect may lie beyond the 'migration front' of oil generated from a source kitchen. It will not receive a charge, since all of the oil expelled into the carrier bed is lost as a residual saturation (and in small traps) in the carrier bed. In order to determine the position of the migration front, accurate calculations of the volumes of petroleum expelled from the kitchen and the rate of loss in the carrier bed need to be made. The errors involved are huge and it would be most unwise for a prospect lying just beyond the calculated migration front to be severely downgraded for this reason. The calculations may, however, give useful order-of-magnitude estimates that assist risk estimation.

The focusing of oil migration into specific flow routes, which sweep through only a very small volume of carrier rock, is probably a very important contributing factor in enabling huge oil volumes to migrate very large distances (several hundred kilometres) in some foreland basins. The heavy oil/asphalt belts on the gentle flanks of these basins (e.g. western Canada) have formed at enormous distances from their source kitchens.

10.3.2.5 Alteration of petroleum

Changes may take place in the physical and chemical properties of petroleum while it is in the trap. These changes may have an important impact on the recoverable fraction and commercial value of an oil accumulation. Their causes are considered under the following four main headings.

1 Biodegradation

Biodegradation is the bacterial alteration of crude oils. Bacteria use any dissolved oxygen present in formation pore waters, or derive oxygen from sulphate ions, in order to selectively oxidize hydrocarbons. Firstly, the light normal-alkanes are removed, followed by branched (iso) alkanes, cycloalkanes and finally the aromatics. The physical effect of biodegradation is to increase the density and viscosity of the oil.

Biodegradation appears to take place only at temperatures of less than 60 to 70 °C. It also appears to require a supply of meteoric water containing dissolved oxygen and nutrients (primarily nitrates and phosphates). These conditions are frequently met in foreland basins, where meteoric water enters the carrier bed/reservoir system in the bordering uplifted thrust belts. Biodegradation may take place in both the thrust belt and in the gentle foreland flank. An example of the former is the Napo Basin, Ecuador; the Athabaska Tar Sands are an example of the latter. Biodegradation may

also occur at the oil–water contacts of petroleum accumulations, resulting in the formation of tar mats, as in the Burgan field of Kuwait.

2 Water washing

Water washing commonly accompanies biodegradation. Hydrocarbon-undersaturated meteoric waters may dissolve some hydrocarbons from a reservoired petroleum mixture. Light alkanes and low boiling point aromatics (e.g. benzene, toluene and zylene) are the most soluble and preferentially removed. The net result is a change in composition similar to that caused by biodegradation. Water washing takes place at temperatures greater than 70 °C. The only requirement is a continued flow of meteoric water.

In assessing the chances of encountering biodegraded or water washed petroleum in a prospect or play, it is necessary to consider the history of fluid movement in the basin since the time of migration through to the present day. Although *present-day* reservoir temperatures may be greater than 70 °C, this is not a guarantee that biodegradation has not taken place. The effects of strong convective meteoric water flow on geothermal gradients must also be taken into account when reconstructing the thermal history of the petroleum.

3 Deasphalting

Deasphalting is a process whereby the precipitation of the heavy asphaltene compounds in a crude oil takes place as a result of the injection of light $C_1 - C_6$ hydrocarbons. This may occur when an oil accumulation experiences a later gas charge as its source kitchen becomes highly mature.

4 Thermal alteration

The variation in petroleum composition that takes place with increasing thermal maturity of the source rock were described in the Section 10.3.2.2. Similar compositional changes take place in a reservoired petroleum with rising temperature. Heavy compounds are replaced by progressively lighter ones, until only methane is present. At high temperatures (greater than 160 °C), oil cracking reactions proceed so rapidly that an oil accumulation may be destroyed within a geologically short period of time.

It is important when assessing exploration plays to block out parts of the fairway that are considered to be susceptible to the petroleum alteration processes described above.

10.4 THE RESERVOIR

Summary

The primary considerations in the assessment of reservoir potential are the likely reservoir porosity and permeability. Porosity and permeability are influenced by the depositional pore-geometrics of the reservoir sediment and the post-depositional diagenetic changes that take place. The porosity types of carbonate rocks include vuggy (pores larger than grains), intergranular (between grains), intragranular or cellular (within grains) and chalky. Diagenetic changes are extremely important, such as dolomitization, fracturing, dissolution, recrystallization and cementation. Sandstone reservoirs have a depositional porosity and permeability controlled by grain size, sorting and packing of the particulate sediment. Diagenetic changes include the authigenesis of clay minerals in the pore space.

Reservoirs are heterogeneous on a number of scales from the km-scale first-order heterogeneities of stratigraphic packages to the microscopic grain scale. Reservoir heterogeneity is the main concern of the development geologist.

The tectonic setting of a basin may determine the composition of clastic reservoirs and therefore their quality. The evaluation of tectonic setting from sandstone composition is termed *provenance* studies. If basin analysis indicates the likely tectonic make-up of a basin, the composition and geometry of reservoir units can be estimated in very general terms.

A petroleum play is defined initially by the depositional or erosional limit of its gross reservoir unit. A reservoir rock must be porous enough to constitute a 'tank' of petroleum within the trap, and its pores must be sufficiently interconnected to allow the contained petroleum fluids to flow through the rock towards the wellbore. Thus, the primary considerations in the assessment of reservoir potential are the likely reservoir *porosity* and *permeability*.

Reservoir porosity affects the reserve of a prospect or play. Reservoir permeability affects the rate at which petroleum fluids may be drawn off from the reservoir during production. Both of these parameters have a large impact on the commercial attractiveness of an exploration or field development opportunity. Porosity and permeability are discussed in brief in Section 10.4.1. Reservoir rocks may result from deposition in a very wide range of environments. The reservoir rocks range from fractured pelagic limestones to highly porous aeolian sandstones. The depositional systems and facies models discussed in Chapter 7 have clear implications for the occurrence of reservoir rocks in depositional sequences. We do not repeat this information here. Section 10.6 highlights some of their stratigraphic trapping possibilities. Reservoir prediction in lightly explored play fairways, however, is still at a relatively primitive stage. Careful sequence-by-sequence interpretations of sedimentary facies, using available local data from outcrop and wells, integrated into a depositional model, and calibrated against analogous sequences elsewhere, is the best approach to reservoir prediction in these circumstances.

Diagenesis may have a large impact on reservoir quality, especially in carbonate sequences (brief overview in Section 10.4.1). In this book we do not cover the complex subject of mineral diagenesis in any detail: for more detailed sedimentological descriptions of common reservoir rock depositional environments, and of mineral diagenesis, readers are referred to texts such as Reading (1986), Walker (1984a), Bathurst (1975), Blatt (1982).

This section is therefore intended to give a broad overview of reservoir character from the point of view of its presence in a petroleum play.

10.4.1 Porosity and permeability

The pore volume of a sediment can be expressed either as an *absolute porosity* given by

$$\phi_a = \left(\frac{\text{bulk volume} - \text{solid volume}}{\text{bulk volume}} \right) 100 \quad (10.12)$$

or, as an *effective porosity*

$$\phi_e = \left(\frac{\text{interconnected pore volume}}{\text{bulk volume}}\right) 100 \quad (10.13)$$

Effective porosity is normally measured in studies of reservoirs.

Different rock types possess different pore geometries, carbonate rocks (Section 10.4.1.1) being very different to siliciclastic rocks (Section 10.4.1.2) in this respect.

Permeability or hydraulic conductivity measures the ability of a medium to transmit fluids and is defined according to the *Darcy equation* which states

$$Q = KA\,(\mathrm{d}P/\mathrm{d}l) \quad (10.14)$$

where Q is the volume of transmitted flow per unit time (flow rate), A is the cross-sectional area and $\mathrm{d}P/\mathrm{d}l$ is the pressure gradient over distance l, or hydraulic gradient. The value of permeability depends not only on rock properties, but also on the medium being transmitted. The *specific permeability*, k, is defined as

$$Q = \frac{kA\gamma}{\mu}\,(\mathrm{d}P/\mathrm{d}l) \quad (10.15)$$

where γ is the specific weight of the fluid and μ is its absolute viscosity; k is measured in Darcys.

The porosity and permeability therefore describe the 'plumbing' of the reservoir. Porosity and permeability are not, however, simply or directly related. Complex pore geometries may, for example, present highly tortuous paths for transmitted fluids with many dead-ends. This will lower permeability whilst porosity may be largely unaffected. Similarly, particular pore-filling mineral habits may have different effects on porosity and permeability: the stacked pseudohexagonal plates of kaolinite reduce porosity strongly but have a less severe effect on permeability, whereas the sheets or 'curtains' of illite have a highly detrimental effect on permeability and a smaller effect on porosity.

10.4.1.1 Carbonate reservoirs

Carbonate rocks are particularly amenable to the study of the relationship between pore geometry and permeability (e.g. Wardlaw and Cassan 1978, Jardine and Wilshart 1987).

Carbonate reservoirs are characterized by extremely heterogeneous porosity and permeability on a number of scales. These heterogeneities are dependent on the environment of deposition of the carbonate facies and, most importantly, on the subsequent diagenetic alteration of the original rock fabric. Here we concentrate on the pore-scale heterogeneities. Typical carbonate facies and facies models are briefly described in Sections 7.1.3.1 and 7.1.4.2.

The main porosity types are:

1 Vuggy (pores larger than grains).
2 Intergranular (between grains).
3 Intragranular or cellular (within grains, such as shell material).
4 Chalky.

Dolomitization produces vuggy or intercrystalline porosity. Burial commonly results in a *fracture porosity*. Nearshore sediments (sabkha sediments, calcarenite beach deposits) have a predominantly intergranular porosity. With dolomitization a micro-vuggy and fine intercrystalline porosity may be developed. Restricted platform sediments consist mainly of carbonate muds with some chalky porosity but very low permeability. Open platform sediments may be varied, and porosity types vuggy, intragranular and intergranular are common. Vuggy and intercrystalline porosity is commonly developed during dolomitization. Platform margin sediments are mainly sands and reefs, and excellent vuggy, intragranular and intergranular porosity is developed. Dolomitization leads to large vugs and medium to coarse intercrystalline pore networks. Foreslope sediments are highly variable. Basinal carbonate muds may have very high initial porosities but this is lost through burial and compaction. These generalizations are summarized in Fig. 10.35.

Diagenetic events leading to changes in porosity and permeability can be summarized under five headings:

• *Dissolution* (leaching) generally improves porosity and permeability.

• *Dolomitization* may improve porosity by creating larger pores, or may reduce it by the growth of interlocking mosaics of dolomite crystals. Dolomitization often increases permeability dramatically, due to the formation of solution vugs and greater degree of post-burial fracturing.

• *Fracturing* by brecciation, faulting and jointing greatly aids permeability.

• *Recrystallization* by aggrading neomorphism of micrite into larger crystal sizes enhances porosity.

(a)

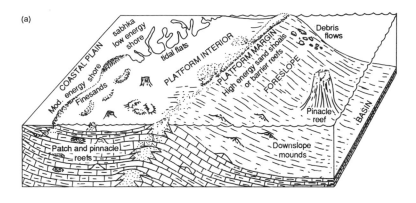

(b)

	Setting	Normal sediments	Associated sediments	Reservoir geometry	Grain size			Limestone φ				Dol. φ		Hydrocarbon significance		
					C	M	F	Vug.	Cell.	I.G.	Chk	Vug.	I.X.	Source	Res'voir	Seal
Platform interior	Near shore	Sands skeletals pellets oolites Muds Tidal flat laminites	Terrigenous clastics Evaporites	Ribbons Sheets										Fair	Good	Good
	Restricted	Muds Minor sands pellets skeletals	Terrigenous clastics Evaporites	Stacked lenses										Fair	Fair	Good
	Open	Sands skeletals pellets oolites Small platform reefs	Terrigenous clastics	Sheets Ribbons Small mounds										Poor	Good	Fair
	Platform margin	Sands oolites skeletals Organic reef	Terrigenous clastics Lime muds	Ribbons										Poor	Good	Poor
	Foreslope	Sands Muds Mud mounds Debris flows Pinnacle reefs	Fine terrigenous clastics Evaporites	Areally small mounds Sheets										Good	Good	Good
	Basin	Muds	Fine terrigenous clastics Chalks Salt	Sheets										Good	Fair	Good
	Submerged platform	Large atolls organic reef reef detritus sands muds laminites	Fine terrigenous clastics Lime muds	Large thick mounds										Good	Good	Good

Fig. 10.35. Synopsis of the main carbonate environments and deposits and their reservoir properties. Vug. = vuggy, cell. = cellular or intragranular, I. G. = intergranular, CHK = chalky, I.X. = intercrystalline. After Jardine and Wilshart (1987).

- *Cements* decrease porosity and permeability, the latter especially when they occur at pore throats.

Wardlaw and Cassan (1978) estimated the efficiency of recovery of hydrocarbons from the various pore types. In general, high intercrystalline porosities are most favourable since they have highly interconnected voids giving good permeability. Vuggy rocks may have rather low microscopic recovery efficiencies since the solution vugs may be only partially connected. The reader is referred to texts such as Bathurst (1975) for fuller treatments of carbonate facies and diagenesis.

10.4.1.2 Sandstone reservoirs

The primary porosity and permeability of sandstones are dependent on the grain size, sorting and packing of the particulate sediment (see summary in Pettijohn 1975, pp. 72–79) and are therefore easier to predict than in carbonate reservoirs.

The porosity of artificially packed natural sand is independent of grain size for sand of the same sorting, but porosity varies strongly with sorting. Average wet-packed porosity for a range of sorting groups varies between 28 per cent for very poorly sorted sand to over 42 per cent for extremely well-sorted sand.

Permeability of unconsolidated sand decreases as grain size becomes finer and as sorting becomes poorer. For wet-packed samples permeability varies over more than an order of magnitude between extremely well-sorted sands and very poorly sorted sands. Permeability varies over 2 orders of magnitude between very fine and coarse sand. The dual effects of sorting (expressed as standard deviation of grain size) and grain size (expressed as mean) on permeability is shown in Fig. 10.36 (after Krumbein and Monk 1942). The probable effect of low sphericity (grain shape) and high angularity (grain roundness) is to increase porosity and permeability of unconsolidated sand.

Grain size analyses of unconsolidated reservoir core samples of Miocene age were carried out by Morrow *et al.* (1969). A predictive scheme was then developed for estimating porosity and permeability from the grain size parameters. The core plugs were from clean, unconsolidated upper Miocene sand from a relatively narrow depth range of 2440 to 2890 m. Porosity and permeability were determined for these core plugs. The best correlation between porosity (ϕ) and grain size distribution was given by a plot of

$$\phi = \log [f(\text{kurtosis, fine-end skewness})]$$

That is, porosity increased as the peakedness (kurtosis) of the grain size distribution increased, but was independent of absolute particle size. For the reservoir core samples there was a definite

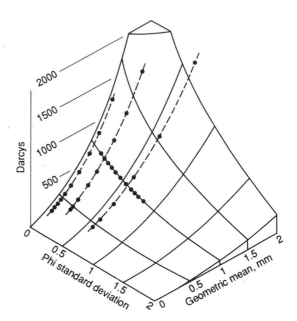

Fig. 10.36. Relation of permeability to grain size and sorting (after Krumbein and Monk 1942, Fig. 5). The vertical axis is the permeability in darcys. The grid lines on the 'permeability surface' are parabolas parallel to the Darcy-size plane, and negative exponentials parallel to the Darcy-standard deviation plane.

tendency for reduction in porosity with increase in the fractional weight of fine material.

The correlation for permeability was more complex. Permeability was found to be a function of particle size, peakedness of distribution (sorting) and fine-end skewness. Qualitatively this is reasonable since permeability is known to increase in coarser, better sorted, clay-poor sands.

Sand *shape fabric* also has some effect on permeability. Potter and Mast (1963) and Mast and Potter (1963) demonstrated that sand shape fabric is determined by palaeocurrent direction (defined by dip of cross-stratification or azimuth of primary current lineation). *Statistically*, the long axes of sand grains lie parallel to current direction and are imbricated up-current, generally at angles of 10° to 25°; a second mode existed of long axes orientated at right angles to the palaeocurrent, representing bedload rolling. Weber *et al.* (1972) conducted a field experiment in a Holocene distributary channel fill in the Netherlands and showed that a grain-alignment heterogeneity was indeed present.

Many siliciclastic reservoirs have a strong diagenetic overprinting that modifies the depositional porosities and permeabilities. Diagenesis invariably has a detrimental effect on reservoir porosity and permeability (see also Section 9.4.3). Many of the advances in studying diagenesis in siliciclastic reservoirs has come through the use of the scanning electron microscope (SEM), X-ray diffraction (XRD), electron microprobe analysis, cathodoluminescence and stable isotope analysis in addition to traditional petrography. Poor reservoir quality often results from the presence of authigenic clay minerals in the pore space (see Stalder (1975) for an early appreciation, Burley, Kantorowicz and Waugh (1985) for a recent review). It is beyond our scope to discuss the complexities of diagenetic processes and geochemistry.

Calculations of reservoir porosity (see Section 8.2.1) and qualitative indications of permeability can be obtained from interpretation of the wireline logs recorded in petroleum boreholes. Porosity and permeability can also be measured from core material. The reader is referred to publications by Schlumberger (1974) and Dresser-Atlas (1982) for manuals on the interpretation of lithology, porosity, clay content, fluid content and other parameters of a reservoir. Additionally, texts such as Tittman (1986) give an account of the physical basis of the routine logging measurements, and Asquith (1982) and Rider (1986) give a general text with many worked examples.

10.4.2 Basic ideas on reservoir heterogeneity

So far we have emphasized the factors influencing porosity and permeability on the grain to microscopic scale. All sedimentary deposits, however, have an inhomogeneity caused by the distribution in time and space of sedimentary facies ('architecture'), and by compaction, deformation, cementation and the nature of pore-filling fluids.

A classification system of reservoir heterogeneities can be based on their size, origin and influence on fluid flow (e.g. Weber 1986, p. 489, Pettijohn, Potter and Siever 1973) (Fig. 10.37). One of the major factors in any classification is the size of the heterogeneities, as illuminated in the scheme that follows:

1 *First-order heterogeneities* (1 to 10 km scale), including the presence of scaling faults and the boundaries of major genetic units of sediment such as fluviatile channel-belts.

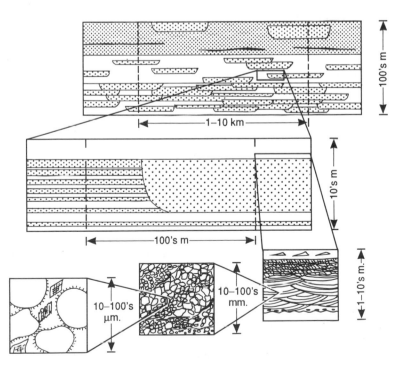

Fig. 10.37. Example of various levels of heterogeneity in a fluviatile reservoir from the kilometre scale down to the microscopic scale. (After Weber 1986).

2 *Second-order heterogeneities* (cm to tens of m scale) which represent variation in permeability with the larger genetic units. The interbedding of shales at channel margins, or the permeability contrasts caused by the coarse-fine stratification of a point bar are examples.

3 *Third-order heterogeneities* (mm to m scale) result from the geometrical arrangements of individual depositional units such as bedforms producing cross-stratification. The inclined foresets of cross-sets also produce a finer scale alternation of moderate and high permeabilities, whilst the toe-sets of bedforms introduce essentially horizontal low permeability 'breaks' or 'baffles' into the reservoir (Weber 1987). *Fractures* and *stylolites* also fall within this physical scale of heterogeneity.

4 *Fourth-order heterogeneities* (μm to mm scale) are represented by variations in grain size and sorting, and *microscopic heterogeneities* caused by the way in which pores are interconnected or blocked.

Large scale heterogeneities of well-spacing size can often be analysed on the basis of detailed well log correlations and by the use of sedimentological models derived from core descriptions. For smaller scale heterogeneities, cores are indispensable, since they provide information on bed thickness, style of cross-stratification, grain size and microscopic features. The correct identification of the environment of deposition of the sediment greatly helps in an assessment of heterogeneity.

A thorough study of the various scales of heterogeneity is essential to the efficient recovery of hydrocarbons from a reservoir but is of less direct concern to the basin analyst.

10.4.3 Provenance of reservoir sediment

The basin analyst is able to provide information on the plate tectonic setting of a basin and its subsidence and thermal history. It is therefore possible to predict the likely source areas or *provenance* of sediment for the basin.

Generally provenance studies involve the sampling of sedimentary rocks (surface exposure, cores, drill cuttings), the identification of the minerals present and their proportions, and the interpretation of the geology of the source region that yielded them. Dickinson (1980), Dickinson and Suczek (1979), Ingersoll and Suczek (1979), Dickinson and Valloni (1980) and Lash (1987) are examples of

this kind of approach. Petrographic components of the framework grains are plotted on a ternary diagram with apices representing quartz (Q), felspar (F) and lithic fragments (L). Variations on the standard QFL diagram are used by distinguishing between, for example, polycrystalline and other forms of quartz, or sedimentary versus volcanic rock fragments. These ternary diagrams are divided into fields representing distinct plate tectonic settings for the sandstone sample, but overlapping of some of the fields severely limits their uses. Mack (1984) gives a useful critique of the methodology of using QFL plots to determine tectonic environment.

The effects of *transport* of sediment on the resulting composition of a sandstone is critical to the correct interpretation of provenance. Sediments in deep sea trenches, for example, may be transported great distances along the ocean floor (Chough and Hesse 1976, Thornburg and Kulm 1987) and their composition more likely reflects the tectonic and compositional nature of the source area rather than that of the site of deposition in the trench (Velbel 1985, Lash 1985). A detailed palaeo-current and sedimentological analysis is necessary to reveal such effects. Transport in terrestrial systems invariably modifies the composition of the sand, but the effects differ markedly according to climatic zone and type of river system. Franzinelli and Potter (1983) provide an elegant study of the Amazon drainage system in this respect. Large river systems in humid and hot climates are optimal for the chemical weathering of unstable grains such as lithic fragments. As a result, the sand at the mouth of the Amazon is dominated by quartz, despite the fact that considerable proportions of lithic grains were contributed to the drainage system at the source (Fig. 10.38). The efficiency of the Amazon system can be contrasted with other terrestrial systems by comparing the composition of sands sampled from beaches around the South American continent (Potter, 1978, 1984). The beaches along the Andean coast have sands dominated by lithics (mostly plutonic and volcanic), the sands along the Brazilian coast and at the Amazon delta are dominated by quartz, whereas the sands of the Patagonian coast of Argentina, fed by moderate river systems in a cold climate, have intermediate compositions with considerable quantities of lithic fragments remaining (Fig. 10.39). In summary, these transport effects

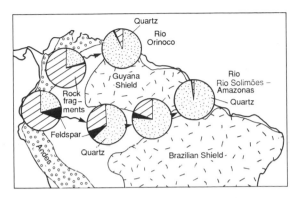

Fig. 10.38. Downstream increase in compositional maturity of sands in the Orinoco and Amazon river systems. The composition of the sands is determined by (a) the geology of the drainage area, Andean terranes providing many unstable rock fragments, and the Precambrian shields and Palaeozoic terranes providing sands dominated by quartz, and (b) transport processes in the humid, large river systems, whereby the unstable rock fragments are broken down by tropical weathering, so that sands near the mouths of the Orinoco and Amazon are quartz-rich, irrespective of source terrane (Franzinelli and Potter 1981).

must be integrated into any provenance study using discriminant plots of sandstone composition to interpret tectonic setting.

The likelihood of the presence of reservoir units of adequate quality in different basin types can be assessed from a knowledge of the hinterland geology and sediment dispersal systems (Section 7.1) (Kingston, Dishroon and Williams 1983 a, b). In broad outline, continental sags typically contain extensive shallow marine, fluviatile, aeolian and lacustrine reservoirs. Rifts may contain early areally restricted and volcanic-rich reservoir units of poor quality, and younger, more extensive, fluviatile, deltaic and shallow marine good quality reservoirs. Passive margins have very extensive shallow marine and deltaic sands or thick carbonate reservoirs. Strike-slip basins have a composition of sedimentary infill determined by the nature of the adjacent plates. Ocean–ocean contacts generally provide poor reservoirs because of contamination by pelagic and volcanogenic material; continent–ocean and continent–continent zones have more chance of providing sources of sand.

Fig. 10.39. The major mineral associations of beach sands sampled from around the South American continent (Potter 1984). The 'leading' (Andean) and 'trailing' (Atlantic) edges of the continent have markedly different sand compositions. The samples along Argentinian beaches are relatively rich in unstables because of the reduced chemical weathering of rock fragments in the cooler Patagonian climate compared to that of the Amazon–Orinoco systems in the north of the continent.

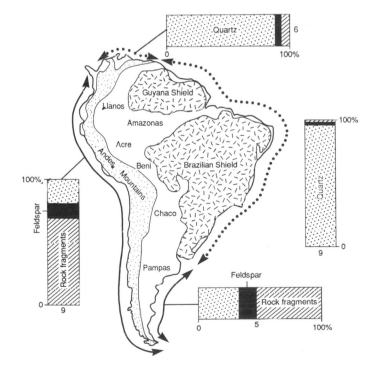

Forearc and trench sediments contain large amounts of volcanogenic material, and porosity and permeability are severely reduced during diagenesis. It must be emphasized, however, that these are broad generalizations and the practising basin analyst should in some sense treat every basin on its own merits after a collation of all available data.

10.5 THE REGIONAL TOPSEAL

Summary

The existence of a petroleum play depends on the presence of an effective regional caprock or topseal.

The basic physical principles governing the effectiveness of petroleum caprocks are the same as those that control secondary migration of petroleum. A caprock is effective if its capillary or displacement pressure exceeds the upward buoyancy pressure exerted by an underlying hydrocarbon column. The capillary pressure of the caprock is largely a function of its pore size. This may be laterally very variable.

The buoyancy pressure is determined by the density of the hydrocarbons and the hydrocarbon column height. A caprock of extremely small pore size is required to prevent the buoyant rise of a tall underlying gas column. Hydrodynamics also affect caprock effectiveness. Loss of gas through caprocks may take place through the process of diffusion.

A worldwide survey of caprocks indicates that the most effective caprock lithologies are fine-grained clastics and evaporites. Ductility is also an important requirement, particularly in tectonically disturbed areas. Salt and anhydrite are the most ductile, followed by organic-rich shales. Caprocks do not need to be thick to be effective, as long as they are laterally persistent. Similarly, depth of burial does not appear to be critical – seals may be effective at all depths.

A good example of a regional caprock is the Upper Jurassic Hith Anhydrite in the Arabian Gulf area, which seals in excess of 100 billion barrels of oil in the underlying Arab reservoirs.

The conditions required for the development of regionally extensive effective caprocks in association with reservoir rocks are frequently met in two particular depositional settings. One of these is where marine shales transgress over gently sloping clastic shelves. An example is the Miocene Telisa Formation shales of the Central Sumatra basin. The other is where evaporites in regressive sabkhas regress over shallow marine carbonate reservoirs, as in the case of the Hith Anhydrite of the Arabian Gulf.

10.5.1 Introduction

The regional caprock or topseal is one of the three essential ingredients of the petroleum play. The nature of the caprock determines not only the efficiency of the subsurface trapping system, but influences the migration routes taken by petroleum fluids on leaving the petroleum source rock. The continuity of the regional topseal largely determines whether the basin has a laterally- or vertically-focused migration system.

Compared to the enormous literature devoted to reservoir rocks, and, over the last twenty years, the massive effort devoted to understanding the geochemistry of source rocks, virtually nothing has been written on caprocks. Downey (1984) and Grunau (1987), however, provide important recent reviews.

10.5.2 The mechanics of sealing

The basic physical principles governing the effectiveness of petroleum caprocks are the same as those controlling secondary migration, which are discussed in Section 10.3.2.4. Only a very brief reminder is given here. The physical principles themselves are relatively well understood, though much has still to be learned about seals in general. Of the papers written on the mechanics of secondary migration and sealing, Schowalter's (1976) is the most recent.

In Section 10.3.2.4, we divided the forces which control secondary migration into those that drive secondary migration, and those that restrict it. The main driving force is *buoyancy*, caused by the fact that petroleum fluids are generally less dense than formation pore waters.

The main restricting force to the movement of a globule or slug of petroleum through a porous rock is its *capillary* or *displacement pressure.* This depends primarily on the size (radius) of the pore throats. A rock will seal an underlying petroleum accumulation if the displacement pressure of its *largest* pore throats equals or exceeds the upwardly directed buoyancy pressure of the petroleum column (Fig. 10.40). The seal potential or capacity of a caprock

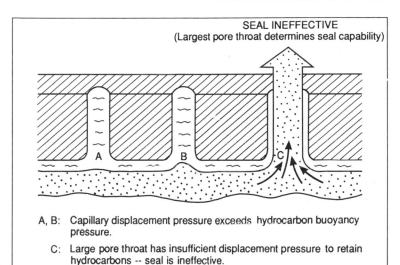

Fig. 10.40. Diagrammatic illustration of the effect of the largest pore throat size on sealing capacity of caprocks (modified from Downey 1984).

A, B: Capillary displacement pressure exceeds hydrocarbon buoyancy pressure.

C: Large pore throat has insufficient displacement pressure to retain hydrocarbons -- seal is ineffective.

can be expressed as the maximum petroleum column height that it will support without leakage. Owing to subsurface density differences of oil and gas, caprocks can support much larger oil columns than gas columns, other things being equal.

The displacement pressure of a piece of caprock can be measured directly in the laboratory by mercury-injection techniques, or can be estimated from the rock porosity and permeability.

These data are useful in evaluating seals, but a very severe limitation is caused by the doubtful representativeness of the analysed sample with respect to the entire sealing surface of the trap. The difference in scale is enormous. Downey (1984) gives a graphic example. For a closure area of 6400 acres (26 km^2), a 4-inch diameter core sample of the caprock would provide a ratio between area of caprock sample to area of caprock surface of 1 to about 3.5 billion! Pore sizes are likely to vary considerably over the lateral extent of the caprock, so a core sample tells us little about the sealing capacity of the caprock as a whole. The existence of large pore networks is critical; these represent the weakest points of the seal.

As a result of these difficulties, seal capacity calculations are subject to wide ranges of error.

10.5.2.1 The effect of hydrodynamics and overpressure

Under hydrodynamic conditions, the driving forces to migration or leakage are modified. Hy-drodynamic flow may either increase or decrease the driving pressure against seals, thus modifying the petroleum column heights the seal can support. When the hydrodynamic force has an upward vector, it acts in support of buoyancy; when it is downward it diminishes the effect of buoyancy on the seal. Hydrodynamic effects on seal capacity may, however, for all practical purposes, be ignored except in those basins with clear evidence of hydrodynamic conditions operating at the present day. In the Powder River Basin of Wyoming, Berg (1975) has shown that hydrodynamic downdip flow has assisted the lateral sealing of the Recluse Muddy and Kitty Muddy fields.

The existence of overpressure in a shale caprock may create a local pore pressure gradient that greatly assists its capacity to seal adjacent normally pressured reservoirs. Studies in the Niger delta (Weber *et al.* 1978) and Gulf Coast (Stuart, 1970) provide field examples.

10.5.2.2 Loss of petroleum through caprocks by diffusion

The work of Leythaeuser, Schaefer and Yukler (1982) has shown that gas may diffuse through water-filled caprocks over geological time-scales. In a study of the 68 billion cubic feet Harlingen gas field in Holland, they estimated that half of the accumulation would be lost by diffusion through the 400 m thick shale caprock in 4.5 million years. Thus, gas fields overlain by water-saturated shale caprocks are likely to be ephemeral phenomena unless continuously topped up by active generation in the area. If losses by diffusion are as severe as Leythaeuser *et al.* suggest, there are some difficulties in explaining the existence of gas fields charged and reservoired in very old sequences, such as for example in the Lower Palaeozoic.

10.5.3 Factors affecting caprock effectiveness

The effectiveness of caprocks worldwide may be examined in terms of their:

- Lithology
- Ductility
- Thickness
- Lateral Continuity
- Burial Depth.

10.5.3.1 Lithology

Caprocks need small pore sizes, so the vast majority of caprocks are fine-grained clastics (clays, shales), evaporites (anhydrite, gypsum, halite) and organic-rich rocks. Other lithologies such as argillaceous limestones, tight sandstones and conglomerates,

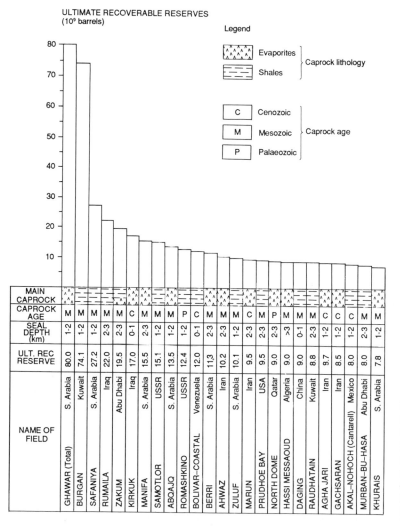

Fig. 10.41. Caprock lithologies, depth range, age and recoverable reserves of the world's 25 largest oil fields. Lithologies are split roughly equally between shales and evaporites. Most of the evaporite caprocks occur in oil fields in the Middle East (modified after Grunau 1987).

cherts and volcanics may also seal, but they are globally far less important, and are frequently of poor quality and geographically of limited extent.

Grunau (1987) compiled information on the caprock lithologies of the world's 25 largest oil and 25 largest gas fields (Figs. 10.41 and 10.42). There was a roughly equal split between shales (13) and evaporites (12) for the 25 oil fields. For the gas fields, shales (16) predominated over evaporites (9).

About 40 per cent of the ultimately recoverable oil reserves from the world's giant oil fields are capped by evaporites, and 60 per cent by shales (Grunau, 1987). For gas, the corresponding volumes are 34 per cent for evaporite caprocks and 66 per cent for shales. The majority of giant *oil fields* with evaporite caprocks are located in the Middle East and North Africa, while shale caprocks to giant oil fields are more ubiquitous, and include Alaska, western Canada, California, the Gulf Coast, Mexico, Venezuela, the North Sea, the Soviet Union, Indonesia and Brunei. Evaporite caprocks to giant *gas fields* are geographically more

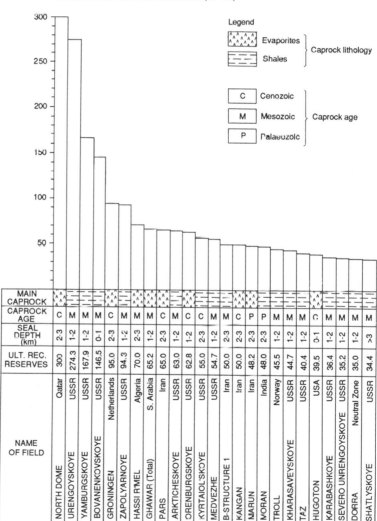

ULTIMATE RECOVERABLE RESERVES (10¹² scf)

Fig. 10.42. Caprock lithologies, depth range, age and recoverable reserves (in standard cubic feet) of the world's 25 largest gas fields. Sixteen of the 25 giant gas fields shown here are capped with shales. The data set is dominated by the shale caprocks of most of the giant gas fields of the Soviet Union (modified after Grunau 1987).

widely distributed, and apart from the Middle East and North Africa, include the Soviet Union, Netherlands/southern North Sea and Brazil.

Nederlof and Mohler (1981), in a statistical analysis of 160 reservoirs/seals, found that caprock lithology was of considerable importance in influencing seal capacity.

10.5.3.2 Caprock ductility

Ductile caprock lithologies are less prone to faulting and fracturing than brittle lithologies. Caprocks are placed under substantial stress during periods of structural deformation, including the deformation responsible for trap formation. During the formation of a simple anticline, for example, fractures may occur in brittle caprocks in the crestal parts of the fold which are undergoing tension. Ductility is, therefore, a particularly important requirement of caprocks in strongly deformed areas such as fold and thrust belts.

The most ductile lithologies are evaporites, and the least ductile cherts. This may explain the extraordinary success of evaporites as caprocks. Table 10.5 lists the main caprock lithologies in order of ductility:

A high kerogen content appears to enhance the ductility of shale caprocks. Many source rocks, therefore, also serve as seals.

Ductility is also a function of temperature and pressure. Evaporites may be brittle at shallow depths, but very ductile at depths of over 1 km.

10.5.3.3 Caprock thickness

A small thickness of fine-grained caprock may have sufficient displacement pressure to support a large hydrocarbon column. Thin caprocks, however, tend to be laterally inpersistent; thus a thick

Table 10.5 Ductility of caprocks (Downey, 1984)

Caprock lithology	Ductility
Salt	Most ductile
Anhydrite	↑
Organic-rich shales	
Shales	
Silty shales	
Calcareous mudstones	
Cherts	Least ductile

caprock substantially improves the chances of maintaining a seal over the entire prospect, or even over the entire play fairway or basin. Typical caprock thicknesses range from tens of metres to hundreds of metres (Grunau, 1987). Very large volumes of petroleum may be sealed by relatively modest thicknesses of caprock. Grunau (1987) gives examples of the 30 metre thick Ahmadi shales which seal the 74 billion barrel Burgan field in Kuwait; the 20 metre thick Arab C-D Anhydrite which seals the Arab D-Jubaila main reservoir of the Ghawar field in Saudi Arabia, the world's largest oil field (approx 80 billion barrels); the 33 metre thick Cap-Rock Anhydrite which seals the Asmari oilfields of the Iranian Zagros Fold Belt.

For gas reservoirs, a thick caprock reduces the risk of substantial losses by diffusion.

10.5.3.4 Lateral seal continuity

In order to provide good regional seals, caprocks need to maintain stable lithological character (and hence capillary pressure and ductility characteristics) and thickness over broad areas. Most prolific petroleum provinces contain at least one of these regional seals. The search for petroleum in these basins may be focused on the base of the regional seal, rather than on any particular reservoir horizon. The lateral variability of the regional seal may be studied using wireline log information and seismostratigraphic analysis.

Some depositional environments and basin settings are more conducive to the establishment of thick and effective regional caprocks than others. Two of these will be discussed in Section 10.5.4. Of particular importance are the distinctive depositional environments that give rise to evaporite deposition (Section 7.1.3.2).

10.5.3.5 Burial depth of caprocks

The present burial depth of caprocks does not appear to be an important factor in influencing seal effectiveness. Grunau (1987) presents histograms showing the seal depths of the world's giant oil and gas fields (Figs 10.43 and 10.44). Almost half of the ultimately recoverable reserves of oil occurs in the 1000 to 2000 m depth range, and 31 per cent in the 2000 to 3000 m range. The world totals are, of course, strongly influenced by the Middle East. There are, however, important regional variations.

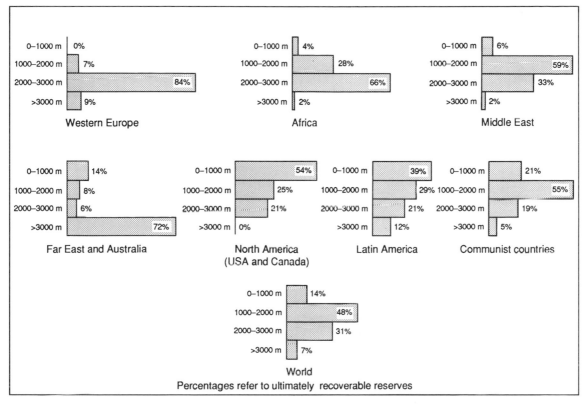

Fig. 10.43. Seal depths of the world's giant oil fields. Almost half of the ultimately recoverable reserves in the world's giant oil fields are sealed at depths of 1 to 2 km. There are important regional variations. In North America, seal depths tend to be shallower, whereas in western Europe, the Far East and Australia they tend to be deeper (after Grunau 1987).

The most striking variations are North America, where over half of the ultimately recoverable oil reserves in giant fields are at depths of less than 1000 m, and the Far East and Australasia, where, in contrast, 72 per cent is at depths greater than 3000 m.

For giant gas fields, the depth distribution is similar at the world level, and regional variations are slightly less strong. As Grunau points out, deep gas is probably much more abundant in nature than Fig. 10.44 indicates, but in many cases it is uneconomic to explore for or to develop.

The overall impression, therefore, is that seals may be effective at all depths. The requirement always is that a unit of high displacement pressure and ductility is present over wide areas. This has got little to do with *present-day* depth of occurrence. However, we know that shale pore diameters do decrease with burial, particularly over the first

2 km (see also Section 8.2.1 on porosity–depth relations). The *maximum attained depth of burial* of shale caprocks is, therefore, likely to have an influence on sealing capability. Many shallow oil accumulations occur in structures that have undergone significant uplift, bringing well-compacted caprocks close to the surface. Provided these caprocks retain their ductility and avoid brittle deformation through the uplift period, there is no reason why they should not be effective caprocks. In the *Duri* field (3.9 billion (10^9) barrels recoverable) of the Central Sumatra Basin, Indonesia, the oil-bearing lower Miocene Bekasap Formation sands occur at depths of only 100 m below ground surface over the crest of the structure, sealed above by lower to mid-Miocene Telisa Formation shales. Furthermore, the overlying Duri Formation reservoirs in the field occur at depths of less than 30 m. The *Minas* field (4.3 billion (10^9) barrels recoverable),

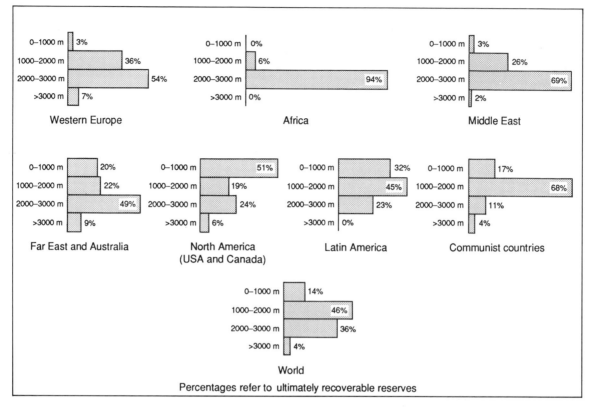

Fig. 10.44. Seal depths of the world's giant gas fields. For the world, the depth distribution is similar to that for oil. Deep gas may, however, be under-represented relative to its occurrence in nature, since in many cases it is uneconomic to explore for deep gas (after Grunau 1987).

in the same basin, is deeper with the lower Miocene Sihapas A-1 sand at approximately 700 m. This reservoir is sealed from younger water-bearing sands in the Upper Telisa by 70 m of Telisa Formation shales.

10.5.3.6 Case study: the Hith Anhydrite regional topseal of the central Arabian Gulf

A classic example of the importance of regional seals on the location of oil and gas is provided by the Middle East. Murris (1980) describes the relationship of oil and gas habitat to the stratigraphic evolution of the area. Figures 10.45 and 10.46 show the stratigraphic development of the central Gulf area, around the Qatar Peninsula, and the relationship of the two main regional caprocks – The Tithonian Hith Anhydrite and the Albian Nahr Umr Shale – to oil and gas accumulations.

The Hith Anhydrite is the regional topseal to the prolific Upper Jurassic Arab reservoirs, sealing well in excess of 100×10^9 (billion) barrels of recoverable oil reserves. Not only do prolific accumulations in the underlying Arab reservoirs demonstrate the effectiveness of the Hith Anhydrite as a seal, but so also does the general *lack* of accumulations in the overlying Lower Cretaceous where the Hith is present. The Hith is occasionally breached by faulting, allowing upward migration of oil from the upper Oxfordian–lower Kimmeridgian Hanifa source rock (for example, at Idd el Shargi in Qatar). Oils in Arab reservoirs have been confidently traced back to Hanifa source rocks, using a broad range of geochemical parameters. In the east of the central Gulf area, where, at and beyond the Upper Jurassic shelf edge the Hith is depositionally absent (Fig. 10.46), hydrocarbons generated in the Hanifa have penetrated upwards into Lower

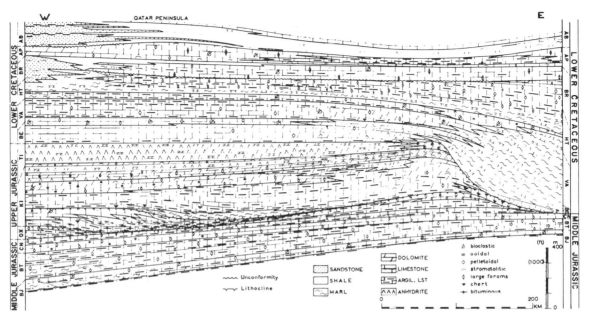

Fig. 10.45. Schematic regional stratigraphic development of the Middle Jurassic to Albian sequences, central Gulf area, Middle East (after Murris, 1980). The Hith Anhydrite is an extensive regional topseal at the top of the Upper Jurassic. It terminates by facies change towards the Upper Jurassic shelf edge in the east of the area. (Reproduced with permission from Murris, 1980.)

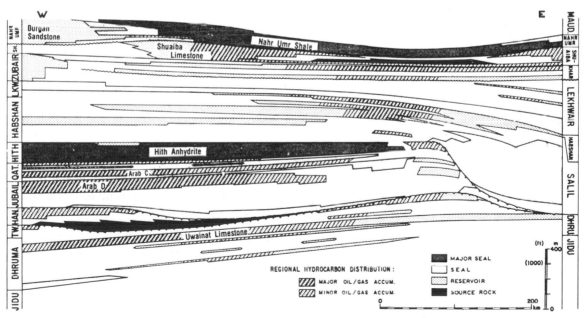

Fig. 10.46. Relationship of oil and gas accumulations to the regional topseals in the Middle Jurassic to Albian sequences, central Gulf area, Middle East (after Murris, 1980). The Upper Jurassic Hith Anhydrite seals the prolific underlying Arab reservoirs, which contain over 100×10^9 (billion) barrels of recoverable oil reserves. The oil has been sourced from the Upper Jurassic Hanifa. (Reproduced with permission from Murris, 1980.)

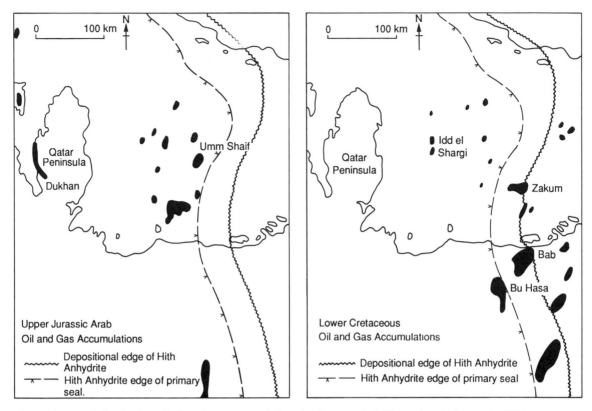

Fig. 10.47. Areal distribution of oil and gas accumulations in the central Gulf area in relation to the Hith Anhydrite caprock (adapted from Murris 1980). Arab accumulations are closely related to the occurrence of the Upper Jurassic Hith topseal. Lower Cretaceous accumulations also occur above the Hith where it is locally breached.

Cretaceous reservoirs. Figure 10.47 shows the areal distribution of Upper Jurassic and Lower Cretaceous accumulations in the central Gulf area, in relation to the main seal edge and depositional edge of the Hith Anhydrite. An unknown but probably high proportion of the Lower Cretaceous reservoired oil in the east of the area has been sourced, however, from the Shuaiba source rock, which is mature only in this eastern area. This contribution, therefore, is quite independent of the presence or absence of the Hith Anhydrite.

10.5.4 The depositional settings of caprocks

We have seen in the previous pages that the requirements for good regional caprocks are the maintenance of stable lithology and ductility over broad areas. A stratigraphic unit is not a caprock unless it seals an underlying reservoir; thus the ideal regional caprock maintains its sealing characteristics over wide areas but also occurs in stratigraphic association with reservoirs. These conditions are liable to be met particularly in two types of depositional setting (see Section 7.1):

1 *As transgressive marine shales on gently sloping clastic shelves.* These regionally extensive shales may form an excellent seal to basal transgressive sandstone reservoirs. This petroleum play, occurring in the *wedge-base position* of the depositional sequence, as described by White (1980), is a very successful one throughout the world, as will be discussed in Chapter 11. The marine transgression may flood wide areas of low-lying coastal flats, and isolate the marine shelf from supplies of coarse clastics. Thus the *transgressive systems tract*

described in Section 6.6, extending from the time at which the palaeo-shelf begins to be onlapped, to the time of sea level highstand, is frequently an ideal location for the development of regional caprocks. However, sandy transgressive systems tracts which line the lower depositional sequence boundary may act as 'thief-zones', allowing the migration undip of petroleum. The correct prediction of shales in the transgressive systems tract is therefore essential.

An example of an excellent and extensive transgressive marine shale caprock is the Early to Mid-Miocene Telisa Formation of the highly productive Central Sumatra Basin, Indonesia. Over 12 billion (10^9) barrels of recoverable oil reserves have been discovered in this basin, much of it as a result of the development of the Telisa shale regional topseal. The Telisa is a marine shelf sequence deposited as a result of the flooding of lower Miocene sand-rich deltas (for example, the Bekasap-Duri and Sihapas deltas). In this way, a prolific reservoir-topseal doublet was formed.

2 *As evaporitic deposits on regressive supratidal sabkhas and in evaporitic interior basins.* In clastic systems, the regressive wedge-top deposits (*sensu* White, 1980), or the late highstand systems tract or shelf margin wedge systems tract (see Section 6.6) are of generally poor seal quality, comprising shallow marine and coastal sands and non-marine deposits. These may form excellent reservoirs, but they do not make good seals. In carbonate systems, however, extensive evaporitic sabkhas may gently prograde across flat marine carbonate platforms, providing extensive and excellent quality seals. The Tithonian Hith Anhydrite of the Arabian Gulf is a good example (Murris, 1980). As the climate became arid in the Tithonian, the shallow carbonate platform in which the prolific Arab reservoirs were deposited was replaced by an extensive sabkha.

Evaporites develop through the interaction of palaeoclimate and palaeogeography. They may develop on supratidal sabkhas (Section 7.1.3.2), in evaporitic continental interior basins (Sections 7.1.2.3, 7.2.1.1), or in wide rifted basins during the early stages of sea floor spreading (Section 7.2.1.2) (e.g. the Albian salt of the South Atlantic). Clear palaeoclimatic and palaeogeographic controls can normally be determined.

Shale caprocks may be deposited in a wider range of depositional environments, ranging from lacustrine (Section 7.1.2.3) through marine shelf (Section 7.1.4.1) to bathyal (Section 7.1.5).

Implications for caprock development over specific play fairways can be interpreted from an understanding of the depositional systems and facies present in a basin; these were discussed in Chapter 7.

10.6 THE TRAP

Summary

The final requirement for the operation of an effective petroleum play is the presence of traps within the play fairway.

A trap represents the location of a subsurface obstacle to the migration of petroleum towards the Earth's surface, which causes a local concentration of petroleum. The petroleum exploration industry is primarily concerned with the recognition of these sites of petroleum accumulation.

Traps are classified into *structural*, stratigraphic and *hydrodynamic* traps.

Structural traps are those caused by tectonic, diapiric, gravitational and compactional processes, and represent the habitat of the bulk of the world's already discovered petroleum resources. The development of most structural traps can be understood in terms of basin-forming mechanics and the ensuing burial history of the basin fill. Examples are the contractional folds of the Zagros fold belt of Iran and Wyoming–Idaho fold/thrust belt, the inversion anticlines of Sumatra, the extensional tilted fault blocks of the North Sea, the extensional rollovers and fault traps of the Niger Delta and US Gulf Coast, the compactional drape anticlines of the North Sea, and the salt domes and related diapiric structures of the Gulf Coast. *Stratigraphic traps* are a diverse group in which the trap geometry is essentially inherited from the original depositional morphology of, or discontinuities in, the basin-fill, or from subsequent diagenetic effects. Much of the world's future *undiscovered* resources are likely to be found in stratigraphic traps, and their discovery will require a very high level of geological expertise. Examples of stratigraphic traps are the fluvial channels and barrier bars in the Cretaceous basins lying along the east flank of the Rockies, the Tertiary submarine fans of the North Sea, the carbonate reefs of the western Canadian

Devonian, southern Mexico and Arabian Gulf, the subunconformity truncation traps exemplified by the Prudhoe Bay (Alaska) and East Texas fields, and the subunconformity palaeotopographic traps of the Gulf of Valencia, Spain. Diagenetic traps include those formed by mineral diagenesis, petroleum tar mat formation, and permafrost and gas hydrate formation.

Hydrodynamic traps are those formed by the movement of interstitial fluids through basins and, in a worldwide context, tend to be relatively uncommon. Hydrodynamic effects, however, are important in some foreland basins.

The majority of the world's giant oil fields have so far been found in anticlinal traps.

Not only must a sealed trap geometry be present for the existence of a petroleum trap, but the timing of its development must also be considered. The geometry must be present prior to the petroleum charge in order to trap petroleum. Thus, an understanding of the history of individual trap growth together with the burial and thermal history of the basin, is essential to the evaluation of petroleum prospects.

10.6.1 The formation of traps for petroleum: introduction

The final requirement for the operation of an effective petroleum play is the presence of traps within the play fairway. Quantitative estimation of the petroleum volumes potentially trapped in the fairway, and in specific, defined prospects is discussed in Chapter 11.

A trap exists where subsurface conditions cause the concentration and accumulation of petroleum. After petroleum is generated and expelled from source rocks, it will move from sites of high potential energy to sites of low potential energy. This process ultimately leads to the loss of the petroleum at the Earth's surface. Subsurface traps *en route* may be considered local (and temporary) potential energy minima. In these places, the migration route of petroleum is obstructed.

The commercial exploitation of petroleum resources depends on the concentration and accumulation of petroleum in traps. The petroleum industry has been dominated by exploration for specific subsurface geometries that are diagnostic of the presence of a trap. The recognition of these geometries, frequently on seismic sections, has been the goal of most explorers for decades. In the earlier days of exploration, the trap geometries pursued were generally large and obvious (for example, large anticlines, often with surface outcrop expression). More recently, the emphasis has shifted towards exploration for the subtle trap, many of them stratigraphic rather than structural in origin.

The same basic physical principles apply to trapping as to secondary migration and seals. A trap is formed where the capillary displacement pressure of a seal exceeds the upward-directed buoyancy pressure of petroleum in the adjoining porous and permeable reservoir rock. A discussion of the mechanics is given in Sections 10.3.2.4 and 10.5.2.

Both oil and gas may occur in a trap; the gas lies above the oil because it is less dense. If a trap is charged first with oil, and then with gas (for example, as a result of increasing source rock maturity), the expanding gas-cap may displace oil downwards past the spill-point(s) of the trap. The oil may then migrate updip to the next available trap. Thus, traps may contain greater proportions of oil relative to gas as the distance from the source kitchen increases. This is the so-called Gussow principle.

Figure 10.48 shows some terms commonly used in the description of traps. Note that the trap

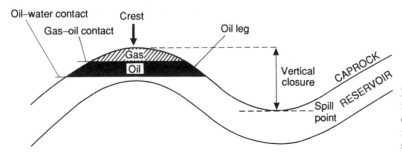

Fig. 10.48. Terms commonly used in the description of traps. A gas cap overlies an oil leg, and the trap in this case is not filled down to its structural spill point.

illustrated is not full-to-spill. A *gas-cap* overlies an *oil leg*, but the (structural) *spill point* is some distance below the *oil–water contact*.

10.6.2 Trap types

The main purpose of trap classification is to allow comparison between one prospect, or one play, and another. A particular trap type in a basin may be characterized by a distinctive field size distribution and drilling success ratio. Trap classification more readily allows the drawing of geological analogies which may be useful in estimation of prospect and play petroleum volumes and risk. This topic is discussed in Chapter 11.

Table 10.6 shows a broad classification of traps. The main subdivision is between *structural traps*, in which the majority of the world's petroleum resources have been found, and *stratigraphic traps*. The classification is based essentially on the *process* causing the formation of the trap, rather than its geometry. If the geological processes operating in a basin are known, therefore, a particular suite of

Table 10.6 Trap types

Structural	Tectonic	Extensional
		Contractional
	Compactional	Drape structures
	Diapiric	Salt movement
		Mud movement
	Gravitational	
Stratigraphic	Depositional	Reefs
		Pinch-outs
		Channels
		Bars
	Unconformity	Truncation
		Onlap
	Diagenetic	Mineral
		Tar mats
		Gas hydrates
		Permafrost
Hydrodynamic		

Note: This broad classification of trap types is based essentially on the process causing trap formation, rather than trap geometry. The main groups of traps are structural, stratigraphic and hydrodynamic.

traps may be predicted. Particular structural traps, for example, can be related to tectonic setting and basin-forming mechanics (Chapters 3 to 5). The detection of stratigraphic traps, on the other hand, is dependent on a good understanding of basin evolution and the stratigraphy of the basin-fill (Chapters 6 and 7).

Structural traps are those caused by tectonic, diapiric, gravitational and compactional processes. The essential point is that movement has occurred in the basin-fill some time after its deposition. *Stratigraphic traps* are those in which the trap geometry is inherited from the original depositional morphology of the basin-fill, or from diagenetic changes that took place subsequently. The best-known stratigraphic traps are caused by facies change or related to unconformities, but we have also included here traps sealed by the undip clogging of pore-space by biodegraded oil, gas hydrates or permafrost. *Hydrodynamic* traps are caused by the flow of water through a reservoir/carrier bed. They may be important in some basins, but are generally rare. More than one process may contribute towards the formation of a trap. Examples are hydrodynamic closures developed on structural noses, onlap and pinch-out traps combined with structural deformation, and channel sands developed on unconformity surfaces. Furthermore, different trap types may be genetically related. A reef, for example, may be overlain by a compactional (drape) anticline.

10.6.2.1 Structural traps

Structural traps formed by tectonic processes

CONTRACTIONAL FOLDS AND THRUST-FAULT STRUCTURES

Contractional folds occur in areas undergoing tectonic compression, and are generally associated with convergent plate boundaries, particularly where continent–continent collision has taken place (Chapter 4). They may also develop where transpression occurs along strike-slip boundaries (Chapter 5).

From a petroleum viewpoint, the most prolific zone of contractional folding is the external zone of the *Zagros Mountains in Iran* (Falcon 1958, Hull and Warman, 1970). Figure 10.49 shows some

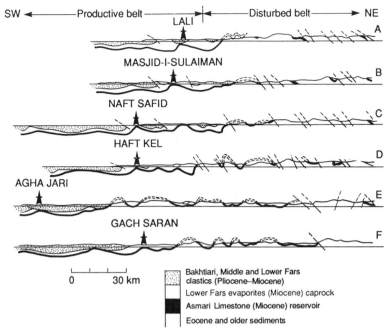

Fig. 10.49. Cross-sections across the Zagros foldbelt of Iran (after Falcon 1958). The folds are large and, at surface, relatively simple. The main producing reservoir is the lower Miocene Asmari Limestone; it has been tectonically fractured, and is sealed by the ductile Lower Fars Group evaporites.

structural cross-sections across the area. The main producing reservoir is the lower Miocene Asmari Limestone, which owes its prolific productivity to tectonically-induced fracturing. The brittle limestone reservoir is overlain by the ductile evaporite caprocks of the Miocene Lower Fars Group. The folds are up to 60 km long and relatively simple where dramatically exposed in outcrop. At depth, however, they are considered to be tighter and

associated with thrust faults which sole out onto a basal detachment, possibly in the Hormuz Salt (Fig. 10.50).

Another example of an area of productive contractional fold structures is the *Wyoming thrust belt, USA*. The Painter reservoir field, discovered in 1977, has been described by Lamb (1980). It is a large overturned anticline developed in the hanging wall of the Absaroka Thrust (Fig. 10.51). The

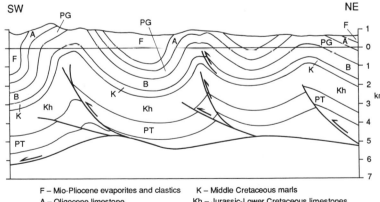

Fig. 10.50. Interpretation of the relationship of Zagros folds to structure at depth, showing the soling out of the thrust faults onto a basal detachment (after Bailey and Stoneley 1981). Similar listric fault styles typify the thrust belts of the western United States (e.g. Idaho–Wyoming) and the Canadian Rockies of Alberta and British Columbia. Exploration for deep traps in these areas may be considerably more difficult than at relatively shallow depths, where structure is simpler.

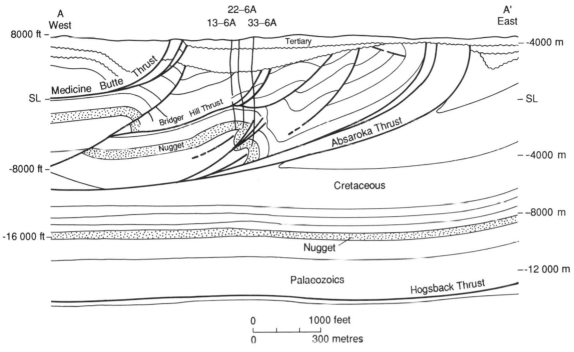

Fig. 10.51. Interpretation of the structure of the Painter Reservoir Field, Idaho–Wyoming thrust belt. The Painter structure has no surface expression, but was identifiable from seismic data. The producing reservoir is the Triassic–Jurassic Nugget Sandstone.

producing reservoir is the Triassic–Jurassic Nugget Sandstone, which has been thrust over the Cretaceous. The field was found in a seismically-defined structure along trend and 16 km south of the previously discovered Ryckman Creek field, which also produces from the Nugget. As can be seen from Fig. 10.51, the Painter structure has no surface expression because it is overlain by the Bridger Hill detachment and a Cretaceous unconformity. It owes its discovery to an improvement in seismic processing techniques in the mid-1970s, which allowed subsurface closure to be detected.

Surface outcrop lithology and terrain have a large influence on the quality of seismic data in fold/thrust belts. Subsurface accumulations may be very difficult to find in those areas where surface structure bears little or no relation to subsurface structure, and where surface or subsurface conditions (karstified limestone, volcanics, rugged terrain) prevent the acquisition of good-quality seismic data.

Anticlines may also develop in areas of local contraction along strike-slip systems. The Wilming-

ton field in the Los Angeles Basin, California, is such an example (Mayuga, 1970). It is developed along the San Andreas fault system. Transpressional anticlines tend to be arranged *en echelon* and are very strongly faulted: they depend on the presence of thick caprocks to seal reservoirs across fault planes. Other examples of highly complicated faulted anticlines that have developed along predominantly strike-slip fault systems are probably the large Seria and Champion fields of Brunei. These fields are located on the relatively proximal part of the Miocene to Recent Baram Delta, and are not only subdivided into a large number of fault compartments but also contain numerous stacked deltaic reservoirs. Some component of shale diapirism is present in the development of Champion.

Some contractional anticlines have developed as a result of the reversal of movement along old extensional faults. These are known as *inversion anticlines*. The evidence for the earlier extensional history is usually the thickening of sediments towards the fault plane during its period of growth.

Good examples are the so-called 'Sunda' folds of some Indonesian basins, for example, Central Sumatra. Inversions tend to occur in areas where relatively subtle changes in the regional stress field cause reversal of movement along faults, and are therefore frequently associated with fault systems that have a strong strike-slip component. Petroleum charge into inversion anticlines may be a problem if a closure did not exist at the time of extension. Migration will tend to be away from the site of the inversion anticline during the extensional phase, and inversion of the basin tends to stop further petroleum generation. Charge may result from the post-inversion re-migration of petroleum.

Traps may develop along contractional faults without any element of folding (Fig. 10.52). These may be in the hangingwall or footwall of the fault, and depend for closure on the juxtaposition of sealing lithologies, or on the sealing of the fault zone itself, for example as a result of a finely powdered fault gouge or cemented zone. To complete the trap, closure is also needed in the third dimension, i.e. into the plane of the paper in Fig. 10.52. This is frequently produced by slight curvature or angularity in the map view of the fault plane, or by the intersection by further faults.

The subthrust (footwall) has sometimes provided an exploration target in thrust belts. These traps are very difficult to define, mainly due to velocity variations caused by the presence of the overthrust sheet, which make seismic interpretation very difficult. Many dry holes have been drilled on velocity 'pull-ups' in subthrust positions – mere illusions of the presence of a trap.

EXTENSIONAL STRUCTURES

Extensional structures form a very important group of traps, being responsible for many of the fields discovered in basins that have experienced a phase of rifting in their geological history (Chapter 3). We will deal in this section only with structural traps resulting from extension of the basement, that is, in stretched rift basins. The extensional structures occurring, for example, in delta sequences that developed in the post-rift stage on passive margins will be covered in the section on gravitational structures.

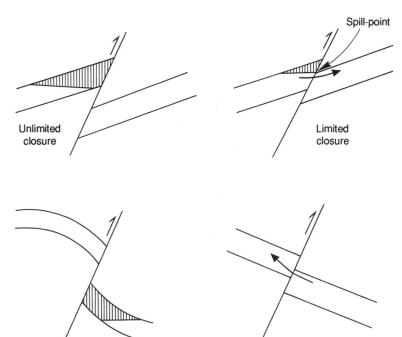

Fig. 10.52. Traps formed by high-angle reverse (contractional) faults. Juxtaposition of permeable bed limits closure. For maximum closure, fault throw needs to be large in relation to reservoir thickness. All of the illustrated trap types also require closure in the third dimension, that is, into the plane of the paper. (After Bailey and Stoneley 1981).

Rollover anticlines may develop in association with basement-controlled growth faults. The Vicksburg flexure in south Texas is an example. Very large sediment thickness changes occur across the fault zone, particularly in the Oligocene section. Large quantities of oil and gas are trapped in rollover anticlines, fault traps, and stratigraphic pinch-outs.

The most prolific play in the East Shetland Basin of the North Sea province occurs in *extensional tilted fault blocks*. The giant Statfjord field (approx 3 billion barrels recoverable), the largest in the North Sea, contains Lower Jurassic Statfjord Formation sandstone reservoirs in a large, westward tilted fault block that has been eroded at its crest to produce a series of Late Jurassic so-called 'Kimmerian' unconformities (Kirk, 1980) (Fig. 10.53). Kimmerian block faulting is responsible for the uplift and erosion of the east flank of Statfjord. The fault block is onlapped by Upper Jurassic Kim-

meridgian source rocks, which, together with Lower Cretaceous shales, form the caprocks to the field. The source rocks were deposited in a restricted basin to the west that was bounded to the east by the uplifted Statfjord block, and to the west by the Hutton/Murchison block. The east flank of the field has been complicated by subsidiary downfaulted blocks.

The main down-to-the-east bounding fault on the east flank of Statfjord has a total displacement of over 1800 m at Statfjord Formation level, and also controls the location of the giant Brent field, only 20 km and on trend to the south. The Statfjord field has total areal extent of 81 km^2.

The trap is a result of truncation of the Brent reservoir at the Kimmerian unconformity surface, and could be considered as stratigraphic. The Statfjord reservoir, however, relies on fault closure. The field is therefore a huge combination structural-stratigraphic trap.

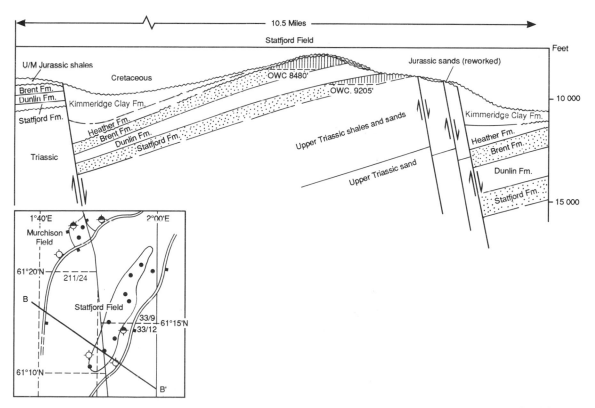

Fig. 10.53. Schematic structural cross-section across the Statfjord field, North Sea (after Kirk 1980). Statfjord is a large westward-tilted fault block that has been strongly eroded at its crest. The trap at Brent Formation level is formed by seal at the erosional unconformity, but the deeper Statfjord Formation reservoir depends on fault closure.

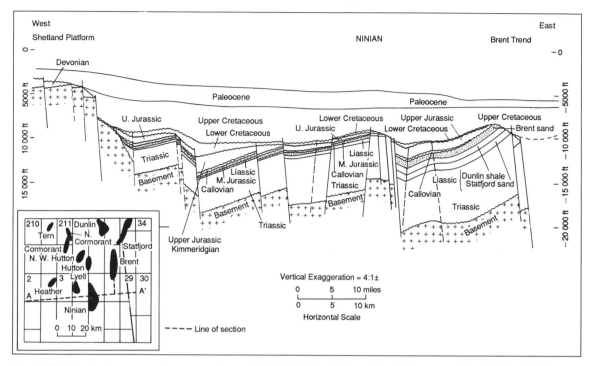

Fig. 10.54. Structural cross-section across the eroded, tilted fault blocks of the Ninian area, East Shetland Basin, North Sea (after Albright, Turner and Williamson 1980). This is a common and successful trap type in the North Sea.

The Ninian field to the southwest is another of the many examples in the East Shetland Basin of this style of trap – the eroded rotated fault block (Albright, Turner and Williamson 1980). As in the Brent Formation reservoir in the Statfjord field, the trap at Ninian is produced primarily by the truncation of Mid-Jurassic reservoirs at Upper Jurassic and Cretaceous unconformities. Figure 10.54 shows a cross-section across the Ninian area of the North Sea.

The giant Piper field of the eastern Moray Firth basin of the North Sea is an example of a more complex faulted structure, comprising three main tilted fault blocks (Maher, 1980). The deepest of the blocks has a separate oil–water contact. The southwestern flank of the Piper structure is a complex series of *fault terraces* stepping down to the southwest into the Witch Ground graben (Fig. 10.55).

Several varieties of extensional fault trap are shown in Fig. 10.56; closure is dependent on the juxtaposed lithology. It is clearly advantageous for the throw of the fault to exceed the gross reservoir thickness. The sealing or non-sealing properties of the fault plane itself are discussed in the section on gravitational structures. It is important to consider the juxtaposed lithology over the whole length of the fault trap. An 'Allan' fault plane map (Allan, 1980) shows the intersection of both footwall and hangingwall lithostratigraphic units onto the fault plane. Figure 10.57 shows a very simple example, from which it is clear that the spill point of the foot wall reservoir into the hangingwall is at 1720 m; closure may have been mapped down to 1740 m if the geometry of the hangingwall had been ignored, resulting in an overestimation of the size of the trap. A much more complicated Allan fault plane map is shown in Fig. 10.58; in this example, petroleum migrates several times across the fault plane to higher structural levels.

Gravitational structures

The most important gravitational structure that forms petroleum traps is the *rollover anticline* into listric growth faults, occurring particularly in delta

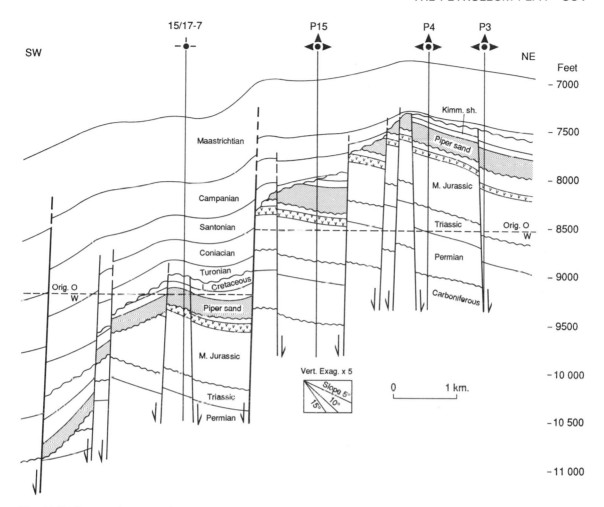

Fig. 10.55. Structural cross-section across the Piper field, North Sea, showing a complex series of fault terraces stepping down to the southwest (after Maher 1980). The deepest of the fault blocks in the southwest has a separate accumulation with its own oil–water contact.

sequences. These structures are not caused by extension in the basement, but are due to instability in the sedimentary cover and its movement under gravity. They are most prone to form where a level of undercompacted (overpressured) clays or ductile salt occurs at depth, into which the growth faults sole out, and which is overlain by a thick sequence of more competent rocks. These conditions are commonly created in thick, progradational delta sequences. Growth faulting and rollover anticlines, therefore, are particularly characteristic of delta sequences. The Gulf of Mexico and Niger Delta are examples. Figure 10.59 shows some types of rollover anticlinal traps in the Niger Delta.

The rollover anticline is the least risky trap for petroleum. Growth faulting may also give rise to *fault traps*. The integrity of the trap depends on the juxtaposition of a shale seal across the fault plane, so these traps tend to be effective on those parts of the delta where the sand/shale ratio is relatively low (say, less than 50 per cent). Even here, many of the sands may be water-bearing owing to cross-fault leakage; production may be obtained from relatively few sands, interspersed with the water-bearing zones, and distributed over a large gross vertical interval.

The fault zone itself may or may not seal. Weber *et al.* (1978) have shown that slivers of sand tend to get caught up in the fault zones on the Niger Delta,

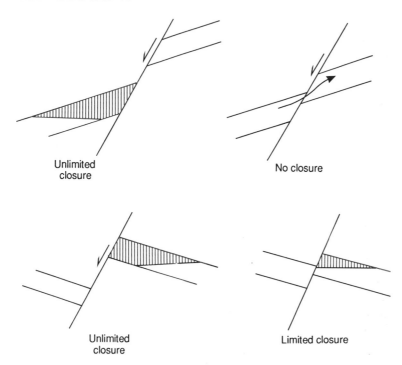

Fig. 10.56. Traps formed by extensional faults (after Bailey and Stoneley 1981). Seal depends on the lithologies juxtaposed against the reservoir across the fault plane. Ideally, fault throw should exceed gross reservoir thickness.

allowing vertical leakage of petroleum. Although destroying fault traps at deeper levels, this leakage may allow a petroleum charge into shallower reservoirs. Other faults off the Niger delta are sealing because of clay smears along the fault plane (Weber *et al.* 1978). Smith (1980) has investigated fault seal in the Louisiana Gulf Coast, and found that some faults seal even when sandstones are juxtaposed across the fault zone, as long as the sands are of different ages. This is due to the presence of fault-zone material that has formed as

a result of mechanical or chemical processes directly or indirectly related to the faulting. Where parts of the *same* sandstone are juxtaposed, the fault tends *not* to seal. Where sand is juxtaposed against shale, a seal is produced.

At shallow depths (a few hundred metres), extensional faults may form open conduits for petroleum; at greater depths, they are likely to be forced closed by overburden pressure.

In the absence of well-understood local circumstances, it is best to evaluate a fault trap on the

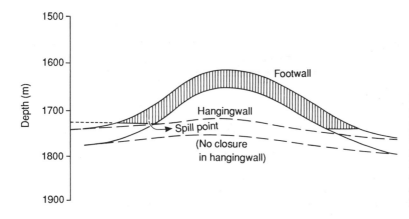

Fig. 10.57. Simple 'Allan' fault plane map showing intersection of permeable lithological units in footwall and hangingwall onto fault plane. The actual spill point may lie at a position above the mapped closure in the footwall, and is controlled by the geometry of the hangingwall.

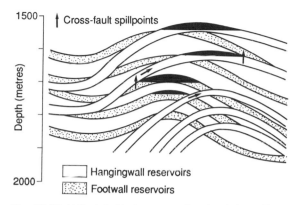

Fig. 10.58. 'Allan' fault plane map showing intersection of reservoir strata in footwall and hanging wall to the fault plane. These intersections determine the spill points at each reservoir level. Petroleum may migrate several times across the fault plane, each time moving to higher structural levels.

basis of the juxtaposed lithology alone, that is, assuming the fault zone allows lateral but not vertical migration.

A further trap that may form as a result of gravitational processes is the *ramp anticline* at the front of *gravitational thrust sheets*. When sheets of detached sediment slide downslope, they may pile up at local obstructions, forming contractional anticlinal features. These occur particularly on deep water delta slopes.

Compactional structures

The most important trap type formed by compactional processes is the drape anticline. It is caused by differential compaction, as illustrated in Fig. 10.60. The presence of a basement horst (effectively non-compactible) causes significant thickness

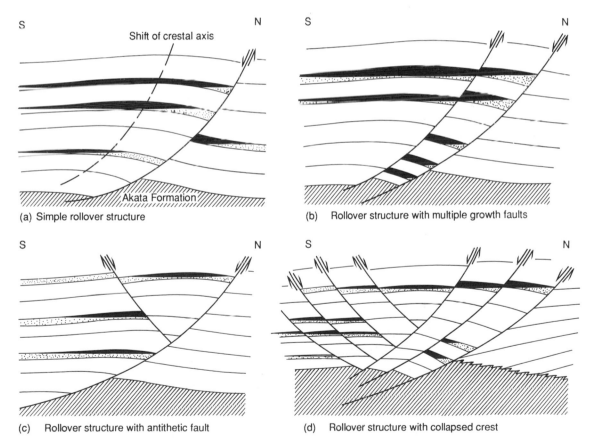

Fig. 10.59. Varieties of rollover structure forming petroleum traps in the Niger delta area (after Weber *et al.* 1980). Note: Only a few reservoir sands are shown in the schematic sections and the sand thickness has been enlarged.

(a) HORST BLOCK

1 Basement palaeotopography

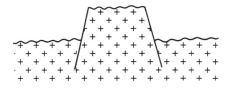

2 Palaeotopography filled-in

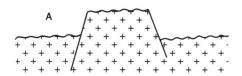

3 Interval A 50% compacted

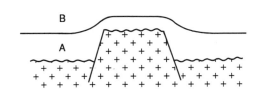

4 Interval A 70% compacted, B 50% compacted

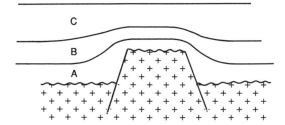

(b) TILTED FAULT BLOCK

1 Rift sequence forms in active half-grabens

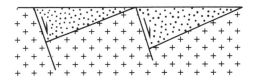

2 First stage of post-rift

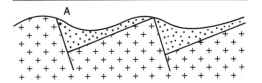

3 Interval A 50% compacted

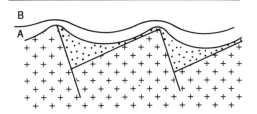

4 Interval A 70% compacted, B 50% compacted

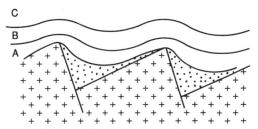

Fig. 10.60. Formation of drape anticline by differential compaction. Relief of anticline increases with depth. Basement topography causes thickness variations in the compactible sediment column. As these sediments compact, drape anticlines are formed.

variations in the overlying highly-compactible sediments. As these compact, drape features are formed because the absolute amount of compaction is greatest where the sediment is thickest.

If the area above the horst remains elevated relative to surrounding areas, shallower-water sedimentary facies may develop which are less compactible than the surrounding muds. This will exaggerate the differential compaction. The sedimentary facies and diagenetic history of the reservoir unit may be quite different over the crest of the drape anticline than off its flanks.

Drape anticlines form a very successful trap type. They are frequently simple features formed without tectonic disturbance (unless basement faults are reactivated) and frequently persist over a long period of geological time, from shortly after the time of reservoir deposition through to the present day. They are therefore available to trap a petroleum charge over a long time span, and are very forgiving of inaccuracy in estimation of charge timing.

Owing to dependence on the existence of basement topography, drape anticlines commonly form in the passive sedimentary cover to rifted supersequences (megasequences), particularly over the relatively elevated parts of tilted fault blocks in the pre-rift (Fig. 10.60b). In these settings, a petroleum charge is frequently needed from syn-rift or very early post-rift source rocks, since the later post-rift is commonly devoid of a source. Communication of the reservoir with the source is often, therefore, the critical factor for this play. A second critical factor may also be the presence of a reservoir in the post-rift. Since a considerable amount of crustal thinning is the cause of the basement topography that forms the drape anticline, these areas commonly subside rapidly due to thermal contraction, leading to deep water conditions. The only reservoirs present may be deep-sea fans. The Lower Tertiary submarine fans of the North Sea are an example.

Examples of drape anticlines are the Forties and Montrose fields of the North Sea, which occur within very large (90 km^2 and 181 km^2 respectively) low relief domal closures at Palaeocene level formed by drape over deeper fault blocks. In the Frigg gas field, the present-day closure on the Eocene submarine fan reservoir is due not only to compaction processes but also to the rejuvenation of Jurassic faults controlling the deeper structure,

and to original depositional topography on top of the fan (Blair, 1975).

Diapiric traps

Diapiric traps result from the movement of salt or overpressured clay.

At depths in excess of 600 to 1000 metres, salt is less dense than its overburden, and liable to upward movement through buoyancy. Salt can flow at surprisingly low temperatures and over long periods of geological time.

Once a density inversion is present, heterogeneities in either the mother layer of salt or clay, or in the overburden, are sufficient to trigger upward movement. Examples of heterogeneities are lateral changes in thickness, density, viscosity or temperature. These changes may be essentially depositional or may be imposed as a result of faulting or folding. In extensional faulted zones, diapirs tend to form through buoyancy where overburden load is most reduced in the footwall. The *salt rollers* of the US Gulf Coast are examples of this triggering mechanism (Fig. 10.61). Faulting may be basement-involved, or thin-skinned, usually soling out in the ductile layer. This layer may also provide a zone of detachment in contractional areas. Differential loading of a salt layer by thick overlying

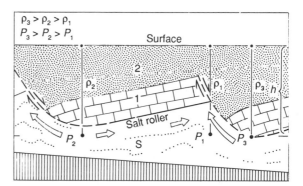

Fig. 10.61. The formation of a salt roller by extensional faulting may trigger the buoyant rise of salt (after Jackson and Galloway 1984). ρ_1, ρ_2 and ρ_3 are sediment column densities, P_1, P_2 and P_3 are pressures at three locations, h is the maximum overburden height (in the hangingwall of the fault) on the top of the salt (S). Overburden pressure is relaxed in the footwall (position P_1) as a result of the extension.

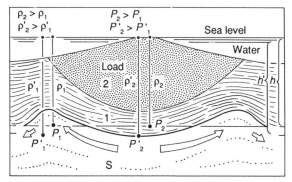

Fig. 10.62 Triggering of salt movement as a result of differential loading by a body of dense sediment. ρ_1, and ρ_2 are the sediment densities prior to movement, exerting pressures P_1 and P_2 at the flank and at the centre of the load respectively. Increased pressure on the salt under the load (P'_2) relative to the pressure at the flank (P'_1) causes lateral and upward displacement towards the lateral diapirs. h and h' are the initial and subsequent depths at which pressures are considered. These conditions commonly occur in young delta sequences (after Jackson and Galloway 1984).

sediments is a powerful triggering mechanism of diapirism in young shallow delta sequences (Fig. 10.62).

STAGES IN THE GROWTH OF SALT STRUCTURES

Salt structures pass through three stages of growth (Fig. 10.63):

1 *The pillow stage*, characterized by the thinning of sediments over the crest of the pillow, and thickening into the adjacent *primary peripheral sink*. No piercement or intrusion of the overlying sediments has taken place. Depositional facies are affected by pillow growth, with higher energy facies, perhaps reefs, developing over the crest. Traps formed at this stage are typically broad domes, while sediments that have been channelled into the topographically low peripheral sink may pinch out pillow-wards, forming stratigraphic traps.

2 *The diapir stage*, when the salt body pierces the overburden. These structures are known as salt piercement structures. As the pillow withdraws to form the diapir, *secondary peripheral sinks* may develop close to the diapir, and inside the earlier primary peripheral sinks of the pillow stage. *Turtle structures*, representing thick lenses of sediment that accumulated in the primary sinks subse-

quently tilted during diapir growth may form petroleum traps. Thick clastics may again pinch out onto the flanks of the diapir, forming stratigraphic traps.

3 *The post-diapir stage.* As the diapir grows, a point is reached where the underlying reservoir of salt is depleted, and it can only continue to rise by thinning of its lower trunk or by complete detachment from the mother salt. Overhangs commonly develop. The geometry of the salt diapir and surrounding strata beneath an overhang is generally poorly known. This is due to the fact that the area is in a seismic shadow, and is relatively infrequently drilled. The typical thickness of a diapir stem is largely unknown. Piercement of salt may take place through to the surface, forming salt domes that are particularly noticeable on Landsat images and aerial photographs.

TRAPPING POTENTIAL

Our understanding of the petroleum entrapment possibilities of salt pillows and diapirs has been developed largely through exploration of the Gulf Coast area of the USA. After the Spindletop discovery in 1901, traps associated with salt diapirs became one of the most important and prolific plays in this outstandingly successful hydrocarbon province. The piercement salt diapirs of the Houston Salt Basin were formed by the loading of Jurassic salt with a thick sequence of Mesozoic and Tertiary sediments. Most of the diapirs have intruded to shallow depths, and have created complexly faulted structures. Radial fault patterns are common over the diapir flanks. Reservoirs are commonly broken into a very large number of separate fault compartments. Syndepositional diapir growth caused substantial thickness and facies changes, as well as local erosion. Stratigraphic pinch-out and unconformity traps were formed, and local reef limestone reservoirs developed.

Most production comes from Eocene to Pliocene age sediments overlying and surrounding the salt, and from the diagenetic salt-dome caprock, which directly overlies and is in contact with the salt. The Cap Rock is a calcite deposit with zones of gypsum and anhydrite, formed by the solution of anhydrite-bearing salt by ground waters. Secondary porosity may be very high (> 40 per cent), but may be irregularly distributed. The Spindletop (Fig. 10.64) and Sour Lake diapirs have produced 60 to

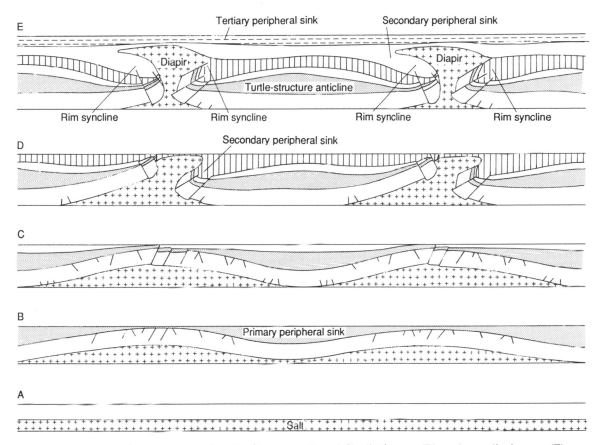

Fig. 10.63. Evolution of salt structures through pillow stage (B and C), diapir stage (D), and post-diapir stage (E). Note the lateral migration of peripheral sinks with each stage (from Seni and Jackson 1983). Turtle structures represent the preserved fill of the peripheral sink.

81 million barrels respectively from Cap Rock reservoirs.

The petroleum trapping possibilities in sediments above and around a mature diapir deposited during its post-diapir stage are shown schematically in Fig. 10.65. These include:

1 Simple domal trap above the diapir with relatively simple or no associated faulting.

2 Domal trap faulted into graben structures.

3 Diapir caprock reservoir (as previously described).

4 Stratigraphic undip pinch-out caused by facies change into the peripheral sink.

5 and **6** Reservoirs sealed against the diapir wall; 5 is beneath an overhang.

7 Unconformity trap formed by erosion towards the diapir crest.

8 and **9** Fault traps in the flanking area.

A variety of further traps, mainly stratigraphic, may develop in the diapir- and pillow-stage sediments (Fig. 10.66). These include structural and combined structural-stratigraphic traps on the crests of turtle structures. A variety of other fault, pinch-out and unconformity traps are also shown.

In summary, diapiric structures may give rise to a very wide range of petroleum traps of both structural and stratigraphic origin. Although the detection of diapirs on seismic sections is not difficult, trap development above and around the diapir is frequently very complex, and individual reservoir units may be quite small, owing to rapid lateral facies changes and complex faulting.

Salt diapirism is likely to occur in quite distinct geological settings. Thick salt deposits may develop in enclosed basins subject to cycles of flooding and desiccation. These may occur in the rift and early

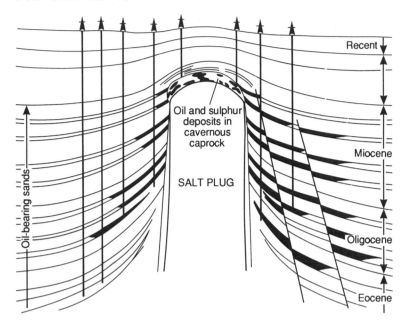

Fig. 10.64. Schematic cross-section through the Spindletop Dome, Texas, showing the distribution of oil reservoirs in both the caprock and on the flanks of the salt plug (after Halbouty 1979).

post-rift stages of continental margin development. Examples are the Jurassic salt of the Gulf Coast and Albian salt of the South Atlantic. The occurrence of salt substantially enhances the petroleum resource potential of these provinces.

MUD DIAPIRISM

Mud diapirism is most likely to develop in pro-delta clay sequences underneath the thick,

rapidly-deposited regressive sandy sequences of modern and Tertiary deltas. Examples are the Baram (Borneo), Niger (West Africa), Mississippi (USA) and Mackenzie (Arctic Canada) deltas. Excess pore pressure builds up in these clay sequences because low permeability prevents the expulsion of sufficient pore fluids as the sediment undergoes compaction. The overpressure lowers the strength of the sediment and promotes ductile flow.

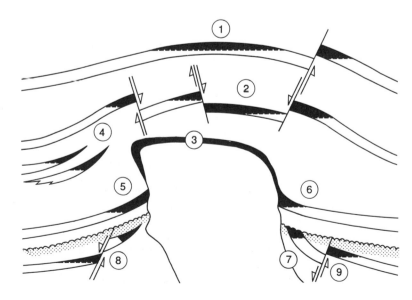

Fig. 10.65. Potential petroleum traps associated with salt diapirs (after Halbouty 1979). A wide variety of stratigraphic and structural traps may develop above and around the diapir, and in its diagenetic caprock.

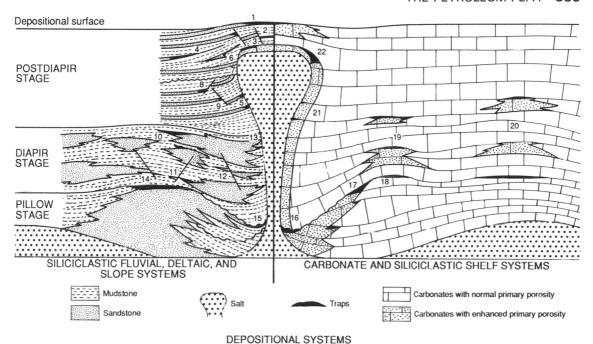

Fig. 10.66. Potential petroleum traps in pillow, diapir and post-diapir stage sediments (after Jackson and Galloway 1984). Faulting, stratigraphic pinch-outs and unconformities may produce a wide variety of traps (numbered 1–22).

Although traps associated with salt diapirs have been better studied, the indications are that mud diapir structures offer broadly similar trapping possibilities. Structures may be very complexly faulted, with reservoirs broken into a multitude of separate units.

Mud diapirs can be distinguished from salt diapirs on seismic sections on the basis of seismic velocity. Salt has a much higher velocity than overpressured shale. Furthermore, mud diapirs tend to lack the well-developed rim synclines that typically surround salt diapirs. Factors may be the shorter-lived process of mud diapirism, and the smaller area from which mud withdrawal appears to take place (Harding and Lowell, 1979).

10.6.2.2 Stratigraphic traps

A stratigraphic trap is primarily caused by some variation in the stratigraphy of the basin-fill. This variation may be essentially inherited from the original depositional characteristics of the fill, or may result from subsequent diagenetic changes to it.

The detailed stratigraphic trap classification of Rittenhouse (1972) is reproduced as Table 10.7, in order to emphasize the tremendous diversity present amongst stratigraphic traps. The broad classification of petroleum trap types given in Table 10.6 incorporates a slightly modified and simplified version of the Rittenhouse classification.

Stratigraphic traps may be classified as either *depositional* (in which the trap geometry is related to sedimentary facies changes), related to *unconformity* surfaces (either above or below) and *diagenetic*. Among the diagenetic traps, there are not only those caused by mineral diagenesis such as dolomitization, but also biodegradation of petroleum (tar mats) and phase changes to petroleum gas (gas hydrates) and interstitial water (permafrost).

A large number and wide range of stratigraphic traps have been discovered by over a century of exploration, since the drilling of the first exploration well by Colonel Drake at Titusville, Pennsylvania in 1857. Many of these stratigraphic discoveries, however, have been made by complete accident or at least by incorrect geological reasoning (Halbouty, 1982). The giant East Texas field for example, a five billion barrel unconformity trap, was drilled as an anticlinal trap. As structural traps of any material size are becoming fewer and fewer, except in frontier basins, an increasingly large

Table 10.7 Rittenhouse's stratigraphic trap classification (Rittenhouse, 1972)

Not Adjacent to Unconformities

I Facies-change traps
  A Current-transported reservoir rock
    1 Eolian
      a Dune (coastal, inland)
      b Eolian-sheet
    2 Alluvial-fan
    3 Alluvial-valley
      a Braided-stream
      b Channel-fill
      c Point-bar
    4 Deltaic (lacustrine, bay)
      a Distributary-mouth bar
      b Deltaic-sheet
      c Distributary channel-fill
      d Finger-bar
    5 Nondeltaic coastal (lacustrine, bay)
      a Beach
      b Barrier-bar
      c Spit, hook, etc.
      d Tidal-delta
      e Tidal-flat
    6 Shallow-marine
      a Tidal-bar
      b Tidal-bar belt
      c Sand-belt
      d Washover
      e Shelf-edge
      f Shallow-winnowed-crestal
      g Shallow-winnowed-flank
      h Shallow-turbidite
    7 Deep-marine
      a Marine-fan
      b Deep-turbidite
      c Deep-winnowed-crestal
      d Deep-winnowed-flank
  B Reservoir rock not current-transported
    1 Gravity
      a Slump
    2 Biogenic carbonate
      a Stratigraphic reef
        1 Shelf-margin
        2 Mound (patch-reef, mud, algal, etc.)
      b Blanket (crinoidal, tidal-flat, lagoonal, etc.)
II Diagenetic traps
  A Nonreservoir to reservoir rock
    1 Replacement (and leached)
      a Dolomitized shelf-margin
      b Dolomitized mound (patch-reef, mud, algal, etc.)
      c Dolomitized blanket (crinoidal, tidal-flat, etc.)

d Dolomitized current-transported deposit (facies or lithologic type)
    2 Leached
      a Leached shelf-margin
      b Leached mound (patch-reef, mud, algal, etc.)
      c Leached blanket (crinoidal, tidal-flat, etc.)
      d Leached current-transported deposit (facies or lithologic type)
    3 Brecciated
    4 Fractured (lithologic type)
  B Reservoir to nonreservoir rock
    1 Compaction
      a Physical compaction
      b Chemical compaction
    2 Cementation

Adjacent to Unconformities

III Traps below unconformities
  A Seal above unconformity
    1 Topography young
      a Valley-flank
      b Valley-shoulder
    2 Topography mature
      a Crestal
      b Dip-slope
      c Escarpment
      d Valley
    3 Topography old
      a Beveled
  B Seal below unconformity
    1 Mineral cement (anhydrite, calcite, etc.)
    2 Tar-seal
    3 Weathering product (weathered-feldspar, weathered-tuff, etc.)
IV Traps above unconformities
  A Reservoir location unconformity-controlled
    1 Two sides
      a Valley-fill
      b Canyon-fill
      c Blowout-fill
    2 One side (buttress)
      a Lake-cliff
      b Coastal-cliff (fault-coastal-cliff)
      c Valley-side (fault-valley-side)
      d Hill-flank (fringing-reef, mound, blanket, etc.)
      e Structure-flank (fringing reef, mound, blanket, etc.)
  B Reservoir location not unconformity-controlled (transgressive)

---

proportion of the world's undiscovered resources will be found in relatively subtle, obscure stratigraphic traps. Halbouty's assertion is that geologists, geophysicists and exploration management must now engage in a *'deliberate search for the subtle trap'*.

The detection of stratigraphic traps requires a high level of geological expertise. Great emphasis must be placed on an understanding of the stratigraphic evolution of the basin, through a detailed sequence-by-sequence analysis. Of particular importance is the understanding of palaeogeography and sedimentary facies for each sequence and sub-sequence.

Pinch-outs

Whatever the geometry and origin of the entire reservoir unit, any porous facies may pinch out laterally, and, when combined with regional structural dip, may give rise to a stratigraphic pinch-out trap. Such traps were previously described on the flanks of salt diapirs.

Pinch-out traps may be extensive and of very low dip. When the depositional pinch out is gradational, an undip transitional 'waste-zone' of very poor quality reservoir rock may be present (Fig. 10.67). Much or all of the petroleum charge may leak into the poorly producible waste-zone, forming a non-commercial accumulation. Ideally, the up-dip facies change should be rapid and complete in order for an effective lateral seal to be produced (Downey, 1984).

Depositional traps

Traps may develop in a wide range of depositional environments, ranging from aeolian dune to submarine fan. The reader is referred to Chapter 7 for more details. A selection of just three of the more common depositional traps is briefly discussed below. These three examples are representative of the variability and complexity of depositional traps.

Fluvial channels Many channel traps have been described from the Cretaceous basins along the eastern flanks of the Rockies (Selley, 1985). An example is the South Glenrock field of the Powder River Basin, Wyoming (Curry and Curry, 1972) (Fig. 10.68), which clearly shows the meandering channel geometry of the productive sand. The South Glenrock example illustrates two important

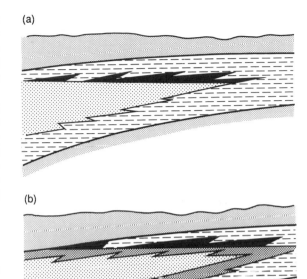

Porous and permeable sand
Tight siltstone
Shale

Fig. 10.67. Location of petroleum accumulation in facies-change traps. When the change is abrupt (a), the petroleum is trapped in good quality reservoir rock. When the change is gradual (b), the petroleum is trapped in an up-dip 'waste zone' of non-reservoir rock (after Downey 1984).

features of many channel traps: that reservoir sand thicknesses are typically small, limiting the reserves of these accumulations, and that the channel-fill may not be reservoir rock but clay.

The geometry of the trap clearly depends on the geometry of the channel. Braided, meandering, anastomosing, delta distributary and tidal channels would have different geometries. Channels may sometimes be detected as a result of differential compaction relative to the surrounding shales. Some degree of structural closure may be needed to produce the trap (for example, a regional tilt or structural nose).

Owing to the isolated nature of channel sands, they may not receive a petroleum charge. It is important, therefore, that source rocks are developed within the same depositional sequence.

The effectiveness of a channel trap clearly depends on the lithology into which the channel is

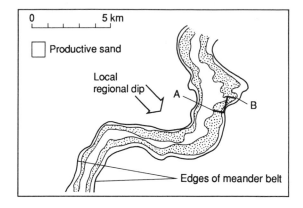

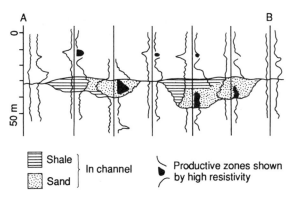

Fig. 10.68. Map and cross-section of the South Glenrock field, Powder River Basin, Wyoming. The trap is a meandering channel stratigraphic trap (after Curry and Curry 1972). These traps typically contain only very small reserves.

incised. Thus, although the channel itself may have a very distinct, sharp, lateral boundary, leakage may occur into the adjoining fluvial sediments.

Clay-filled channels may provide a lateral seal to reservoir sands in the adjacent incised sequence. Examples are present in the Sacramento Valley of California (Garcia, 1981) (Fig. 10.69), and the Pennsylvanian Minnelusa sandstones of the Powder River Basin, Wyoming (Van West, 1972).

Submarine fans These sediment bodies may be developed on a very much larger scale, and give rise to large petroleum accumulations. An understanding of the sediment-transport system is required for accurate prediction of submarine fans: a sequence-by-sequence basin analysis is particularly critical. We have previously seen that submarine fans tend to be developed at particular stages

(lowstands) and locations (base of slope) in the depositional sequence (see especially Sections 6.6.1 and 6.6.2).

The Tertiary of the North Sea is beginning to provide numerous examples of submarine fan reservoirs. An example described in the literature is the Balder oilfield of the Norwegian sector (Sarg and Skjold 1982). This field is a series of Palaeocene-age sand-rich suprafan lobes, with the trap geometry provided by depositional topography and subsequent submarine erosion. A hemipelagic shale caprock seals the suprafan complex. The discovery well was drilled by Esso in 1967 as a test of a structural prospect. The area was reinterpreted in the 1970s using seismostratigraphic principles and the field found to be a stratigraphic lowstand systems tract trap.

Reefs Carbonate build-ups (Section 7.1.4.2) may provide high-relief stratigraphic traps. Isolated pinnacle reefs may be completely encased in younger marine shale. Barrier reefs on carbonate shelves generally need to be sealed updip by tight back-reef facies. As noted for pinch-out traps, there is a risk that the back-reef facies will constitute a vaste zone of nonproductive reservoir rock. Although the relief on many reef traps may be considerable, the size of the accumulation is dependent on the presence or absence of permeable strata that terminate against the reef body. Careful seismostratigraphic analysis may be required to detect these zones of leakage.

The distribution of reservoir units within reef complexes may be variable and unpredictable, not only owing to depositional facies changes but also to the effects of diagenesis.

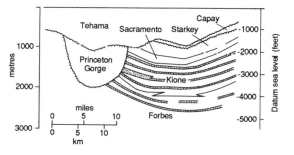

Fig. 10.69. Reservoirs truncated and sealed by an incised impermeable clay channel-fill (after Garcia 1981). This example is from the Sacramento Valley of California.

Reefs have formed very successful petroleum traps around the world, including in the Devonian of the Western Canada Basin, in the Sirte Basin of Libya, in the Tertiary of the Salawati and North Sumatra Basins of Indonesia, in the Miocene of Sarawak, in the Permian Basin of west Texas, and in southern Mexico and the Arabian Gulf.

Sophisticated exploration techniques, primarily in geophysical data acquisition, processing and interpretation, may be needed to explore successfully for reefs in mature provinces. The remaining, smaller reefs in these provinces are extremely subtle features, their detection requiring a very high level of geological and geophysical expertise.

Unconformity traps

A variety of traps may develop at unconformities, both immediately above and immediately below the unconformity surface.

Many of the depositional stratigraphic traps previously described may also develop on unconformities. An example is a pinch-out trap. Overstepping marine shales frequently provide a topseal to shallow shelf or shorezone sands. These traps develop on margins undergoing marine onlap and transgression, for example, in the transgressive systems tract of sequence stratigraphic terminology (Section 6.6).

Supra-unconformity sands may be localized by topography in the unconformity surface. Thus, incisive channels in the Type 1 unconformity surface (Section 6.6, Fig. 6.34) formed at a relative low stand may become sand-filled during rising sea level. Valleys developed along the strike of the outcropping pre-unconformity strata may be the location of the first post-unconformity fluvial sediment. In both cases, a stratigraphic trap at the unconformity surface may be formed, particularly if regional structural dip is in a suitable direction.

Traps developed beneath an unconformity by *truncation* of reservoir beds may give rise to giant fields. Examples have been previously described of eroded extensional fault blocks in the East Shetland Basin of the North Sea (e.g. the Brent reservoir in the Statfjord field, Fig. 10.53). A further example is the Fortescue–Halibut field of the Gippsland Basin, Australia. There may be several elements to the closure developed in subunconformity traps, including the topography present at the erosional unconformity, and the

structural geometry of the pre-unconformity beds. Cross-faulting, for example, may provide closure in the strike direction. Topography on the unconformity may have been provided in the first instance by fault scarps.

The lithology of the post-unconformity sediments is critical to the effectiveness of subunconformity truncation traps. If, for example, a thin marine basal transgressive sand is present above the unconformity, leakage may occur. This sand may be so thin as to be below seismic resolution. Careful seismostratigraphic interpretation and basin analysis is required in order to understand the causes of the unconformity and the evolution of the sequences above and below it, before accurate predictions may be made of sedimentary facies impinging on the unconformity surface. Ideal conditions are created where the unconformity subsides rapidly into deep water depths, perhaps as a result of rapid initial fault-controlled or thermal subsidence, and the topography is passively infilled by deep marine muds. Traps developed at unconformities that have been onlapped by shelf, shorezone or non-marine sequences may not be effectively sealed (cf. Section 10.5.4 regarding caprocks).

When the subunconformity strata are carbonates, exposure at surface may have very important implications for reservoir development in the subunconformity trap. The Casablanca field in the Gulf of Valencia, offshore southern Spain, was formed by Miocene faulting, erosion and leaching of a Mesozoic sequence of tight, dense limestones (Watson, 1981). The faulting formed the palaeotopography and fractured the brittle carbonates, allowing the penetration of meteoric waters and the development of secondary porosity to depths of up to 150 m. The eroded, elongate limestone ridge is overlain by mid-Miocene Alcanar Formation organic-rich marls, which charge and cap the accumulation. A similar example is the Angila field in the Sirte basin of Libya (Fig. 10.70); in this case it is weathered granite that forms the reservoir in the unconformity trap (Williams, 1968).

Probably the two best-known North American examples of unconformity traps are the Prudhoe Bay field in Alaska, and the East Texas field. At East Texas, the productive Woodbine sandstone reservoir is sandwiched between two unconformities. The trap is formed by truncation beneath the upper of the two (Fig. 10.71).

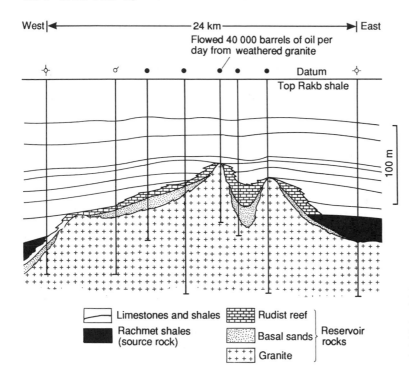

Fig. 10.70. Cross-section through the subunconformity trap of the Angila field, Libya. Some of the production is from fractured and weathered subunconformity granitic basement (after Williams 1968).

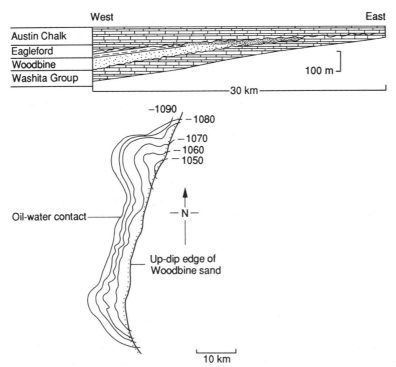

Fig. 10.71. Cross-section and map of the East Texas field. This is a giant accumulation trapped by truncation of the Woodbine Sandstone beneath an unconformity (after Minor and Hanna 1941).

Diagenetic traps

There are a number of traps in which diagenesis has played a significant part. Examples are where cementation (e.g. of dolomite or calcite) has provided an updip seal to an accumulation, or where leaching has produced a local reservoir in an otherwise impermeable sequence. These trapping mechanisms are usually established after the main phase of oil generation, and have relatively low predictive importance.

An interesting form of diagenetic trap is where cementation has taken place below the oil–water contact of an existing accumulation (diagenesis is frequently inhibited by the presence of petroleum in the pore space). This may seal in the accumulation despite the subsequent removal of the original trapping mechanism, for example by tectonic activity.

At temperatures of less than about 70 °C, and in the presence of meteoric water, we have seen that bacterial degradation of oil may take place (see Section 10.3.2.5). This can form an impermeable updip tar mat that seals subsequently migrated oil. Examples have been quoted from the Californian San Joaquin Valley (Wilhelm, 1945) and Russian Volga–Ural region (Vinogradov Aver'yanov and Nigmati 1983).

Similarly, a change of phase from gas or liquid to solid may also provide an unusual kind of trap. At high latitudes, permafrost may provide an updip seal to petroleum accumulations. At particular pressure-temperature conditions (low temperature, high pressure) petroleum gases may form *solid hydrates*. These are solid crystalline precipitates of gas and water. Not only is gas trapped in the hydrates themselves, but may accumulate in reservoirs underlying the zone of hydrate formation (Downey, 1984). Hydrates are most likely to form in shallow onshore reservoirs in permafrost areas (examples have been quoted at depths of up to 200 m in Siberia), and in cold, deep sea areas (the Glomar Challenger cored gas hydrates at subsea depths of 600 m on the Blake Plateau).

10.6.2.3 Hydrodynamic traps

Hydrodynamics was briefly introduced in Section 10.3.2.4 in the context of the movement of petroleum fluids through basins. There are relatively few basins worldwide where hydrodynamics is known to have a significant impact on the entrapment of petroleum. These are typically foreland basins where porous and permeable carried beds have been uplifted and exposed in adjoining fold/thrust belts, allowing the influx of meteoric water (see Section 9.3.3). If an outlet for the water is available elsewhere in the 'plumbing system' of the basin, hydrodynamic flow of the basin fluids may take place. Once the necessary conditions are known to be established at the basin scale, individual exploration prospects can be evaluated with a view to hydrodynamic effects.

Under conditions of strong hydrodynamic flow, petroleum–water contacts may be inclined rather than horizontal, petroleum may be completely flushed from structural or stratigraphic closures, or hydrodynamic closures may be produced (for example on structural noses) where there is no other form of closure present. Each prospect needs to be evaluated individually, since prospectivity will depend on a host of regional and local factors that may be difficult to assess. Readers are referred to Hubbert (1953) for the theory of hydrodynamic trapping.

Clearly, an understanding of the structural and stratigraphic evolution of the basin is required for an assessment of the impact, if any, of hydrodynamic conditions on petroleum entrapment. Generally, however, hydrodynamic traps and hydrodynamic effects appear to be relatively rare.

10.6.2.4 Worldwide statistics on trap types

Moody (1975) has presented statistics on the numbers of giant (> 500 million barrels) oil fields and ultimately recoverable oil reserves distributed amongst the various trap types. His total sample was 198 giant oil fields discovered up to the early 1970s. Gas is not included. The results are shown in Fig. 10.72. Although the classification used by Moody does not relate in detail to the breakdown presented in this book, the statistics are revealing. By far the greatest proportion of the world's giant oil fields have been found in anticlinal structural traps, with the second biggest grouping being traps in which a combination of factors are involved. Stratigraphic traps are probably under-represented. Many stratigraphic traps are probably still to be found, or have been found since the production of the Moody statistics. Many stratigraphic

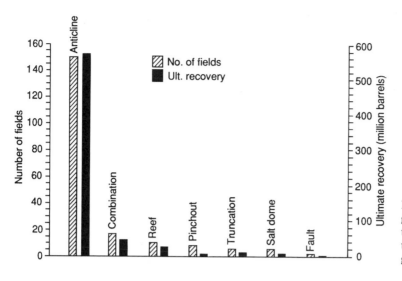

Fig. 10.72. Worldwide statistics on giant oil fields according to trap type (after Moody 1975). Anticlinal traps dominate the world's existing giant oil discoveries.

accumulations also tend to be sub-giant, and not included in the Moody dataset.

10.6.3 Timing of trap formation

An understanding of the mechanism of trap formation, and therefore the timing of trap formation, is essential to prospect evaluation. A trap that developed too late to receive a petroleum charge will be dry. Each of the structural, stratigraphic, and hydrodynamic trap types discussed in this section has implications for trap timing that are obvious. Depositional and unconformity traps are very early, dating from the time the sealing units became effective. Thus these traps are ready to receive a charge from a very early stage. Some structural traps, however, are very late in relation to petroleum charge. Each trap needs to be individually evaluated.

The timing problem is perhaps best illustrated by considering a currently active fold/thrust belt. The fold/thrust belt forms as a result of shortening, at least in the sedimentary cover, which results in uplift. Uplift results in cooling of the overthrust sheets, which switches off petroleum generation. Thus a timing problem may exist, unless generation is maintained in subthrust positions by loading of the allochthonous sheet, or unless earlier-trapped petroleum re-migrates into the new fold structures.

A similar problem exists for inversion structures. Prior to inversion, migration is usually directed away from the site of the future trap. Inversion may switch off new generation, thus the charge into the inversion trap must be from re-migrated oil.

Structures formed in areas of continuous subsidence also need careful evaluation, so that the growth of the structure may be closely related to the timing and volume of petroleum charge. The sedimentary section may need to be backstripped (Chapter 8), to the time of first trap formation, and the charge system geochemically modelled.

11 Quantification of undiscovered potential

'I have often admired the mystical way of Pythagoras, and
the secret magic of numbers'

(Sir Thomas Browne (1605–1682))

SUMMARY

Quantitative estimates of the undiscovered potential of petroleum plays are required by petroleum exploration companies in order to evaluate exploration investment opportunities and guide long-term strategic plans. A range of geological, geochemical and statistical techniques have been developed over the years to estimate undiscovered resources. A play assessment method is described in this chapter that combines these techniques, and is built upon the geological interpretations of depositional sequences made through the process of basin analysis. There are four stages to the approach.

The first stage is the development of the play concept and the mapping out of the play fairway. Play fairway maps show the key geological controls on the play, namely those controlling the presence of an effective reservoir, petroleum charge, regional topseal and traps. Play fairways are primarily reservoir-defined, and may be subdivided into a number of *common-risk segments*, defined by lateral variations in play chance, anticipated field sizes or drilling success ratios.

The second stage is estimation of the *number and sizes of undiscovered fields* in each common-risk segment. Ranges of possible field sizes must be consistent with the sizes of already identified prospects in the segment, and be calibrated against already discovered field sizes in the same or analogous play. The potential recoverable petroleum volumes present in a prospect may be calculated as the prospect trap capacity through the 'pore-volume equation', or as the volume of petroleum charge, if this is less. The charge volume calculation is subject to large errors, owing to difficulties in estimating petroleum volumes lost during migration. If field sizes are borrowed from an analogue play, care should be taken that the analogue is truly valid.

Assessment or risk is the third stage. There are three elements of risk. *Model risk* is the risk on the validity of the geological model that underpins the entire play assessment, specifically the geometry and sedimentary facies of the depositional sequences involved in the play. Given the model is valid, *conditional play risk* is the chance that an effective play is produced, that is, a sealed producible reservoir with a petroleum charge. Model risk and conditional play risk together constitute *play chance,* the overall chance of the play working. Play chance can be calibrated with worldwide statistics on the success rates of plays in various parts of the 'facies cycle wedge' or depositional sequence. Sandstone plays in wedge-base (transgressive systems tract) positions and carbonate wedge-top (highstand systems tract) plays are amongst the most successful worldwide. *Prospect-specific risk* is particular to individual prospects, and relates mainly to the validity of the individual trap. Prospect-specific risk exists even in proven play fairways, and is caused largely by unpredictable heterogeneities in geology over the extent of the fairway. Prospect-specific risks can be calibrated by drilling success ratios in proven analogous plays. The chance of success in an individual prospect (*prospect success chance*) is the product of each of these three elements of risk, and is the probability of encountering petroleum volumes in the prospect within the range predicted.

Stage 4 is the calculation of *assessment curves,* which show the range of petroleum volumes that may be found in the play or prospect, together with their probability of occurrence. These volumes should be compared with geochemical petroleum charge volumes, so that no greater resource is placed into the play than can realistically be charged from the source rock.

11.1 INTRODUCTION

Quantitative estimates of the undiscovered potential of plays are required by petroleum exploration companies in order to evaluate exploration investment opportunities and guide long-term strategic plans. Estimates of the likely size and timing of future discoveries, and hence future petroleum supply, also form an important element in the planning studies of government organizations.

A number of different techniques have been developed over the years for the estimation of undiscovered petroleum resources. White and Gehman (1979) and Miller (1986) have provided recent reviews. Techniques fall generally into four main groups:
• *Subjective methods*, relying on the personal experience and ability of the assessor. These estimates

can be highly biased. The *Delphi* technique provides a means of collecting together a range of personal opinions held by individual assessors.
• *Basin statistics* comprise the historic field sizes and drilling success ratios for a play, and provide a means of calibrating volume and risk estimates against the reality of past experience. Basin statistics can be drawn from within the same play as the one being assessed, or can be borrowed from an analogue play elsewhere.
• *Statistical modelling of historical discovery data* Statistical 'discovery process models' that characterize the process by which previous oil and gas discoveries have been made, are applied to predict undiscovered field sizes. The models require statistical information on the sizes and timing of previous discoveries. Geological expertise is needed to define the play.
• *Geochemical modelling* uses the physical and chemical principles of petroleum generation, migration and entrapment to calculate the volumes generated, expelled, lost and available to charge traps. The approach is limited by the enormous uncertainties surrounding volumes lost during migration and as a result of leakage from traps.

Combinations of these techniques are normally employed in practice. The technique described in this chapter is such a combination approach, and is appropriate for plays that have been assessed via the discipline of basin analysis. It is a four-step approach:
1 Development of the play according to the concepts described in Chapter 10, and the mapping out of the play fairway.
2 Estimation of the numbers and sizes of undiscovered fields in the fairway, from the evaluation of individual identified prospects, already discovered field sizes, and from the use of field size statistics borrowed from a geological *analogue* to the play being assessed.
3 Estimation of play fairway risks. These are subjectively estimated, and based primarily on our confidence in the interpreted sedimentary facies distributions in each of the depositional sequences involved in the play, and the implications of these facies for the presence of reservoir rocks, petroleum charge, and regional topseals. The estimates can be calibrated by reference to the success rates of analogous plays.
4 Calculations of undiscovered potential and cali-

bration with charge volumes calculated from geochemical modelling.

This approach is a modified version of that described by Baker *et al.* (1986).

11.2 DEFINITION AND MAPPING OF THE PLAY FAIRWAY

The first stage in play assessment is the definition of the play and the mapping of its fairway.

Play fairway maps show the geographical distribution of the key geological controls on the play fairway. We have seen in Chapter 10 that these geological controls are those that determine:
- the presence of an effective reservoir
- a petroleum charge into the reservoir
- a regional topseal to the reservoir
- the presence of traps
- the timely relationship of the above factors.

Gerard Demaison (1984) introduced the 'generative basin concept', and described how petroleum generative kitchens could be mapped out. White (1988) has extended this concept by including the reservoir, topseal and trapping controls on the play.

Plays are essentially reservoir-defined. Hence, fairways at different stratigraphic levels in a basin may be stacked vertically. Within a single play, all prospects and discovered fields share a common geological mechanism for petroleum occurrence. Petroleum accumulations, discovered or undiscovered, within a single play fairway, can be considered to constitute a naturally occurring population or family of geological phenomena. Thus, each play can be characterized by a specific distribution of field sizes. Furthermore, the drilling success ratio within a play may also be a characteristic feature. These assumptions underpin the play assessment method.

The concepts of 'unproven plays' and 'play chance' were introduced in Chapter 10. Owing to the interplay of the critical geological factors over the extent of the fairway, it is normal for play chance to vary areally. This variation in play chance may be due to hard evidence of adverse geology in different parts of the fairway (for example, determined from well or seismic data), or to variations in the quantity or quality of the data base, which allows greater or less confidence in the interpretations made. As a result, an unproven fairway may be subdivided into a mosaic of segments; within each segment play chance is constant. These segments can be termed *common-risk segments*.

The fairway can also be subdivided into segments if there are strong reasons for believing field sizes are likely to be significantly different (for example, as a result of differing structural development in different parts of the fairway), or drilling success ratios are likely to vary significantly. Thus the three factors that define fairway segments are variations in:
- play chance
- expected field sizes
- expected drilling success ratios

In this way, reservoir-defined plays are sometimes subdivided by trap type. Examples are the Kapuni Overthrust segment and the Kapuni Inversion segment of the Eocene Kapuni Coal Measures Play of the Taranaki Basin, New Zealand.

Figure 11.1 is an example of a simple play map, modified from White (1988). The following items of geological information are contained on the map:
1 the depositional or erosional limits of the reservoir unit
2 the distribution of reservoir facies within the gross reservoir unit
3 the areas where a source rock is present
4 the area where it is mature (the kitchen)
5 a migration zone around the kitchen; together with (4) it represents the area receiving a petroleum charge
6 areas where there is an effective regional seal
7 areas where traps are present (structural or stratigraphic)
8 oil and gas fields, dry holes, and untested prospects, leads and notional prospects
9 drilling success ratios for specific parts of the fairway.

The drilling success ratio is the ratio of the number of *technical successes* to the number of *valid tests* of the fairway.

A *technical success* is an exploration well that flows petroleum to surface or in which the presence of petroleum in drill-stem or wireline formation tests convincingly demonstrates the presence of a pool of recoverable petroleum. It carries no implication of commerciality. (A commercial success is a well that discovers an economically developable petroleum accumulation).

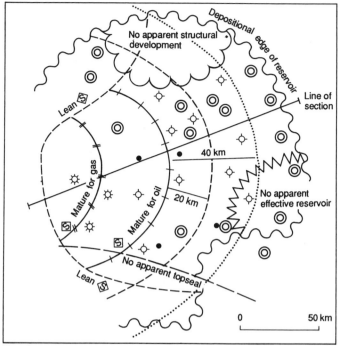

Line of section

No apparent structural development

Depositional edge of reservoir

Lean Ⓢ

Mature for gas

Mature for oil

40 km

20 km

No apparent effective reservoir

No apparent topseal

Lean Ⓢ

0   50 km

X section

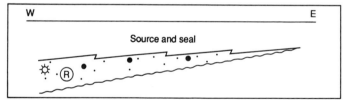

W   E

Source and seal

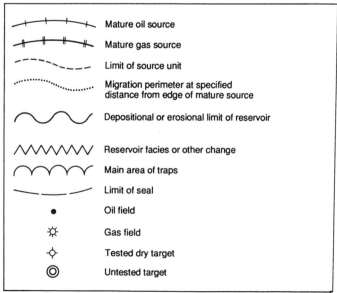

⊥⊥⊥ (curve)	Mature oil source
⊥⊥⊥ (curve)	Mature gas source
– – –	Limit of source unit
⋯⋯	Migration perimeter at specified distance from edge of mature source
∿	Depositional or erosional limit of reservoir
WWW	Reservoir facies or other change
⌣⌣⌣	Main area of traps
——	Limit of seal
●	Oil field
☼	Gas field
◇	Tested dry target
◎	Untested target

Fig. 11.1. Example of a play map (after White 1988). The play map should show all of the important geological factors likely to control hydrocarbon accumulation in the fairway. The fairway is initially reservoir-defined, then the extent of charge, topseal and likely trap development are added.

A *valid test* is a well which penetrated the play fairway, and intended to test an exploration target in the play fairway. In basins with vertically stacked fairways, there may be many more penetrations of a fairway than there are valid tests. Valid tests are not only those wells that tested valid traps in the fairway.

There are normally far more dry holes than technical successes in a play. Within a proven play, dry holes are caused by local geological variations, such as the absence of a lateral seal in a faulted prospect, the existence of a migration shadow, or local diagenetic destruction of reservoir porosity.

These factors contribute to prospect-specific risk, which will be discussed further in Section 11.4.

In Fig. 11.1, an area of mature source in the west of the area has experienced the highest drilling success ratio (75 per cent). As we move outwards, firstly a migration distance of 20 km, and then a migration distance of 40 km, the success ratios drop to 28 per cent and then 20 per cent. This is interpreted to be caused by more and more prospects failing to lie on migration routes as the distance from the kitchen increases. Despite these variations in success ratio, however, the play is effectively proven in these areas. Beyond the 40 km migration perimeter, there are no existing discoveries and the play is unproven. The unproven part of the fairway is divided into the following *common risk segments,* illustrated in Fig. 11.2:

• B: an area of no apparent structural development in the north (no traps). The seismic grid, however, is coarse and there may be stratigraphic traps, so the possibility of traps cannot be ruled out (play chance assessed at 0.20).

• C: an area beyond the 40 km migration perimeter in the east. Although unlikely, a very focused migration path could charge a prospect in this segment (play chance is assessed at 0.50).

• D: an area beyond the 40 km migration perimeter where reservoir effectiveness may be destroyed by diagenesis (play chance assessed at 0.30).

• E: an area in the northwest, updip from a lean source. However, the geochemical database is in

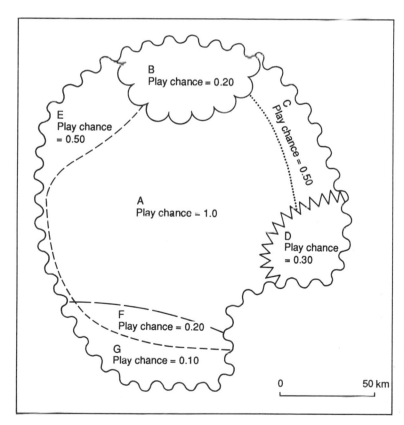

Fig. 11.2. Subdivision of the play fairway into common-risk segments. These are controlled by the distribution of reservoir charge, topseal and likely trap development shown on Fig. 11.1. A play chance is assessed for each common-risk segment. In this illustration, segment A is considered already proven (play chance = 1.0).

fact quite poor and source rock interpretations unreliable (play chance is assessed at 0.5).
• F: an area in the south with no apparent topseal (play chance assessed at 0.20).
• G: an area in the south updip from a lean source and also with no apparent regional topseal (play chance assessed at 0.10).

The assessment of play chance probabilities will be discussed in Section 11.4. Beyond the depositional or structurally controlled limits of the gross reservoir unit, there is no play at all. This is the outer limit of the fairway.

11.3 NUMBERS AND SIZES OF UNDISCOVERED FIELDS

The next step, having defined the play and mapped the fairway, is to estimate how many fields there could be in the fairway, and how big they are. These estimates can be represented as a frequency distribution of undiscovered field sizes. The integral of the predicted field-size distribution, or sum of all the predicted fields, is an estimate of the undiscovered potential of the play.

11.3.1 Field sizes

Undiscovered field sizes can be estimated in a number of ways. Much depends on the amount and quality of data available. When a large data base is available, and it is felt that all prospects have been identified, undiscovered field size estimates may be built up from evaluations of each of the individual prospects. Calculation of prospect petroleum volumes is described below.

This approach, which is rarely possible for the whole play and is time-consuming, places great faith in the reliability of the individual prospect evaluations, which should therefore be calibrated against field sizes already discovered in the play.

It is normally desirable, therefore, to estimate field sizes in alternative ways. Options include:
• Using the existing field size frequency distribution for the play as a guide to future field sizes. This approach assumes that both the discovered and undiscovered fields are random samples from the population of all fields. This is extremely unlikely since the biggest fields tend to be found relatively early. As a result, the approach will tend to overestimate undiscovered field sizes.

• Borrowing a field size frequency distribution from an anlaogue play. An example, from Baker et al. (1986), is shown in Fig. 11.3, and comprises 68 fields in rollover traps in the Tertiary of the US Gulf Coast (a more refined definition of the play is, however, advisable).
• Carrying out a Monte Carlo simulation of the input parameters (area of trap, porosity, etc.) for a typical prospect volume calculation, using a range of values for each parameter that represents the full possible variation that may exist across the fairway as a whole.

A combination of these approaches will normally yield the most realistic and reliable estimates. The following two requirements should always be satisfied, whichever approach is adopted:
1 That estimates are calibrated against known field sizes in the same or an analogous play;
2 That they are consistent with the predicted sizes of any identified prospects in the play.

Great care should be taken to ensure that the field size distribution does not extend to unrealistically large values. The largest few fields generally have an overwhelming impact on the resources of the play, so it is important to get them right. Baker et al. (1986) show that 80 per cent of the total amount of oil in the approximately 14 000 fields found up to 1970 in the conterminous United States occurs in the biggest 3 per cent of the fields (> 50 million barrels) (Fig. 11.4). It may be sensible to tie the upper limit of the field size distribution approximately to the size of the biggest known prospect.

A sensible, practical lower limit to the field size distribution should also be set. This should be at least as low as the minimum *economic* field size. There will almost certainly be a large number of fields smaller than this limit that contribute to the overall resource endowment, but they should, for practical reasons, be ignored in the play assessment.

If a field size distribution is borrowed from an analogous play, it is essential that the analogue is valid. It is very easy to be overconfident in the validity of an analogue. To a large extent, every play is unique. Use of an analogue, however, allows a realistic resource estimate to be made in the absence of local data. An analogue should be valid in terms of:
• charge volumes (source volumetrics)
• reservoir characteristics

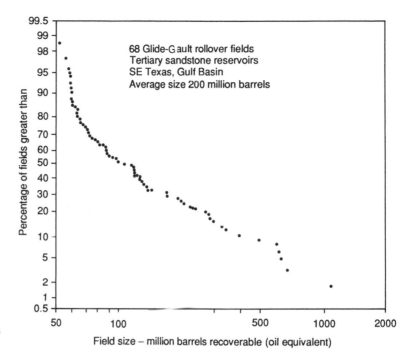

Fig. 11.3. Field size distribution of rollover traps at down-to-basin faults, south Texas, Gulf Basin (after Baker *et al.* 1986). This distribution may be borrowed in order to assess likely field sizes in a geologically analogous play.

- the migration 'plumbing' system, including the retention characteristics of the seals
- the structural style, which controls the density and sizes of traps.

Analogues based on these factors are valid in terms of the physical and chemical processes of petroleum generation, migration and entrapment,

and are preferred to analogues drawn on the basis of geological basin classification schemes (Kingston, Dishroon and Williams 1983a, b, Klemme 1980).

Kingston *et al.* (1983a) have described a global basin classification scheme based on the tectonics that formed the basin, the nature of the contained

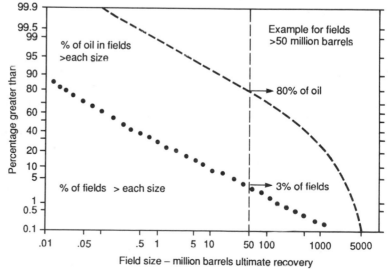

Fig. 11.4. Field size distribution for 13 985 oil fields for the conterminous United States (after Baker *et al.* 1986). 80 per cent of the total volume of oil is found in the biggest 3 per cent of the fields (those greater than 50 million barrels). It is particularly important, therefore, to estimate accurately the sizes of the biggest fields in a play, since these have an overwhelming impact on the ultimate resource.

depositional sequences, and the tectonics that modified it (cf. Chapter 1). This classification is used as the basis for predicting diagnostic suites of hydrocarbon plays in each basin type (Kingston *et al.* 1983b). While this scheme provides an extremely useful broad geological framework that may be used to understand the overall basin development and the development of broad hydrocarbon plays, it tends to be too coarse as a resource assessment tool, unless the data base is extremely poor indeed.

Perhaps the most versatile approach to the use of analogues in play assessment is to assemble a data base comprising geological descriptions and field size and success ratio statistics for as large and representative a selection of the world's important play systems as possible, and to 'search' this data base using whatever geological criteria are known or are considered important for the play being assessed. Thus, parameters such as:
• megasequence (or supersequence) type (e.g. extensional syn-rift, contractional foreland basin)
• gross structural style (e.g. contractional thin-skinned thrusting, gravity tectonics (growth faults))
• gross depositional environment (e.g. non-marine lacustrine, basinal submarine fan)
• gross lithology (e.g. sandstone)
• trap type (e.g. extensional rotated fault blocks, contractional ramp anticline)
may be used as search criteria, depending on the knowledge level for the play being assessed. Clearly, retrievals of potential analogues can be based on either very broad or very narrow criteria, which makes the approach applicable to plays at widely varying stages of exploration maturity.

11.3.2 Calculation of prospect volumes

11.3.2.1 Trap capacity

The potential recoverable reserves of a prospect can be calculated using the simple 'pore-volume equation'. For an oil prospect:

Recoverable Reserves

$$\text{(STB)} = \frac{\text{BRV} \times \text{N/G} \times \phi \times \text{Shc} \times \text{RF} \times 6.29}{\text{FVF}}$$

(11.1)

where BRV = Bulk Rock Volume in m^3

N/G = Net/Gross ratio of the reservoir rock body making up the BRV
ϕ = average reservoir porosity
Shc = average hydrocarbon saturation
RF = Recovery Factor (the fraction of the in-place petroleum expected to be recovered to surface)
6.29 = factor converting m^3 to barrels
FVF = Formation Volume Factor of oil. This is the amount that the oil volume shrinks on moving from reservoir to surface
STB = Stock Tank Barrels, i.e. barrels at standard conditions of 60 °F and 14.7 psia.

A similar calculation can be made for gas, with the oil shrinkage factor in the denominator replaced by a gas expansion factor in the numerator.

The Bulk Rock Volume can be estimated in a number of ways. If the geometry of the gross reservoir body approximates to a slab, BRV is calculated as the product of the trap area and the average thickness of the gross reservoir body. If the reservoir geometry is not a slab, it is more accurate to construct a trap area versus depth plot (see Fig. 11.5). Area/depth curves should be drawn for the mapped seismic reflector, the top of the reservoir, and the base of the reservoir. The BRV is simply the area between the top reservoir and base reservoir curves, and the spill point of the trap.

Because there is uncertainty over the exact value for each of the parameters in the pore-volume equation, they should be expressed as *probability distributions*. These represent the geologist's perception of the range of values a parameter may have, with their associated relative probability of occurrence. Some commonly used probability distributions are shown on Fig. 11.6. The probability distributions for each of the parameters may then be multiplied together using the very common *Monte Carlo simulation* procedure, to yield a probability distribution of petroleum volumes. The mechanics of this procedure are outlined on Fig. 11.7. A minimum and a maximum petroleum volume could be arbitrarily defined as the 5th and 95th percentiles of this distribution, but other values could be chosen. The probability distribution of petroleum volumes may be presented as a cumulative probability distribution or 'assessment curve' to show the chances of volumes exceeding given values, as in Fig. 11.8. In this figure, there is

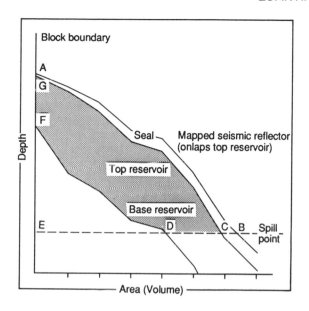

ABE = Total volume of rock enclosed by mapped horizon
ABC = Volume of overlying seal
FDE = Volume of underlying non-reservoir (or non-petroleum bearing) rock
GCDF = Volume of the reservoir (petroleum bearing rock)

Fig. 11.5. Area versus depth plots for calculation of Bulk Rock Volume. Curves should be drawn for the mapped seismic reflector and top and bottom of the gross reservoir unit.

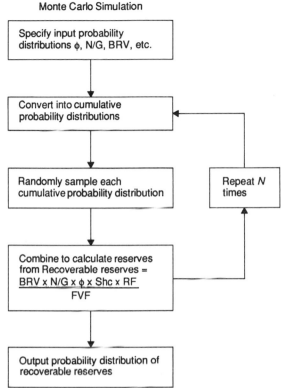

Fig. 11.7. Monte Carlo simulation for the calculation of prospect volumes. Typically, a thousand or more iterations are made to yield a probability distribution of potential petroleum volumes.

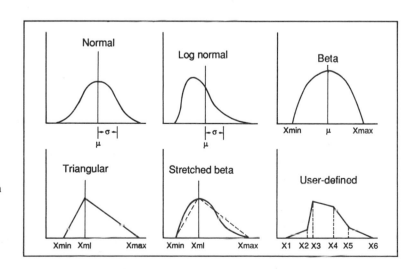

Fig. 11.6. Common types of continuous probability distribution used in petroleum volume calculations. These represent uncertainty in the value for parameters such as porosity and hydrocarbon saturation.

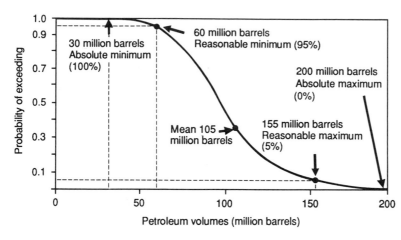

Fig. 11.8. Unrisked prospect assessment curve. This shows the chance of exceeding specific values of petroleum volume. Volumes could lie anywhere between 30 and 200 million barrels. There is a 50 per cent chance of exceeding 100 million barrels, and a 5 per cent chance of exceeding 155 million barrels.

a 50 per cent chance of volumes exceeding 100 million barrels. The 5 per cent and 95 per cent points, which may be taken as a reasonable maximum and reasonable minimum respectively, are 155 million barrels and 60 million barrels. The absolute minimum is about 30 million barrels.

11.3.2.2 Charge volumes

The petroleum volume of a prospect calculated in this way is in effect only its trap capacity. If the charge is inadequate, the trap will not be full. The trap capacity should in these circumstances be discounted by a Degree of Fill factor (DOF), or the charge volumes from geochemical modelling used directly as the limiting petroleum volume for the prospect.

Mackenzie and Quigley (1988) have described an approach to the calculation of prospect charge volumes (Fig. 11.9) (see also Section 10.3.2.2). First, the mass of petroleum expelled from that part of the source kitchen that provides a drainage or catchment area for the prospect being evaluated, is calculated. This is achieved by dividing the kitchen into a series of 'isomaturity slabs', and for each slab making the following calculation:

$$M_E = (P_0) \, (PGI) \, (PEE_n) \, (\rho_{rock}) \, (y) \, (AREA)$$
$$(11.2)$$

where M_E = mass of petroleum expelled from the slab

P_0 = average initial petroleum potential

PGI = Petroleum Generation Index (the fraction of petroleum-prone material that has been transformed into petroleum)

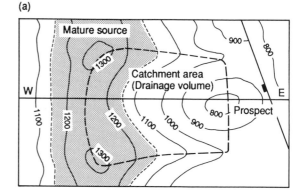

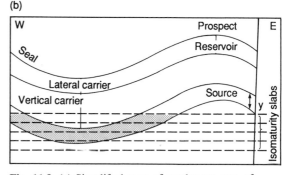

Fig. 11.9. (a) Simplified map of catchment area of hypothetical prospect for petroleum fluids. (b) E–W cross-section through region and prospect. Masses of petroleum expelled are calculated for each 'isomaturity slab' of source rock within the catchment area of the prospect. These are summed and converted to volumes expelled. The charge volume to the prospect is then the volume expelled minus the volume lost along the migration pathway. Volumes lost during secondary migration are a function of the volume of carrier bed through which the petroleum passes (after Mackenzie and Quigley 1988).

PEE$_n$ = net Petroleum Expulsion Efficiency (the fraction of petroleum formed in the source that has been expelled towards the prospect)

ρ_{rock} = bulk density of the source rock

y = mean thickness of source rock

AREA = area of the isomaturity slab in the catchment area of the prospect.

The masses of petroleum expelled from each slab are then summed. Masses can be converted to volumes expelled (V_E).

Second, an estimate of the petroleum volumes lost along the migration pathway is calculated:

$$V_L = \phi f V_D \qquad (11.3)$$

where V_L = volume of petroleum lost

ϕ = porosity of the rock through which the petroleum flows

f = residual hydrocarbon saturation in this rock

V_D = volume of this rock

The charge volume (V_C) for the prospect is then:

$$V_C = V_E - V_L \qquad (11.4)$$

Some of the parameters in these equations are very difficult to estimate, particularly those concerning volumes lost. As a result, charge volumes can normally be predicted to an accuracy of ±50 per cent at best.

11.3.3 Number of fields

The number of undiscovered fields may be estimated in at least three ways:

• by *counting* the number of identified and notional prospects greater than the minimum size on the field size distribution, and dividing by an anticipated drilling success ratio (which may be borrowed from an analogue play).

• by borrowing a *prospect density* (number of prospects per unit area) from an analogue, scaling the number of prospects to the area of the play, and dividing by the drilling success ratio.

• by directly borrowing a *field density* from an analogous play. This obviates the need to count prospects (which may be impossible or impractical) or to estimate drilling success ratio.

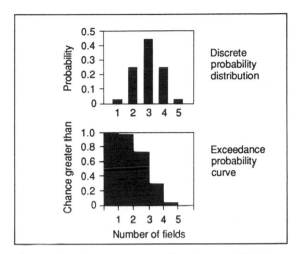

Fig. 11.10. Estimated number of undiscovered fields in a play (after Baker *et al.* 1986). Between 1 and 5 fields may occur in the play fairway, and the most likely number is 3. The probability distribution and exceedance curve are shown in this example.

The number of undiscovered fields should be considered as a discrete probability distribution, as shown in Fig. 11.10. In this example, somewhere between one and five fields are expected, and three is the most likely number. Rather than consider the number of undiscovered fields as a probability distribution, it is frequently adequate to make a single-point estimate (for example, three).

11.4 PLAY FAIRWAY RISK ASSESSMENT

There are three elements of risk in play fairway analysis which are an outcome of the process of conceiving a play, which was described in Sections 10.1 and 10.2. These are:

• model risk ⎫
• conditional play risk ⎬ play chance
• prospect-specific risk ⎭

The relationship between these risks is illustrated in Fig. 11.11.

Model risk Model risk is the risk on the validity of the geological model which underpins the play assessment. Specifically, this means that the sequence geometry and sedimentary facies must occur as predicted. Model risk is largely a function of the adequacy of the data types (outcrop, wells

Model risk
The chance that the interpreted geological model for the fairway is valid, specifically the presence of the predicted sedimentary facies.

Conditional play risk
Given this model, the chance that an effective play is produced, specifically a timely petroleum charge into a producible reservoir with an effective regional topseal.

Prospective-specific risk
Given this play, the chance that a specific prospect contains a volume of technically recoverable hydrocarbons within a predicted range.

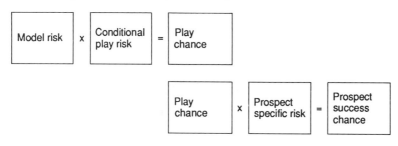

Fig. 11.11. Three elements of risk in prospect and play assessment.

and seismic) in constraining the stratigraphic interpretation.

Figure 11.12 shows three possible geological models that may be interpreted to explain the observational data on a seismic section. Each has different implications for play assessment. Model 1 is of a regressive platform sequence unconformably overlying a non-marine sequence, and overlain by a thick sequence of prograding clastics. These three sequences provide the ingredients of a petroleum play, as indicated. Model 2 is quite different – a volcanic plateau – with no chance of developing a play. Model 3 is a variation on model 1, but a basal trangressive sand is present at the base of sequence 3, which causes a problem with topseal over much of the fairway. This sand is below seismic resolution. Each model has a probability of being valid, which is its model risk.

Conditional play risk   Given that the geological model is valid, conditional play risk is the chance that an effective play is produced, that is, a sealed producible reservoir with a petroleum charge. If, for example, an element of model risk was on the occurrence of lacustrine mudstones within a rifted half-graben sequence, the corresponding element of conditional play risk is on whether these lacustrine mudstones, if present, contain mature source rocks.

11.4.1 Play chance

Model risk and conditional play risk together constitute the absolute risk on a play being present. This is known as *play chance.* A play chance estimate is built up by considering the charge, reservoir, topseal and trap elements of the play, and relates to the chances that these elements will combine somewhere in the fairway to produce the required number and size of fields. Table 11.1 is an example of a play chance assessment for the same play as illustrated in Section 11.3.

If an element of the geological model or conditional play is absent, the entire play is invalidated. Typical values of play chance are shown in Table 11.2.

A means of *calibrating* play chance against worldwide experience is provided by White (1980). The basis for White's study was a compilation of 1150 plays from 200 'facies cycle wedges' in 80 basins worldwide. The facies cycle wedge may be considered equivalent to a depositional sequence (Chapter 6). White (1980) collected statistics on the occurrence of both successful and unsuccessful plays according to position of the play within the 'facies cycle wedge'. Each facies cycle wedge comprises a fully or partly complete transgressive-regressive sequence, bounded above and below by

Fig. 11.12. (*Opposite*) Schematic illustration of three possible geological models that explain the observable seismic data. Each has different implications for play assessment. The chance of each model being valid is known as model risk.

(a) THE DATA

(b) THE INTERPRETATION

Sequence

Model 1

| Prograding shelf |
| Carbonate platform |
| Lacustrine syn-rift |

(c) THE INTERPRETATION

Sequence

Model 2

| Fluvial plain |
| Volcanic plateau |
| Meta-sediments |

(d) THE INTERPRETATION

Sequence

Model 3

| Prograding shelf |
| Shallow marine |
| Carbonate platform |
| Lacustrine syn-rift |

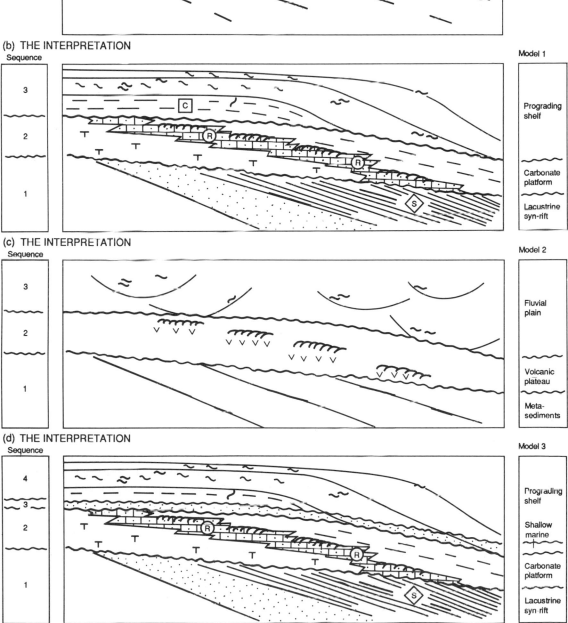

Table 11.1 Example of a play chance assessment, broken down into model and conditional play risk

	Model risk	Conditional play risk	Play chance
Reservoir	1.0	0.9	0.9
Charge	0.9	0.8	0.7
Topseal	0.9	0.9	0.8
Traps	1.0	1.0	1.0
Overall	0.8	0.65	0.5

Note: In this example, there is a 50 per cent chance that a reservoir, charge, topseal and traps will combine somewhere in the fairway to produce the specified number and sizes of fields.

regional unconformities, and containing characteristic facies associations (Figs. 11.13 and 11.14).

White defined five main play types, representing wedge-base, middle-wedge, wedge-top, wedge-edge, and subunconformity positions. These are in effect systems tracts within deposition sequences (Chapter 6). Different risk (play chance) is associated with each of these different wedge positions, as a consequence of different reservoir/charge/seal relationships.

Figure 11.15 shows the chance of a major (> 50 million barrels of oil) field being present in each of the play types. We can see that if we are confident of a wedge-base position and a sandstone lithology, the chance of the play being productive is 60 per cent. This is the least risky play. In contrast, sandstone wedge-top and wedge-edge plays are the riskiest, with only 15 per cent chance of being productive. In carbonate wedges, the wedge-top is the least risky with about a 45 per cent chance of being productive. Overall, 35 per cent of the plays studied (outside the Middle East) were productive, indicating an overall worldwide play chance of

0.35. Sandstone wedge plays are less risky (38 per cent productive) than carbonate wedge plays (30 per cent productive).

These statistics may be used, therefore, as a rough calibration to a subjectively estimated play chance.

It is useful to relate the statistics to reservoir, source, topseal and trap development in each of the wedge positions.

Wedge-base plays

Wedge-base sandstones are frequently shoreline deposits directly overlying an unconformity over which a marine transgression has taken place. They therefore are equivalent to transgressive systems tracts. Pinch-outs caused by landward onlap of the basal reservoir beds may provide stratigraphic traps. Overlying transgressive marine shales and micritic limestones are usually the source beds, as well as a good regional topseal. Sand distribution may be related to topography on the underlying unconformity, which may also contribute to the

Table 11.2 Typical values for each element of risk

Type of risk		Play chance	Prospect-specific risk	Prospect success chance
Typical values	Low risk	1 in 1 to 1 in 2	< 1 in 3	< 1 in 5
	Moderate risk	1 in 2 to 1 in 4	1 in 3 to 1 in 6	1 in 5 to 1 in 10
	High risk	> 1 in 4	> 1 in 6	> 1 in 10
Calibration		Over 1000 plays studied by White (1980)	Drilling success ratios in *proven* play systems Ensure analogue is valid	

Note: Ideally, play chance and prospect-specific risk should be separately calibrated using different datasets. Play chance is the more difficult to calibrate, but prospect-specific risk can usually be calibrated using drilling success ratios in analogous proven play systems.

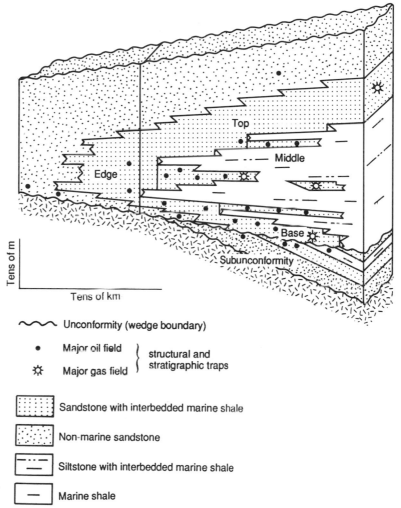

Fig. 11.13. Distribution of facies in the sand–shale wedge of White (1980). Wedge-base, -edge, -middle, -top and -subunconformity positions may be identified.

formation of structural (drape) closures. All the ingredients, therefore, for a productive play are frequently present. Examples of wedge-base plays include the Cretaceous Woodbine–Tuscaloosa of the US Gulf Coast, Miocene Sihapas of Central Sumatra, and Oligocene Oficina of eastern Venezuela.

Wedge-base carbonate reefs and sheets are usually sealed above by fine-grained facies such as shale. Again, topographic features on the underlying unconformity probably control the location of wedge-base reefs. Examples are the Devonian Leduc of western Canada, the Pennsylvanian Canyon reefs of west Texas and Oligocene Porquero reefs of northern Colombia.

Wedge-middle plays

Wedge-middle reservoirs are often both underlain and overlain by fine-grained lithologies. Reservoirs may be isolated lenses or tongues extending out from the base, edge or top of the wedge, and are frequently controlled by the nature of the depositional slope. Wedge-middle reservoirs are predominantly base-of-slope submarine fans (i.e. low stand fans, low stand wedges), but include nearshore shelf sands. Cherty or calcareous fractured shales (as in the Monterey Formation of the Los Angeles basin) may also form reservoirs. The predominance of fine-grained lithologies usually ensures a seal to reservoirs where present, but they tend to be

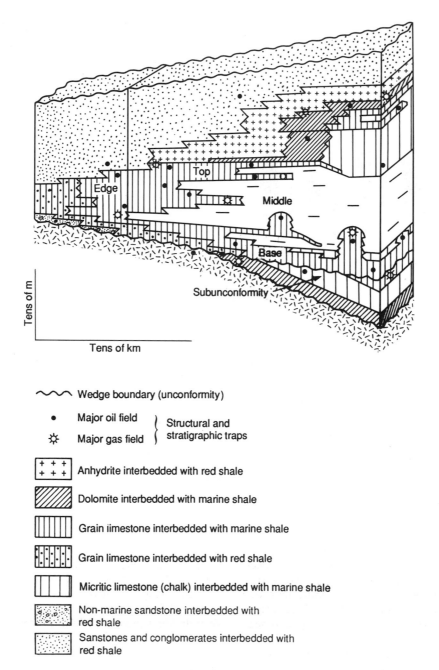

Tens of m

Tens of km

〜〜〜 Wedge boundary (unconformity)

●   Major oil field   } Structural and
☼   Major gas field   } stratigraphic traps

Anhydrite interbedded with red shale

Dolomite interbedded with marine shale

Grain iimestone interbedded with marine shale

Grain limestone interbedded with red shale

Micrilic limestone (chalk) interbedded with marine shale

Non-marine sandstone interbedded with
red shale

Sanstones and conglomerates interbedded with
red shale

Fig. 11.14. Distribution of facies in carbonate–shale wedge of White (1980). A prominent feature is the thick evaporite unit in the wedge-top. This may form an excellent regional topseal.

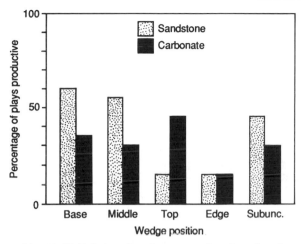

Fig. 11.15. Relative play chance as a function of wedge position for 1150 studied plays in 80 productive basins (after White 1980). The most frequently productive sandstone wedge position is wedge-base (60 per cent productive; play chance of 0.60), whereas the wedge-top is most frequently productive in carbonate wedges (45 per cent productive; play chance of 0.45). Overall, sandstone wedge plays are slightly more frequently productive (38 per cent) than carbonate wedge plays (30 per cent).

of patchy distribution. Source rocks may also not be present in the middle-wedge; thus charge may also be a problem. The Permian Basin Spraberry and Cretaceous Viking and Cardium of western Canada are examples of wedge-middle plays.

Wedge-top plays

Regressive wedge-top reservoirs are normally overlain by poorly developed non-marine anhydrites and shales or coarse clastics. Regional topseal is therefore a major risk. Underlying fine-facies shales are frequently organic-poor but land-plant source rocks may be present. Wedge-top sandstone reservoirs are often very thick, and include major regressive delta sequences. Thus, charge and seal problems explain the poor success rate (15 per cent) of wedge-top clastic plays. Wedge-top sediments are equivalent to the progradational parts of the highstand systems tract or shelf margin systems tract.

In contrast to the relatively risky wedge-top sandstone plays, wedge-top carbonate plays are highly productive, particularly in the Middle East. Carbonate reservoirs in wedge-top positions may be leached, and sealed by overlying non-marine facies anhydrite. Large reef complexes may be located on the edges of mature carbonate shelf developments (e.g. Permian Leonard and Guadalupe of west Texas). Other examples of wedge-top carbonate plays include the Jurassic Smackover of the Gulf Coast, Jurassic Arab of Saudi Arabia, and Miocene Asmari of Iran.

Wedge-edge plays

Wedge-edge plays comprise coarse clastic reservoirs underlain and overlain by non-marine facies. As a result, good seals are usually absent, unless anhydrite is well developed in the non-marine facies. Charge is also a problem unless the wedge-edge is close to adjacent fine-grained source rocks. Successful wedge-edge plays are often where there is some structural control on sedimentation, as in active extensional lacustrine basins.

Subunconformity plays

It is difficult to generalize about subunconformity plays, since topseal depends on lithologies at the base of the overlying wedge. Sources may be either below or above the unconformity. The Cretaceous Woodbine of the East Texas field is an example.

11.4.2 Prospect-specific risk

Whereas play chance refers to those factors that could wipe out the entire play, prospect-specific risk refers to those uncertainties specific to individual prospects. Prospect-specific risk is independent from prospect to prospect. It therefore relates mainly to the presence and effectiveness of the trap, though there may also be prospect-specific risks on the prospect charge system or reservoir. The following list gives an idea of the local geological factors that may result in the failure of an individual prospect within a proven play fairway. The drilling success ratios achieved in proven play fairways are a function of prospect-specific risk.

Trap/seal

• Caprock becomes sandy and leaks;
• lateral seal in ineffective, or fault plane seal is absent in a fault block prospect;

• trap geometry is illusory, for example due to seismic lateral velocity changes, or seismic miscorrelation.

Charge

• Source quality and/or thickness significantly deteriorated in the prospect drainage area, causing inadequate petroleum charge;
• source rock was immature in the prospect drainage area;
• there was no migration route into the particular prospect, as a result of discontinuous or poor quality carrier bed;
• heavy migration losses resulted in the prospect lying beyond the migration front of the petroleum charge.

Reservoir

• Shaled out or eroded at the prospect;
• impermeable, perhaps as a result of local diagenesis.
    This list is by no means comprehensive.

11.4.2.1 Geological factors affecting prospect-specific risk

Prospect-specific risk is related primarily to the *unpredictability of changes in geology* across the fairway. Simple stratigraphy and structuring will generally allow a very good drilling success ratio of say, 1 in 2 (and often relatively large fields). Extremely variable stratigraphy combined with a complex style and history of structuring will generally give a very poor success ratio, of say, worse than 1 in 10.

An example of relatively simple structuring and stratigraphy is the *Oriente Basin of Ecuador*, where success ratios are 1 in 2 (technical) and 1 in 3 (commercial). The high success ratio is primarily a result of ubiquitous reservoirs, good carrier bed development and extensive intraformational seals in the Cretaceous back-arc megasequence. These have allowed the wide distribution of oils generated in the west from the back-arc megasequence as a result of burial under Tertiary foreland basin sediments. Most dry wells have resulted primarily from uncertain trap definition (for example, in seismic depth conversion of small, low relief time closures) and from the effects of hydrodynamics.

Examples of complex basin histories are the *Taranaki Basin of New Zealand*, and the *Sabah Basin of Malaysia*.

The Taranaki Basin is a Late Cretaceous to Tertiary extensional basin modified by mid-Miocene transpression. The very complex structural history makes a good understanding of the play systems very difficult. The technical success ratio is 1 in 6, but for commercial discoveries it is 1 in 13.

In the Tertiary Sabah Basin, long-lived active deformation has resulted in complicated structures with elements of growth faulting, phases of compressional wrenching and diapirism. This has produced strong lateral lithological and facies changes at many stratigraphic levels. As a result, apart from complex trap geometries, reservoir and source development and cross-fault seals (sand/shale ratio) are variable and unpredictable. The technical success ratio is 1 in 10.

11.4.2.2 The effect of the data base on prospect-specific risk

The amount and quality of the available data can have a major effect on drilling success ratio, because to a large extent it determines our ability to detect local geological variations that are critical to the success of the well. Overall, a poor data base tends to result in inaccurate geological perceptions of the prospect. As a result, for example, wells may be drilled on imagined rather than real trap geometries.

While high technology may be applied to enhance poor data quality in some areas, in others the geological conditions are overwhelmingly unsuitable for good data acquisition. Examples are where surface terrain or outcrop (carbonates, evaporites or volcanics) adversely affect seismic acquisition.

In the Papua New Guinea fold belt, a rugged karstified terrain and a thick outcropping Tertiary limestone sequence preclude acquisition of good quality seismic, and a great premium must be placed on the application of structural geological models and surface geological mapping.

In the Gulf of Suez, Egypt, there is a marked contrast between the success ratios achieved in the Miocene Kareem and Rudeis plays (1 in 5 and 1 in 4), which lie just below and conformable with the highly reflective Zeit salt, and the pre-Miocene

Nubian–Matulla play (1 in 10), which is deeper and unconformable with the easily mapped base Salt reflector. Structure cannot therefore be resolved seismically in the pre-Miocene, and drilling success ratios are poor. There are fewer dry wells in the Kareem and Rudeis because structure can be defined; most dry wells are due to patchy reservoir development which cannot be detected by seismic interpretation in advance of drilling.

11.4.2.3 The effect of exploration maturity on prospect-specific risk

Drilling success ratios often change through time as exploration proceeds. Care should be taken, therefore, that a drilling success ratio is selected in order to calibrate prospect-specific risk that is appropriate for the exploration maturity of the play being assessed. In very mature plays of finite size, a point is reached where the success ratio begins to decay as it becomes more and more difficult to find fields of the minimum economic size. It is important to distinguish between a *cumulative* or *historic success ratio*, based on all previously drilled valid tests of the play, and a *short-term success ratio* calculated from a particular segment of the exploration history.

The North Sumatra example (Fig. 11.16) shows a decay in success ratio after the highly successful initial exploration period of the first 40 or 80 wells. Note that the short-term success ratio has dropped to zero for fields greater than 1 million barrels, since there are effectively no more fields of this minimum size to be found. The historic (cumulative) success ratio, however, has barely dropped below 1 in 10, owing to the prolific successes of the early years. The short-term success ratio dropped due to economic constraints. Without economic filtering, success ratios may continue for a long time at a level, although field sizes are likely to drop dramatically. This process is illustrated by field sizes and success ratios in the Permian Basin of west Texas (Fig. 11.17).

Prospect-specific risk, therefore, can be estimated after careful evaluation of the charge, reservoir and trap of an individual prospect, with reference to the known or likely heterogeneity of the fairway, the quality of the data base, and the stage of exploration maturity of the play. Prospect-specific risk can be calibrated with drilling success ratios drawn from analogous proven play systems.

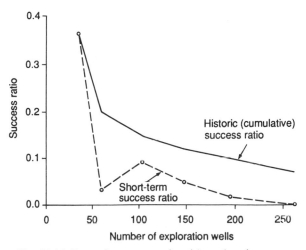

Fig. 11.16. Decay in success ratio with exploration maturity (for fields of greater than 1 million barrels) in the North Sumatra basin, Indonesia. After 250 wells the short-term success ratio had dropped effectively to zero, while the cumulative success ratio still indicated about 1 in 10.

Typical values of prospect-specific risk are shown in Table 11.2.

Prospect success chance

The chance of success in an individual prospect is the product of all of the three preceding elements of risk (Fig. 11.11). It refers to the probability of encountering petroleum volumes within the range predicted. It is related directly, therefore, to the probability distribution of reserves calculated for the prospect (Section 11.3).

11.5 ASSESSMENT CURVES

Assessment curves illustrate the range of petroleum volumes that may be found in a play or prospect, together with their probability of occurrence.

11.5.1 Play assessment curves

Using the field size distribution in Fig. 11.3, and the field number distribution in Fig. 11.10, an unrisked assessment curve of the undiscovered potential of the play can be calculated by Monte Carlo simulation (Fig. 11.18). Firstly, the field number distribution is randomly sampled. If a field number of three

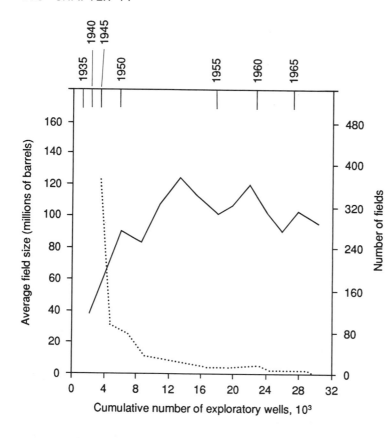

Fig. 11.17. The number of discoveries of oil and gas fields and their average size for 14 successive drilling intervals of approximately 2000 exploratory wells each in the Permian Basin, from 1921 to 1974 (after Root and Drew 1979). Although average field sizes (dotted line) declined significantly as exploration proceeded, success ratios (solid line) have remained steady. Economic truncation has not yet had a major impact on drilling success ratios.

is selected, the field size distribution is sampled three times, and the selected field sizes summed. This represents one possible value of undiscovered potential. This process is repeated a large number of times, say, several thousand, and an undiscovered potential probability distribution produced. In Figure 11.18, we see that the mean is 600 million barrels, which represents the most likely number of fields (3) at the average size (200). The minimum assessment of 50 million barrels represents 1 field of the minimum size (50) being selected.

Using the play chance, a risked assessment curve can be calculated from the unrisked one (Fig. 11.18).

If the play fairway comprises a number of common-risk segments, this process must be repeated for each segment, and the risked assessment curves added to produce an expectation for the play as a whole. Similarly, multiple plays may be added to provide an assessment for a whole basin or country.

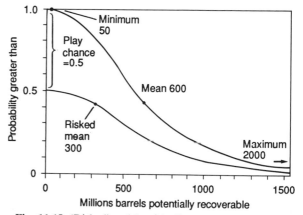

Fig. 11.18. 'Risked' and 'unrisked' assessment curves. The unrisked curve is produced by combining the field number and field size distributions, and illustrates the possible petroleum volumes given the play is productive. The risked curve is produced after multiplication by the play chance. Together, these curves present vitally important information about the play being assessed.

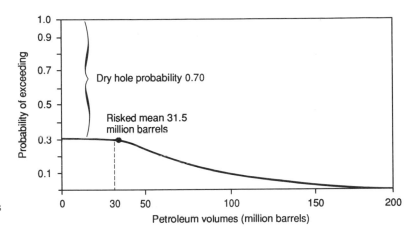

Fig. 11.19. Risked prospect assessment curve, based on Fig. 11.8, and using a prospect success chance of 0.30.

If a Monte Carlo simulation computer program is not available, undiscovered potential can be calculated manually by sampling regularly the field size distribution a number of times equal to the predicted number of undiscovered fields in the play, and summing the selected field sizes. This approach will, of course, yield only a single-point estimate of future potential, but a range of realistic field sizes will be provided.

The unrisked potential of each segment should be compared with geochemical volumetric calculations, made in the same way as described in Section 11.3 for individual prospects.

11.5.2 Prospect assessment curves

At the prospect level, the unrisked assessment curve (Fig. 11.8) must be risked using the prospect success chance in order to produce a risk assessment curve (Fig.11.19). In the example shown the prospect success chance is 0.30.

There is therefore a 70 per cent chance that petroleum volumes will be less than the absolute minimum of about 30 million barrels on the unrisked probability distribution. It is convenient to consider this event a dry hole outcome, though it may include any volumes up to 30 million barrels. The risked mean (31.5 million barrels) is 0.3 times the unrisked mean of 105 million barrels. The risked mean is useful for comparison with other prospects. Different prospects can be ranked on the basis of their risked mean petroleum volumes. The unrisked mean is the expected petroleum volume in the prospect given it is not dry, and forms the basis for engineering and commercial evaluation of the prospect.

References

Aharon, P. (1983) 140 000 yr isotope climatic record from raised coral reef in New Guinea. *Nature*, **304**, 720–723.

Ahnert, F. (1970) Functional relationships between denudation relief and uplift in large mid-latitude drainage basins. *Am. J. Sci.*, **270**, 243–263.

Aihara, A. (1980) Formation and organic metamorphism of the Palaeogene coal deposits in the Japanese islands. Congres géologique international (Paris 1980), *Rev. Ind. minér.*, 'Les Techniques', suppl. June 1980, 307–314.

Albright, W.A., Turner, W.L. and Williamson, K.R. (1980) Ninian field, UK sector, North Sea. In: *Giant Oil and Gas Fields of the Decade 1968–1978* (Ed. by M.T. Halbouty), 173–194, *Am. Assoc. petrol. Geol. Mem.* **30**.

Alexander, J. and Leeder, M.R. (1987) Active tectonic control on alluvial architecture. In: *Recent Developments in Fluvial Sedimentology* (Ed. by F.G. Ethridge, R.M. Flores, and M.D. Harvey), 243–252, *Spec. Publ. Soc. econ. Paleont. Mineral.* **39**.

Allan, U.S. (1980) A model for the migration and entrapment of hydrocarbons. *American Association of petroleum Geologists Research Conference* on *Seals for Hydrocarbons*, Keystone, Colorado, Sept. 14–17, 1980, unpublished book of abstracts.

Allen, G.P., Laurier, D. and Thouvenin, J. (1979) Etude sédimentologique du delta de la Mahakam. *Notes et Mémoires,* **15**, Compagnie Française des Pétroles, Paris.

Allen, P.A. and Collinson, J.D. (1986) Lakes. In: *Sedimentary Environments and Facies* (Ed. by H.G. Reading), 63–94, Blackwell Scientific, Oxford, 615 pp.

Allen, P.A., Cabrera, L., Colombo, F. and Matter, A. (1983) Variations in alluvial style on the Eocene–Oligocene alluvial fan of the Scala Dei Group, SE Ebro Basin, Spain. *J. geol. Soc. London*, **140**, 133–146.

Allen, P.A., Homewood, P. and Williams, G.D. (1986) Foreland basins: an introduction. In: *Foreland Basins* (Ed. by P.A. Allen and P. Homewood), 3–12, *Spec. Publ. int. Assoc. Sedimentol.*, **8**, Blackwell Scientific, Oxford.

Allen, P.A., Mange-Rajetzky, M.A., Matter, A. and Homewood, P. (1985) Dynamic palaeogeography of the open Burdigalian seaway, Swiss Molasse Basin. *Eclog. géol. Helv.*, **78**, 351–381.

Anderson, R.Y. and Kirkland, D.W. (1960) Origin, varves and cycles of Jurassic Todilto Formation, New Mexico. *Bull. Am. Assoc. petrol. Geol.*, **44**, 37–52.

Anderton, R. (1976) Tidal shelf sedimentation: an example from the Scottish Dalradian. *Sedimentology*, **23**, 429–458.

Andrews-Speed, C.P., Oxburgh, E.R. and Cooper, B.A. (1984) Temperatures and depth-dependent heat flow in western North Sea. *Bull. Am. Assoc. petrol. Geol.*, **68**, 1764–1781.

Arthur, M.A. and Schlanger, S.O. (1979) Cretaceous 'oceanic anoxic events' as causal factors in development of reef reservoired giant oil fields. *Bull. Am. Assoc. petrol. Geol.*, **63**, 870–885.

Artyushkov, E.V. (1973) Stresses in the lithosphere caused by crustal thickness inhomogeneities. *J. geophys. Res.*, **78**, 7675–7708.

Ashby, M.F. and Verall, R.A. (1977) Micromechanisms of flow and fracture and their relevance to the rheology of the upper mantle. *Phil. Trans. R. Soc. London*, **A288**, 59–95.

Asquith, D.O. (1970) Depositional topography and major marine environments, Late Cretaceous, Wyoming. *Bull. Am. Assoc. petrol. Geol.*, **54**, 1184–1224.

Asquith, G. (1982) *Basic Well Log Analysis for Geologists.* Methods in Exploration Series, American Association of Petroleum Geologists, Tulsa, Oklahoma, 216 pp.

Athy, L.F. (1930) Density, porosity and compaction of sedimentary rocks. *Bull. Am. Assoc. petrol. Geol.*, **14**, 1–24.

Atkinson, C.D. (1983) *Comparative sequences of ancient fluviatile deposits in the Tertiary, South Pyrenean Basin, northern Spain.* Unpublished PhD thesis, University of Wales, 350 pp.

Atwater, G.T. and Miller, E.E. (1965) The effect of increase in porosity with depth on future development of oil and gas reserves in South Louisiana (abstract). *Bull. Am. Assoc. petrol. Geol.*, **49**, 334.

Atwater, T. (1970) Implications of plate tectonics for the Cenozoic tectonic evolution of western North America. *Bull. geol. Soc. Am.*, **81**, 3513–3536.

Aubouin, J. (1965) Geosynclines. *Developments in Geotectonics*, Elsevier, Amsterdam, 335 pp.

Aydin, A. and Nur, A. (1982) Evolution of pull-apart basins and their scale independence. *Tectonics*, **1**, 91–105.

Aydin, A. and Nur, A. (1985) The types and roles of stepovers in strike-slip tectonics. In: *Strike-Slip Deformation, Basin Formation and Sedimentation* (Ed. by K.T. Biddle and N. Christie-Blick), *Spec. Publ. Soc. econ. Paleont. Mineral.* **37**, 35–44.

Badley, M.E., Price, J.D., Rambech Dahl, C. and Agdestein, T. (1988) The structural evolution and the northern Viking Graben and its bearing upon extensional modes of basin formation. *J. geol. Soc. London*, **145**, 455–472.

Bailey, R.J. and Stoneley, R. (1981) Petroleum: entrapment and conclusions. In: *Economic Geology and Geotectonics* (Ed. by D.H. Tarling), 73–97, Blackwell Scientific, Oxford, 213 pp.

Baker, B.H. and Morgan, P. (1981) Continental rifting: progress and outlook. *Eos, Trans. Am. geophys. Union*, **62**, 585–586.

Baker, B.H., Mohr, P.A. and Williams, L.A.J. (1972) Geology of the Eastern Rift System of Africa. *Spec. Pap. geol. Soc. Am.*, **136**.

Baker, R.A., Gehman, H.M., James, W.R. and White, D.A. (1986) Geologic field number and field size assessments of oil and gas plays. In: *Oil and Gas Assessment – Methods and Applications* (Ed. by D.D. Rice), 25–32, *Am. Assoc. petrol. Geol. Studies in Geology*, **21**.

Bally, A.W. (1975) A geodynamic scenario for hydrocarbon occurrences. *Proc. 9th World Petrol. Congr., Tokyo*, 33–44, Vol. 2 (Geology). Applied Science Publishers Barking.

Bally, A.W. (1982) Musings over sedimentary basin evolution. *Phil. Trans. R. Soc. London*, **A305**, 325–338.

Bally, A.W. and Oldow, J.S. (1984) *Plate Tectonics, Structural Styles, and the Evolution of Sedimentary Basins*. Unpublished course notes, 238 pp.

Bally, A.W. and Snelson, S. (1980) Realms of subsidence. In: *Facts and Principles of World Petroleum Occurrence* (Ed. by A.D. Miall), *Can. Soc. petrol. Geol. Mem.* **6**, 9–75.

Bally, A.W., Bernoulli, D., Davis, G.A. and Montadert, L. (1981) Listric normal faults. *Oceanologica Acta*, 87–102.

Bamford, D. (1979) Seismic constraints on the deep geology of the Caledonides of northern Britain. In: *The Caledonides of the British Isles – Reviewed* (Ed. by A. Harris, C.H. Holland and B.E. Leake). *Spec. Publ. geol. Soc. London* **8**, 93–96.

Banks, R.J. and Swain, C. (1978) The isostatic compensation of East Africa. *Proc. R. astr. Soc.*, **A364**, 331–352.

Barazangi, M. and Dorman, J. (1969) World seismicity map compiled from ESSA Coast and Geodetic Survey epicentre data, 1961–1967. *Bull. seismol. Soc. Am.*, **59**, 369–380.

Bartlett, W.L., Friedman, M. and Logan, J.M. (1981) Experimental folding and faulting of rocks under confining pressure. Part IX. Wrench faults in limestone layers. *Tectonophysics*, **79**, 255–277.

Barton, P. and Wood, R. (1984) Tectonic evolution of the North Sea basin: crustal stretching and subsidence. *Geophys. J. R. astr. Soc.*, **79**, 987–1022.

Bathurst, R.G.C. (1971) Carbonate Sediments and their Diagenesis. *Developments in Sedimentology*, **12**, Elsevier, Amsterdam, 620 pp.

Bathurst, R.G.C. (1975) Carbonate Sediments and their Diagenesis (2nd edition). *Developments in Sedimentology* **12**, Elsevier, Amsterdam, 658 pp.

Beach, A., Bird, T. and Gibbs, A. (1987) Extensional tectonics and crustal structure: deep seismic reflection data from the northern North Sea Viking Graben. In: *Continental Extensional Tectonics* (Ed. by M.P. Coward, J.F. Dewey and P.L. Hancock), *Spec. Publ. geol. Soc. London,* **28**, 467–476.

Beadle, L.C. (1974) *The Inland Waters of Tropical Africa*. Longman, London, 365 pp.

Beard, J.H., Sangree, J.B. and Smith, L.A. (1982) Quaternary chronology, palaeoclimate, depositional sequences, and eustatic cycles. *Bull. Am. Assoc. petrol. Geol.*, **66**, 158–169.

Beaumont, C. (1978) The evolution of sedimentary basins on a viscoelastic lithosphere: theory and examples. *Geophys. J. R. astr. Soc.*, **55**, 471–497.

Beaumont, C. (1981) Foreland basins. *Geophys. J. R. astr. Soc.*, **65**, 291–329.

Beaumont, C., Keen, C.E. and Boutilier, R. (1982) On the evolution of rifted continental margins; comparison of models and observations for Nova Scotian margin. *Geophys. J. R. astr. Soc.*, **70**, 667–715.

Beaumont, C., Quinlan, G. and Hamilton, J. (1988) Orogeny and stratigraphy: numerical models of the Palaeozoic in the eastern interior of North America. *Tectonics*, 7, 389–416.

Bechtel, T., Forsyth, D. and Swain, C. (1987) Mechanisms of isostatic compensation in the vicinity of the East African rift, Kenya. *Geophys. J. R. astr. Soc.*, **90**, 445–465.

Beerbower, J.R. (1964) Cyclothems and cyclic depositional mechanisms in alluvial plain sedimentation. In: *Symposium on Cyclic Sedimentation* (Ed. by D.F. Merriam), 31–42, *Bull. Kansas Geol. Surv.* **169**, Vol. 1.

Begin, A.B., Ehrlich, A. and Nathan, Y. (1974) Lake Lisan, the Pleistocene precursor of the Dead Sea. *Bull. geol. Surv. Israel*, **63**, 30 pp.

Belderson, R.H., Johnson, M.A. and Kenyon, N.H. (1982) Bedforms. In: *Offshore Tidal Sands, Process and Deposits* (Ed. by A.H. Stride), 27–57, Chapman and Hall, London.

Belknap, D.F. *et al.* (1987) Late Quaternary sea level changes in Maine. In: *Sea Level Fluctuation and Costal Evolution* (Ed. by D. Nummedal, O.H. Pilkey and J.D. Howard), *Spec. Publ. Soc. econ. Paleont. Mineral.* **41**, 71–85.

Bellamy, D. (1978) *Botanic Man.* Hamlyn, London.

Ben-Avraham, Z., Almagor, G. and Garfunkel, Z. (1979) Sediments and structure of the Gulf of Elat (Aqaba) – northern Red Sea. *Sedim. Geol.*, **23**, 239–267.

Berg, R.R. (1975a) Depositional environment of Upper Cretaceous Sussex Sandstone, House Creek Field, Wyoming. *Bull. Am. Assoc. petrol. Geol.*, **59**, 2099–2110.

Berg, R.R. (1975b) Capillary pressures in stratigraphic traps. *Bull. Am. Assoc. petrol. Geol.*, **59**, 939–956.

Berggren, W.A., Kent, D.V., Flynn, J.J. and Van Couvering, J.A. (1985) Cenozoic geochronology. *Bull. geol. Soc. Am.*, **96**, 1407–1418.

Bernard, H.A., LeBlanc, R.J. and Major, C.F. Jr (1962) Recent and Pleistocene geology of southeast Texas. *Geology of Gulf Coast and Central Texas and Guidebook of Excursion*, 175–225, Houston Geological Society.

Berné, S., Auffret, J.-P. and Walker, P. (1988) Internal structure of subtidal sandwaves revealed by high-resolution seismic reflection. *Sedimentology*, **35**, 5–20.

Bernoulli, D. and Jenkyns, H.C. (1974) Alpine, Mediterranean and Central Atlantic Mesozoic facies in relation to the early evolution of the Tethys. In: *Modern and Ancient Geosynclinal Sedimentation* (Ed. by R.H. Dott and R.H. Shaver), 129–160, *Spec. Publ. Soc. econ. Paleont. Mineral.* **19**.

Biddle, K.T., Uliana, M.A., Mithchum, R.M. Jr, Fitzgerald, M.G. and Wright, R.C. (1986) The stratigraphic and structural evolution of the central and eastern Magallanes Basin, southern South America. In: *Foreland Basins* (Ed. by P.A. Allen and P. Homewood), *Spec. Publ. int. Assoc. Sedimentol.*, **8**, 41–61.

Bischoff, J.L. and Henyey, T.L. (1974) Tectonic elements of the central part of the Gulf of California. *Bull. geol. Soc. Am.*, **85**, 1893–1904.

Blackwell, D.D. and Chockalingam, S. (1981) Heat flow and crustal evolution of rift provinces in the western United States: Snake River Plain region and Basin and Range Province. *Conference on Processes of Planetary Rifting*, Lunar and Planetary Institute, Houston, 212–215.

Blair, D.G. (1975) Structural styles in North Sea oil and gas fields. In: *Petroleum and the Continental Shelf of NW Europe*, Vol. 1 (Ed. by A.W. Woodland), 327–338, Applied Science Publishers, Barking.

Blatt, H. (1982) *Sedimentary Petrology.* W.H. Freeman, San Francisco, 564 pp.

Blatt, H., Middleton, G.V. and Murray, R.C. (1972) *Origin of Sedimentary Rocks.* Prentice-Hall, Englewood Cliffs, New Jersey, 634 pp.

Boles, J.R. (1981) Clay diagenesis and effects on sandstone cementation (case histories from the Gulf Coast Tertiary). In: *Clays and the Resource Geologist* (Ed. by F.J. Longstaffe), *Min. Ass. Can. Short Course Handbook*, **7**, 148–168.

Bond, G. (1978) Speculations on real sea level changes and vertical motions of continents at selected times in the Cretaceous and Tertiary periods. *Geology*, **6**, 247–250.

Bond, G.C. and Kominz, M.A. (1984) Construction of tectonic subsidence curves for the early Paleozoic miogeocline, southern Canadian Rocky Mountains: implications for subsidence mechanisms, age of breakup, and crustal thinning. *Bull. geol. Soc. Am.*, **95**, 155–173.

Bostick, N.H. (1979) Microscopic measurement of the level of catagenesis of solid organic matter in sedimentary rocks to aid exploration for petroleum and to determine former burial temperatures – a review. In: *Aspects of Diagenesis* (Ed. by P.A. Scholle and P.R. Schluger), 17–43, *Spec. Publ. Soc. econ. Paleont. Mineral.* **26**.

Bostick, N.H. and Alpern, B. (1977) Principles of sampling, preparation and constituent selection for microphotometry in measurement of maturation of sedimentary organic matter. *J. Microscopy*, **109**, 41–47.

Bott, M.H.P. (1980) Mechanism of subsidence at passive continental margins. In: *Dynamics of Plate Interiors. Am. Geophysical Union Geodynamics Series*, **1**, 27–32.

Bott, M.H.P. (1981) Crustal doming and the mechanism of continental rifting. *Tectonophysics*, **73**, 1–8.

Bott, M.H.P. (1982) *The Interior of the Earth: its Structure, Constitution and Evolution,* 2nd edition. Elsevier, Amsterdam, 403 pp.

Bott, M.H.P. and Kusznir, N.J. (1979) Stress distribution associated with compensated plateau uplift structures with application to the continental splitting mechanism. *Geophys. J. R. astr. Soc.*, **56**, 451–459.

Bouma, A.H. (1962) *Sedimentology of some Flysch Deposits: a Graphic Approach to Facies Interpretation.* Elsevier, Amsterdam, 168 pp.

Boyer, B.W. (1982) Green River laminites: does the playa-lake model really invalidate the stratified lake model? *Geology*, **10**, 321–324.

Boyles, J.M. and Scott, A.J. (1982) A model for migrating shelf-bar sandstones in Upper Mancos Shale (Campanian), northwestern Colorado. *Bull. Am. Assoc. petrol. Geol.*, **66**, 491–508.

Brace, W.H. and Kohlstedt, D.L. (1980) Limits on

lithospheric stress imposed by laboratory experiments. *J. geophys. Res.*, **85**, 6248–6252.

Bradley, W.H. (1964) Geology of the Green River Formation and associated Eocene rocks in southwestern Wyoming and adjacent parts of Colorado and Utah. *Prof. Pap. U.S. geol. Surv.* **496–A**.

Bradley, W.H. and Eugster, H.P. (1969) Geochemistry and palaeolimnology of the trona deposits and associated authigenic minerals of the Green River Formation of Wyoming. *Prof. Paper U.S. geol. Surv.* **469–B**, 71 pp.

Brenchley, P.J. and Williams, B.P.J. (1985) *Sedimentology: Recent Developments and Applied Aspects. Spec. Publ. Geol. Soc. London*, **18**, Blackwell Scientific, Oxford, 342 pp.

Brewer, J.A., Good, R., Oliver, J.E., Brown, L.D. and Kaufman, S. (1983) COCORP profiling across the southern Oklahoma aulacogen: overthrusting of the Wichita Mountains and compression within the Anadarko Basin. *Geology*, **11**, 109–114.

Bridge, J.S. and Leeder, M.R. (1979) A simulation model of alluvial stratigraphy. *Sedimentology*, **26**, 617–644.

Bridges, P.H. (1982) Ancient offshore tidal deposits. In: *Offshore Tidal Sands. Processes and Deposits* (Ed. by A.H. Stride), 172–192, Chapman and Hall, London.

Brown, L.F. and Fisher, W.L. (1977) Seismic-stratigraphic interpretation of depositional systems. In: *Seismic Stratigraphy – Applications to Hydrocarbon Exploration* (Ed. by C.E. Payton), *Am. Assoc. petrol. Geol. Mem.* **26**, 213–248.

Bruhn, R.L., Stern, C.R. and De Wit, M.J. (1978) Field and geochemical data bearing on the development of a Mesozoic volcano-tectonic rift zone and back-arc basin in southernmost South America. *Earth planet. Sci. Letters*, **41**, 32–46.

Buck, W.R. (1984) *Small-scale convection and the evolution of the lithosphere.* Unpublished PhD thesis, MIT, 256 pp.

Buck, W.R. (1986) Small-scale convection induced by passive rifting: the cause for uplift of rift shoulders. *Earth planet. Sci. Letters*, **77**, 362–372.

Buck, W.R. Steckler, M.S. and Cochran, J.R. (1988) Thermal consequences of lithospheric extension: pure and simple. *Tectonics*, **7**, 213–234.

Burchfiel, B.C. and Davis, G.A. (1972) Structural framework and evolution of the southern part of the Cordilleran orogen, western United States. *Am. J. Sci.*, **272**, 97–118.

Burchfiel, B.C. and Royden, L. (1982) Carpathian foreland fold and thrust belt and its relation to Pannonian and other basins. *Bull. Am. Assoc. petrol. Geol.*, **66**, 1179–1195.

Burchfiel, B.C. and Stewart, J.H. (1966) 'Pull-apart' origin of the central segment of Death Valley, California. *Bull. geol. Soc. Am.*, **77**, 439–442.

Burke, K. (1972) Longshore drift, submarine canyons, and submarine fans in development of Niger delta. *Bull. Am. Assoc. petrol. Geol.*, **56**, 1975–1983.

Burke, K. (1976a) The Chad Basin: an active intra-continental basin. *Tectonophysics*, **36**, 197–206.

Burke, K. (1976b) Development of graben associated with the initial ruptures of the Atlantic Ocean. In: *Sedimentary Basins of Continental Margins and Cratons* (Ed. by M.H.P. Bott), *Tectonophysics*, **36**, 93–112.

Burke, E. (1977) Aulacogens and continental break-up. *Ann. Rev. Earth planet. Sci.*, **5**, 371–396.

Burke, K. and Dewey, J.F. (1973) Plume generated triple junctions: key indicators in applying plate tectonics to old rocks. *J. Geol. Chicago*, **81**, 406–433.

Burke, K. and Wilson, J.T. (1972) Is the African plate stationary? *Nature*, **239**, 387–390.

Burke, K., Dessauvagie, T.F.J. and Whiteman, A.J. (1972) Geological history of the Benue Valley and adjacent areas. In: *African Geology* (Ed. by A.J. Whiteman and T.F.J. Dessauvagie), 187–206, Ibadan.

Burley, S.D., Kantorowicz, J.D. and Waugh, B. (1985) Clastic diagenesis. In: *Sedimentology: Recent Developments and Applied Aspects* (Ed. by P.J. Brenchley and B.P.J. Williams), 189–226, *Spec. Publ. geol. Soc. London*, **18**.

Butler, G.P., Harris, P.M., and Kendall, G.G. St. C. (1982) Recent evaporites from the Abu Dhabi coastal flats. In: *Deposition and Diagenetic Spectra of Evaporites* (Ed. by C.R. Handford, R.G. Loucks and G.R. Davies), 33–64, *Soc. econ. Mineral. Paleont. Core Workshop No. 3*, Calgary, 1982.

Cant, D.J. and Walker, R.G. (1978) Fluvial processes and facies sequences in the sandy braided South Saskatchewan River, Canada. *Sedimentology*, **25**, 625–648.

Cardott, B.J. and Lambert, M.W. (1985) Thermal maturation by vitrinite reflectance of Woodford Shale, Anadarko Basin, Oklahoma. *Bull. Am. Assoc. petrol. Geol.*, **69**, 1982–1998.

Carey, S.W. (1958) The tectonic approach to continental drift. In: *Continental Drift: a Symposium on the Present Status of the Continental Drift Hypothesis*, held in the Geology Department of the University of Tasmania in March 1956, 177–355, 363 pp.

Carey, S.W. (1976) *The Expanding Earth. Developments in Geotectonics*, **10**, Elsevier, Amsterdam, 488 pp.

Case, J.E. and Holcombe, T.L. (1980) Geologic-tectonic map of the Caribbean region. *U.S. geol. Surv. Misc. Invest. Ser., map* **I–1100**.

Catacosinos, P.A. (1973) Cambrian lithostratigraphy of the Michigan Basin. *Bull. Am. Assoc. petrol. Geol.*, **57**, 2404–2418.

Cathles, L.M. (1975) *The Viscosity of the Earth's Mantle.* Princeton University Press.

Cercone, R.K. (1984) Thermal history of Michigan Basin. *Bull. Am. Assoc. petrol. Geol.*, **68**, 130–136.

Cermak, V. (1979) Review of heat flow measurements in Czechoslovakia. In: *Terrestrial Heat Flow in Europe*

(Ed. by V. Cermak and L. Rybach), New York, Springer-Verlag, 152–160.

Chapple, W.M. (1978) Mechanics of thin-skinned fold-and-thrust belts. *Bull. geol. Soc. Am.*, **89**, 1189–1198.

Chapple, W.M. and Forsyth, D.W. (1979) Earthquakes and bending of plates at trenches. *J. geophys. Res.*, **84**, 6729–6749.

Chase, T.E., Menard, H.W. and Mammerickx, J. (1970) *Bathymetry of the North Pacific*, Chart 8 of 10, Scripps Institution of Oceanography and Institute of Marine Resources.

Cheadle, M.J., Czuchra, B.L., Byrne, T., Ando, C.J., Oliver, J.E., Brown, L.D., Kaufman, S., Malin, P.E. and Phinney, R.A. (1985) The deep crustal structure of the Mojave desert, California, from COCORP seismic reflection data. *Tectonics*, **5**, 293–300.

Chen, W.-P. and Molnar, P. (1983) Focal depths of intracontinental and interplate earthquakes and their implications for the thermal and mechanical properties of the lithosphere. *J. geophys. Res.*, **88**, 4183–4214.

Chough, S. and Hesse, R. (1976) Submarine meandering thalweg and turbidity currents flowing for 4000 km in the Northwest Atlantic Mid-Ocean Channel, Labrador Sea. *Geology*, **4**, 529–533.

Christie, P.A.F. and Sclater, J.G. (1980) An extensional origin for the Witchground/Buchan graben in the northern North Sea. *Nature*, **283**, 729–732.

Christie-Blick, N. and Biddle, K.T. (1985) Deformation and basin formation along strike-slip faults. In: *Strike-slip Deformation, Basin Formation and Sedimentation* (Ed. by K.T. Biddle and N. Christie-Blick), *Spec. Publ. Soc. econ. Paleont. Mineral.* **37**, 1–34.

Clark, J.A., Farrell, W.E. and Peltier, W.R. (1978) Global changes of post-glacial sea level: a numerical calculation. *Quaternary Research*, **9**, 265–287.

Clemmensen, L.B. and Abrahamsen, K. (1983) Aeolian stratification and facies association in desert sediments, Arran Basin (Permian), Scotland. *Sedimentology*, **30**, 311–339.

Clemmensen, L.B. and Blakey, R.C. Erg deposits in the Lower Jurassic Wingate Sandstone, northeastern Arizona: oblique dune sedimentation. *Sedimentology*, **36**, 449–470.

Clifton, H.E., Hunter, R.E. and Phillips, R.L. (1971) Depositional structures and processes in the non-barred, high energy nearshore. *J. sedim. Petrol.*, **41**, 651–670.

Cloetingh, S., McQueen, H. and Lambeck, K. (1985) On a tectonic mechanism for regional sea level variations. *Earth planet. Sci. Letters*, **75**, 157–166.

Cloos, E. (1955) Experimental analysis of fracture patterns. *Bull. geol. Soc. Am.*, **66**, 241–256.

Cochran, J.R. (1983) Effects of finite extension times on the development of sedimentary basins. *Earth planet. Sci. Letters*, **66**, 289–302.

Cohee, G.V. and Landes, K.K. (1958) Oil in the Michigan Basin. In: *Habitat of Oil* (Ed. by L.G. Weeks), 473–493, American Association of Petroleum Geologists, Tulsa, Oklahoma.

Cole, R.D. and Picard, M.D. (1981) Sulfur isotope variations in marginal lacustrine rocks of the Green River Formation, Colorado and Utah. In: *Recent and Ancient Nonmarine Depositional Environments; Models for Exploration* (Ed. by F.G. Ethridge and R.M. Flores), 261–275, *Spec. Publ. Soc. econ. Paleont. Mineral.* **31**.

Colella, A. (1981) Preliminary core analysis of Crati submarine fan deposits (Ionian Sea). *Abstr. int. Assoc. Sedimentol. 2nd Eur. Mtg, Bologna*.

Coleman, J.M. (1969) Brahmaputra River: channel processes and sedimentation. *Sedim. Geol.*, **3**, 129–239.

Cooles, G.P., Mackenzie, A.S. and Quigley, T.M. (1986) Calculations of masses of petroleum generated and expelled from source rocks. In: *Advances in Organic Geochemistry* (Ed. by D. Leythaeuser and J. Rullkotter), 235–246, Pergamon Press, Oxford.

Connan, J. (1974) Time-temperature relation in oil genesis. *Bull. Am. Assoc. petrol. Geol.*, **58**, 2516–2521.

Covey, M. (1986) The evolution of foreland basins to steady state: evidence from the western Taiwan foreland basin. In: *Foreland Basins* (Ed. by P.A. Allen and P. Homewood), 77–90, *Spec. Publ. int. Assoc. Sedimentol.* **8**, Blackwell Scientific, Oxford.

Coward, M.P. (1986) Heterogeneous stretching, simple shear and basin development. *Earth planet. Sci. Letters*, **80**, 325–336.

Cox, A. (Editor) (1973) *Plate Tectonics and Geomagnetic Reversals*. W.H. Freeman, San Francisco, 702 pp.

Cox, A. and Hart, R.B. (1986) *Plate Tectonics: How it Works*. Blackwell Scientific Publications, Palo Alto, California, 392 pp.

Crittenden, M.D. Jr. (1963) Effective viscosity of the Earth derived from isostatic loading of Pleistocene Lake Bonneville. *J. geophys. Res.*, **68**, 5517–5530.

Crough, S.T. (1978) Thermal origin of mid-plate hot-spot swells. *Geophys. J.R. astr. Soc.*, **55**, 451–469.

Crough, S.T. (1979) Hotspot epeirogeny. *Tectonophysics*, **61**, 321–333.

Crough, S.T. (1983) Rifts and swells: geophysical constraints on causality. *Tectonophysics*, **94**, 23–37.

Crowell, J.C. (1954) *Geology of the Ridge Basin area, California*. California Division of Mines Bulletin 170, Map Sheet 7.

Crowell, J.C. (1974a) Sedimentation along the San Andreas Fault, California. In: *Modern and Ancient Geosynclinal Sedimentation* (Ed. by R.H. Dott Jr. and R.H. Shaver), *Spec. Publ. Soc. econ. Paleont. Mineral.* **19**, 292–303.

Crowell, J.C. (1974b) Origin of late Cenozoic basins in southern California. In: *Tectonics and Sedimentation* (Ed. by W.R. Dickinson), *Spec. Publ. Soc. econ. Paleont. Mineral.* **22**, 190–204.

Crowell, J.C. (1975) The San Gabriel Fault and Ridge Basin. In: *San Andreas Fault in Southern California*

`(Ed. by J.C. Crowell), 208–233, California Division of Mines and Geology Special Report **118**.

Crowell, J.C. (1981) Juncture of the San Andreas system and the Gulf of California rift. *Oceanologica Acta, Proc. 26th int. geol. Congr.*, Paris, 137–141.

Crowell, J.C. and Link, M.H. (1982) Ridge Basin, Southern California. In: *Geologic History of Ridge Basin, Southern California* (Ed. by J.C. Crowell and M.H. Link), 1–4, Society of Economic Paleontologists and Mineralogists Pacific Section. 304 pp.

Curray, J.R. (1964) Transgressions and regressions. In: *Papers in Marine Geology* (Ed. by R.L. Miller), 175–203, Macmillan, New York.

Curray, J.R. (1965) Late Quaternary history, continental shelves of the United States. In: *The Quaternary of the United States* (Ed. by H.E. Wright and D.G. Frey), 723–735, Princeton University Press, New Jersey.

Curry, W.H. and Curry, W.H. III (1972) South Glennock oilfield, Wyoming: a pre-discovery thinking and post-discovery description. In: *Stratigraphic Oil and Gas Fields* (Ed. by R.E. King), 415–427, *Am. Assoc. petrol. Geol. Mem.* **16**.

Curray, J.R., Emmel, F.J. and Crampton, P.J.S. (1969) Holocene history of a strandplain, lagoonal coast, Nayarit, Mexico. In: *Coastal Lagoons – A Symposium* (Ed. by A.A. Castanares and F. B. Phleger), 63–100, Universidad Nacional Autónoma, Mexico.

Dahlen, F.A. (1984) Noncohesive critical Coulomb wedges: an exact solution. *J. geophys. Res.*, **89**, 10 125–10 133.

Dahlstrom, C.D.A. (1970) Structural geology in the eastern margin of the Canadian Rocky Mountains. *Bull. Can. petrol. Geol.*, **18**, 332–406.

Daly, A.R. and Lilly, D.H. (1985) Thermal subsidence and generation of hydrocarbons in the Michigan Basin: Discussion. *Bull. Am. Assoc. petrol. Geol.*, **69**, 1181–1184.

Dalziel, I.W.D., De Wit, M.J. and Palmer, K.F. (1974) A fossil marginal basin in the southern Andes. *Nature*, **250**, 291–294.

Damuth, J.E. and Embley, R.W. (1981) Mass transport processes on the Amazon Cone: western Equatorial Atlantic. *Bull. Am. Assoc. petrol. Geol.*, **65**, 629–643.

Davidson-Arnott, R.G.D. and Greenwood, B. (1976) Facies relationships on a barred coast, Kouchibouguac Bay, New Brunswick, Canada. In: *Beach and Nearshore Sedimentation* (Ed. by R.A. Davis Jr. and R.L. Ethington), 149–168, *Spec. Publ. Soc. econ. Paleont. Mineral.* **24**.

Davies, D.K. (1968) Carbonate turbidites, Gulf of Mexico. *J. sedim. Petrol.*, **38**, 1100–1109.

Davis, D., Suppe, J. and Dahlen, F.A. (1983) Mechanics of fold-and-thrust belts and accretionary wedges. *J. geophys. Res.*, **88**, 1153–1172.

De Bremaecker, J.-C. (1983) Temperature, subsidence, and hydrocarbon maturation in extensional basins: a finite element model. *Bull. Am. Assoc. petrol. Geol.*, **67**, 1410–1414.

Demaison, G. (1984) The generative basin concept. In: *Petroleum Geochemistry and Basin Evaluation* (Ed. by G. Demaison and R.J. Murris), 1–14, *Am. Assoc. petrol. Geol. Mem.* **35**.

Demaison, G.J. and Moore, G.T. (1980) Anoxic environments and oil source bed genesis. *Bull. Am. Assoc. petrol. Geol.*, **64**, 1178–1209.

Dever, G.R., Hoge, H.P., Hester, N.C. and Ettensohn, F.R. (1977) *Stratigraphic Evidence for Late Paleozoic Tectonism in Northeastern Kentucky*. Field Trip Guide: Lexington. Kentucky Geological Survey.

Dewey, J.F. (1977) Suture zone complexities: a review. *Tectonophysics*, **40**, 53–67.

Dewey, J.F. (1980) Episodicity, sequence, and style at convergent plate boundaries. In: *The Continental Crust and its Mineral Deposits* (Ed. by D.W. Strangeway), 553–573, *Spec. Paper geol. Assoc. Canada*, **20**.

Dewey, J.F. (1982) Plate tectonics and the evolution of the British Isles. *J. geol. Soc. London*, **139**, 371–414.

Dewey, J.F. and Bird, J.M. (1970) Mountain belts and the new global tectonics. *J. geophys. Res.*, **75**, 2625–2647.

Dewey, J.F. and Pindell, J.L. (1985) Neogene block tectonics of eastern Turkey and northern South America: continental applications of the finite difference method. *Tectonics*, **4**, 71–83.

Dibblee, T.W. Jr (1977) Strike-slip tectonics of the San Andreas fault and its role in Cenozoic basin development. In: *Late Mesozoic and Cenozoic Sedimentation and Tectonics in California*, San Joaquin Geological Society Short Course, 26–38.

Dickinson, W.R. (1974) Plate tectonics and sedimentation. In: *Tectonics and Sedimentation* (Ed. by W.R. Dickinson), 1–27, *Spec. Publ. Soc. econ. Paleont. Mineral.*, **22**, Tulsa, Oklahoma.

Dickinson, W.R. (1980) Plate tectonics and key petrologic associations. In: *The Continental Crust and its Mineral Deposits* (Ed. by D.W. Strangeway), 341–360, *Spec. Paper geol. Assoc. Canada*, **20**, J.T. Wilson Volume.

Dickinson, W.R. (1981) Plate tectonics and the continental margin of California. In: *The Geotectonic Development of California* (Ed. by W.G. Ernst). 1–28, Prentice-Hall, Englewood Cliffs, New Jersey.

Dickinson, W.R. and Seely, D.R. (1979) Structure and stratigraphy of forearc regions. *Bull. Am. Assoc. petrol. Geol.*, **63**, 2–31.

Dickinson, W.R. and Suczek, C.A. (1979) Plate tectonics and sandstone compositions. *Bull. Am. Assoc. petrol. Geol.*, **63**, 2164–2182.

Dickinson, W.R. and Valloni, R. (1980) Plate tectonics and provenance of sands in modern ocean basins. *Geology*, **8**, 82–86.

Dietz, R.S. (1963) Collapsing continental rises, an actualistic concept of geosynclines and mountain building. *J. Geol. Chicago*, **71**, 314–333.

Dingle, R.V. (1980) Large allochthonous sediment masses and their role in the construction of the continental slope and rise off southwestern Africa. *Mar. Geol.*, **37**, 333–354.

Dominguez, L.L., Mullins, H.T. and Hine, A.C. (1988) Cat Island platform, Bahamas: an incipiently drowned Holocene carbonate shelf. *Sedimentology*, **35**, 805–819.

Donovan, D.T. and Jones, E.J.W. (1979) Causes of world-wide changes in sea level. *J. geol. Soc. London*, **136**, 187–192.

Dott, R.H. Jr (1983) Episodic sedimentation: how normal is average? How rare is rare? Does it matter? *J. sedim. Petrol*; **53**, 5–23.

Dow, W.G. (1977) Kerogen studies and geological interpretations. *J. geochem. Exploration*, **7**, 79–99.

Dow, W.G. and O'Connor, D.I. (1982) Kerogen maturity and type by reflected light microscopy applied to petroleum exploration. In: *How to Assess Maturation and Paleotemperatures. Soc. econ. Paleont. Mineral. Short Course Notes*, **7**, 133–157.

Dowdle, W.L. and Cobb, W.M. (1975) Static formation temperature from well logs – an empirical method. *J. petrol. Tech.*, Nov. 1975, 1326.

Downey, J.S. (1984) Geohydrology of the Madison and associated aquifers in parts of Montana, North Dakota, South Dakota, and Wyoming. *Prof. Paper U.S. geol. Surv.*, **1273-G**, 47 pp.

Downey, M.W. (1984) Evaluating seals for hydrocarbon accumulations. *Bull. Am. Assoc. petrol. Geol.*, **68**, 1752–1763.

Dresser-Atlas (1982) *Well Logging and Interpretation Techniques*. Dresser Industries Inc.

Duff, P. McL. D., Hallam, A. and Walton, E.K. (1967) *Cyclic Sedimentation. Developments in Sedimentology*. Elsevier, Amsterdam.

Dunbar, C.O. and Rogers, J. (1957) *Principles of Stratigraphy*. Wiley , New York, 356 pp.

Durand, B., Alpern, B., Pittion, J.L. and Pradier, B. (1986) Reflectance of vitrinite as a control of thermal history of sediments. In: *Thermal Modeling in Sedimentary Basins* (Ed. by J. Burrus), 441–474, *lst IFP Exploration Research Conference, Carcans, France, June 3–7, 1985*. Editions Technip, Paris, 600 pp.

Egyed, L. (1956) The change in the Earth's dimensions determined from palaeogeographical data. *Geofisica Pura e Applicata*, **33**, 42–48, Milano, Italy (in English).

Elliott, D. (1976) The motion of thrust sheets. *J. geophys. Res.*, **81**, 949–963.

Ellis, P.G. and McClay, K.R. (1988) Listric extensional fault systems – results of analogue model experiments *Basin Research*, **1**, 55–70.

Emter, D. (1971) *Ergebnisse seismischer Untersuchungen der Erdkruste und des obersten erdmantels in Südwestdeutschland*. Dissertation, Universität Stuttgart, 108 pp.

Emery, K.O. (1977) Structure and stratigraphy of divergent continental margins. In: *Geology of Continental Margins* (Ed. by H. Yarborough *et al.*), *Am. Assoc. petrol. Geol. Continuing Education Course Notes Series*, **5**, B1–B20, Washington, D.C.

England, P.C. (1983) Constraints on extension of continental lithosphere. *J. geophys. Res.*, **88**, 1145–1152.

England, P.C. and Houseman, G.A. (1988) Uplift and extension of the Tibetan plateau. *J. geophys. Res.*, **326**, 301–320.

England, P.C. and McKenzie, D.P. (1982) A thin viscous sheet model for continental deformation. *Geophys. J. R. astr. Soc.*, **70**, 295–322.

Espitalié, J., Laporte, J.L., Madec, M., Marquis, F, Leplat, P., Paulet, J. and Boutefeu, A. (1977) Méthode rapide de caractérisation des roches mères, de leur potentiel pétrolier et de leur degré d'évolution. *Rev. Fr. Pét.*, **32**, 23–42.

Etheridge, M.A., Branson, J.C., Stuart-Smith, P.G., and Schere, A.S. (1984) The geometry of extensional structures in the Bass basin. *Geol. Soc. Aust. Abstr.*, **12**, 164–165.

Ettensohn, F.R. (1981) Mississippian–Pennsylvanian boundary in northeastern Kentucky. *Geol. Soc. Am. Cincinnati '81 Field Trip Guidebook*, 195–257.

Eugster, H.P. (1967) Hydrous sodium silicates from Lake Magadi, Kenya: precursors of bedded chert. *Science*, **157**, 1177–1180.

Eugster, H.P. (1986) Lake Magadi, Kenya: a model for rift valley hydrochemistry and sedimentation? In: *Sedimentation in the African Rifts* (Ed. by L.E. Frostick *et al.*), 177–189, *Spec. Publ. geol. Soc. London*, **25**, Blackwell Scientific, Oxford.

Eugster, H.P. and Hardie, L.A. (1975) Sedimentation in an ancient playa-lake complex: the Wilkins Peak member of the Green River formation of Wyoming. *Bull. geol. Soc. Am.*, **86**, 319–334.

Eugster, H.P. and Kelts, K. (1983) Lacustrine chemical sediments. In: *Chemical Sediments and Geomorphology* (Ed. by A.S. Goudie and K. Pye), 321–368, Academic Press, London.

Fairbanks, R.G. and Matthews, R.R. (1978) The marine oxygen isotope record in Pleistocene corals, Barbados, West Indies. *Quaternary Research*, **10**, 181–196.

Fairbridge, R.W. (1961) Eustatic changes in sea level. In: *Physics and Chemistry of the Earth* (Ed. by L.H. Ahrens), **4**, 99–185, Pergamon Press, London.

Falcon, N.L. (1958) Position of oil fields of southwest Iran with respect to relevant sedimentary basins. In: *Habitat of Oil* (Ed. by L.G. Weeks), 1279–1293, Symposium Volume, American Association of Petroleum Geologists, Tulsa, 1384 pp.

Falvey, D.A. (1974) The development of continental margins in plate tectonic theory. *J. Aust. pet. Explor. Assoc.*, **14**, 95–106.

Falvey, D.A. and Middleton, M.F. (1981) Passive continental margins: evidence for a prebreakup deep crustal metamorphic subsidence mechanism. In: *Colloquium on Geology of Continental Margins* (C3, Paris, 7–17 July 1980), *Oceanologica Acta*, **4** (Supplement), 103–114.

Fasel, J.-M. (1986) *Sédimentologie de la Molasse d'Eau Douce Subalpine entre Le Léman et la Gruyère*. Unpubl. PhD thesis, University of Fribourg (Switzerland), 143 pp.

Feinstein, S. (1981) Subsidence and thermal history of southern Oklahoma aulacogen: implications for petroleum exploration. *Bull. Am. Assoc. petrol. Geol.*, **65**, 2521–2533.

Fischer, A.G. (1964) The Lofer cyclothems of the Alpine Triassic. In: *Symposium on Cyclic Sedimentation* (Ed. by D.F. Merriam), *Bull. Kansas State Geological Survey*, **169**, 107–149.

Fischer, A.G. (1975) Origin and growth of basins. In: *Petroleum and Global Tectonics* (Ed. by A.G. Fischer and S. Judson), Princeton University Press, 322 pp.

Fischer, J.H. and Barratt, M.W. (1985) Exploration in Ordovician of Central Michigan Basin. *Bull. Am. Assoc. petrol. Geol.*, **69**, 2065–2076.

Fischer, W.L., Brown, L.F., Scott, A.J. and McGowen, J.H. (1969) Delta systems in the exploration for oil and gas. *Bur. econ. Geol. Univ. Texas, Austin*, 78 pp.

Fisk, H.N. (1944) *Geological Investigations of the Alluvial Valley of the Lower Mississippi River*. Mississippi River Commission, Vicksburg, Miss., 78 pp.

Fisk, H.N. (1947) *Fine Grained Alluvial Deposits and their Effects on Mississippi River Activity*. Mississippi River Commission, Vicksburg, Miss., 78 pp.

Flemings, P.B. and Jordan, T.E. (1989) A synthetic stratigraphic model of foreland basin development. *J. geophys. Res.*, **94**, 3851–3866.

Flemming, B.W. (1978) Sand transport patterns in the Agulhas current (south-east African continental margin). *Sedim. Geol.*, **26**, 179–205.

Flint, R.F. (1971) *Glacial and Quaternary Geology*. Wiley, New York, 892 pp.

Franzinelli, E. and Potter, P.E. (1983) Petrology, chemistry and texture of modern river sands, Amazon River system. *J. Geol. Chicago*, **91**, 23–39.

Freund, R. (1970) Rotation of strike-slip faults in Sistan, southeast Iran. *J. Geol. Chicago*, **78**, 188–200.

Freund, R. (1971) The Hope Fault, a strike-slip fault in New Zealand. *Bull. N.Z. geol. Surv.*, **86**, 1–49.

Friend, P.F. (1983) Towards the field classification of alluvial architecture or sequence. In: *Modern and Ancient Fluvial Systems* (Ed. by J.D. Collinson and J. Lewin), 345–354, *Spec. Publ. int. Assoc. Sedimentol.*, **6**, Blackwell Scientific, Oxford.

Friend, P.F., Slater, M.J. and Wiliams, R.C. (1979) Vertical and lateral building of river sandstone bodies, Ebro Basin, Spain. *J. geol. Soc. London*, **136**, 39–46.

Frostick, L.E. and Reid, I. (1986) Evolution and sedimen-

tary character of lake deltas fed by ephemeral rivers in the Turkana basin, northern Kenya. In: *Sedimentation in the African Rifts* (Ed. by L.E. Frostrick *et al.*), 113–125, *Spec. Publ. geol. Soc. London*, **25**.

Frostick, L.E., Reid, I. and Layman, J.T. (1983) Changing size distribution of suspended sediment in arid-zone flash floods. In: *Modern and Ancient Fluvial Systems* (Ed. by J.D. Collinson and J. Lewin), 97–106, *Spec. Publ. int. Assoc. Sedimentol.*, **6**, Blackwell Scientific, Oxford.

Forel, F.A. (1892) *Le Léman: Monographie Limnologique*, Vol. 1, Géographie, Hydrographie, Géologie, Climatologie, Hydrologie. F. Rouge, Lausanne, 543 pp.

Foucher, J.-P., Le Pichon, X. and Sibuet, J.-C. (1982) The ocean-continent transition in the uniform stretching model: the role of partial melting in the upper mantle. *Phil. Trans.R. Soc. London*, **A305**, 27–43.

Fountain, D.M. and Salisbury, M.H. (1981) Exposed cross-sections through the continental crust: implications for crustal structure, petrology and evolution. *Earth planet. Sci. Letters*, **56**, 263–277.

Fowler, J.H. and Kuenzi, W.D. (1978) Keweenawan turbidites in Michigan (deep borehole red beds): a foundered basin sequence developed during evolution of a protoceanic rift system. *J. geophys. Res.*, **83**, 5833–5843.

Fryberger, S.G. and Schenk, C. (1981) Wind sedimentation tunnel experiments on the origins of aeolian strata. *Sedimentology*, **28**, 805–821.

Gagliano, S.M. and Van Beek, J.L. (1970) Geologic and geomorphic aspects of deltaic processes, Mississippi delta system. In: *Hydrologic and Geologic Studies of Coastal Louisiana*, Report No. 1, Centre for Wetland Resources, Louisiana State University, 140 pp.

Galloway, W.E. (1975) Process framework for describing the morphologic and stratigraphic evolution of deltaic depositional systems. In: *Deltas, Models for Exploration* (Ed. by M.L. Broussard), 87–98, Houston Geological Society, Houston.

Garcia, R. (1981) Depositional systems and their relation to gas accumulation in Sacramento Valley, California. *Bull. Am. Assoc. petrol. Geol.*, **65**, 653–673.

Garfunkel, Z. (1981) Internal structure of the Dead Sea leaky transform (rift) in relation to plate kinematics. *Tectonophysics*, **80**, 81–108.

Garfunkel, Z. and Bartov, Y. (1977) The tectonics of the Suez rift. *Bull geol. Surv. Israel*, **71**, 44 pp.

Garfunkel, Z., Zak, I. and Freund, R. (1981) Active faulting in the Dead Sea Rift. *Tectonophysics*, **80**, 1–26.

Garland, G.D. (1971) *Introduction to Geophysics: Mantle, Core and Crust*. W.B. Saunders, Toronto, 420 pp.

Garner, D.L. and Turcotte, D.L. (1984) The thermal and mechanical evolution of the Anadarko Basin. *Tectonophysics*, **107**, 1–24.

Gibling, M.R., Charn Tantisukrit, Wutti Utamo, Theerapongs Thanasuthipitak and Mungkorn Haraluck (1985a) Oil shale sedimentology and geochemistry in Cenozoic Mae Sot basin, Thailand. *Bull. Am. Assoc. petrol. Geol.*, **69**, 767–780.

Gibling, M.R., Yongyut Ukakimaphan and Suthep Srisuk (1985b) Oil shale and coal in intermontane basins of Thailand. *Bull. Am. Assoc. petrol. Geol.*, **69**, 760–766.

Giltner, J.P. (1988) Application of extensional models to the northern Viking Graben. *Norsk geol. Tidsskrift.*

Ginsburg, R.N. (1971) Landward movement of carbonate mud: new model for regressive cycles in carbonates. *Abstr. Bull. Am. Assoc. petrol. Geol.*, **55**, 340.

Ginsburg, R.N. (1975) *Tidal Deposits: a Casebook of Recent Examples and Fossil Counterparts.* Springer-Verlag, Berlin, 428 pp.

Ginsburg, R.N. and James, N.P. (1974) Holocene carbonate sediments of continental shelves. In: *The Geology of Continental Margins* (Ed. by C.A. Burk and C.L. Drake), 137–155, Springer-Verlag, Berlin.

Girdler, R.W. (1970) A review of Red Sea heat flow. *Phil. Trans. R. Soc. London*, A267, 191–203.

Glaeser, J.D. (1978) Global distribution of barrier islands in terms of tectonic setting. *J. Geol. Chicago*, **86**, 283–297.

Goetze, C. (1978) The mechanisms of creep in olivine. *Phil. Trans. R. Soc. London*, A288, 99–119.

Goetze, C. and Evans, B. (1979) Stress and temperature in the bending lithosphere as constrained by experimental rock mechanics. *Geophys. J. R. astr. Soc.*, **59**, 463–478.

Goldhammer, R.K., Dunn, P.A. and Hardie, L.A. (1987) High-frequency glacio-eustatic sealevel oscillations with Milankovitch characteristics recorded in Middle Triassic platform carbonates in northern Italy. *Am. J. Sci.*, **287**, 853–892.

Goodwin, P.W. and Anderson, E.J. (1985) Punctuated aggradation cycles: a general hypothesis of episodic stratigraphic accumulation. *J. Geol. Chicago*, **93**, 515–533.

Gordon, M.B. and Hempton M.R. (1986) Collision-induced rifting: the Grenville orogeny and the Keweenawan rift of North America. *Tectonophysics*, **127**, 1–25.

Gosnold, W.D. and Fischer, D.W. (1986) Heat flow studies in sedimentary basins. In: *Thermal Modeling in Sedimentary Basins* (Ed. by J. Burrus), 199–218, lst IFP Exploration Research Conference, Carcans, France, June 3–7, 1985. Editions Technip, Paris, 600 pp.

Green, A.R. (1977) The evolution of the earth's crust and sedimentary basin development. In: *The Earth's Crust* (Ed. by J. Heacock), 1–17, American Geophysical Union, *Geophysical Monograph* 10.

Gressly, A. (1938) Observations géologiques sur le Jura Soleurois. *Neue Denkschr. allg. schweiz, Ges. Naturw.*, **2**, 1–112.

Griggs, D.T., Turner, F.J. and Heard, H.C. (1960) Deformation of rocks at 500 to 800 °C. In: *Rock Deforma-*

tion (Ed. by D.T. Griggs and J. Handin), 21–37, *Geol. Soc. Am. Mem.* 79, 382 pp.

Grocott, J. and Watterson, J. (1980) Strain profile of a boundary within a large ductile shear zone. *J. struct. Geol.*, **2**, 111–117.

Grotzinger, J.P. (1986a) Cyclicity and paleoenvironmental dynamics, Rocknest platform, northwest Canada. *Bull. geol. Soc. Am.*, **97**, 1208–1231.

Grotzinger, J.P. (1986b) Upward-shallowing platform cycles: a response to 2.2 billion years of low-amplitude, high-frequency (Milankovitch band) sea level oscillations. *Paleoceanography*, **1**, 403–416.

Grotzinger, J.P. (1989) Introduction to Precambrian reefs. In: *Reefs, Canada and Adjacent Area* (Ed. by H.H.J. Geldsetzer, N.P. James and G.E. Tebbutt), 9–12, *Can. Soc. petrol. Geol. Mem.* 13.

Grove, A.T. and Warren, A. (1968) Quaternary landforms and climate on the south side of the Sahara. *Geogr. J. London*, **134**, 194–208.

Grow, J.A., Mattick, R.E. and Schlee, J.S. (1979) Multichannel seismic depth sections and interval velocities over outer continental shelf and upper slope between Cape Hatteras and Cape Cod. In: *Geological Investigations of Continental Margins* (ed. by J.S. Watkins, L. Montadert and P.W. Dickerson), *Am. Assoc. petrol. geol. Mem.* **29**, 65–83.

Grunau, H.R. (1987) A worldwide look at the caprock problem. *J. petrol. Geol.*, **10**, 245–266.

Gussow, W.C. (1954) Differential entrapment of oil and gas: a fundamental principle. *Bull. Am. Assoc. petrol. Geol.*, **38**, 816–853.

Gust, D.A., Biddle, K.T., Phelps, D.W. and Uliana, M.A. (1985) Associated Middle to Late Jurassic volcanism and extension in southern South America. *Tectonophysics*, **116**, 223–253.

Haack, U. (1982) Radioactivity of rocks. In: *Physical Properties of Rocks* (Ed. by G. Angenheister), 433–481, Vol. 1b, Springer-Verlag, Berlin.

Hagemann, H.W. (1974) Petrographische und Palynologische Untersuchung der organischen Substanz (Kerogen) in den Liassischen Sedimenten Luxemburgs. In: *Advances in Organic Geochemistry* (Ed. by B. Tissot and F. Bienner), 29–37, Technip, Paris.

Halbouty, M.T. (1979) *Salt Domes, Gulf Region, United States and Mexico (Second Edition).* Gulf Publishing, Houston, Texas, 561 pp.

Halbouty, M.T. (1982) The time is now for all explorationists to purposely search for the subtle trap. In: *The Deliberate Search for the Subtle Trap* (Ed. by M.T. Halbouty), 1–10, *Am. Assoc. petrol. Geol. Mem.* 32.

Halbouty, M.T., King, R.E., Klemme, H.D., Dott, R.H. Sr and Meyerhoff, A.A. (1980) World's giant oil and gas fields, geologic factors affecting their formation and basin classification. In: *Geology of Giant Petroleum Fields* (Ed. by M.T. Halbouty), *Am. Assoc. petrol. Geol. Mem.* **14**, 502–555.

Hall, T. and Al-Haddad, F.M. (1976) Seismic velocities in the Lewisian metamorphic complex, north-west

Britain – *in situ* measurements. *Scot. J. Geol.*, **12**, 305–314.

Hallam, A. (1963) Major epeirogenic and eustatic changes since the Cretaceous and their possible relationship to crustal structure. *Am. J. Sci.*, **261**, 397–423.

Hallam, A. (1977) Secular changes in marine inundation of USSR and North America through the Phanerozoic. *Nature*, **269**, 769–772.

Ham, W.E., Denison, R. E, and Merritt, C.A. (1964) Basement rocks and structural evolution of southern Oklahoma. *Bull. Oklahoma geol. Surv.*, **73-3**, 61 pp.

Hamblin, A.P. and Walker, R.G. (1979) Storm-dominated shallow marine deposits; the Fernie-Kootenay (Jurassic) transition, southern Rocky Mountains. *Can. J. Earth Sci.*, **16**, 1673–1690.

Hamilton, W. (1979) Tectonics of the Indonesian region. *Prof. Paper U.S. geol. Surv.*, **1078**.

Hanks, T.C. (1971) The Kuril trench–Hokkaido rise system: large shallow earthquakes and simple models of deformation. *Geophys. J. R. astr. Soc.*, **23**, 173–189.

Haq, B.U., Hardenbol, J. and Vail, P.R. (1987) Chronology of fluctuating sea levels since the Triassic (250 Myr ago to present). *Science*, **235**, 1156–1167.

Harding, T.P. and Lowell, J.D. (1979) Structural styles, their plate tectonic habitats and hydrocarbon traps in petroleum provinces. *Bull. Am. Assoc. petrol. Geol.*, **63**, 1016–1058.

Harris, L.B. and Cobbold, P.R. (1984) Development of conjugate shear bands during simple shearing. *J. struct. Geol.*, **7**, 37–44.

Haxby, W.F. Turcotte, D.L. and Bird, J.M. (1976) Thermal and mechanical evolution of the Michigan Basin. *Tectonophysics*, **36**, 57–75.

Hayes, M.O. (1979) Barrier island morphology as a function of tidal and wave regime. In: *Barrier Islands – from the Gulf of St. Lawrence to the Gulf of Mexico* (Ed. by S.P. Leatherman), 1–27, Academic Press, New York.

Hays, J.D. and Pitman, W.C. (1973) Lithospheric plate motion, sea level changes, and climatic and ecological consequences. *Nature*, **246**, 18–22.

Hays, J.D. Imbrie, J. and Shackleton, N.J. (1976) Variations in the Earth's orbit: pacemaker of the ice ages. *Science*, **194**, 2212–2232.

Heard, H.C. (1960) Transition from brittle fracture to ductile flow in Solenhofen limestone as a function of temperature, confining pressure, and interstitial fluid pressure. In: *Rock Deformation* (Ed. by D.T. Griggs and J. Handin), 193–226, *Geol. Soc. Am. Mem.* **79**, 382 pp.

Hedberg, H.D. (1936) Gravitational compaction of clays and shales. *Am. J. Sci.*, **31**, 241–287.

Heezen, B.C. and Hollister, C.D. (1964) Deep sea current evidence from abyssal sediments. *Mar. Geol.*, **1**, 141–174.

Heling, D. and Teichmüller, M. (1974) Die Grenze Montmorillonit/Mixed Layer Minerale und ihre Beziehung zur Inkohlung in der grauen Schichtenfolge des Oligozäns im Oberrheingraben. *Fortschritte in der Geologie von Rheinland und Westfalen, Krefeld*, **24**, 113–128.

Hellinger, S.J. and Sclater, J.G. (1983) Some comments on two-layer extensional models for the evolution of sedimentary basins. *J. geophys. Res.*, **88**, 8251–8270.

Henyey, T.L. and Bischoff, J.L. (1973) Tectonic elements of the northern part of the Gulf of California. *Bull. geol. Soc. Am.*, **84**, 315–330.

Héroux, Y., Chagnon, A. and Bertrand, R. (1979) Compilation and correlation of major thermal maturation indicators. *Bull. Am. Assoc. petrol. Geol.*, **63**, 2128–2144.

Hinze, W.J. and Merritt, D.W. (1969) Basement rocks of the Michigan Basin. In: *1969 Annual Field Excursion* (Ed. by H. B. Stonehouse), 28–59, Michigan Basin Geological Society, Lansing, Mich.

Hinze, W.J., Kellog, R.L. and O'Hara, N.W. (1975) Geophysical studies of basement geology of the southern peninsula of Michigan. *Bull. Am. Assoc. petrol. Geol.*, **59**, 1562–1584.

Hirn, A. (1976) Sondages sismiques profonds en France. *Bull. Soc. géol. France*, **23**, 1065–1071.

Hirn, A. and Perrier, G. (1974) Deep seismic sounding in the Limagne graben. In: *Approaches to Taphrogenesis* (ed. by J.H. Iliies and K. Fuchs). Stuttgart, E. Schweizerbartsche Verlagsbuchhandlung, 329–340.

Hirst, J.P.P. and Nichols, G.J. (1986) Thrust tectonic controls on Miocene alluvial distribution patterns, southern Pyrenees. In: *Foreland Basins* (Ed. by P.A. Allen and P. Homewood), 247–258, *Spec. Publ. int. Assoc. Sedimentol.* **8**, Blackwell Scientific, Oxford.

Hobbs, B.E., Means, W.D. and Williams, P.F. (1976) *An Outline of Structural Geology*. Wiley International, New York, 571 pp.

Hoffman, P. (1974) Shallow and deep water stromatolites in Lower Proterozoic platform-basin facies change, Great Slave Lake, Canada. *Bull. Am. Assoc. petrol. Geol.*, **58**, 856–867.

Hoffman, P. (1976) Stromatolite morphologies in Shark Bay, Western Australia. In: *Stromatolites* (Ed. by M.R. Walter), 261–272, Elsevier, Amsterdam.

Holeman, J.N. (1968) Sediment yield of major rivers of the world. *Water Resources Res.*, **4**, 737–747.

Holmes, A. (1965) *Principles of Physical Geology*. Nelson, London, 1288 pp.

Homewood, P., Allen, P.A. and Williams, G.D. (1986) Dynamics of the Molasse Basin of western Switzerland. In: *Foreland Basins* (Ed. by P.A. Allen and P. Homewood), 199–217, *Spec. Publ. int. Assoc. Sedimentol.*, **8**, Blackwell Scientific, Oxford.

Hood, A., Gutjahr, C.C.M. and Heacock, R.L. (1975) Organic metamorphism and the generation of petroleum. *Bull. Am. Assoc. petrol. Geol.*, **59**, 986–996.

Houbolt, J.J.H.C. (1968) Recent sediments in the southern bight of the North Sea. *Geol. Mijnbouw*, **47**, 254–273.

Houseman, G. and England, P.C. (1986) A dynamical model of lithosphere extension and sedimentary basin formation. *J. geophys. Res.*, **91**, 719–729.

Howell, D.G. and Vedder, J. (1981) Structural implications of stratigraphic discontinuities across the southern California borderland. *J. sedim. Petrol.*, **49**, 517–540.

Howell, D.G., Crouch, J.K., Greene, H.G., McCulloch, D.S. and Vedder, J.G. (1980) Basin development along the late Mesozoic and Cainozoic California margin: a plate tectonic margin of subduction, oblique subduction and transform tectonics. In: *Sedimentation in Oblique Slip Mobile Zones* (Ed. by P.F. Ballance and H.G. Reading), 43–62, *Spec. Publ. int. Assoc. Sedimentol.*, **4**, Blackwell Scientific, Oxford.

Hower, J. (1981) Shale diagenesis. In: *Clays and the Resource Geologist* (Ed. by F.J. Longstaffe), *Min. Ass. Can. Short Course Handbook*, **7**, 60–80.

Hsü, K.J. (1979) Thin-skinned plate tectonics during neo-Alpine orogenesis. *Am. J. Sci.*, **279**, 353–366.

Hsü, K.J. (1983) *Mountain Building Processes*. Academic Press, Orlando, Florida.

Hsü, K.J., Ryan, W.B.F. and Cita, M.B. (1973) Late Miocene desiccation of the Mediterranean. *Nature*, **242**, 240–244.

Hubbard, R.J. (1988) Age and significance of sequence boundaries on Jurassic and Early Cretaceous rifted continental margins. *Bull. Am. Assoc. petrol. Geol.*, **72**, 49–72.

Hubbard, R.J., Pape, J. and Roberts, D.G. (1985) Depositional sequence mapping as a technique to establish tectonic and stratigraphic framework and evaluate hydrocarbon potential on a passive continental margin. In: *Seismic Stratigraphy II: an Integrated Approach* (Ed. by O.R. Berg and D. Woolverton), *Am. Assoc. petrol. Geol. Mem.* **39**, 79–91.

Hubbert, M.K. (1953) Entrapment of petroleum under hydrodynamic conditions. *Bull. Am. Assoc. petrol. Geol.*, **37**, 1954–2026.

Hubbert, M.K. and Rubey, W.W. (1959) Role of fluid pressure in mechanics of overthrust faulting. *Bull. geol. Soc. Am.* **70**, 115–166.

Huff, K.F. (1978) Frontiers of world oil exploration. *Oil and Gas Journal*, **76**, No. 40, 214–220.

Hull, C.E. and Warman, H.R. (1970) Asmari oil fields of Iran. In: *Geology of Giant Petroleum Fields*, 428–437, *Am. Assoc. petrol. Geol. Mem.* **14**,

Hunt, J.M. (1961) Distribution of hydrocarbons in sedimentary rocks. *Geochim. Cosmochim. Acta*, **22**, 37–49.

Hunt, J.M. (1979) *Petroleum Geochemistry and Geology*. W. H. Freeman, San Francisco, 617 pp.

Hunt, J.M. (1980) Application of Geochemistry to Petroleum Exploration. *Proc. American Association of Petroleum Geologists Fall Education Conference, Houston, Texas*.

Hunter, R.E. (1977) Basic types of stratification in small aeolian dunes. *Sedimentology*, **24**, 361–388.

Hunter, R.E. (1981) Stratification styles in eolian sandstones: some Pennsylvanian to Jurassic examples from the western Interior, USA. In: *Modern and Ancient Nonmarine Depositional Environments; Models for Exploration* (Ed. by F.G. Ethridge and R.M. Flores), 315–329, *Spec. Publ. Soc. econ. Paleont. Mineral.* **31**.

Hunter, R.E. and Rubin, D.M. (1983) Interpreting cyclic cross-bedding with an example from the Navajo Sandstone. In: *Eolian Sediments and Processes* (Ed. by M.E. Brookfield and T.S. Ahlbrandt), 429–454, Elsevier, Amsterdam.

Illich, H.A. and Grizzle, P.L. (1985) Thermal subsidence and generation of hydrocarbons in Michigan Basin: Discussion. *Bull. Am. Assoc. petrol. Geol.*, **69**, 1401–1403.

Imbrie, J. (1982) Astronomical theory of the Pleistocene ice ages: a brief historical review. *Icarus*, **50**, 408–422.

Ingersoll, R.V. (1988) Tectonics of sedimentary basins. *Bull. geol. Soc. Am.*, **100**, 1704–1719.

Ingersoll, R.V. and Suczek, C.A. (1979) Petrology and provenance of Neogene sand from Nicobar and Bengal fans, DSDP 211 and 218. *J. sedim. Petrol.* **49**, 1217–1228.

Irving, E. (1979) Paleopoles and paleolatitudes of North America and speculations about displaced terrains. *Can. J. Earth Sci.*, **16**, 669–694.

Isacks, B., Oliver, J. and Sykes, L. (1968) Seismology and the new global tectonics. *J. geophys. Res.*, **73**, 5855–5899.

Jacob, H. and Kuckelhorn, K. (1977) Das Inkohlungsprofil der Bohrung Miesbach 1 und seine erdölgeologische Interpretation. *Erdöl-Erdgas Z.*, **4**, 115–124.

Jackson, J.A. (1987) Active normal faulting and crustal extension. In: *Continental Extensional Tectonics* (Ed. by M.P. Coward, J.F. Dewey and P.L. Hancock), 3–18, *Spec. Publ. geol. Soc. London* **28**.

Jackson, J.A. and McKenzie D.P. (1983) The geometrical evolution of normal fault systems. *J. struct. Geol.*, **5**, 471–482.

Jackson, M.P.A. and Galloway, W.E. (1984) Structural and depositional styles of Gulf Coast Tertiary Continental Margins: application to hydrocarbon exploration. *Am. Assoc. petrol. Geol. Continuing Education Course Notes Series*, **25**.

James, N.P. (1983) Reef environment. In: *Carbonate Depositional Environments* (Ed. by P.A. Scholle, D.G. Bebout and C.H. Moore), 346–440, *Am. Assoc. petrol. Geol. Mem.* **33**.

James, N.P. (1984a) Shallowing-upward sequences in carbonates. In: *Facies Models* (Ed. by R.G. Walker), 213–228, *Geoscience Canada Reprint Series* **1**.

James, N.P. (1984b) Reefs. In: *Facies Models* (Ed. by R. G. Walker), 229–244, *Geoscience Canada Reprint Series*, **1**.

Jansa, L.F. and Wade, J.A. (1975) Geology of the continental margin off Nova Scotia and Newfoundland. In: *Offshore Geology of Eastern Canada* (Ed. by W.J.M. Van der Linden and J.A. Wade), *Geol. Surv. Canada Pap. 74-30*, **2**, 51–105.

Jardine, D. and Wilshart, J.W. (1987) Carbonate reservoir description. In: *Reservoir Sedimentology* (Ed. by R.W. Tillman and W.J. Weber), 129–152, *Spec. Publ. Soc. econ. Paleont. Mineral.* **40**.

Jarvis, G.T. and McKenzie, D.P. (1980) The development of sedimentary basins with finite extension rates. *Earth planet. Sci. Letters,* **48**, 42–52.

Jenkyns, H.C. (1986) Pelagic environments. In: *Sedimentary Environments and Facies* (Ed. by H.G. Reading), 343–397, Blackwell Scientific, Oxford.

Johnson, H.D. and Baldwin, C.T. (1986) Shallow siliciclastic seas. In: *Sedimentary Facies and Environments* (Ed. by H.G. Reading), 229–282, Blackwell Scientific, Oxford.

Johnson, K.S., Branson, C.C., Curtis, N.M. Jr, Ham, W.E., Harrison, W.E., Marcher, M.V. and Roberts, J.F. (1972) Geology and earth resources of Oklahoma – an atlas of maps and cross-sections. *Oklahoma geol. Surv. Educational Publ.* **1**, 8 pp.

Jordan, T.E. (1981) Thrust loads and foreland basin evolution, Cretaceous, western United States. *Bull. Am. Assoc. petrol. Geol.*, **65**, 2506–2520.

Junger, A. (1976) Tectonics of the southern California borderland. In: *Aspects of the Geologic History of the California Continental Borderland* (Ed. by D.G. Howell), 486–498, *Misc. Publ. Am. Assoc. petrol. Geol. Pacific Section*, **24**.

Kachel, N.B. and Smith, J.D. (1986) Geological impact of sediment transporting events on the Washington continental shelf. In: *Shelf Sands and Sandstones* (Ed. by R.J. Knight and J.R. McLean), 145–162, *Can. Soc. petrol. Geol. Mem.* **11**, Calgary, 347 pp.

Kantsler, A.J., Smith, G.C. and Cook, A. C (1978) Lateral and vertical rank variation: implications for hydrocarbon exploration. *J. Aust. pet. Explor. Assoc.*, **18**, 143–156.

Karig, D.E. and Moore, G.F. (1975) Tectonically controlled sedimentation in marginal basins. *Earth planet. Sci. Letters,* **26**, 233–238.

Karig, D.E., Suparka, S., Moore, G.F. and Hehanussa, P.E. (1978) Structure and Cenozoic evolution of the Sunda Arc in the central Sumatra region. In: *Geological and Geophysical Investigations of Continental Margins* (Ed. by J.S. Watkins, I. Montadert and P.W. Dickerson), 223–237, *Am. Assoc. petrol. Geol. Mem.* **29**.

Karner, G.D. (1986) Effects of lithospheric in-plane stress on sedimentary basin stratigraphy. *Tectonics*, **5**, 573–588.

Karner, G.D. and Watts, A.B. (1983) Gravity anomalies and flexure of the lithosphere at mountain ranges. *J. geophys. Res.*, **88**, 10 449–10 477.

Karner, G.D. Steckler, M.S. and Thorne, J.A. (1983) Long-term thermo-mechanical properties of the lithosphere. *Nature*, **304**, 250–252.

Kay, M. (1947) Geosynclinal nomenclature and the craton. *Bull. Am. Assoc. petrol. Geol.*, **31**, 1289–1293.

Kay, M. (1951) *North American Geosynclines. Geol. Soc. Am. Mem.* **48**, 143 pp.

Keen, C.E. (1985) The dynamics of rifting: deformation of the lithosphere by active and passive driving forces. *Geophys. J. R. astr. Soc.*, **80**, 95–120.

Keen, C.E., Beaumont, C. and Boutilier, R. (1983) A summary of thermo-mechanical model results for the evolution of continental margins based on three rifting processes. *Am. Assoc. petrol. Geol. Mem.* **34**, 725–728.

Kelley, P.A., Bissada, K.K., Burda, B.H., Elrod, L.W. and Pheifer, R.N. (1985) Petroleum generation potential of coals and organic-rich deposits: significance in Tertiary coal rich basins. *Proc. Indonesian Petroleum Association, 14th Annual Convention, 1985.*

King, B.C. and Williams, L.A.J. (1976) The East African rift system. In: *Geodynamics; Progress and Prospects* (Ed. by C.L. Drake), American Geophysical Union, Washington D.C., 63–74.

Kinghorn, R.R.F. (1983) *An Introduction to the Physics and Chemistry of Petroleum.* Wiley, Chichester, England, 420 pp.

Kingston, D.R., Dishroon, C.P. and Williams, P.A. (1983a) Global basin classification. *Bull. Am. Assoc. petrol. Geol.,* **67**, 2175–2193.

Kingston, D.R., Dishroon, C.P. and Williams, P.A. (1983b) Hydrocarbon plays and global basin classification. *Bull. Am. Assoc. petrol Geol.*, **67**, 2194–2198.

Kinsman, D.J.J. (1975) Rift valley basins and sedimentary history of trailing continental margins. In: *Petroleum and Global Tectonics* (Ed. by A.G. Fischer and S. Judson), Princeton University Press, 83–126.

Kirby, S.H. (1983) Rheology of the lithosphere. *Rev. Geophys. Space Phys.*, **21**, 1458–1487.

Kirk, R.H. (1980) Statfjord field: a North Sea giant. In: *Giant Oil and Gas Fields of the Decade 1968–1978* (Ed. by M.T. Halbouty), 95–116, *Am. Assoc. petrol. Geol. Mem.* **30**.

Kite, G.W. (1972) An engineering study of crustal movement around the Great Lakes. *Inland Waters Directorate, Dept Environment, Tech. Bull.*, **63**, Ottawa, 57 pp.

Klemme, H.D. (1980) Petroleum basins – classification and characteristics. *J. petrol. Geol.*, **3**, 187–207.

Klemperer, S.L. (1988) Crustal thinning and nature of extension in the northern North Sea from deep seismic reflection profiling. *Tectonics*, **7**, 803–821.

Kocurek, G. (1981a) Significance of interdune deposits and bounding surfaces in aeolian dune sands. *Sedimentology*, **28**, 753–780.

Kocurek, G. (1981b) Erg reconstruction: the Entrada Sandstone (Jurassic) of northern Utah and Colorado.

Palaeogeogr., Palaeoclimat., Palaeoecol., **36**, 125–153.

Koide, H. and Bhattacharji, S. (1977) Geometric patterns of active strike-slip faults and their significance as indicators for areas of energy release. In: *Energetics of Geological Processes* (Ed. by S.K. Saxena), 46–66, Springer-Verlag, New York.

Krumbein, W.C. and Monk, G.D. (1942) Permeability as a function of the size parameters of unconsolidated sand. *Am. Inst. Min. Metall. Eng., Tech. Publ.* **1492**, 11 pp.

Kübler, B., Pittion, J.-L., Héroux, Y., Charolais, J. and Weidmann, M. (1979) Sur le pouvoir réflecteur de la vitrinite dans quelques roches du Jura, de la Molasse et des Nappes préalpines, helvétiques et penniques (Suisse occidentale, Haute Savoie). *Eclog. géol. Helv.*, **72**, 347–373.

Kusznir, N.J. and Karner, G.D. (1985) Dependence of the flexural rigidity of the continental lithosphere on rheology and temperature. *Nature*, **316**, 138–142.

Kusznir, N.J. and Park, R.G. (1987) The extensional strength of the continental lithosphere: its dependence on geothermal gradient, and crustal composition and thickness. In: *Continental Extensional Tectonics* (Ed. by M.P. Coward, J.F. Dewey and P.L. Hancock), *Spec. Publ. geol. Soc. London* **28**, 35–52.

Kusznir, N.J., Karner, G.D. and Egan, S. (1987) Geometric, thermal and isostatic consequences of detachments in continental lithosphere extension and basin formation. In: *Sedimentary Basins and Basin-Forming Mechanisms* (Ed. by C. Beaumont and A.J. Tankard), 185–203, *Can. Soc. petrol. Geol. Mem.* **12**.

Lachenbruch, A.H. and Sass, J.H. (1980) Heat flow and energetics of the San Andreas fault zone. *J. geophys. Res.*, **85**, 6185–6222.

Lamb, C.F. (1980) Painter reservoir field: giant in the Wyoming thrust belt. *Bull. Am. Assoc. petrol. Geol.*, **64**, 638–644.

Langford, R.P. (1989) Fluvial-aeolian interactions. Part I: modern systems. *Sedimentology*, **36**, 1023–1035.

Langford, R.P. and Chan, M.A. (1989) Fluvial-aeolian interactions. Part II: ancient system. *Sedimentology*, **36**, 1037–1051.

Larson, R.L. and Pitman, W.C. (1972) Worldwide correlation of Mesozoic magnetic anomalies and its implications. *Bull. geol. Soc. Am.*, **83**, 3645–3662.

Lash, G.G. (1985) Recognition of trench-fill in orogenic flysch sequences. *Geology*, **13**, 867–870.

Lash, G.G. (1987) Longitudinal petrographic variations in a Middle Ordovician trench deposit, central Appalachian orogen. *Sedimentology*, **34**, 227–235.

Le Pichon, X., and Angelier, J. (1979) The Hellenic arc and trench system: a key to the neotectonic evolution of the eastern Mediterranean area. *Tectonophysics*, **60**, 1–42.

Le Pichon, X. and Francheteau, J. (1978) A plate tectonic analysis of the Red Sea – Gulf of Aden area. *Tectonophysics*, **46**, 369–406.

Le Pichon, X. and Sibuet, J.-C. (1981) Passive margins: a model of formation. *J. geophys. Res.*, **86**, 3708–3720.

Le Pichon, X., Francheteau, J. and Bonnin, J. (1973) *Plate Tectonics.* Elsevier, New York, 300 pp.

Leeder, M.R. (1982) *Sedimentology: Process and Product.* Allen and Unwin, London, 344 pp.

Leeder, M.R. and Alexander, J. (1987) The origin and tectonic significance of asymmetrical meander-belts. *Sedimentology*, **34**, 217–226.

Leeds, A.R., Knopoff, L. and Kausel, E.G. (1974) Variations of upper mantle structure under the Pacific Ocean. *Science*, **186**, 141–143.

Lejay, A. (1988) *Géométrie et architecture des corps sédimentaires tidaux en basin d'avant-pays.* Master's thesis, Université Louis Pasteur, Strasbourg, 31 pp.

Leopold, L.B. and Wolman, M.G. (1957) River channel patterns: braided, meandering and straight. *Prof. Pap. U.S. geol. Surv.*, **282-B**, 85 pp.

Leythaeuser, D., Schaefer, R.G. and Yukler, A. (1982) Role of diffusion in primary migration of hydrocarbons. *Bull. Am. Assoc. petrol. Geol.*, **66**, 408–429.

Liggett, M.A. and Childs, J.F. (1974) Crustal extension and transform faulting in the southern Basin Range Province. *Argus Explor. Co. Rept. of Inv.*, NASA-CR-137256, E74-10411, 28 pp.

Liggett, M.A. and Ehrenspeck, H.E. (1974) Pahranagat shear system, Lincoln County, Nevada. *Argus Explor. Co., Rept. of Inv.*, NASA-CR-136388, E74-10206, 10 pp.

Link, M.H. (1982) Provenance, palaeocurrents and palaeogeography of Ridge Basin, southern California. In: *Geologic History of Ridge Basin, Southern California* (Ed. by J.C. Crowell and M.H. Link), 265–276, Society of Economic Paleontologists Mineralogists, Pacific Section.

Link, M.H. and Osborne, R.H. (1978) Lacustrine facies in the Pliocene Ridge Basin Group: Ridge Basin, California. In: *Modern and Ancient Lake Sediments* (Ed. by A. Matter and M.E. Tucker), 169–187, *Spec. Publ. int. Assoc. Sedimentol.* **2**, Blackwell Scientific, Oxford.

Lopatin, N.V. (1971) Temperature and geologic time as factors in coalification (in Russian). *Akad. Nauk SSSR Izvestiya, Seriya Geologicheskaya*, **3**, 95–106.

Loutit, T.S., Hardenbol, J., Vail, P.R. and Baum, G.R. (1988) Condensed sections: the key to age dating of continental margin sequences. In: *Sea Level Changes: an Integrated Approach* (Ed. by Wilgus, B.S. Hastings, H.W. Posamentier, J.C. van Wagoner, C.A. Ross and G.C. St.C. Kendall), *Spec. Publ. Soc. econ. Paleont. Mineral.* **42**, 183–213.

Lowe, D.R. (1988) Suspended load fall-out rate as an independent variable in the analysis of current structures. *Sedimentology*, **35**, 765–776.

Lowell, J.D. and Genik, G.J. (1972) Sea floor spreading and structural evolution of southern Red Sea. *Bull. Am. Assoc. petrol. Geol.*, **56**, 247–259.

Luheshi, M.N. and Jackson, D. (1986) Conductive and convective heat transfer in sedimentary basins. In: *Thermal Modeling in Sedimentary Basins* (Ed. by J. Burrus), 219–234, *lst IFP Exploration Research Conference, Carcans, France, June 3–7, 1985*. Editions Technip, Paris, 600 pp.

Lyon-Caen, H. and Molnar, P. (1985) Gravity anomalies, flexure of the Indian plate, and the structure, support and evolution of the Himalaya and Ganga Basin. *Tectonics*, **4**, 513–538.

Lyon-Caen, H. and Molnar, P. (1989) Constraints on the deep structure and dynamic processes beneath the Alps and adjacent regions from an analysis of gravity anomalies. *Geophys. J. Int.*, **99**, 19–32.

McConnell, R.B. (1977) East African rift system dynamics in view of Mesozoic apparent polar wander. *J. geol. Soc. London*, **134**, 33–39.

McConnell, R.B. (1980) A resurgent taphrogenic lineament of Precambrian origin in eastern Africa. *J. geol. Soc. London*, **137**, 483–489.

McCubbin, D.G. (1982) Barrier island and strand plain facies. In: *Sandstone Depositional Environments* (Ed. by P.A. Scholle and D. Spearing), 247–279, American Association of Petroleum Geologists, Tulsa, Oklahoma.

Mack, G.H. (1984) Exceptions to the relationship between plate tectonics and sandstone composition. *J. sedim. Petrol.*, **54**, 212–220.

Mackenzie, A.S. (1984) Applications of biological markers to petroleum geochemistry. In: *Advances in Petroleum Geochemistry* (Ed. by J. Brooks and D.H. Welte), Vol. 1, 115–214, Academic Press, London.

Mackenzie, A.S. and Quigley, T.M. (1988) Principles of geochemical prospect appraisal. *Bull. Am. Assoc. petrol. Geol.*, **72**, 399–415.

McKenzie, D.P. (1977) Surface deformation, gravity anomalies and convection. *Geophys. J. R. astr. Soc.*, **48**, 211–238.

McKenzie, D.P. (1978a) Some remarks on the development of sedimentary basins. *Earth planet. Sci. Letters.*, **40**, 25–32.

McKenzie, D.P. (1978b) Active tectonics of the Alpine-Himalayan belt: the Aegean and surrounding regions. *Geophys. J.R. astr. Soc.*, **55**, 217–254.

McKenzie, D.P., Roberts, J.M. and Weiss, N.O. (1974) Convection in the Earth's mantle: towards a numerical simulation. *J. fluid Mech.*, **62**, 465–538.

McNutt, M.K. and Kogan, M.G. (1987) Isostasy in the USSR, 2, Interpretation of admittance data. In: *The Composition, Structure, Dynamics of the Lithosphere-Aesthenosphere System* (Ed. by K. Fuchs and C. Froideveaux), *Geodyn. Ser.*, **16**, 309–327, American Geophysical Union, Washington D.C.

McNutt, M.K., Diament, M. and Kogan, M.G. (1988) Variations of elastic plate thickness at continental thrust belts. *J. geophys. Res.*, **93**, 8825–8838.

McWhae, J.R.H. (1981) Structure and spreading history of the northwestern Atlantic region from the Scotian shelf to Baffin Bay. In: *Geology of the North Atlantic Borderlands* (Ed. by J.W. Kerr and A.J. Fergusson), 299–332, *Can. Soc. petrol. Geol. Mem.* **7**.

Magara, K. (1976) Thickness of removed sedimentary rocks, paleopore pressure and paleotemperatures, southwestern part of western Canada Basin. *Bull. Am. Assoc. petrol. Geol.*, **60**, 554–565.

Maher, C.E. (1980) Piper oil field. In: *Giant Oil and Gas Fields of the Decade 1968–1978* (Ed. by M.T. Halbouty), 131–172, *Am. Assoc. petrol. geol. Mem.* **30**.

Majorowicz, J.A. and Jessop, A.M. (1981) Present heat flow and a preliminary geothermal history of the central Prairies Basin, Canada. *Geothermics*, **10**, 81–93.

Majorowicz, J.A., Jones, F.W., Lam, H.L. and Jessop, A.M. (1984) The variability of heat flow both regional and with depth in southern Alberta, Canada: effect of groundwater flow? *Tectonophysics*, **106**, 1–29.

Mann, P., Hempton, M.R., Bradley, D.C. and Burke, K. (1983) Development of pull-apart basins. *J. Geol. Chicago*, **91**, 529–554.

Mann, P., Draper, G. and Burke, K. (1985) Neotectonics of a strike-slip restraining bend system, Jamaica. In: *Strike-Slip Deformation, Basin Formation, and Sedimentation* (Ed. by K.T. Biddle and N. Christie-Blick), *Spec. Publ. Soc. econ. Paleont. Mineral.* **37**, 211–226.

Manspeizer, W. (1985) The Dead Sea rift: impact of climate and tectonism on Pleistocene and Holocene sedimentation. In: *Strike-Slip Deformation, Basin Formation, and Sedimentation* (Ed. by K.T. Biddle and N. Christie-Blick), 143–158, *Spec. Publ. Soc. econ. Paleont. Mineral.* **37**.

Mareschal, J.-C. (1983) Mechanisms of uplift preceding rifting. *Tectonophysics*, **94**, 51–66.

Marzo, M., Nijman, W. and Puigdefabregas, C. (1988) Architecture of the Castissent fluvial sheet sandstones, Eocene, South Pyrenees, Spain. *Sedimentology*, **35**, 719–738.

Mast, R.F. and Potter, P.E. (1963) Sedimentary structures, sand shape fabrics, and permeability, II. *J. Geol. Chicago*, **70**, 548–565.

Mayne, S.J., Nicholas, E., Bigg-Wither, A.L., Rasidi, J.S. and Raine, M.J. (1974) Geology of the Sydney basin – a review. *Bull. Bur. Min. Resources Aust.*, **149**.

Mayuga, M.N. (1970) Geology and development of California's giant Wilmington oil field. In: *Geology of Giant Petroleum Fields*, 158–184, *Am. Assoc. petrol. Geol. Mem.* **14**.

Menard, H.W. (1964) *Marine Geology of the Pacific*. McGraw-Hill, New York, 271 pp.

Miall, A.D. (1981) Alluvial sedimentary basins: tectonic setting and basin architecture. In: *Sedimentation and Tectonics in Alluvial Basins* (Ed. by A.D. Miall), 1–33, *Spec. Pap. geol. Ass. Can.* **23**.

Miall, A.D. (1984) *Principles of Sedimentary Basin Analysis*. New York, Springer-Verlag, 490 pp.

Miall, A.D., Balkwill, H.R. and Hopkins, W.S. Jr (1980) Cretaceous and Tertiary sediments of Eclipse Trough, Bylot Island area, Arctic Canada, and their regional setting. *Geol. Surv. Canada Paper*, 79–23.

Middleton, G.V. (1978) Facies. In: *Encyclopaedia of Sedimentology* (Ed. by R.W. Fairbridge and J. Bourgeois), 323–325, Dowden, Hutchinson and Ross, Stroudsburg, Penn.

Middleton, M.F. (1982) Tectonic history from vitrinite reflectance. *Geophys. J.R. astr. Soc.*, **68**, 121–132.

Middleton, M.F. and Schmidt, P.W. (1982) Palaeothermometry of the Sydney basin. *J. geophys, Res.*, **87**, 5351–5359.

Mike, K. (1975) Utilization of the analysis of ancient river beds for the detection of Holocene crustal movements. In: *Recent Crustal Movements* (Ed. by N. Pavoni and R. Green), *Tectonophysics*, **29**, 359–368.

Miller, B.M. (1986) Resource appraisal methods: choice and outcome. In: *Oil and Gas Assessment – Methods and Applications* (Ed. by D.D. Rice), *Am. Assoc. petrol. Geol., Studies in Geology* **21**, 1–24.

Milliman, J.D. and Meade, R.H. (1981) World-wide delivery of river sediment to the oceans. *J. Geol. Chicago*, **91**, 1–21.

Minor, H.E. and Hanna, M.A. (1941) East Texas oil field. In: *Stratigraphic Type Oil Fields*, 600–640, American Association of Petroleum Geologists, Tulsa. Oklahoma.

Minster, J.B., Jordan, T.H., Molnar, P. and Haines E. (1974) Numerical modelling of instantaneous plate tectonics. *Geophys. J. R. astr. Soc.*, **36**, 553–562.

Mitchum, R.M. Jr, Vail, P.E. and Thompson, S. III (1977) The depositional sequence as a basic unit for stratigraphic analysis. In: *Seismic Stratigraphy – Applications to Hydrocarbon Exploration* (Ed. by C.E. Payton), *Am. Assoc. petrol. Geol. Mem.* **26**, 53–62.

Molnar, P. (1988) Continental tectonics in the aftermath of plate tectonics. *Nature*, **335**, 131–137.

Molnar, P. and Atwater, T. (1978) Interarc spreading and Cordilleran tectonics as alternates related to the age of subducted oceanic lithosphere. *Earth planet. Sci. Letters*, **41**, 330–340.

Molnar, P. and Tapponnier, P. (1975) Cenozoic tectonics of Asia: effects of a continental collision. *Science*, **189**, 419–426.

Monnier, F. (1982) Thermal diagenesis in the Swiss molasse basin: implications for oil generation. *Can. J. Earth Sci.*, **19**, 328–342.

Montadert, L., Roberts, D.G., Auffret, G., Bock, W., Du Peuple, P.A., Hailwood, E.A., Harrison, W., Kagami, H., Lumsden, D.N., Mueller, C., Schnitker, D., Thompson, T.L. and Timofeev, P.P. (1977) Rifting and subsidence on passive continental margins in the north east Atlantic. *Nature*, **268**, 305–309.

Montadert, L., Roberts, D.G., de Charpal, O. and Guennoc, P. (1979) Rifting and subsidence of the northern continental margin of the Bay of Biscay. *Init.*

Rep. Deep Sea Drilling Proj., **48**, 1025–1060.

Moody, J.D. (1975) Distribution and geological characteristics of giant oil fields. In: *Petroleum and Global Tectonics* (Ed. by A.G. Fischer and S. Judson), 307–319, Princeton University Press.

Moore, D.G. (1969) Reflection profiling studies of the California continental borderlands. *Spec. Pap. geol. Soc. Am.* **107**, 138 pp.

Moore, G.F. and Karig, D.E. (1976) Development of sedimentary basins on the lower trench slope. *Geology*, **4**, 693–697.

Moore, G.F. and Karig, D.E. (1980) Structural geology of Nias Island, Indonesia: implications for subduction zone tectonics. *Am. J. Sci.*, **280**, 193–223.

Moore, W.S. (1982) Late Pleistocene sea level history. In: *Uranium Series Disequilibrium: Application to Environmental Problems* (Ed. by M. Ivanovich and R.S. Harmon), 481–496, Clarendon Press, Oxford.

Morelli, C., Gantar, G. and Pisani, M. (1975) Bathymetry, gravity and magmatism in the Strait of Sicily and in the Ionian Sea. *Bolletino di Geofisica, Teorica ed Applicata*, **11**, 3–190.

Moretti, I. and Turcotte, D.L. (1985) A model for erosion, sedimentation, and flexure with application to New Caledonia. *J. Geodynamics*, **3**, 155–168.

Morgan, P. (1983) Constraints on rift thermal processes from heat flow and uplift. *Tectonophysics*, **94**, 277–298.

Morgan, P. and Baker, B.H. (1983) Introduction-processes of continental rifting. *Tectonophysics*, **94**, 1–10.

Morgan, W.J. (1981) Hotspot tracks and the early rifting of the Atlantic. In: *Papers presented to the Conference on the Processes of Continental Rifting, Lunar Planet. Inst.*, Houston, 1–4.

Morgan, W.J. (1982) Hotspot tracks and the opening of the Atlantic and Indian Oceans. In: *The Sea* (Ed. by C. Emiliani), Vol. 7, Wiley, New York.

Morrow, N.R., Huppler, J.D. and Simmons, A.B. (1969) Porosity and permeability of unconsolidated, Upper Miocene sands from grain-size analysis. *J. sedim. Petrol.*, **39**, 312–321.

Morton, R.A. and Price, W.A. (1987) Late Quaternary sea level fluctuations and sedimentary phases of the Texas coastal plain and shelf. In: *Sea Level Fluctuation and Coastal Evolution* (Ed. by D. Nummedal, O.H. Pilkey and J.D. Howard), *Spec. Publ. Soc. econ. Paleont. Mineral.* **41**, 181–198.

Mueller, S. Peterschmitt, E., Fuchs, K., Emter, D. and Ansorge, J. (1973) Crustal structure of the Rhinegraben area. *Tectonophysics*, **20**, 381–391.

Multer, H.G. (1971) *Field Guide in some Carbonate Rock Environments, Florida Keys and Western Bahamas.* Fairleigh Dickinson University, Madison, New Jersey, 158 pp.

Murris, R.J. (1980) Middle East: stratigraphic evolution

and oil habitat. *Bull. Am. Assoc. petrol. Geol.*, **64**, 597–618.

Mutti, E. (1979) Turbidites et cones sous-marins profonds. In: *Sédimentation Détritique* (Ed. by P. Homewood), 353–419, *Inst. Géol. Univ. Fribourg (Switzerland), Short Course (Troisième Cycle) Notes.*

Mutti, E. (1985) Turbidite systems and their relations to depositional sequences. In: *Provenance of Arenites* (Ed. by G.G. Zuffa), 65–93, *NATO-ASI Series*, Reidel Publishing, Dordrecht, The Netherlands.

Mutti, E. and Ghibaudo, G. (1972) Un esempio di torbiditi di conoide sottomarina esterna: le Arenarie di San Salvatore (Formazione di Bobbio, Miocene) nell' Apennino di Piacenza. *Memorie dell'Accademia delle Scienze di Torino, Classe di Scienze Fisiche, Matematiche e Naturali, Ser.*, **4**, No. 16, 40 pp.

Mutti, E. Luterbacher, H.P., Ferrer, J. and Rosell, J. (1972) Schema stratigrafico e lineamenti di facies del Paleogene marino della Zona Centrale Sud-Pirenaica tra Tremp (Catalogna) e Pamplona (Navarra). *Mem. Soc. geol. ital.* **11**, 391–416.

Mutti, E., Remacha, E., Sgavetti, M., Rosell, J., Valloni, R. and Zamorrano, M. (1985) Stratigraphy and facies characteristics of the Eocene Hecho Group turbidite systems, south-central Pyrenees. In: *Field Trip Guidebook of the 6th I.A.S. Eur. Mtg, Lleida, Spain, Excursion 12*, 579–600.

Nakiboglu, S.M. and Lambeck, K. (1983) A reevaluation of the isostatic rebound of Lake Bonneville. *J. geophys. Res.*, **88**, 10 439–10 448.

Nash, D.B. (1984) Morphological dating of fluvial terrace scarps and fault scarps near West Yellowstone, Montana. *Bull. geol. Soc. Am*, **95**, 1413–1424.

Nederlof, M.H. and Mohler, H.P. (1981) Quantitative investigation of trapping effect of unfaulted caprock (abstr.). *Bull. Am. Assoc. petrol. Geol.*, **65**, 964.

Neugebauer, H.J. (1978) Crustal doming and the mechanism of rifting. Part 1, Rift formation. *Tectonophysics*, **45**, 159–186.

Neugebauer, H.J. (1983) Mechanical aspects of continental rifting. *Tectonophysics*, **94**, 91–108.

Nicholson, C., Seeber, L., Williams, P. and Sykes, L.R. (1985a) Seismicity and fault kinematics through the eastern Transverse Ranges, California: block rotation, strike-slip faulting and shallow-angle thrusts. *J. geophys. Res.*, **91**, 4891–4908.

Nicholson, C., Seeber, L., Williams, P.L. and Sykes, L.R. (1985b) Seismic deformation along the southern San Andreas fault, California: implications for conjugate slip rotational block tectonics. *Tectonics*, **5**, 629–648.

Niemitz, J.W. and Bischoff, J.L. (1981) Tectonic elements of the southern part of the Gulf of California. *Bull. geol. Soc. Am.*, **92**, 360–407.

Nienhuis, P.H. (1981) Distribution of organic matter in living marine organisms. In: *Marine Organic Chemistry* (Ed. by E.K. Duursma and R. Dawson), 31–69, Elsevier, Amsterdam, 521 pp.

Nijman, W. and Puigdefabregas, C. (1978) Coarse-grained point bar structure in a molasse-type system, Eocene Castissent Sandstone Formation, South Pyrenean basin. In: *Fluvial Sedimentology* (Ed. by A.D. Miall), 487–510, *Can. Soc. petrol. Geol. Mem.* **5**.

Nijman, W. and Nio, S.D. (1975) The Eocene Montañana delta (Tremp-Graus Basin of Lerida and Huesca, southern Pyrenees, Spain). In: *The Sedimentary Evolution of the Palaeogene South Pyrenean Basin*, 1–20, *9th int. Congr. I.A.S., Nice, France, Part B*.

Nilsen, T.H. and McLaughlin, R.J. (1985) Comparison of tectonic framework and depositional patterns of the Hornelen strike-slip basin of Norway and the Ridge and Little Sulphur Creek strike-slip basins of California. In: *Strike-Slip Deformation, Basin Formation and Sedimentation* (Ed. by K.T. Biddle and N. Christie-Blick), 79–103, *Spec. Publ. Soc. econ. Paleont. Mineral.*, **37**.

Nittrouer, C.A. and Sternberg, R.W. (1981) The formation of sedimentary strata in an allochthonous shelf environment: the Washington continental shelf. *Marine Geology.* **42**, 201–232.

Nittrouer, C.A., DeMaster, D.J., Kuehl, S.A. and McKee, B.A. (1986) Association of sand with mud deposits accumulating on continental shelves. In: *Shelf Sands and Sandstones* (Ed. by R.J. Knight and J.R. McLean), 17–25, *Can. Soc. petrol. Geol. Mem.*, **11**, Calgary, 347 pp.

Normark, W.R. (1970) Growth patterns of deep sea fans. *Bull. Am. Assoc. petrol. Geol.*, **54**, 2170–2195.

Normark, W.R. (1978) Fan valleys, channels and depositional lobes on modern submarine fans: characters for recognition of sandy turbidite environments. *Bull. Am. Assoc. petrol. Geol.*, **62**, 912–931.

Normark, W.R. and Dickson, F.H. (1976) Sublacustrine fan morphology in Lake Superior. *Bull. Am. Assoc. petrol. Geol.*, **60**, 1021–1036.

Normark, W.R. and Gutmacher, C.E. (1988) Sur submarine slide, Monterey Fan, central California. *Sedimentology*, **35**, 629–648.

Normark, W.R., Piper, D.J.W., and Hess, G.R. (1979) Distributary channels, sandy lobes and mesotopography of Navy Submarine Fan, California Borderland with application to ancient fan sediments. *Sedimentology*, **26**, 749–774.

Nunn, J.A. and Sleep, N.H. (1984) Thermal contraction and flexure of intracratonic basins: a three dimensional study of the Michigan Basin. *Geophys. J.R. astr. Soc.*, **76**, 587–635.

Nunn, J.A., Sleep, N.H. and Moore, W.E. (1984) Thermal subsidence and generation of hydrocarbons in Michigan Basin. *Bull. Am. Assoc. petrol. Geol.*, **68**, 296–315.

Odonne, F. and Vialon, P. (1983) Analogue models of folds above a wrench fault. *Tectonophysics*, **99**, 31–46.

Ogniben, L., Parotto, M. and Praturion, A. (Eds) (1975) *Structural Model of Italy.* Consiglio Nazionale Delle Ricerche, **90**, 502 pp.

Ori, G.G. and Friend, P.F. (1984) Sedimentary basins, formed and carried piggyback on active thrust sheets. *Geology,* **12**, 475–478.

Ori, G.G., Roveri, M. and Valloni, F. (1986) Plio-Pleistocene sedimentation in the Apenninic-Adriatic foredeep (central Adriatic Sea, Italy). In: *Foreland Basins* (Ed. by P.A. Allen and P. Homewood), 183–198, *Spec. Publ. int. Assoc. Sedimentol.,* **8**, Blackwell Scientific, Oxford.

Orowan, E. (1967) Seismic damping and creep in the mantle. *Geophys. J. R. astr. Soc.,* **14**, 191–218.

Palciauskas, V.V. (1986) Models for thermal conductivity and permeability in normally compacting basins. In: *Thermal Modeling in Sedimentary Basins* (Ed. by J. Burrus), 323–336, *1st IFP Exploration Research Conference, Carcans, France, June 3–7, 1985.* Editions Technip, Paris, 600 pp.

Parrish, J.T., Ziegler, A.M. and Scotcse, C.R. (1982) Rainfall patterns and the distributions of coals and evaporites in the Mesozoic and Cenozoic. *Palaeogeography, Palaeoclimatology, Palaeoecology,* **40**, 67–101.

Parkinson, N. and Summerhayes, C. (1985) Synchronous global sequence boundaries. *Bull. Am. Assoc. petrol. Geol.,* **69**, 685–687.

Parsons, B. (1982) Causes and consequences of the relation between area and age of the ocean floor. *J. geophys. Res.,* **87**, 289–302.

Parsons, B. and Sclater, J.G. (1977) An analysis of the variation of ocean floor bathymetry with heat flow and age. *J. geophys. Res.,* **82**, 803–827.

Paterson, M.S. (1958) Experimental deformation and faulting in Wombeyan marble. *Bull. geol. Soc. Am.,* **69**, 465–476.

Payton, C.E. (Editor) (1977) *Seismic Stratigraphy–Applications to Hydrocarbon Exploration. Am. Assoc. petrol. Geol. Mem.* **26**, 516 pp.

Peltier, W.R. (1980) Models of glacial isostasy and relative sea level. In: *Dynamics of Plate Interiors* (Ed. by A.W. Ball, P.L. Bender, T.R. McGetchin and R.I. Walcott), *Am. geophys. Union Geodynamics Ser.,* **I**, 111–127.

Petters, S.W. (1978) Stratigraphic evolution of the Benue trough and its implications for the Upper Cretaceous palaeogeography of West Africa. *J. Geol. Chicago,* **86**, 311–322.

Pettijohn, F.J. (1975) *Sedimentary Rocks* (3rd Edition). Harper and Row, New York. 628 pp.

Pettijohn, F.J., Potter, P.E. and Siever, R. (1973) *Sand and Sandstone.* Springer-Verlag, New York, 618 pp.

Philippi, G.T. (1965) On the depth, time and mechanism of petroleum generation. *Geochim. et Cosmochim. Acta,* **29**, 1021–1049.

Pilkey, O.H., Locker, S.D. and Cleary, W.J. (1980) Comparison of sand-layer geometry on flat floors of ten modern depositional basins. *Bull. Am. Assoc. petrol. Geol.,* **64**, 841–856.

Pitman, W.C. (1978) Relationship between eustacy and stratigraphic sequences of passive margins. *Bull. geol. Soc. Am.,* **89**, 1389–1403.

Pitman, W.C. (1979) The effect of eustatic sea level changes on stratigraphic sequences at Atlantic margins. In: *Geological and Geophysical Investigations of Continental Margins. Am. Assoc. petrol. Geol. Mem,* **29**, 453–460.

Pitman, W.C. and Andrews, J.A. (1985) Subsidence and thermal history of small pull-apart basins. In: *Strike-Slip Deformation, Basin Formation and Sedimentation* (Ed. by K.T. Biddle and N. Christie-Blick), *Spec. Publ. Soc. econ. Paleont. Mineral.* **37**, 45–49.

Platt, J.P. (1986) Dynamics of orogenic wedges and the uplift of high-pressure metamorphic rocks. *Bull. geol. Soc. Am.* **97**, 1037–1053.

Poag, C.W. (1980) Foraminiferal stratigraphy, palaeoenvironments and depositional cycles in the Outer Baltimore Canyon Trough. In: *Geological Studies of the COST B-3 Well, US Mid-Atlantic Continental Slope Area* (Ed. by P.A. Scholle), *US Geol. Surv. Circular* **833**, 44–65.

Poag, C.W. (1985) Depositional history and stratigraphic reference section for Central Baltimore Canyon Trough. In: *Geological Evolution of the US Atlantic Margin* (Ed. by C.W. Poag), 217–264, Van Nostrand Reinhold, New York, 383 pp.

Pollock, H.N. and Chapman, D.S. (1977) On the regional variation of heat flow, geotherms, and lithospheric thickness. *Tectonophysics* **38**, 279–296.

Pollock, H.N. and Watts, D. (1976) Thermal profile of the Michigan Basin. *Trans. Am. geophys. Union,* **57**, 595.

Ponte, F.C. Fonseca, J. and Carozzi, A.V. (1980) Petroleum habitats in the Mesozoic–Cenozoic of the continental margin of Brazil. In: *Facts and Principles of World Petroleum Occurrence* (Ed. by A.D. Miall), 857–886, *Can. Soc. petrol. Geol. Mem.* **6**.

Porter, M.L. (1987) Sedimentology of an ancient erg margin: the Lower Jurassic Aztec Sandstone, southern Nevada and southern California. *Sedimentology,* **34**, 661–680.

Posamentier, H.W. and Vail, P.R. (1988) Eustatic controls on clastic deposition, II, sequence and systems tract models. In: *Sea Level Changes: An Integrated Approach* (Ed. by C.K. Wilgus et al.), 125–154, *Spec. Publ. Soc. econ. Paleont. Mineral.* **42**.

Posamentier, H.W., Jervey, M.T. and Vail, P.R. (1988) Eustatic controls on eustatic deposition, I, conceptual framework. In: *Sea Level Changes: An Integrated Approach* (Ed. by C.K. Wilgus, B.S. Hastings, C.G. St.C. Kendall, H.W. Posamentier, C.A. Ross and J.C. Van Wagoner), 109–124, *Spec. Publ. Soc. econ. Paleont. Mineral.,* **42**.

Potter, P.E. (1978) Petrology and chemistry of modern big river sands. *J. Geol. Chicago*, **86**, 423–449.

Potter, P.E. (1984) South American modern beach sand and plate tectonics. *Nature*, **311**, 645–648.

Potter, P.E. and Mast, R.F. (1963) Sedimentary structures, sand shape fabrics, and permeability, I. *J. Geology Chicago*, **71**, 441–471.

Powell, T.G., Foscolos, A.E., Gunther, P.R. and Snowdon, L.R. (1978) Diagenesis of organic matter and fine clay minerals: a comparative study. *Geochim. et Cosmochim. Acta*, **42**, 1181–1197.

Price, R.A. (1973) Large-scale gravitational flow of supracrustal rocks, southern Canadian Rockies. In: *Gravity and Tectonics* (Ed. by K. de Jong and R. Scholten), 491–502, Wiley, New York.

Price, R.A. (1981) The Cordilleran foreland thrust and fold belt in the southern Canadian Rocky Mountains. In: *Thrust and Nappe Tectonics* (Ed. by K.R. McClay and N.J. Price), 427–448, *Spec. Publ. geol. Soc. London* **9**, Blackwell Scientific, Oxford.

Price, R.A. and Hatcher, R.D. (1983) Tectonic significance of similarities in the evolution of the Alabama–Pennsylvania Appalachians and the Alberta–British Columbia Canadian Cordillera. In: *Contributions to the Tectonics and Geophysics of Mountain Chains* (Ed. by R.D. Hatcher, H. Williams and I. Zietz), *Geol. Soc. Am. Mem.* **158**, 149–160.

Pryor, W.A. (1973) Permeability–porosity patterns and variations in some Holocene sand bodies. *Bull. Am. Assoc. petrol. Geol.*, **57**, 162–189.

Purcell, W.R. (1949) Capillary pressures – their measurement using mercury and the calculation of permeability therefrom. *AIME Petroleum Trans.*, **186**, 39–48.

Purdy, E.G. (1963a) Recent calcium carbonate facies of the Great Bahama Bank.I. Petrography and reaction groups. *J. Geol. Chicago*, **71**, 334–355.

Purdy, E.G. (1963b) Recent carbonate facies of the Great Bahama Bank. II. Sedimentary facies. *J. Geol. Chicago*, **71**, 472–497.

Quinlan, G.M. and Beaumont, C. (1984) Appalachian thrusting, lithospheric flexure, and the Paleozoic stratigraphy of the eastern interior of North America. *Can. J. Earth Sci.*, **21**, 973–996.

Ragotzkie, R.A. (1978) Heat budgets of lakes. In: *Lakes: Chemistry, Geology, Physics* (Ed. by A. Lerman), 1–20, Springer-Verlag, Berlin.

Ramsay, J.G. and Huber, M.I. (1983) *The Techniques of Modern Structural Geology*, Vol. 1 *Strain Analysis*. Academic Press, London, 307 pp.

Rawson, P.F. and Riley, L.A. (1982) Latest Jurassic–Early Cretaceous events and the 'late Cimmerian unconformity' in the North Sea. *Bull. Am. Assoc. petrol. Geol.*, **66**, 2628–2648.

Read, J.F., Grotzinger, J.P., Bova, J.A. and Koerschner, W.F. (1986) Models for generation of carbonate cycles. *Geology*, **14**, 107–110.

Reading, H.G. (1980) Characteristics and recognition of strike-slip fault systems. In: *Sedimentation in Oblique-Slip Mobile Zones* (Ed. by P.F. Ballance and H.G. Reading), *Spec. Publ. int. Assoc. Sedimentol.* **4**, 7–26.

Reading, H.G. (1982) Sedimentary basins and global tectonics. *Proc. geol. Assoc.*, **93**, 321–350.

Reading, H.G. (Ed.) (1986) *Sedimentary Environments and Facies*, 2nd edition. Blackwell Scientific, Oxford, 615 pp.

Reasenberg, P. and Ellsworth, W.L. (1982) Aftershocks of the Coyote Lake, California earthquake of August 6, 1979: a detailed study. *J. geophys. Res.*, **87**, 10 637–10 655.

Reineck, H.-E. and Singh, I.B. (1975) *Depositional Sedimentary Environments*. Springer-Verlag, Berlin, 439 pp.

Reinson, G.E. (1984) Barrier island and associated strandplain systems. In: *Facies Models* (Ed. by R.G. Walker), 119–140, Geoscience Canada Reprint Series, **1**.

Ricci Luchi, F. (1986) The Oligocene to Recent foreland basins of the northern Apennines. In: *Foreland Basins* (Ed. by P.A. Allen and P. Homewood), 105–140, *Spec. Publ. int. Assoc. Sedimentol.* **8**, Blackwell Scientific, Oxford.

Ricci Lucchi, F., Colella, A., Gabbianelli, G., Rossi, S. and Normark, W.R. (1984) The Crati Submarine Fan, Ionian Sea. *Geo-Mar. Letters*, **3**, 71–78.

Rider, M.H. (1986) *The Geological Interpretation of Well Logs*. Blackie, Glasgow, 175 pp.

Riedel, W. (1929) Zur Mechanik geologischer Brucherscheinungen. *Zentralblatt fur Mineralogie, Geologie und Palaeontologie*, **1929B**, 354–368.

Risk, M.J. and Rhodes E.G. (1985) From mangroves to petroleum precursors: an example from tropical N.E. Australia. *Bull. Am. Assoc. petrol. Geol.*, **69**, 1230–1240.

Rittenhouse, G. (1972) Stratigraphic trap classification. In: *Stratigraphic Oil and Gas Fields – Classification, Exploration Methods and Case Histories* (Ed. by R.E. King), 14–28, *Am. Assoc. petrol. Geol. Mem.* **16**.

Robert, P. (1988) *Organic Metamorphism and Geothermal History*. Elf-Aquitaine and Reidel Publishing, Dordrecht, 311 pp.

Rodgers, D.A. (1980) Analysis of pull-apart basin development produced by en echelon strike-slip faults. In: *Sedimentation in Oblique Slip Mobile Zones* (Ed. by P.F. Ballance and H.G. Reading), *Spec. Publ. int. Assoc. Sedimentol.*, **4**, 27–41.

Romankevich, E.A. (1984) *Geochemistry of Organic Matter in the Ocean*. Springer-Verlag, Berlin, 334 pp.

Rona, P.A. (1982) Evaporites at passive margins. In: *Dynamics of Passive Margins* (Ed. by R.A. Scrutton), 116–132, American Geophysical Union and Geological Society of America Geodynamics Series, **6**.

Rosendahl, B.R., Reynolds, D.J., Lorber, P.M., Burgess, C.F., McGill, J., Scott, D., Lambiase, J.J. and Derksen, S.J. (1986) Structural expressions of rifting: lessons from Lake Tanganyika, Africa. In: *Sedimentation in the African Rifts* (Ed. by L.E. Frostick *et al.*), 29–43, *Spec. Publ. geol. Soc. London* 25.

Rowley, D.B. and Sahagian, D. (1986) Depth-dependent stretching: a different approach. *Geology* 14, 32–35.

Royden, L.H. (1985) The Vienna Basin: a thin-skinned pull-apart basin. In: *Strike-Slip Deformation, Basin Formation and Sedimentation* (Ed. by K.T. Biddle and N. Christie-Blick), *Spec. Publ. Soc. econ. Paleont. Mineral.* 37, 319–338.

Royden, L. and Karner, G.D. (1984) Flexure of the continental lithosphere beneath Apennine and Carpathian foredeep basins: evidence for an insufficient topographic load. *Bull. Am. Assoc. petrol. Geol.*, 68, 704–712.

Royden, L. and Keen, C.E. (1980) Rifting processes and thermal evolution of the continental margin of eastern Canada determined from subsidence curves. *Earth planet. Sci. Letters*, 51, 343–361.

Royden, L., Patacca, E. and Scandone, P. (1987) Segmentation and configuration of subducted lithosphere in Italy; an important control on thrust-belt and foredeep-basin evolution. *Geology*, 15, 714–717.

Royden, L., Sclater, J.G. and Von Herzen, R.P. (1980) Continental margin subsidence and heat flow: important parameters in formation of petroleum hydrocarbons. *Bull. Am. Assoc. petrol. Geol.*, 64, 173–187.

Royden, L. Horváth, F., Nagymarosy, A. and Stegena, L. (1983) Evolution of the Pannonian Basin system, 2. Subsidence and thermal history. *Tectonics*, 2, 91–137.

Royse, F., Warner, M.A. and Reese, D.L. (1975) Thrust belt structural geometry and related stratigraphic problems, Wyoming–Idaho–northern Utah. *Rocky Mt. Assoc. Geol., 1975, Guidebook*, 41–54.

Rubin, D.M. and Hunter, R.E. (1983) Reconstructing bedform assemblages from compound cross-bedding. In: *Eolian Sediments and Processes* (Ed. by M.E. Brookfield and T.S. Ahlbrandt), 407–427, Elsevier, Amsterdam.

Rubey, W.W. and Hubbert, M.K. (1960) Role of fluid pressure in mechanics of overthrust faulting, II, Overthrust belt in geosynclinal area of western Wyoming in light of fluid-pressure hypothesis. *Bull. geol. Soc. Am.*, 60, 167–205.

Ruddiman, W.F. and McIntyre, A. (1981) Oceanic mechanisms for amplification of the 23 000-year ice-volume cycle. *Science*, 212, 617–626.

Russell, K.L. (1968) Oceanic ridges and eustatic changes in sea level. *Nature*, 218, 861–862.

Rust, B.R. (1978) Depositional models for braided alluvium. In: *Fluvial Sedimentology* (Ed. by A.D. Miall), 605–625, *Can. Soc. petrol. Geol. Mem.*, 5, Calgary.

Rutter, E.H. (1976) The kinetics of rock deformation by pressure solution. *Phil. Trans. R. Soc. London*, A283, 203–219.

Rutter, E.H. (1983) Pressure solution in nature, theory and experiment. *J. geol. Soc. London*, 140, 725–740.

Rybach, L. (1986) Amount and significance of radioactive heat sources in sediments. In: *Thermal Modelling in Sedimentary Basins* (Ed. by J. Burrus), 311–322, *lst IFP Exploration Research Conference, Carcans, France, June 3–7, 1985*. Editions Technip, Paris, 600 pp.

Rybach, L. and Cermak, V. (1982) Radioactive heat generation in rocks. In: *Physical Properties of Rocks* (Ed. by G. Angenheister), 353–371. Vol. 1b, Springer-Verlag, Berlin.

St John, B., Bally, A.W. and Klemme, H.D. (1984) *Sedimentary Provinces of the World – Hydrocarbon Productive and Non-productive.* American Association of Petroleum Geologists, Tulsa, Oklahoma.

Salveson, J.O. (1976) Variations in the oil and gas geology of rift basins. *Egyptian General Petroleum Corp., 5th Explor. Sem., Cairo, Egypt, November 15–17, 1976.*

Salveson, J.O. (1978) Variations in the geology of rift basins – a tectonic model. Paper presented at *Rio Grande Rift Symposium*, Santa Fe, New Mexico.

Sarg, J.F. and Skjold, L.J. (1982) Stratigraphic traps in Palaeocene sands in the Balder area, North Sea. In: *The Deliberate Search for the Subtle Trap* (Ed. by M.T. Halbouty), 197–206, *Am. Assoc. petrol. Geol. Mem.* 32.

Schlanger, S.O. and Jenkyns, H.C. (1976) Cretaceous oceanic anoxic events: causes and consequences. *Geol. Mijnbouw*, 55, 179–184.

Schluger, P.R. (Ed.) (1979) *Diagenesis as it Affects Clastic Reservoirs. Spec. Publ. Soc. econ. Paleont. Mineral.* 26, 443 pp.

Schlumberger (1974) *Well Evaluation Conference, North Sea* (Ed. by R. Campbell *et al.*), Schlumberger, France, 171 pp.

Scholle, R.A. (1977) Chalk diagenesis and its relation to petroleum exploration: oil from chalks, a modern miracle. *Bull. Am. Assoc. petrol. Geol.*, 61, 982–1009.

Schowalter, T.T. (1976) The mechanics of secondary hydrocarbon migration and entrapment. *Wyoming Geol. Assoc. Earth Science Bull*, 9, 1–43.

Schubert, C. (1982) Origin of Cariaco Basin, southern Caribbean Sea. *Mar. Geol.*, 47, 345–360.

Schumm, S.A. (1963) A tentative classification of alluvial river channels. *U.S. geol. Surv. Circ.* 477.

Schumm, S.A. (1977) *The Fluvial System.* Wiley, New York, 338 pp.

Sclater, J.G. and Christie, P.A.F. (1980) Continental stretching: an explanation of the post Mid-Cretaceous subsidence of the central North Sea basin. *J. geophys. Res.*, 85, 3711–3739.

Sclater, J.G., Jaupart, C. and Galson, D. (1980) The heat flow through oceanic and continental crust and the heat loss of the Earth. *Rev. Geophys. Space Physics*, 18, 269–311.

Sclater, J.G., Parsons, B. and Jaupart, C. (1981) Oceans and continents: similarities and differences in the mechanisms of heat loss. *J. geophys. Res.*, **86**, 11 535–11 552.

Sclater, J.G., Royden, L., Horvath, F., Burchfiel, B.C., Semken, S. and Stegena, L. (1980) Formation of the intra-Carpathian basins as determined from subsidence data. *Earth planet. Sci. Letters*, **51**, 139–162.

Sears, S.O. and Lucia, F.J. (1979) Reef-growth model for Silurian pinnacle reefs, northern Michigan reef trend. *Geology*, **3**, 299–302.

Sears, S.O. and Lucia, F.J. (1980) Dolomitization of northern Michigan Niagara reefs by brine refluxion and fresh water/sea water mixing. In: *Concepts and Models of Dolomitization* (Ed. by D. H. Zenger, J.B. Dunham and R.C. Ethington), 215–235, *Spec. Publ. Soc. econ. Mineral. Paleont.* **28**.

Seeber, L., Armbruster, J.G. and Quittmeyer, R.C. (1981) Seismicity and continental subduction in the Himalayan arc. In: *Zagros, Hindu Kush, Himalaya Geodynamic Evolution* (Ed. by H.K. Gupta and F.M. Delany), American Geophysical Union Geodynamics Series, **3**, 215–242.

Selley, R.C. (1978) Porosity gradients in North Sea oil-bearing sandstones. *J. Geol. Soc. London*, **135**, 119–132.

Selley, R.C. (1985) *Elements of Petroleum Geology.*W. H. Freeman, New York, 449 pp.

Sellwood, B.W. (1986) Shallow marine carbonate environments. In: *Sedimentary Environments and Facies*, 2nd Edition (Ed. by H. G. Reading), 283–342, Blackwell Scientific, Oxford, 615 pp.

Sengör, A.M.C. (1976) Collision of irregular continental margins: implications for foreland deformation of Alpine-type orogens. *Geology*, **4**, 779–782.

Sengör, A.M.C. and Burke, K. (1978) Relative timing of rifting and volcanism on Earth and its tectonic implications. *Geophys. Res. Letters*, **5**, 419–421.

Sengör, A.M.C., Burke, K. and Dewey, J.F. (1978) Rifts at high angles to orogenic belts: tests for their origin and the Upper Rhine Graben as an example. *Am.J. Sci.*, **278**, 24–40.

Sengor, A.M.C., Gorur, N. and Saroglu, F. (1985) Strike-slip faulting and related basin formation in zones of tectonic escape: Turkey as a case study. In: *Strike-Slip Deformation, Basin Formation and Sedimentation* (Ed. by K.T. Biddle and N. Christie-Blick), *Spec. Publ. Soc. econ. Paleont. Mineral.* **37**, 227–264.

Seni, S.J. and Jackson, M.P.A. (1984) Sedimentary record of Cretaceous and Tertiary salt movement, East Texas Basin. *Bur. econ. Geol., University of Texas, Rept. of Invest.* **139**.

Servant, M. and Servant, S. (1970) Les formations lacustres et les diatomées du quaternaire récent du fond de la cuvette tschadienne. *Rev. Géogr. phys. Géol. dynam.*, **13**, 63–76.

Shackleton, N.J. (1977) Oxygen isotope and palaeomagnetic evidence for early northern hemisphere glaciation. *Nature*, **270**, 216–219.

Shackleton, N.J. and Opdyke, N.D. (1973) Oxygen isotope and palaeomagnetic stratigraphy of equatorial Pacific core V28–V2381: oxygen isotope temperatures and ice volumes on a 10^5 year and 10^6 year scale. *Quaternary Research*, **3**, 339–355.

Shanmugam, G. (1985) Significance of coniferous rain forests and related organic matter in generating commercial quantities of oil, Gippsland Basin, Australia. *Bull. Am. Assoc. petrol. Geol.*, **69**, 1241–1254.

Sharp, R.V. (1976) Surface faulting in Imperial Valley during the earthquake swarm of January–February, 1975. *Bull. seismol. Soc. Am.*, **66**, 1145–1154.

Sheridan, R.E. (1974) Atlantic continental margin of North America. In: *The Geology of Continental Margins* (Ed. by C.A. Burk, and C.L. Drake), Springer-Verlag, New York, 391–408.

Sheridan, R.E., Crosby, J.T., Kent, K.M., Dillon, W.P. and Paull, C.K. (1981) The geology of the Blake Plateau and Bahamas regions. In: *Geology of the North Atlantic Borderlands* (Ed. by J.W. Kerr and A.J. Fergusson), 95–117, *Can. Soc. petrol. Geol. Mem.* 7.

Shibaoka, M. and Bennett, A.J.R. (1977) Patterns of diagenesis in some Australian sedimentary basins. *J. Aust. pet. Explor. Assoc*, **17**, 58–63.

Shibaoka, M., Bennett, A.J.R. and Gould, K.W. (1973) Diagenesis of organic matter and occurrence of hydrocarbons in some Australian sedimentary basins. *J. Aust. pet. Explor. Assoc.*, **17**, 58–63.

Shinn, E.A. (1983) Birdseyes, fenestrae, shrinkage pores and loferites; a reevaluation. *J. sedim. Petrol.*, **53**, 619–629.

Shor, G.G. and Pollard, D.D. (1964) Mohole site selection studies north of Maui. *J. geophys. Res.*, **69**, 1627–1637.

Shumskij, P.A. (1969) In: *International Symposium on Antarctic Glaciological Exploration.* Scientific Committee on Antarctic Research, Cambridge, England.

Sibson, R.H. (1983) Continental fault structure and the shallow earthquake source. *J. geol. Soc. London*, **140**, 741–768.

Siebold, E and Hinz, K. (1974) Continental slope construction and destruction, West Africa. In: *The Geology of Continental Margins* (Ed. by C.A. Burk and C.L. Drake) 179–196, Springer, Heidelberg.

Siedler, E. and Jacoby, W.R. (1981) Parameterized rift development and upper mantle anomalies. *Tectonophysics*, **73**, 53–68.

Sinclair, H.D., Coakley, B., Allen, P.A. and Watts, A.B. (1990) Simulation of foreland basin stratigraphy using a diffusion model of mountain belt erosion: an example from the Alps of eastern Switzerland. *Tectonics* (in review).

Sleep, N.H. (1971) Thermal effects of the formation of Atlantic continental margins by continental break-up. *Geophys. J. R. astr. Soc.*, **24**, 325–350.

Sleep, N.H. and Snell, N.S. (1976) Thermal contraction

and flexure of midcontinent and Atlantic marginal basins. *Geophys. J. R. astr. Soc.*, **45**, 125–154.

Sleep, N.H. Nunn, J.A. and Chou, L. (1980) Platform basins. *Annual Reviews of Earth and Planetary Science*, **8**, 17–34.

Sleep, N.H. and Sloss, L.L. (1978) A deep borehole in the Michigan Basin. *J. geophys. Res.*, **83**, 5815–5819.

Sloss, L.L. (1950) Paleozoic stratigraphy in the Montana area. *Bull. Am. Assoc. petrol. Geol.*, **34**, 423–451.

Sloss, L.L. (1962) Stratigraphic models in exploration. *J. sedim. Petrol.*, **32**, 415–422.

Sloss, L.L. (1963) Sequences in the cratonic interior of North America. *Bull. geol. Soc. Am.*, **74**, 93–114.

Sloss, L.L. (1979) Global sea level change: a view from the craton. *Am. Assoc. petrol. Geol. Mem.* **29**, 461–467.

Sloss, L.L. (1988) Forty years of sequence stratigraphy. *Bull. geol. Soc. Am.*, **100**, 1661–1665.

Sloss, L.L. and Speed, R.C. (1974) Relationships of cratonic and continental-margin tectonic episodes. In: *Tectonics and Sedimentation* (Ed. by W.R. Dickinson), 98–119, *Spec. Publ. Soc. econ. Paleont. Mineral.* **22**.

Smith, A.G. (1976) Plate tectonics and orogeny: a review. *Tectonophysics*, **33**, 215–285.

Smith, D.A. (1980) Sealing and non-sealing faults in Louisiana Gulf Coast Salt Basin. *Bull. Am. Assoc. petrol. Geol.*, **64**, 145–172.

Smith, D.G. and Smith, N.D. (1980) Sedimentation in anastomosed river systems: examples from alluvial valleys near Banff, Alberta. *J. sedim. Petrol.*, **50**, 157–164.

Smith, L. and Chapman, D.S. (1983) On the thermal effects of groundwater flow, 1. Regional scale systems. *J. geophys. Res.*, **88**, 593–608.

Smith, L.A. (1965) Paleoenvironmental variation curves and palaeoeustatics. *Trans. Gulf Coast Assoc. geol. Sci.*, **15**, 47–60.

Smoot, J.P. (1983) Depositional subenvironments in an arid closed basin; the Wilkins Peak Member of the Green River Formation (Eocene), Wyoming, USA. *Sedimentology*, **30**, 801–828.

Snyder, D.B. and Barazangi, M. (1986) Deep crustal structure and flexure of the Arabian plate beneath the Zagros collisional mountain belt as inferred from gravity observations. *Tectonics*, **5**, 361–373.

Solli, M. (1976) *En seismisk Skorpeundersokelse Norge-Shetland.* MSc thesis, University of Bergen, Norway, 155 pp.

Sonder, L.J. and England, P.C. (1989) Effects of a temperature-dependent rheology on large-scale continental extension. *J. geophys. Res.* **94**, 7603–7619.

Spohn, L. and Schubert, G. (1983) Convective thinning of the lithosphere: a mechanism for rifting and mid-plate volcanism on Earth, Venus and Mars. *Tectonophysics*, **94**, 67–90.

Stach, E., MacKowsky, M.-Th. Teichmuller, M., Teich-

müller, R., Taylor, G.H. and Chandra, D. (1982) *Stach's Textbook of Coal Petrology (3rd Edition).* Gebrüder Borntraeger, Berlin-Stuttgart, 535 pp.

Stalder, P.J. (1975) Cementation of Pliocene–Quaternary fluviatile clastic deposits in and along the Ocean Mountains. *Geol. Mijnbouw*, **54**, 148–156.

Steckler, M.S. (1981) *Thermal and mechanical evolution of Atlantic-type margins.* Unpubl. PhD thesis, Columbia University, New York, 261 pp.

Steckler, M.S. (1985) Uplift and extension at the Gulf of Suez: indications of induced mantle convection. *Nature*, **317**, 135–139.

Steckler, M.S. and Watts, A.B. (1978) Subsidence of the Atlantic-type continental margin off New York. *Earth planet. Sci. Letters*, **41**, 1–13.

Steckler, M.S. and Watts, A.B. (1981) Subsidence history and tectonic evolution of Atlantic-type continental margins. In: *Dynamics of Passive Margins* (Ed. by R.A. Scrutton), American Geophysical Union Geodynamics Series **6**, 184–196.

Steckler, M.S., Watts, A.B. and Thorne, J.A. (1988) Subsidence and basin modelling at the U.S. Atlantic continental margin. In: *The Atlantic Continental Margin, U.S.*, Vol. 1–2, *The Geology of North America* (Ed. by R.E. Sheridan and J.A. Grow), 199–416, Geological Society of America.

Stewart, J.H. (1977) Rift systems in the western United States. In: *Tectonics and Geophysics of Continental Rifts* (Ed. by I.B. Ramberg and E.R. Neumann). Reidel Publishing, Dordrecht, 89–110.

Stockmal, G.S., Beaumont, C. and Boutilier, R. (1986) Geodynamic models of convergent tectonics: the transition from rifted margin to overthrust belt and consequences for foreland basin development. *Bull. Am. Assoc. petrol. Geol.*, **70**, 181–190.

Stow, D.A.V. (1985) Deep sea clastics: where are we and where are we going? In: *Sedimentology: Recent Developments and Applied Aspects* (Ed. by P.J. Brenchley and B.P.J. Williams), 67–93, *Spec. Publ. geol. Soc. London* **18**, Blackwell Scientific, Oxford.

Stow, D.A.V. and Lovell, J.P.B. (1979) Contourites: their recognition in modern and ancient sediments. *Earth Sci. Rev.*, **14**, 251–291.

Strasser, A. (1988) Shallowing-upward sequences in Purbeckian peritidal carbonates (lowermost Cretaceous, Swiss and French Jura Mounains). *Sedimentology*, **35**, 369–384.

Stride, A.H. (1963) Current swept floors near the southern half of Great Britain. *Q. J. geol. Soc. London*, **119**, 175–199.

Stride, A.H. (Ed.) (1982) *Offshore Tidal Sands: Process and Deposits.* Chapman and Hall, London, 213 pp.

Stuart, C.A. (1970) Geopressures. *Proc. 2nd Symp. Abnormal Subsurface Pressure, Louisiana State University, Baton Rouge, Louisiana, Supplement*, 121 pp.

Stubblefield, W.L., McGrail, D.W. and Kersey, D.G.

(1984) Recognition of transgressive and post-transgressive sand ridges on the New Jersey continental shelf. In: *Siliciclastic Shelf Sediments* (Ed. by R.W. Tillman and C.T. Siemers), 1–24, *Spec. Publ. Soc. econ. Paleont. Mineral.* **34**.

Sturm, M. and Matter, A. (1978) Turbidites and varves in Lake Brienz (Switzerland): deposition of clastic detritus by density currents. In: *Modern and Ancient Lake Sediments* (ed. by A. Matter and M.E. Tucker), 145–166, *Spec. Publ. int. Assoc. Sedimentol.* **2**, Blackwell Scientific, Oxford.

Surdam, R.C. and Wolfbauer, C.A. (1975) Green River Formation, Wyoming: a playa-lake complex. *Bull. geol. Soc. Am.*, **86**, 335–345.

Surlyk, F., Clemmensen, L.B. and Larsen, H.C. (1981) Post-Palaeozoic evolution of the East Greenland continental margin. In: *Geology of the North Atlantic Borderlands* (Ed. by J.W. Kerr and A.J. Fergusson), 611–646, *Can. Soc. petrol. Geol. Mem.* **7**.

Suter, J.R., Berryhill, H.L. Jr. and Penland, S. (1987) Late Quaternary sea level fluctuations and depositional sequences, southwest Louisiana continental shelf. In: *Sea Level Fluctuation and Coastal Evolution* (Ed. by D. Nummedal, O.H. Pilkey and J.D. Howard), *Spec. Publ. Soc. econ. Paleont. Mineral.* **41**, 199–219.

Swift, D.J.P. (1974) Continental shelf sedimentation. In: *The Geology of Continental Margins* (Ed. by C.A. Burk and C. L. Drake), 117–135, Springer-Verlag, Berlin.

Swift, D.J.P. and Field, M.E. (1981) Evolution of a classic sand ridge field, Maryland sector, North American inner shelf. *Sedimentology*, **28**, 461–482.

Swift, D.J.P., Kofoed, J.W., Saulsbury, F.P. and Sears, P. (1972) Holocene evolution of the shelf surface, south and central Atlantic shelf of North America. In: *Shelf Sediment Transport: Process and Pattern* (Ed. by D.J.P. Swift, D.B. Duane and O.H. Pilkey), 499–574, Dowden, Hutchinson and Ross, Stroudsburg, Pennsylvania.

Swift, D.J.P., McKinney, T.F. and Stahl, L. (1984) Recognition of transgressive and post-transgressive sand ridges on the New Jersey continental shelf: Discussion. In: *Siliciclastic Shelf Sediments* (Ed. by R.W. Tillman and C.T. Siemers), 25–36, *Spec. Publ. Soc. econ. Paleont. Mineral.* **34**.

Sykes, L. (1967) Mechanisms of earthquakes and nature of faulting on the mid-ocean ridges. *J. geophys. Res.*, **72**, 2131–2153.

Sylvester, A.G. (1988) Strike-slip faults. *Bull. geol. Soc. Am.*, **100**, 1666–1703.

Tankard, A.J. (1986) On the depositional response to thrusting and lithospheric flexure: examples from the Appalachian and Rocky Mountain basins. In: *Foreland Basins* (Ed. by P.A. Allen and P. Homewood), *Spec. Publ. int. Assoc. Sedimentol.* **8**, 369–394.

Tankard, A.J., Jackson, M.P.A., Eriksson, K.A., Hobday, D.K., Hunter, D.R. and Minter, W.E.L. (1982) *Crustal Evolution of Southern Africa.* Springer-Verlag, New York. 523 pp.

Tapponnier, P. and Molnar, P. (1976) Slip-line field theory and large-scale continental tectonics. *Nature*, **264**, 319–324.

Tchalenko, J.S. (1970) Similarities between shear zones of different magnitudes. *Bull. geol. Soc. Am.*, **81**, 1625–1640.

Tchalenko, J.S. and Ambraseys, N.N. (1970) Structural analyses of the Dasht-e Baÿaz (Iran) earthquake fractures. *Bull. geol. Soc. Am.*, **81**, 1625–1640.

Teichmüller, M. (1970) Bestimmung des Inkohlungsgrades von kohligen Einschlussen in Sedimenten des Oberrheingrabens: ein Hilfsmittel bei der Klärung geothermischer Fragen. In: *Graben Problems* (Ed. by J.H. Illies and S. Müller), *Upper Mantle Sci. Proj. Sci. Rept.*, **27**, 124–142.

Teichmüller, M. (1982) Fluoreszenz von Liptiniten und Vitriniten in Beziehung zu Inkohlungsgrad und Verkokungsverhalten. *Geol. Landesamt Nordrhein-Westfalen. Spec. Paper*, 119 pp.

Teichmüller, M. and Teichmüller, R. (1975) Inkohlungsuntersuchungen in der Molasse des Alpenvorlands. *Geologica bavar.*, **73**, 123–142.

Terres, R.R. and Sylvester, A.G. (1981) Kinematic analysis of rotated fractures and blocks in simple shear. *Bull. seismol. Soc. Am.*, **71**, 1593–1605.

Thomas, B.M. (1982) Land plant source rocks for oil and their significance in Australian basins. *J. Aust. pet. Explor. Assoc.*, **22**, 164–178.

Thornburg, T.M. and Kulm, L.D. (1987) Sedimentation in the Chile Trench: depositional morphologies, lithofacies and stratigraphy. *Bull. geol. Soc. Am.*, **98**, 33–52.

Tillman, R.W. and Siemers, C.T. (Eds) (1984) *Siliciclastic Shelf Sediments. Spec. Publ. Soc. econ. Paleont. Mineral.* **34**, 268 pp.

Tissot, B. (1969) Premières données sur les mécanismes et le cinétique de la formation du pétrole dans les sédiments; simulation d'un schéma reactionnel sur ordinateur. *Revue de l'Institut Français du Pétrole*, **24**, 470–501.

Tissot, B. (1973) Vers l'évaluation quantitative du pétrole formé dans les bassins sédimentaires. *Revue Assoc. Fr. Tech. Pét.*, **222**, 27–31.

Tissot, B. and Espitalié, J. (1975) L'evolution thermique de la matière organique des sediments: applications d'une simulation mathématique. *Revue de l'Institut Français du Pétrole*, **30**, 743–777.

Tissot, B.P. and Welte, D.H. (1978) *Petroleum Formation and Occurrence: A New Approach to Oil and Gas Exploration.* Springer-Verlag, Berlin, 538 pp.

Tissot, B.P. and Welte, D.H. (1984) *Petroleum Formation and Occurrence* (2nd edition). Springer-Verlag, New York, 699 pp.

Tittman, J. (1986) *Geophysical Well Logging.* Academic Press, New York.

Toksöz, M.N. and Bird, P. (1977) Formation and evolution of marginal basins and continental plateaus. In: *Island Arcs, Deep Sea Trenches and Back arc Basins* (Ed. by M. Talwani and W.C. Pitman), 379–393, *Maurice Ewing Series*, 1, *American Geophysical Union.*

Turcotte, D.L. (1983) Driving mechanisms of mountain building. In: *Mountain Building Processes* (Ed. by K.J. Hsu), 141–146, Academic Press, Orlando, Florida.

Turcotte, D.L. and Bernthal, M.J. (1984) Synthetic coralreef terraces and variations of Quaternary sea level. *Earth planet. Sci. Letters*, **70**, 121–128.

Turcotte, D.L. and Emerman, S.H. (1983) Mechanisms of active and passive rifting. *Tectonophysics*, **94**, 39–50.

Turcotte, D.L. and Kenyon, P.M. (1984) Synthetic passive margin stratigraphy. *Bull. Am. Assoc. petrol. Geol.*, **68**, 768–775.

Turcotte, D.L. and Oxburgh, E.R. (1973) Mid-plate tectonics. *Nature* **244**, 337–339.

Turcotte, D.L. and Schubert, G. (1982) *Geodynamics: Applications of Continuum Mechanics to Geological Problems.* Wiley, New York, 450 pp.

Turcotte, D.L. and Willeman, R.J. (1983) Synthetic cyclic stratigraphy. *Earth planet. Sci. Letters.*, **63**, 89–96.

Uyeda, S. and Kanamori, H. (1979) Back-arc opening and the mode of subduction. *J. geophys. Res.*, **84**, 1049–1061.

Vai, G.B. and Ricci Lucchi, F. (1977) Algal crusts, autochthonous and clastic gypsum in a cannibalistic evaporite basin: a case history from the Messinian of the northern Apennines. *Sedimentology*, **24**, 211–244.

Vail, P.E. Mitchum, R.M. Jr and Thompson, S. III (1977a) Relative changes of sea level from coastal onlap. In: *Seismic Stratigraphy – Applications to Hydrocarbon Exploration* (Ed. by C.E. Payton), *Am. Assoc. petrol. Geol. Mem.* **26**, 63–82.

Vail, P.R., Mitchum, R.M. Jr and Thompson S. (1977b) Seismic stratigraphy and global changes of sea level, Part 4: Global cycles of relative changes of sea level. In: *Seismic Stratigraphy – Applications to Hydrocarbon Exploration* (Ed. by C.E. Payton), *Am. Assoc. petrol. Geol. Mem* **26**, 83–97.

Valentine, J.W. and Moores, E.M. (1970) Plate tectonic regulation of faunal diversity and sea level. *Nature*, **228**, 657–669.

van Gijzel, P. (1982) Characterization and identification of kerogen and bitumen and determination of thermal maturation by means of qualitative and quantitative microscopical techniques. In: *How to Assess Maturation and Paleotemperatures, Soc. econ. Paleont. Mineral. Short Course Notes*, 7, 159–216.

van Hinte, J.E. (1978) Geohistory analysis–application of micropalaeontology in exploration geology. *Bull. Am. Assoc. petrol. Geol.*, **62**, 201–222.

Van Wagoner, J.C., Posamentier, H.W., Mitchum, R.M., Vail, P.R., Sarg, J.F., Loutit, T.S., and Handenbol, J. (1988) An overview of the fundamentals of sequence stratigraphy and key definitions. In: *Sea Level Changes: an Integrated Approach, Spec. Publ. Soc. econ. Paleont. Mineral.*, **42**, 39–45.

Van West, F.P. (1972) Trapping mechanisms of Minnelusa oil accumulations, northeastern Powder River basin, Wyoming. *Mountain Geologist*, **9**, 3–20.

Vassoevich, N.B., Visotskij, I.V., Guseva, A.N. and Olenin, V.B. (1967) Hydrocarbons in the sedimentary mantle of the Earth. *Proc. 7th World Petr. Congr.*, **2**, 37–45.

Veevers, J.J. (1981) Morphotectonics of rifted continental margins in embryo (East Africa), youth (Africa-Arabia), and maturity (Australia). *J. Geol. Chicago*, **89**, 57–82.

Velbel, M.A. (1985) Mineralogy of mature sandstones in accretionary prisms. *J. sedim. Petrol.*, **55**, 685–690.

Vening-Meinesz, F.A. (1941) Gravity over the Hawaiian Archipelago and over the Madeira arca. *Proc. Netherlands Acad., Wetensia*, 44 pp.

Vening-Meinesz, F.A. (1948) *Gravity Expeditions at Sea, 1923–1938.* Netherlands Geodetic Commission, Waltman, Delft.

Vine, F. and Matthews, D.H. (1963) Magnetic anomalies over ocean ridges. *Nature*, **199**, 947–949.

Vinogradov, L.S., Aver'yanov, I.S. and Nigmati, I.S. (1983) Catagenetically sealed oil pools of the Volga–Ural region. *Petroleum Geology*, **19**, 266–268.

Wagner, C.W. and Van der Togt, C. (1973) Holocene sediment types and their distribution in the Southern Persian Gulf. In: *The Persian Gulf: Holocene Carbonate Sedimentation and Diagenesis in a Shallow Epicontinental Sea* (Ed. by B.H. Purser), 123–156, Springer-Verlag, Berlin, 471 pp.

Walcott, R.I. (1970) Flexural rigidity, thickness and viscosity of the lithosphere. *J. geophys. Res.*, **75**, 3941–3954.

Walker, R.G. (Ed.) (1984a) *Facies Models* (2nd Edition). Geoscience Canada Reprint Series, 1, 317 pp.

Walker. R. G. (1984b) Upper Cretaceous (Turonian) Cardium Formation, southern Foothills and Plains, Alberta. In: *Shelf Sands and Sandstone Reservoirs* (Ed. by R.W. Tillman *et al.*), 1–44, Soc. econ. Paleont. Mineral. Short Course, San Antonio, Texas.

Waples, D.W. (1980) Time and temperature in petroleum formation: application of Lopatin's method to petroleum exploration. *Bull. Am. Assoc. petrol. Geol.*, **64**, 916–926.

Waples, D.W. (1981) *Organic Geochemistry for Exploration Geologists.* Burgess Publishing, Minneapolis, Minnesota.

Wardlaw, N.C. and Cassan, J.P. (1978) Estimation of recovery efficiency by visual observation of pore systems in reservoir rocks. *Bull. Can. Petrol. Geol.*, **26**, 572–585.

Watson, H.J. (1981) Casablanca field, offshore Spain, a palaeogeomorphic trap. *Bull. Am. Assoc. petrol. Geol. Abstr.*, **65**, 1005–1006.

Watts, A.B. (1988) Gravity anomalies, crustal structure and flexure of the lithosphere at the Baltimore Canyon Trough. *Earth planet. Sci. Letters,* **89**, 221–238.

Watts, A.B. (1982) Tectonic subsidence, flexure and global changes in sea level. *Nature,* **297**, 469–474.

Watts, A.B. and Cochran, J.R. (1974) Gravity anomalies and flexure of the lithosphere along the Hawaiian–Emperor seamount chain. *Geophys. J. R. astr. Soc.,* **38**, 119–141.

Watts, A.B. and Ryan, W.B.F. (1976) Flexure of the lithosphere and continental margin basins. *Tectonophysics,* **36**, 25–44.

Watts, A.B. and Steckler, M.S. (1979) Subsidence and eustasy at the continental margin of eastern North America. *Am. geophys. Union, Maurice Ewing Ser.,* **3**, 218–239.

Watts, A.B. and Talwani, M. (1974) Gravity anomalies seaward of trenches and their tectonic implications. *Geophys. J. R. astr. Soc.,* **36**, 57–90.

Watts, A.B. and Thorne, J. (1984) Tectonics, global changes in sea level and their relationship to stratigraphic sequences at the US Atlantic continental margin. *Mar. petrol. Geol.,* **1**, 319–339.

Watts, A.B., Karner, G.D. and Steckler, M.S. (1982) Lithospheric flexure and the evolution of sedimentary basins. *Phil. Trans.R. Soc. London,* **A305**, 249–281.

Weber, K.J. (1982) Influence of common sedimentary structures on fluid flow in reservoir models. *J. petrol. Tech.,* March 1982, 665–672.

Weber, K.J. (1986) How heterogeneity affects oil recovery. In: *Reservoir Characterization* (Ed. by L.W. Lake and H.B. Carroll, Jr), 487–544, Academic Press, Orlando, 659 pp.

Weber, K.J. (1987) Computation of initial well productivities in aeolian sandstone on the basis of a geological model, Leman Gas Field, U.K. In: *Reservoir Sedimentology* (Ed. by R.W. Tillmann and K.J. Weber), 333–354, *Spec. Publ. Soc. econ. Paleont. Mineral.* **40**.

Weber, K.J. Eype, R., Leynse, D. and Moens, C. (1972) Permeability distribution in a Holocene distributary channel-fill near Leerdam (The Netherlands) – permeability measurements and *in-situ* fluid flow experiments. *Geol. Mijnbouw,* **51**, 53–62.

Weber, K.J., Mandl, G., Pilaar, W.F., Lehner, F. and Precious, R. G. (1978) The role of faults in hydrocarbon migration and trapping in Nigerian growth fault structures, 2643–2651, *Offshore Tech. Conf., Houston, paper OTC* 3356.

Weber, V.V. and Maximov, S.P. (1976) Early diagenetic generation of hydrocarbon gases and their variations dependent on initial organic composition. *Bull. Am. Assoc. petrol. Geol.,* **60**, 287–293.

Weise, B.R. (1980) Wave-dominated delta systems of the Upper Cretaceous San Miguel Formation, Maverick Basin, South Texas. *Report of Investigations,* **107**, Bur. econ. Geol., University of Texas, Austin, 33 pp.

Wendlandt, R.F. and Morgan, P. (1982) Lithospheric thinning associated with rifting in East Africa. *Nature,* **298**, 734–736.

Wernicke, B. (1981) Low-angle normal faults in the Basin and Range Province: nappe tectonics in an extending orogen. *Nature,* **291**, 645–648.

Wernicke, B. (1985) Uniform-sense normal simple shear of the continental lithosphere. *Can.J. Earth Sci.,* **22**, 108–125.

Wernicke, B. and Burchfiel, B.C. (1982) Modes of extensional tectonics. *J. struct. Geol.,* **4**, 105–115.

Wheeler, H.E. (1958) Time stratigraphy. *Bull. Am. Assoc. petrol. Geol.,* **42**, 1047–1063.

White, D.A. (1980) Assessing oil and gas plays in facies cycle wedges. *Bull. Am. Assoc. petrol. Geol.,* **64**, 1158–1178.

White, D.A. (1988) Oil and gas play maps in exploration and assessment. *Bull. Am. Assoc. petrol. Geol.,* **72**, 944–949.

White, D.A. and Gehman, H.M. (1979) Methods of estimating oil and gas resources. *Bull. Am. Assoc. petrol. Geol.,* **63**, 2183–2191.

White, N. and McKenzie, D.P. (1988) Formation of the steer's head geometry of sedimentary basins by differential stretching of the crust and mantle. *Geology,* **16**, 250–253.

Wilcox, R.E., Harding, T.P. and Seely, D.R. (1973) Basin wrench tectonics. *Bull. Am. Assoc. petrol. Geol.,* **57**, 74–96.

Wilhelm, O. (1945) Classification of petroleum reservoirs. *Bull. Am. Assoc. petrol. Geol.,* **29**, 1537–1580.

Wilkinson, B.R. (1982) Cyclic cratonic carbonates and Phanerozoic calcite seas. *J. Geol. Education,* **30**, 189–203.

Willett, S.D., Chapman, D.S. and Neugebauer, H.J. (1985) A thermo-mechanical model of continental lithosphere. *Nature,* **314**, 520–523.

Williams, D.G. (1984) Correlation of Pleistocene marine sediments of the Gulf of Mexico and other basins using oxygen isotope stratigraphy. In: *Principles of Pleistocene Stratigraphy Applied to the Gulf of Mexico* (Ed. by N. Healy-Williams), 65–118, International Human Resources Development Corporation, Boston.

Williams, H.H., Kelley, P.A., Janks, J.S. and Christensen, R.M. (1985) The Palaeogene rift basin source rocks of Central Sumatra. *Proc. Indonesia Petroleum Association. 14th Annual Convention, October, 1985.*

Williams, J.J. (1968) The stratigraphy and igneous reservoirs of the Angila field, Libya. In: *Geology and Archaeology of Northern Cyrenaica, Libya* (Ed. by T.F. Barr), 197–206.

Williams, R.C. (1975) *Fluvial deposits of Oligo-Miocene age in the southern Ebro Basin, Spain.* Unpublished. PhD thesis, University of Cambridge.

Wilson, J.L. (1975) *Carbonate Facies in Geologic History.* Springer-Verlag, Berlin, 471 pp.

Wilson, J.T. (1966) Did the Atlantic close and then re-open? *Nature*, **211**, 676–681.

Wilson, J.T. and Burke, K. (1972) Two types of mountain building. *Nature*, **239**, 448–449.

Windley, B.F. (1977) *The Evolving Continents.* Wiley, Chichester.

Wise, D.U. (1974) Continental margins; freeboard and volumes of continents and oceans through time. In: *The Geology of Continental Margins* (Ed. by C.A. Burk and C.L. Drake), 45–58, Springer-Verlag, New York.

Woidt, W.-D. and Neugebauer, H.J. (1981) Lithospheric thinning and the dynamics of density instabilities. *Eos, Trans. Am. geophys. Union*, **62**, 814.

Wood, R. and Barton, P. (1983) Crustal thinning and subsidence of the North Sea. *Nature*, **302**, 134–136.

Woodcock, N.H. (1986) The role of strike-slip fault systems at plate boundaries. *Phil. Trans. R. Soc. London*, **A317**, 13–29.

Woodwell, G.M., Whittaker, R.H., Reiners, W.A., Likens, G.E., Delwiche, C.C. and Botkin, D.B. (1978) The biota and the world carbon budget. *Science*, **199**, 141–146.

Wright, L.D. (1977) Sediment transport and deposition at river mouths: a synthesis. *Bull. geol. Soc. Am.*, **88**, 857–868.

Wright, L.D. and Coleman, J.M. (1974) Mississippi River mouth processes; effluent dynamics and morphologic development. *J. Geol. Chicago*, **82**, 751–778.

Wyllie, P.J. (1971) *The Dynamic Earth.* Wiley, New York, 416 pp.

Zak, I. and Freund, R. (1981) Asymmetry and basin migration in the Dead Sea rift. *Tectonophysics*, **80**, 27–38.

Index